Applied Theories of Wave Propagation

Edited by **Ian Nicklaw**

LANRYE
INTERNATIONAL

New Jersey

Published by Clanrye International,
55 Van Reypen Street,
Jersey City, NJ 07306, USA
www.clanryeinternational.com

Applied Theories of Wave Propagation
Edited by Ian Nicklaw

International Standard Book Number: 978-1-63240-069-7 (Hardback)

Printed in the United States of America.

Contents

Preface

Over the recent decade, advancements and applications have progressed exponentially. This has led to the increased interest in this field and projects are being conducted to enhance knowledge. The main objective of this book is to present some of the critical challenges and provide insights into possible solutions. This book will answer the varied questions that arise in the field and also provide an increased scope for furthering studies.

This book discusses the applied theories of wave propagation in a comprehensive manner. A wave is one of the most fundamental physics phenomena discovered by humans since ancient times. The wave is also one of the most-studied phenomena in physics which can be well elucidated by mathematics. Such studies can be the best explanation of the "science" of wave propagation, summarizing the laws of nature by employing human defined symbols, languages and operators. An in-depth understanding of waves and wave propagation can assist us in enhancement of the quality of life and guide us into a direction for future explorations of the universe and nature. The book elucidates appealing theories and applications for readers interested in studying about waves and wave propagations. It also acts as a reference for experts working in the fields described in this book.

I hope that this book, with its visionary approach, will be a valuable addition and will promote interest among readers. Each of the authors has provided their extraordinary competence in their specific fields by providing different perspectives as they come from diverse nations and regions. I thank them for their contributions.

Editor

Tsunami Wave Propagation

Alexey Androsov, Sven Harig, Annika Fuchs, Antonia Immerz, Natalja Rakowsky, Wolfgang Hiller and Sergey Danilov

Additional information is available at the end of the chapter

1. Introduction

Tsunami hazard is connected with loss of human life, flooding of coastal structures, destruction of berthing designs, infrastructures of coastal water areas.

In support of the Tsunami Early Warning System for the Indian Ocean, a finite element model TsunAWI for simulations of wave propagation has been developed. It is part of German Indonesian Tsunami Early Warning System (GITEWS) (www.gitews.de) serving to predict arrival times and expected wave heights. TsunAWI is based on an unstructured grid approach employing finite elements to solve the governing equations.

Finite-element methods are widely used in studies of wave generation and propagation in different fields of fluid dynamics. They are often employed to simulate propagation of long waves such as ocean tides and tsunamis in the ocean in the framework of shallow-water equations [1–3]. The main reason to prefer FE modelling is that the solution is computed over a mesh that can be adapted to cover basins with complex geometries characterized by irregular bottom topography and coastlines.

The purpose of this work is to describe a complex system of tsunami warning, based on the numerical modelling of tsunami events in Indian Ocean. The complexity of the problem stems from insufficient information about the sources generating tsunami, real bottom morphometry and extremely short warning time. In addition, some physical processes, such as interaction of tsunami waves with long tidal waves or nonhydrostatic effects, commonly neglected in models of tsunami wave propagation, may lead to substantial corrections.

Our work, therefore, addresses the influence of tidal dynamics on tsunami wave propagation in coastal areas. A tsunami wave is much shorter than tidal waves which explain why tidal waves are usually ignored in tsunami modelling. There are three approaches to account for the interaction between tsunami waves and tides. The first one presumes that the wave propagation is linear, so that tides reduce or augment (depending on their phase) the amplitude of arriving waves [4]. In this case, having simulated tidal patterns in advance, it would be possible to predict arriving waves as a simple superposition of signals. The second approach assumes that the interaction has a nonlinear character caused by changes in the fluid layer thickness in the shallow area [5]. Finally, the third approach suggests that the main effect is due to nonlinear interactions between tidal and tsunami velocities [6]. The studies performed thus far are of theoretical character and do not involve practical examples.

As a tsunami wave arrives in coastal regions with rough bathymetry, the nonhydrostatic part of pressure, neglected in the standard hydrostatic configuration of TsunAWI, gains in importance. The account for nonhydrostatic effects corrects the wave propagation speed. The impact of nonhydrostatic effects is investigated using the standard benchmark problem of a solitary wave runup on a plane beach [7].

The organization of present work is as follows. Section 2 introduces the basic equations, the numerical implementation of the model and the mesh generation algorithm. Its final part presents the nonhydrostatic pressure correction algorithm. Section 3 describes the architecture of the Tsunami Early Warning System. In section 4, the system is applied to simulate some realistic scenario of tsunami wave propagation in the Indian Ocean. Section 5 deals with tidal-tsunami interactions whereas section 6 illustrates manifestations of nonhydrostatic effects. Section 7 contains the conclusions of the work.

2. Description of the model

2.1. Boundary-value problem in Cartesian coordinates

In the domain $\tilde{\Omega} = \{(x,y) \in \Omega, 0 \leq t \leq T\}$, where Ω is a plane domain with boundary $\partial\Omega$, we consider the vertically averaged equations of motion and continuity

$$\mathbf{v}_t + (\mathbf{v} \cdot \nabla)\mathbf{v} + g\nabla\zeta = \Phi \equiv f\mathbf{k} \times \mathbf{v} - rH^{-1}\mathbf{v}|\mathbf{v}| + H^{-1}\nabla(K_h H \nabla \mathbf{v}), \tag{1}$$

$$\zeta_t + \nabla \cdot (H\mathbf{v}) = 0, \tag{2}$$

where $\mathbf{v} = (u,v)$ is the horizontal velocity vector, $H = h + \zeta$ is the total water depth, $H > 0$, h is the unperturbed water depth, and ζ is the surface elevation, $\nabla = (\partial/\partial x, \partial/\partial y)$ is the gradient operator, f the Coriolis parameter, $\mathbf{k}$ is the unit vector in the vertical direction, r is the bottom friction coefficient, and K_h is the eddy viscosity coefficient. The set of (1) and (2) is known as the rotating shallow water equations.

On the solid part of the boundary, $\partial\Omega_1$, and on its open part, $\partial\Omega_2$, we impose the following boundary conditions

$$v_n|_{\partial\Omega_1} = 0, \quad \Gamma(\mathbf{v},\zeta)|_{\partial\Omega_2} = \Theta_1, \tag{3}$$

where v_n is the velocity normal to $\partial\Omega_1$, Γ is the operator of the boundary conditions and Θ_1 is a vector-function determined by the boundary regime and different for inflow and outflow [8]. In practice, when the full information on the open boundary is unavailable, in place of the second condition (3) one commonly imposes the boundary condition on the elevation $\zeta|_{\partial\Omega_2} = \psi(x,y,t)$ or the radiation boundary condition $\mathbf{v}_n = \mathbf{v} \cdot \mathbf{n} = \sqrt{g/H}\zeta$. The latter provides free linear wave passage through the open boundary (when the Coriolis acceleration plays a small role). Here $\mathbf{n}$ is the outer unit normal to $\partial\Omega_2$. The accuracy of the reduced boundary-value formulation with only the sea level assigned at the open boundary, was analyzed in [9]. The problem (1)-(3) for the combination $\mathbf{u} = (\mathbf{v}, \zeta)$ is solved for given initial conditions: $\mathbf{u}|_{t=0} = \mathbf{u}^0$.

The equation of energy for set (1) and (2) has the form

$$\frac{\partial E}{\partial t} + \nabla \cdot \left[H \left(g\zeta + \frac{1}{2} |\mathbf{v}|^2 \right) \mathbf{v} \right] = -r |\mathbf{v}|^{3/2} + \mathbf{v} \cdot \nabla(K_h H \nabla \mathbf{v}) \tag{4}$$

where

$$E = \frac{1}{2} \left(H |\mathbf{v}|^2 + g\zeta^2 \right) \tag{5}$$

is the total energy per unit area.

2.2. Method

The finite element spatial discretization is based on the approach by Hanert et al. [10] with some modifications like added viscous and bottom friction terms, corrected momentum advection terms, radiation boundary condition and nodal lumping of mass matrix in the continuity equation. The basic principles of discretization follow the paper of [10] and are not repeated here.

Simulation of tsunami wave propagation benefits from using an explicit time discretization. Indeed, numerical accuracy requires relatively small time steps, which reduces the main advantage of implicit schemes. Furthermore, modelling the inundation processes usually requires very high spatial resolution in coastal regions (up to some tens of meters) and consequently large number of nodes, drastically increasing necessary computational resources in case of implicit temporal discretization.

The leap-frog discretization was chosen as a simple and easy to implement method. We rewrite eqs. (1) and (2) in time discrete form,

$$\frac{\mathbf{v}^{n+1} - \mathbf{v}^{n-1}}{2\Delta t} + fk \times \mathbf{v}^n + g\nabla\zeta^n + \frac{r}{H^n} |\mathbf{v}^n| \mathbf{v}^{n+1} - \nabla K_h \nabla \mathbf{v}^{n-1} + (\mathbf{v}^n \nabla) \mathbf{v}^n = 0, \tag{6}$$

$$\frac{\zeta^{n+1} - \zeta^{n-1}}{2\Delta t} + \nabla \cdot (H^n \mathbf{v}^n) = 0. \tag{7}$$

Here Δt is the time step length and n the time index. The leap-frog three-time-level scheme provides second-order accuracy and is neutral within the stability range. This scheme however has a numerical mode which is removed by the standard filtering procedure. Notice that friction and viscosity contributions deviate from the usual leap-frog method.

2.3. Momentum advection scheme

Because of the discontinued character of velocity representation, special care is required with respect to the implementation of momentum advection. Earlier experiments with $P_1^{NC} - P_1$ code revealed problems with spatial noise and instability of momentum advection when the discretization is used in the form described in [10]. A modified implementation without upwinding terms was found to work well, yet when paired with rather high viscous dissipation for removing small-scale noise in the velocity field. In addition, the implementation of the momentum advection for P_1^{NC} velocities involves cycling over edges in the numerical code, in addition to cycling over elements to assemble the elemental (regular) contributions. This reduces numerical efficiency. This lead us to a simpler approach, which provides some smoothing of the velocity field while removing edge contributions.

According to this approach, prior to calculating the advection term in the momentum equation we project the velocity from the P_1^{NC} to the P_1 space in order to smooth it. To make this projection numerically efficient, nodal quadrature (lumped mass matrix) is employed. The projected velocity is then used to estimate the advection term. Finally we proceed as usual by multiplying the result with a P_1^{NC} basis function and integrating over the domain [11]. This results in a very stable behaviour.

2.4. Other implementation details

2.4.1. Wetting and drying, viscosity

For modelling wetting and drying we use a moving boundary technique which utilizes linear least square extrapolation through the wet-dry boundary and into the dry region. We apply "dry node concept" developed in [12] The idea of this concept is to exclude dry nodes from the solution and then to extrapolate elevation to the dry nodes from their wet neighbours. Because the scheme is neutrally stable it demands horizontal viscosity in places of large gradients of the solution. The coefficient of horizontal viscosity is determined by a Smagorinsky parameterization [13].

2.4.2. Code parallelization

Since a large number of scenarios has to be calculated, code optimization and parallelization is crucial. The operational version of TsunAWI is parallelized employing OpenMP. It is therefore limited to shared memory platforms. The parallelization is implemented by defining parallel regions in the numerically most demanding parts of the code and splitting up the corresponding loops, thus sharing the load to the CPUs involved. The remaining part of the code stays serial. The implementation in this ideology needs therefore smaller changes than a full parallel implementation based on MPI, and proves to be numerically efficient.

2.4.3. Mesh generation

The quality of the triangulation of the model domain is crucial for the model results. The meshes used in the following studies were generated by a mesh generator based on the freely available software Triangle by Jonathan Shewchuk [14]. Starting from a model domain defined within a topography/bathymetry data set (in our case GEBCO 30″) Triangle is used to generate a mesh based on a refinement rule depending on the water depth and prescribed by the corresponding wave phase velocity and the CFL criterion. The triangulation will be refined until the edges in all triangles fulfill the criterion

$$\Delta x \leq \min\{c\sqrt{gh}, c_g \frac{h}{\nabla h}\} \tag{8}$$

Since the Triangle output is not yet smooth enough for numerical experiments several iterations of smoothing are applied. Smoothing steps consist of edge swapping, torsion smoothing and linear smoothing. Torsion smoothing tries to equal out angles around each node, linear smoothing acts on the distance between nodes. These strategies are described in [15].

2.5. Nonhydrostatic approach

2.5.1. Limitations of the shallow water model

The shallow water equations are derived on the assumption that the horizontal motion of the fluid dominates over the vertical one. Using the wave length λ and the reference water depth h as characteristical values of horizontal and vertical motion the ratio

$$\delta := \frac{h}{\lambda} \ll 1, \tag{9}$$

must be fulfilled. In this case, terms containing the vertical component of velocity are very small compared with others and a model reduced to the two horizontal dimensions still provides a good approximation of the three-dimensional flow transport. In the course of neglecting the impact of vertical motion, the hydrostatic approximation is approved: The pressure term is limited to static pressure $p_0 = \rho g(\zeta - z)$, as dynamic pressure forces are mainly induced by vertical elevation. With regard to tsunami propagation, condition (9) is satisfied in deep ocean, as the wave length accounts for hundreds of kilometers. When the tsunami reaches coastal regions, the wave length decreases rapidly and the ratio δ as defined above may become less strict, especially in the presence of horizontal inhomogeneities. Waves become dispersive in this limit. They cannot be represented by the standard shallow water equations since in a hydrostatic model the phase velocity $c \approx \sqrt{gh}$ of a wave packet is not affected by the wave length. A more accurate model may be required near the shore. We are seeking how to improve TsunAWI so that the nonhydrostatic effects can be taken into account if required.

2.5.2. The nonhydrostatic approach

In search of approach that corrects the given one, it is useful to look at terms neglected so far. The deviation from the hydrostatic approximation to the real pressure

$$p' = p - p_0 \tag{10}$$

is of particular interest and serves as a starting point. Returning to the momentum equation in vertical direction and considering the missing nonhydrostatic pressure term in (1), the depth-averaged momentum equations result in

$$\partial_t \mathbf{v} + (\mathbf{v} \cdot \nabla)\mathbf{v} + g\nabla\zeta + \frac{1}{\rho H} \int_{-h}^{\zeta} \nabla p' dz = \Phi \tag{11}$$

$$\partial_t w + (\mathbf{v} \cdot \nabla)w \quad + \frac{1}{\rho H} \int_{-h}^{\zeta} \partial_z p' dz = 0 \tag{12}$$

in which w describes the depth-averaged vertical velocity component. Both additional unknowns w and p' are simplified by assuming linear behavior of non-averaged fields in vertical direction. The values at surface and bottom are partly given by the boundary conditions: the kinematic boundary conditions determine the vertical velocities w_ζ and w_{-h} as

$$w_\zeta = \partial_t \zeta + \mathbf{v} \cdot \nabla \zeta, \tag{13}$$

$$w_{-h} = \quad -\mathbf{v} \cdot \nabla h. \tag{14}$$

At the surface, the boundary condition for pressure enforces the nonhydrostatic part of pressure to vanish, just like the hydrostatic counterpart. Hence, p' depends only on nonhydrostatic bottom pressure $q := p'_{-h}$.

2.5.3. Discretization

For solving equations (11) and (12) a two-step procedure is applied, as suggested in [16] in the framework of finite-difference model, used in [17] for wave breaking and run-up issues and realized in a Finite Element/Finite Volume context by [18]. Firstly, the hydrostatic shallow water equations (1), (2) are stepped forward as before, just as the hydrostatic variant of equation (12) with $p' \equiv 0$. The additional unknowns q and w are introduced as vectors containing the values at the nodes of the triangulation in the same way as the sea surface elevation ζ. In a second step the resulting velocity vector $(\tilde{\mathbf{v}}, \tilde{w})$ is corrected by

$$\mathbf{v}^{n+1} = \tilde{\mathbf{v}}^{n+1} + \frac{2\Delta t}{\rho} \left(\nabla \frac{q^{n+1}}{2} - \frac{q^{n+1}}{2} \frac{\nabla(\zeta^n - h)}{H^n} \right), \tag{15}$$

$$w^{n+1} = \tilde{w}^{n+1} + \frac{2\Delta t}{\rho} \frac{q^{n+1}}{H^n}, \tag{16}$$

in which the correction terms depending on q^{n+1} are nothing else but the calculated integral terms of (11) and (12). The factor $2\Delta t$ arises from using the leapfrog time-stepping scheme. This correction is performed at the end of each time step. However, the nonhydrostatic part of bottom pressure q^{n+1} must be estimated before. While p_0 can be calculated explicitly, the computation of q is performed in an implicit manner. By including the equations (15) and (16) in the depth-integrated continuity equation

$$\int_{-h}^{\zeta} \nabla \cdot \mathbf{v} + \partial_z w \, dz = 0, \tag{17}$$

the nonhydrostatic bottom pressure can be determined. Because of finite-element discretization, equation (17) reduces to a system of linear equations with a large sparse matrix whose entries vary with time step. As concerns the CPU time, computing q proves to be rather expensive compared to the time required by the shallow water model.

3. The German-Indonesian Tsunami Early Warning System GITEWS

The German-Indonesian Tsunami Early Warning System GITEWS was founded after the devastating Indian Ocean Tsunami 2004 as a joint project of German research institutes leaded by GFZ and Indonesian institutes and authorities. In 2005, Indonesia and Germany agreed in a joint declaration to develop a warning system coordinated by the UNESCO Intergovernmental Oceanographic Commission. The system was inaugurated in 2008 and evaluated in 2010 by an international commission including the heads of the four operating worldwide tsunami warning centers.

3.1. Architecture of the Indonesian TEWS

The Indonesian warning systems meets the challenge of near field warning with extremely short warning time. As tsunamigenic earthquakes originate close to the shore, the time between a seismic event and the issue of a warning is limited to just a few minutes. The TEWS therefore relies on a repository of tsunami scenarios TSR precomputed with the tsunami model TsunAWI, see section 3.3. Figure 1 shows a schematic overview of the components of

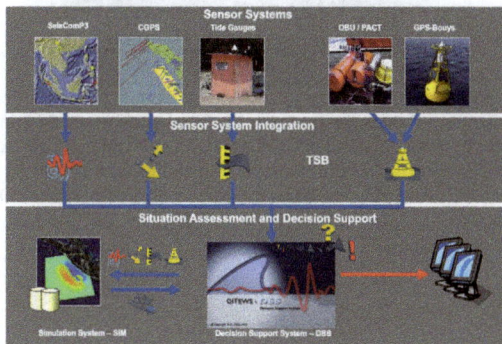

Figure 1. Schematic overview of the early warning process. In case of an earthquake, all available sensor data are gathered by the Decision Support System DSS via the Tsunami Service Bus TSB and sent to the Simulation System SIM. A set of scenarios fitting to the sensor data are delivered and aggregated to an overall perspective.

the TEWS. Data of various sensors, such as the seismic data analyzed by SeisComp, GPS sensors as well as data from tide gauges and buoys is collected via the Tsunami Service Bus TSB [19], and distributed to the Decision Support System DSS [20], which triggers the Simulation System SIM (see section 3.2) with the available sensor data. Based on this data, the SIM delivers a set of best matching scenarios. The DSS performs a worst case aggregation over these scenarios and visualizes expected wave heights and arrival times for the Indonesian coasts. The system is installed at the warning center in Jakarta/Indonesia and assists the Chief Officer on Duty in assessing the potential tsunami risk. He then disseminates warnings to governmental institutions, local disaster management, action forces and media.

3.2. Simulation System (SIM)

In the following the SIM with its components and the algorithm for selecting the best fitting scenarios to an earthquake event shall be introduced in a nutshell. A more detailed overview of the SIM is given in [21], though referring to a prior version of the selection algorithm.

The SIM is written as a Java web application offering web processing WPS and web notification services WNS conforming to the open GIS consortium OGC standard. It is accessible via HTTP-Requests where request and response details are transferred to and from the SIM in XML format.

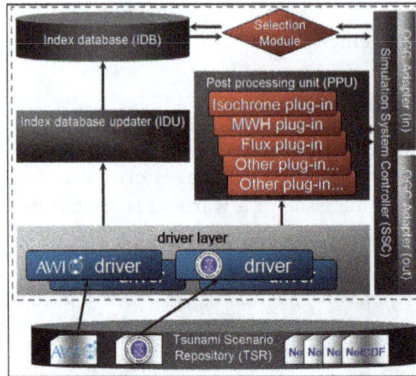

Figure 2. Overview of the SIM software components. Main components are the controller, responsible for overall coordination of processes, the Index Database IDB with data products from the TSR for fast access in case of an event, Postprocessing Unit for extraction of these data products from scenarios, the selection module determining best fitting scenarios for given sensor data, and the OGC-adapter as communication interface.

3.2.1. SIM components

A graphical overview of the software components is given in figure 2.

- Index Database IDB: As mentioned in section 3.1, warning time is crucial for the region of the Sunda Arc. To ensure a response time of the SIM of less than 10 s, significant data products are extracted in advance from each scenario contained in the pre-calculated scenario repository TSR. The IDB contains maximum sea surface height (mwh) and arrivaltimes at coastal forecast zones, isolines of mwh, isochrones, sea surface height time series at tide gauge and buoy locations, and GPS displacement values at GPS station locations.

- Index Database Updater: This component is used to store extracted data products in the index database.

- Postprocessing Unit PPU: The data products are extracted by the post processing plugins compounded in the PPU. The plugins are written in C to employ fast mathematical calculation routines. There task is to extract the information stored in the IDB and to generate SHP-files for visualizing the wave propagation in the DSS. Furthermore, the simulated inundation can be extracted as SHP-files, which are a basis for a priori risk assessment and hazard maps.

- Driver Layer: The driver layer separates the scenario data format from the internal representation of data in the SIM and thus allows to integrate scenarios from different origins and in different formats. One driver per scenario type acts as a translator between scenario data format and the data structure used in the SIM. At the time being, only scenarios calculated with TsunAWI are addressed.

- Scenario Selection Module: The component responsible for the identification of the best fitting scenarios to an earthquake event is described in section 3.2.2.

- Tsunami Scenario Repository TSR: It contains all pre calculated scenarios and is described in more detail in section 3.3

3.2.2. Selection algorithm

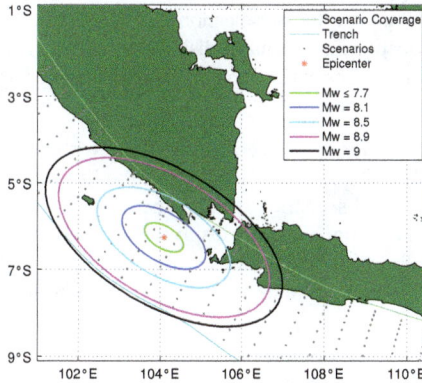

Figure 3. Visualization of the seismic uncertainty ellipse with different size depending on the observed magnitude.

The selection algorithm uses a multi sensor approach combining the different available sensor types to acquire a set of best matching scenarios to an earthquake event. By basing the selection on different sensor data types, uncertainties resulting from inaccurate measurements and errors in the tsunami model are reduced.

In the selection algorithm as described in [21], the different sensor types and even individual sensor stations could be weighted so to factor their individual estimated uncertainty. Initially, the so called matching values generated for each sensor type for a scenario were accumulated in an overall weighted sum of matching values defining a measure of suitability to the current event.

The experiences with real sensor data in the GITEWS project showed that each sensor group has to be regarded separately with its characteristics in mind. The weighted sum over all matching values is therefore replaced by a stepped approach starting with seismic data delivering the most robust values, followed by CGPS data.

The algorithm is specialized for the Sunda Arc taking into advantage that the rupture is often oriented along the trench. The preselection of scenarios based on seismic data is performed by calculating an elliptic area around the measured epicenter within which the selected scenarios lie (figure 3). The dimension of the ellipse depends on the measured magnitude M_W, the long ellipse-axis is given by

$$r_L = 10^{0.5 - [M_W + 0.3] - 1.8} \text{km}.$$

To ensure that at least one scenario is covered, for small magnitudes $\tilde{r}_L = 180$km is chosen. The orientation of the ellipse is derived from the orientation of the trench between the two coordinates found by going $r_L/2$ up an down the trench from the nearest point to the epicenter on the trench.

The second important sensor data class in GITEWS are CGPS dislocation vectors. The sensor data is reliable, and the measurement arrives at the DSS fast, thus allowing for a better estimate of the tsunami hazard in the first few minutes after the earthquake. For each sensor and each scenario (that remains after preselection by the seismic sensor data) the length of the measured and the corresponding scenario dislocation vector are compared. Scenarios for which a defined minimum number of similar GPS values is reached are taken into account for the final set of best fitting scenarios.

If several scenarios are chosen for one epicenter, only those with the largest magnitude within the uncertainty range $[M_w - 0.5, M_w + 0.3]$ is kept in the list, because in DSS processing, a worst case aggregation over all scenarios is performed. For each coastal forecast zone, the shortest arrival time is taken from all scenarios in the list, and the highest maximum wave height. The scenarios with lower magnitudes will not account for this aggregation, hence skipping them reduces the amount of data to be processed without changing the result.

3.3. The Tsunami Scenario Repository

As of March 2012, the GITEWS Tsunami Scenario Repository TSR contains 3470 scenarios for prototypic ruptures (RuptGen2.1 [22]) with magnitudes in the range of Mw=7.2,7.4,...,9.0 on 528 different epicenters, see figure 4. These scenarios for prototypic ruptures in the Sunda

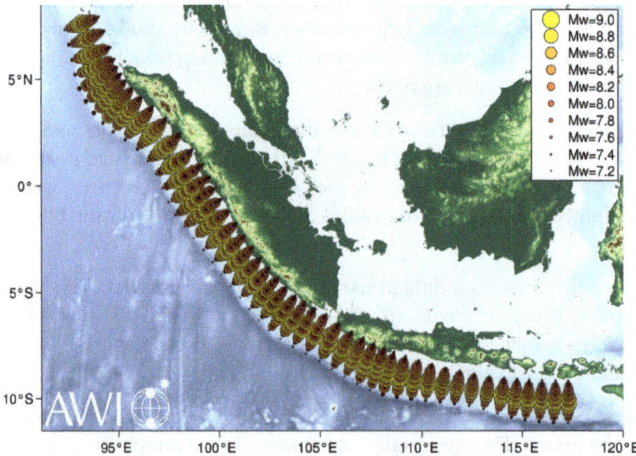

Figure 4. Tsunami scenario repository for GITEWS. Each circle symbolises a scenario with center at the given position. Scenarios for up to ten different magnitudes Mw= 7.2, 7.4, ..., 9.0 are calculated at each of the 528 positions, resulting in 3470 scenarios in total.

Arc are used in the warning situation as well as for a priori risk analysis and hazard maps. The quality of the scenarios, being crucial for both tasks, relies strongly on the numerical implementation of the governing physical equations and on an accurate representation of the bathymetry and topography especially in coastal regions. Therefore, GITEWS included the development and concise validation of the tsunami model TsunAWI, described in detail in section 4.

For the TSR, TsunAWI operates on a grid with the resolution changing seamlessly between 14 km in deep sea, 150 m at the coast and down to 50 m in regions of special interest, e.g., densely populated areas and at tide gauges. The run-up scheme provides a realistic approximation of the inundation hazard. The dense net of scenarios allows to evaluate the

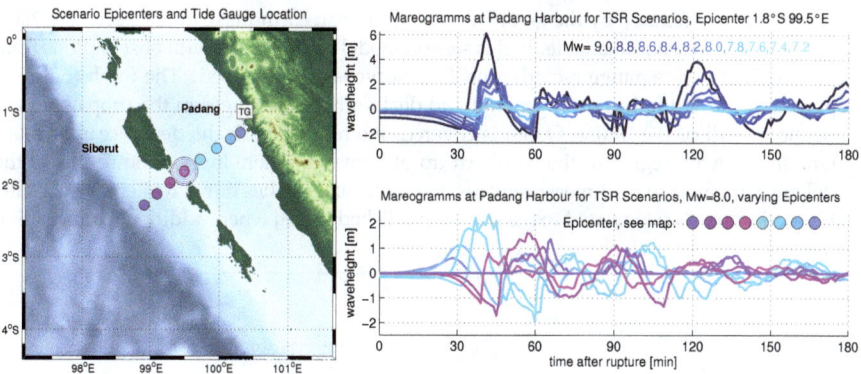

Figure 5. Influence of the magnitude Mw and of the source location on the strength of a tsunami: Mareograms at Padang Harbor for TSR-scenarios originating at the same epicenter with different magnitude (above) and with equal magnitude, but different origin (below).

effect of small changes in epicenter and magnitude as illustrated in figure 5 for artificial mareograms for the harbour of Padang, Sumatra. The logarithmic energy scale of the magnitude is clearly illustrated by the upper mareogram. The lower graph shows more complex features due to varying epicenters on a line perpendicular to the trench. While the magnitude is fixed to Mw=8.0 in this example, the tsunamis originating closest to the coast have the lowest impact. On the one hand, this is due to the initial conditions generated by RuptGen which assumes a source depth of 0 km at the trench increasing up to 100 km under the main islands. A second factor determining the strength of the tsunami is the height of the water column at the origin, and this of course is small in coastal regions.

Epicenters southwest off the coast yield increasing wave heights in Padang, until the Mentawai islands are reached. The pink epicenter marks a turning point where the corresponding tsunami is no longer trapped in reflections between Sumatra and the islands, relieving the situation for Padang.

4. Tsunami modeling with finite elements: applications

TsunAWI is still under development and therefore the performance and model results must be constantly evaluated to ensure a consistent and stable code. Validation needs to be performed on several levels, among these are:

- Consistency / convergence,
- Benchmarks in idealized settings with well defined results,
- Comparisons in real cases with measured data.

This section deals mostly with the last case, section 6 includes some cases with idealized bathymetry and compares hydrostatic and non hydrostatic results.

4.1. Indian Ocean tsunami in December 2004

For the tsunami generated by the great Sumatra-Andaman earthquake on 26 December 2004, a large amount of data is available. In this section model comparisons with observations from satellite altimetry, tide gauge records, and field surveys are discussed. The mesh (figure 6) employed in this study has been generated to deal with all the stages in the propagation of the tsunami. All of the Indian Ocean is covered, the resolution in the deep ocean is about 15 km. In the Aceh region in the Northern tip of Sumatra, where inundation results of the model are compared to field measurements, the mesh size reaches down to 40 m. The results in this section are closely related to the studies published in [23] where additional information can be found.

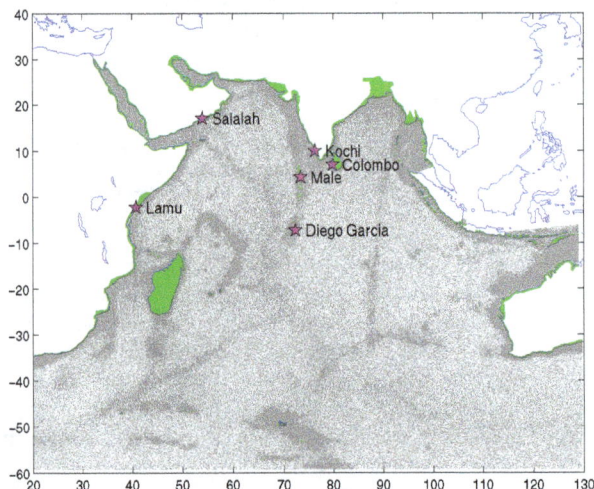

Figure 6. Mesh density in the model domain. The green areas are land nodes contained in the mesh which are initially dry. The resolution ranges from 15 km in the deep ocean to 40 m in the Northern tip of Sumatra.

4.1.1. Available topography and bathymetry data

Topography and bathymetry in the following experiments are based on several data sources. The GEBCO data set [24] is globally available at 30 arc seconds resolution and is used in all meshes as initial topographic and bathymetric information which is replaced by more precise data wherever they are available. Bathymetric data is locally improved by ship measurements and nautical charts. Topographic information is improved by the SRTM data set [25] which is freely available at a resolution of 90 m, however the vertical accuracy is usually not sufficient for model studies of runup in the coastal domain. In the area of Indonesia the SRTM data were additionally processed by the German Aerospace Agency (DLR) and provided at a resolution of 30m. In Aceh region additional bathymetric and topographic data were provided by BPPT Jogyakarta.

4.1.2. Source model

In tsunami modeling it turns out that the exact knowledge of the source, i.e. the initial conditions of the model is of crucial importance for comparisons with data. Usually the source parameters are optimized with respect to certain measurements and normally it's not possible to match model results with different measurements like tide gauges *and* inundation. The source model used in the following studies is based on the results presented in [26]. The objective of that paper is to optimize the subfaults of the earthquake such that an optimal agreement with certain tide gauge records is obtained. The resulting subfaults are shown in figure 7. In order to demonstrate the impact of source parameters with respect to the matching with data the orientation of the Southern faults is modified as indicated in that figure. The strike angle of subfaults A/C is changed from 340/340 degrees as proposed in [26] to 290/320 degrees. All other parameters can be found in in [23, 26].

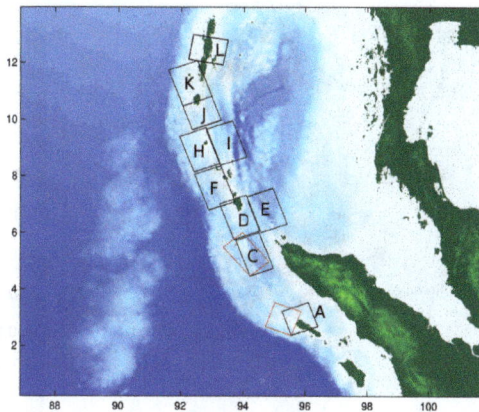

Figure 7. Sub-faults as proposed in [26]. The rupture area has been decomposed into 12 sub-regions. With a rupture speed of 1.7 km/s, the whole rupture process takes 12 min. Faults B and G have zero slip in our experiments and are not displayed. The red faults have different strike angle and show better agreement with satellite altimetry.

4.1.3. Model setup

The mesh used in these experiments consists of 5 million nodes and 10 million triangles. The finest resolution enforces a time step of half a second. The model is integrated for 10 hours. In the positions along the satellite tracks the model state is written to a file every second.

This enables a careful comparison between model and satellite altimetry data. Additional information on model setup and initialization is contained in [23].

4.1.4. Satellite altimetry

For the first time the Indian Ocean tsunami was observed by several Satellite missions. **Jason 1 (J1)** and **TOPEX/POSEIDON (T/P)** were above the bay of Bengal about two hours after the earthquake, whereas **ENVISAT** observed the tsunami about 3h20' after the event. In all cases the tsunami signal was extracted from the altimeter measurements by subtracting the data from consecutive cycles. Table 1 summarizes the cycles that were taken into account in the three missions. Figure 8 contains the groundtracks of **J1** and **T/P** whereas figure 9 displays the extracted tsunami signal for **J1** and **T/P**. The model results are interpolated in time and space and extracted from a Hovmöller diagram as shown in figure 10. The **J1** signal in figure 9 shows clearly a double peak in the position of the leading wave crest which is due to the partial waves generated by the southern subfaults, whereas in the **T/P** measurements these partial waves overlap. This behavior is reproduced by the model however as it turns out the matching depends strongly on the fault parameters. Changing the strike angles in fault A from 340 to 290 and in fault C from 340 to 320 improves the matching and lowers the RMS errors as indicated in table 1.

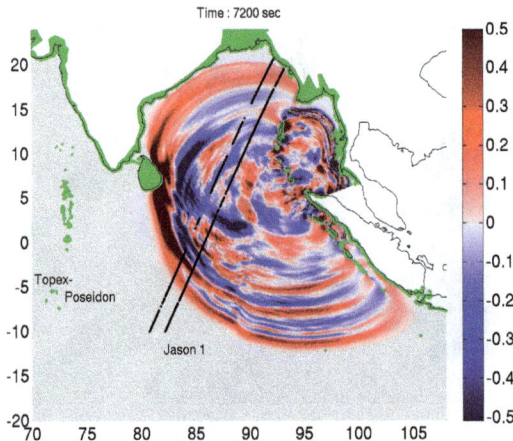

Figure 8. Model snapshot after two hours. Sea surface elevation together with the positions of the satellite tracks Jason 1 and TOPEX/POSEIDON

Mission	Pass	Cycle	Equator time	rms (340/340)	rms (290/320)
Jason 1	129	109	02:55 UTC	0.243 m	0.238 m
Topex/Poseidon	129	452	03:01 UTC	0.223 m	0.164 m

Table 1. Satellite Missions used in this study for data comparison. The last columns quantify the RMS error obtained in scenarios with different strike angles in subfaults A and C (figure 7).

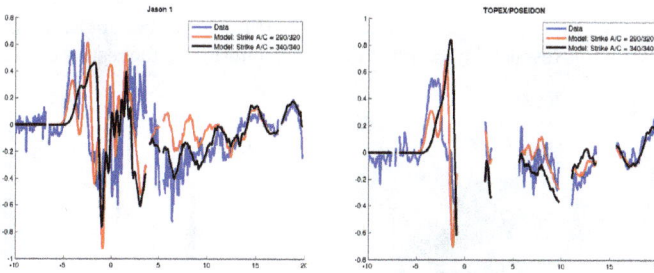

Figure 9. Tsunami signal extracted from satellite tracks Jason 1 and TOPEX/POSEIDON (blue lines) together with model results along the tracks interpolated in space and time (red and black lines).

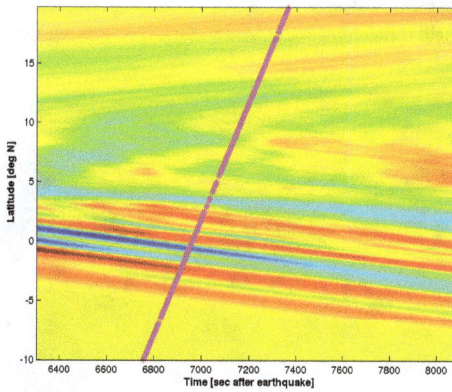

Figure 10. In the location of the satellite tracks (figure 8) the sea surface elevation is written to file every second. From the resulting Hovmöller diagram the model results corresponding to the satellite observations as shown in figure 9 are interpolated in space and time.

4.1.5. Inundation

After the event in December 2004 several field surveys examined the runup and flow depth in the affected regions. In the area of Banda Aceh, which was most heavily hit by the tsunami eight locations with well documented field measurements were chosen to compare model runup to these results. Runup and flow depth depend heavily on the prescribed roughness parameterization. Several experiments with identical initial conditions and varying Manning parameter were conducted. In all cases constant manning number was applied in the whole model domain.

Figure 11 shows the locations as well as a comparison of measurements and model results for different roughness parameters in a bar diagram. These results are obtained with strike angles 290/320 in subfaults A/C. From the corresponding rms errors the best fitting Manning number for experiments with constant parameters can be chosen. Figure 12 shows the flow depth in Aceh region together with a line depicting the boundary of inundation as it was obtained from Satellite images (provided by DLR).

Figure 13 displays the flow depth comparison for different strike angles in the Southern subfaults. Also in this respect the modified orientation improves the results considerably.

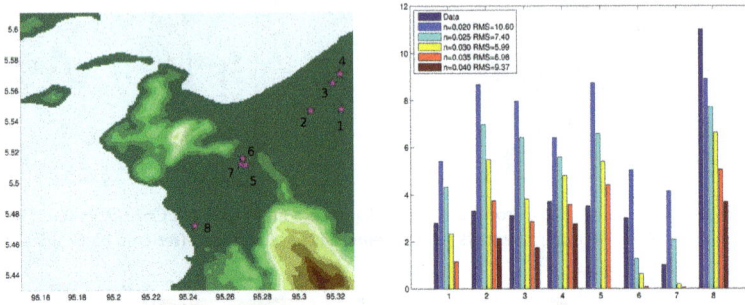

Figure 11. a) Positions of field measurements, b) model results for varying roughness parameters in comparison.

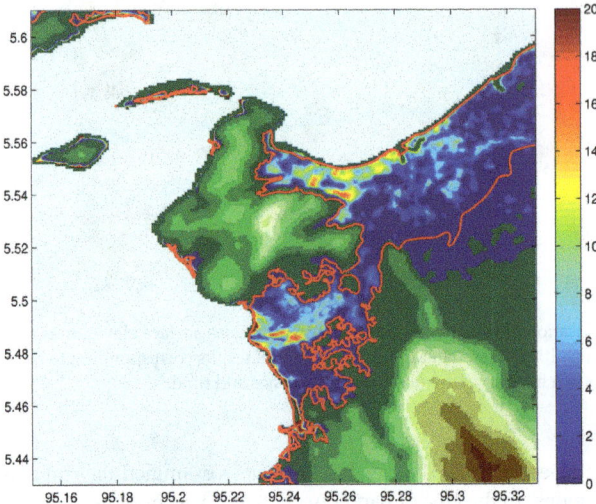

Figure 12. Flowdepth obtained for n=0.025 in Aceh region. The red line indicates the inundated area as it was determined from satellite images and provided by DLR.

4.1.6. Tide gauge records

The Indian Ocean tsunami was observed by tide gauges world wide. Since the arrival time and estimated wave height are among the most important warning products good matching between model results and arrival times and the height of the leading wave crest as they were recorded by tide gauges is desirable. Therefore hindcast experiments with such comparisons are included in the present study as well. Figure 14 summarizes the time series in some of the locations throughout the Indian Ocean where data is available. The comparisons between tide gauge records and model results show generally a good agreement with respect to arrival

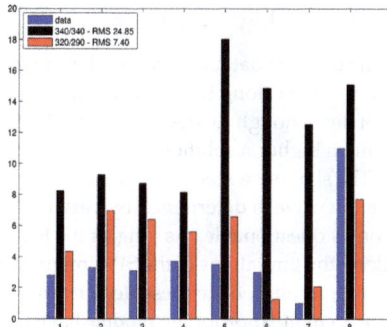

Figure 13. Flowdepth comparison for experiments with strike angles of plates A and C set to 340/340 and 290/320 (compare figure 7). The rms error is considerably improved by the adjustment.

time. The wave height of the leading crest is sometimes underestimated however in most of the locations the agreement is very good. The orientation of the Southern subfaults does not influence the results in far distance as Salalah and Lamu show, the matching in Colombo is slightly better with the uniform values (340/340). This is consistent with the derivation of these strike values as this station was used for optimization in [26].

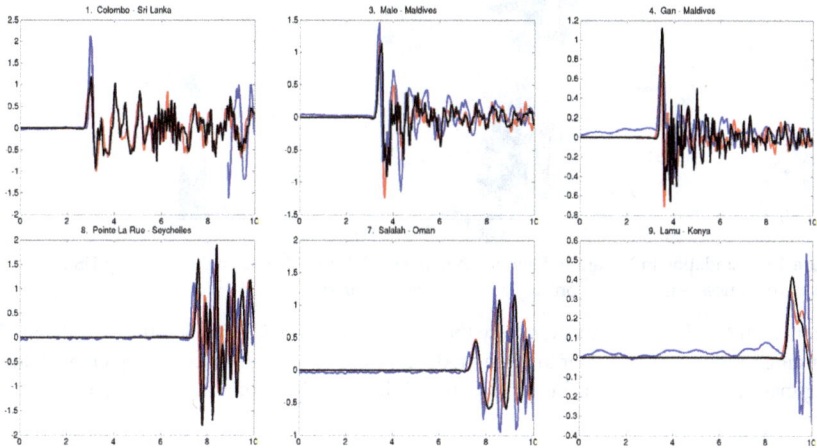

Figure 14. Tide gauge records (blue lines, time in hours, elevation in meter) and corresponding model results in locations displayed in figure 6. The red lines show results for strike angles 290/320 degrees, black lines the corresponding results for 340/340 degrees in subfaults A/C.

4.1.7. Summary of the Indian Ocean study

Summing up, it turns out that TsunAWI is able to reproduce observational results on several scales of wave propagation. Both the large scale propagation and arrival times throughout the Indian Ocean as well as the inundation in a selected area with high resolution is feasible in the approach with unstructured meshes. However the model results depend strongly on the quality of the source model.

4.2. Inundation experiments with high resolution

The scenarios generated for the tsunami database are based on a mesh with resolution ranging from 10 km in the deep ocean to 150 m along the global coastline and 50 m in priority regions. This resolution is certainly not high enough for detailed inundation simulations. Hazard maps are based on data sets with much higher resolution and the corresponding model setup must contain a comparable mesh. The shallow water equations are based on very small aspect ratio defined in (9) and as soon as the wave is determined by very fine scale topographic features the validity of the SWE becomes questionable. As long as the focus of such model studies is put on the inundated area alone the limitations of the SWE might still be acceptable and in the following experiments with high resolution are presented. The actual error however has to be investigated and comparisons to more appropriate models have to be performed. Section 2.5 deals with some of the improvements possible with nonhydrostatic corrections in idealized settings. Within the GITEWS project hazard maps were produced by DLR based on model

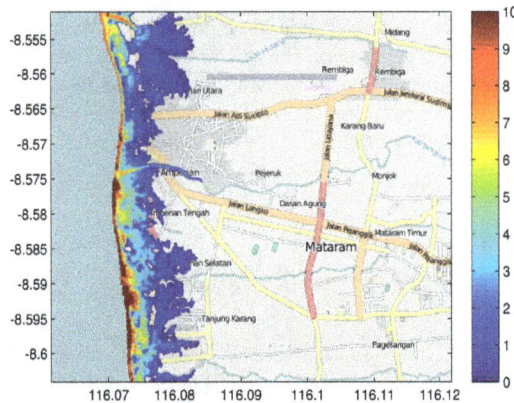

Figure 15. Inundation in Mataram. Flowdepth in the model study based on the Intermap DSM. The bottom roughness is given by a constant Manning value n=0.02.

results with MIKE 21 FM carried out by the German Research Center Geesthacht (GKSS). The Manning roughness parameter in these model runs are varying in space and given by detailed roughness maps. Details on this approach and the model can be found in [27] and citations therein.

The focus of this section is to highlight the importance of the bottom roughness parameter. Simulations of tsunamis with identical initial conditions were carried out in a mesh with resolution up to 5 m. The study area is Mataram, on the island of Lombok, Indonesia. The topography data in this case was based on the Intermap data set and provided by DLR. The resolution of the topography data is 5m and two versions, a digital surface map (DSM) and digital terrain map (DTM) were used. The DSM version is a first-reflection data set and contains elevations of vegetation and buildings. The DTM on the other hand is a model where all these features were removed.

Figure 15 displays the flow deph based on the DSM topography data, whereas figure 16 shows the same quantity for the DTM data with three different values of Manning's n. Additionally,

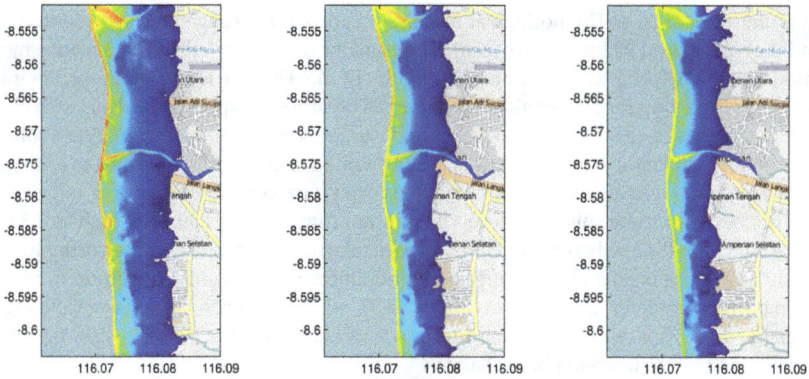

Figure 16. Results as in figure 15, however for Intermap DTM with three different Manning parameters n=0.04, n=0.06, and n=0.08 in the left, middle and right panel.

in case of a tsunami buildings and infrastructure will partly collapse and result in a very heterogenious fluid flow which may be only poorly approximated by a bottom friction parameterization. The result in figure 15 contains small scale features, however since it doesn't take into account the destruction of buildings and vegetation, it may be underestimating the true inundation. [27] suggests the use of DTM data with detailed roughness maps. Results for DTM data and different roughness values are displayed in figure 16. According to [27] the right panel corresponds rather to the situation of an urban area with partly collapsing buildings whereas the left panel corresponds to the situation of the whole area covered by coarse sand.

Given the vast differences displayed in figure 15 and 16 this section and its results is mostly meant to raise the awareness for the dependency of inundation results on the friction parameters. The quality of topography data plays an even bigger role and the best suited data set and model parameterization must be chosen before hazard maps are produced.

5. Tide-tsunami interaction

For investigating the influence of tidal motion on tsunami wave propagation three model runs were carried out: a) pure tidal motion, b) pure tsunami wave propagation and c) tsunami wave propagation on the background tidal motion. As object of modelling we have chosen the Indonesian coast including south part of Java, Bali, Lombok, Sumbawa and Sumba islands and North Sea. The choice of these two objects is not accidental. The first of these is the object with complex morphometry and sharp bathymetry, which contains a large number of islands and straits. The North Sea is a shelf sea with slowly varying bathymetry.

5.1. Interaction between tides and tsunami for Indonesian coast

For calculations, we use the ETOPO 30 sec. morphometry dataset and data on tides in the Indian Ocean derived from the TPXO6.2 dataset of oceanic tides [28]. The calculations were performed on an unstructured mesh (figure 17 a) with 177132 nodes and 347098 elements with a time step of 20 s. The horizontal mesh size varies in the range between 160 m (in the inundation zone) and 29 km (deep ocean). The tsunami wave is initialized by a source model as used in the GITEWS scenario database (figure 17 b). The particular choice of the initial condition is irrelevant for our considerations here. The series of experiments deals with effects of nonlinear interaction between tidal waves and tsunamis in the Indian Ocean. A full solution (propagation of tsunami wave on the tidal background) and composite solution (an arithmetic sum of tsunami wave and tides computed separately) are compared. It turns out that the difference between these solutions is very significant (figure 18) reaching as high as 3.5 m in the coastal region (St.3) where nonlinearity is particularly important. It is noteworthy that the first tsunami wave is only slightly affected by nonlinearity (change in amplitude $\approx 5-8\%$, with max. amplitude of tsunami wave 16 m, St.3), while the second wave is affected more essentially, at $\approx 25\%$ (with max. amplitude second tsunami wave 3.5 m). Also, the strong nonlinear interaction leads to phase changes.

In search of explanation as to why the impact of tide-tsunami interaction is so significant, we repeated the above cases once more but switching off the momentum advection term in the equation of motion. Figure 18 (Right panel) clearly shows that the evolution of potential and kinetic energy is now significantly different from the full nonlinear case. Hence we conclude that in the near-shore regime, non-linear interaction of the wave-induced velocities contributes greatly to the complex behaviour of tide-tsunami wave phenomena. The exact mechanisms still call for a more careful analysis and suggest a topic for future research.

5.2. Tide-tsunami interactions in the North Sea

The last known mega-tsunami to hit rim countries of the North Sea took place over 8,000 years ago. But coastal areas of the North Sea are vulnerable to tsunamis caused by localized underwater landslides. Numerical simulations are performed on the basis of a high-resolution multilayer model to study initial generation of waves by landslides in the Storegga area and with a TsunAWI model to study the wave propagation further in the North Sea.

As before, three simulations have been performed to identify the character of the nonlinear interaction of tide and tsunami waves. First simulation dealt with the M_2 tidal structure in the North Sea. In the second case only tsunami waves excited by a signal from the landslide model were simulated. The third simulation dealt with a joint solution for the tsunami wave on the

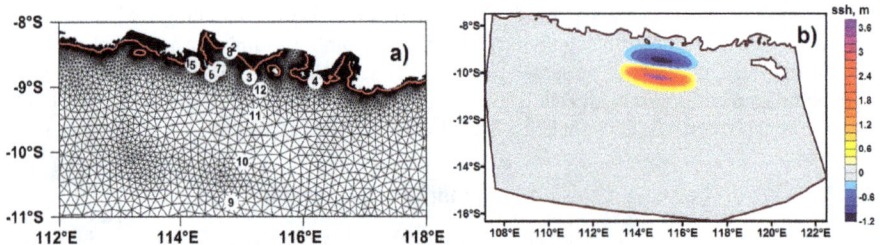

Figure 17. a) Part of the unstructured mesh with control stations. White curve marks coastal line. b) Initial condition for tsunami wave.

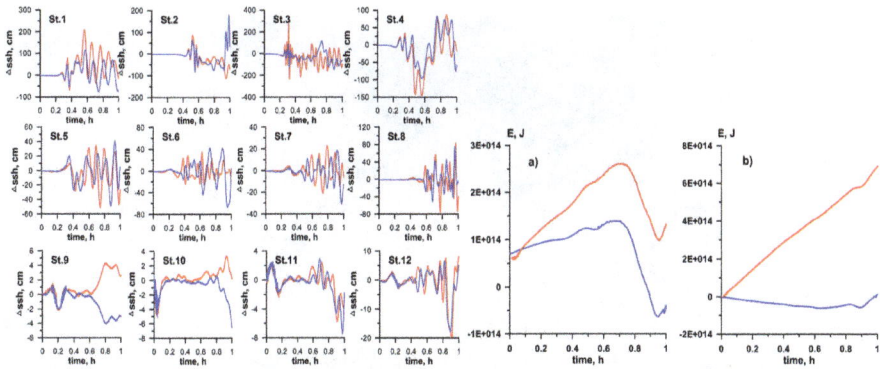

Figure 18. Left panel: Nonlinear interaction for control stations (1-4 - coastal; 5-8 - shelf; 9-12 - deep). Difference in elevation between full solution and composite solution. Red line - high tide; blue line - low tide. Right panel: Difference in potential (a) and kinetic energy (b) between full solution and composite solution. Red line - with advection in momentum equation; blue line - without advection.

background of semidiurnal tide. In this case the boundary information (north boundary) is represented by a superposition of two waves.

A combination of two Kelvin waves (one propagating from the northern open boundary along the coast of England and another through the English Channel) produces three nodal lines: on an output from English Channel, near north coast of Germany, and southern coast of Norway [11]. There is reasonable agreement for semidiurnal tidal elevations with respect to observation (figure 19).

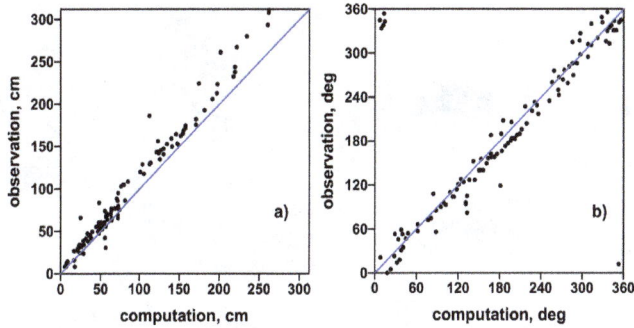

Figure 19. Correlation between the observed and computed values of the amplitudes (a) and phases (b) of the tide wave M_2. The RMS error of the computation is 29.8 cm (for 112 stations).

A two-layer model based on shallow water equations written in a curvilinear coordinate system is applied to simulate waves generated by hypothetical submarine slides at Storegga on the Norwegian continental slope (figure 20). The numerical method is based on composite schemes for split operators [29]. The model is initialized by prescribing submarine slides in the Storegga area off western Norway [30, 31]. The signal at the open boundary of the North Sea

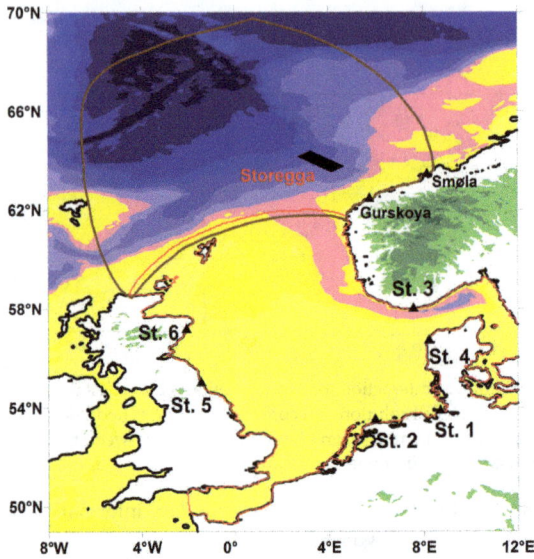

Figure 20. Computational domain with slide positions (Storegga) and control stations. The brown line shows the domain for submarine slides modelling. The red line marks the domain for tsunami wave propagation (TsunAWI model).

simulated by the model serves as the boundary condition for finite-element model (TsunAWI) used to simulate wave propagation further in the North Sea.

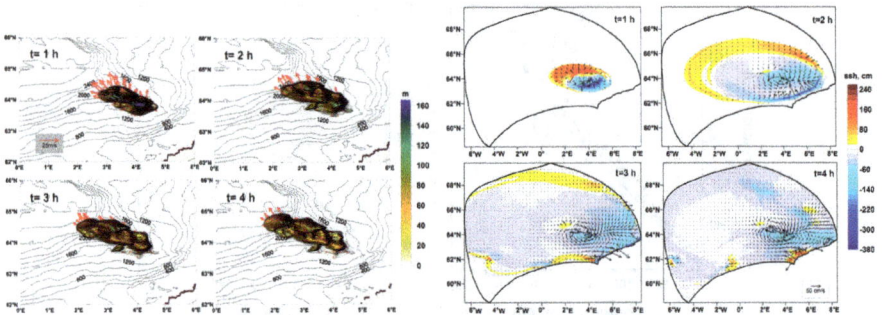

Figure 21. Thickness (m) and velocity (m/s) for Storegga landslide at four subsequent times (Left panel). Sea surface elevation (cm) and velocity (m/s) for Storegga landslide (Right panel) at four time instants.

Figure 21 (Left panel) shows the position, thickness of the submarine slide (Storegga) and velocity field in the bottom layer (landslide) at four different time moments. The slide velocity reaches 25 m/s, in agreement with [30, 31]. The submarine slide leads to long wave on the surface as shown in figure 21 (Right panel) for the same time moments as in figure 21 (Left panel).

A composite solution and full solution are compared. It turns out that the difference between these solutions is fairly large (figure 22, Left panel) reaching as high as 60 cm in the coastal region where nonlinearity is particularly important. Additional experiments showed that the arrival time of tsunami waves in the full solution varies but slightly. In search of explanation

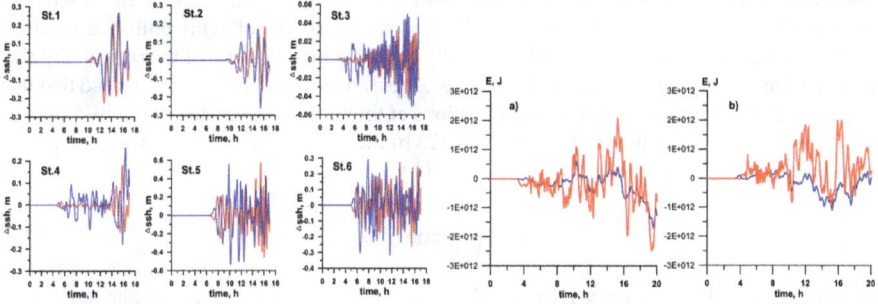

Figure 22. Left panel: Nonlinear interaction. Difference in elevation between a composite solution and full solution. Red line - high tide; blue line - low tide. Right panel: Difference in energy between a solution with advection and solution without (a) momentum advection (blue line) and solution without (b) nonlinearity in the continuity equation (red line).

as to why the impact of tide-tsunami interaction is so significant, we repeated the above cases once more but switching off the momentum advection term in the equation of motion (linear case) or nonlinearity due to change of layer thickness in the equation of continuity (figure 22, Right panel). Here we can see the other mechanism of tide-tsunami nonlinear interaction due to level change.

6. Validation of the nonhydrostatic approach

For comparison of the results delivered by the original shallow water model on the one hand with the nonhydrostatic extension on the other two standard testcases are carried out. The first one is investigated in [18], it describes the behavior of a standing wave within a closed basin in respect of the phase velocity depending on the ratio δ defined in (9). As second application a tank experiment (see [32]) is modeled and the results are compared against observation data at various instants. This example shows dispersive effects that occur when a solitary wave is running up a plane beach.

6.1. Standing wave in a basin

The computational domain Ω of this testcase presents a rectangular basin with fixed length and width $l = 10$ m and $w = 4$ m, respectively, while its depth h vary in different experiments between 0.25 m and 10.5 m. At the walls of the basin, condition (3) for solid boundaries $\partial\Omega_1$ is applied. With a wave length of $\lambda = 2l$ a standing wave with an amplitude of $a = 0.01$ m can be arranged by the initial condition

$$\zeta^0(x, y) = -a \cos(\frac{2\pi x}{\lambda}), \tag{18}$$

as illustrated in Figure 23 a). Both, horizontal and vertical velocity equal zero initially. In the absence of sinks and sources, two waves with same amplitude and frequency emerge. They move in opposite directions and form a standing wave with antinodes at the boundaries, $x \in \{0,l\}$, and a node in the middle of the basin, $x = l/2$. An unstructured mesh with a resolution of about 0.125 m covers the domain Ω. Using a time step size of $\Delta t = 0.005$ s, several experiments with varying basin depths were carried out with both the original shallow water model and the nonhydrostatically corrected one. Because of the periodicity, the propagation speed can be determined. In this example an inviscid fluid is assumed, so that the pressure gradient is the only force. Modifications of the pressure term due to nonhydrostatic effects entail changes of the motion. In Figure 23 b) the results of both models are compared with the reference propagation speed, estimated by

$$c = \sqrt{\frac{g}{k}\tanh{(kh)}}, \tag{19}$$

in which $k = 2\pi/\lambda$ is the wave number. If $\delta \ll 1$, the argument of the hyperbolic tangent in (19) is very small and the approximation $\tanh(x) \approx x$ can be adopted. So, the propagation speed converges to the phase velocity $c_{sw} = \sqrt{gh}$ that is independent of wavelength and characterizes motions in the standard shallow water approximation. It is not surprising that the phase velocity curve based on the shallow water model is congruent with c_{sw} for all δ. More interesting is the result of the nonhydrostatic approach: although it is simplified by using depth-averaged values, it offers a good approximation to the propagation speed up to ratio δ in excess of about 0.4. Consequently, as δ is increased, the nonhydrostatic correction provides a better approximation to the wave propagation speed.

6.2. Solitary wave on a plane beach

In the second testcase the results of the shallow water model and the nonhydrostatic approach are compared to observation data of a tank experiment [32]. The main part of tank is of constant water depth d. Near the shoreline ($x = 0$), the bathymetry gradually ascents with a constant slope of $\tan(\alpha) = 1/19.85$, as depicted in Figure 24. All spatial quantities in the

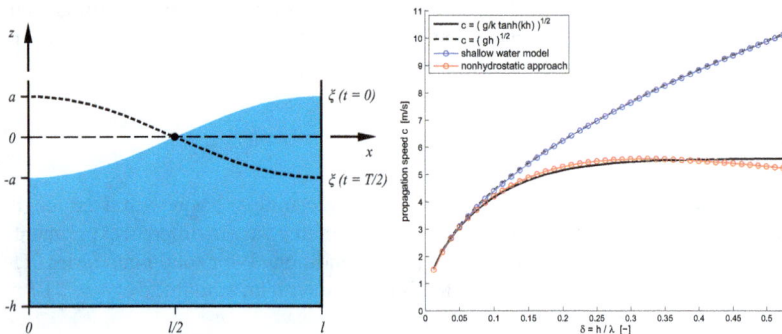

Figure 23. a) Setup of testcase 6.1: Standing wave in a basin. b) Different results in reference to propagation speed c.

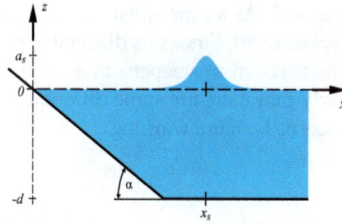

Figure 24. Setup of testcase 6.2: Solitary wave on a plane beach.

model setup are made dimensionless with the tank depth d. To simplify matters, $d = 1$ m is chosen. A solitary wave with maximum height $a_s := a/d$ is given by

$$\zeta^0(x,y) = \frac{a}{d}\text{sech}^2\left(\sqrt{\frac{3a}{4d}}(x - x_s)\right), \quad x_s = \frac{d}{\tan(\alpha)} + \sqrt{\frac{4d}{3a}}\text{arccosh}(\sqrt{20}). \quad (20)$$

with an initial horizontal velocity of $\mathbf{v}^0 = (-\zeta^0\sqrt{g/d}, 0)^T$. The vertical velocity is prescribed by the boundary conditions (13) and (14). While the ramp can be flooded for $x < 0$, boundary condition (3) on $\partial\Omega_1$ are imposed at the solid walls of the tank. An unstructured mesh with $\Delta x \in [0.1, 0.2]$, in which the finer resolution accords to the ascending part, covers the computational domain $\Omega = [-10, 70] \times [-0.5, 0.5]$. The time step is selected as $\Delta\tau = \Delta t\sqrt{g/d} = 0.004$. With the help of a Manning factor of $n = 0.01$ the low friction inside of the tank is approximated. By setting the maximum wave height $a_s = 0.0185$, the shallow water model and the nonhydrostatic approach provide similar results which agree very well with the observation data. More interesting is the case of breaking wave with $a_s = 0.3$. Figure 25 illustrates different stages of the flow evolution in four snapshots. It is apparent that the nonhydrostatic approach approximates the shape of the solitary wave much better than

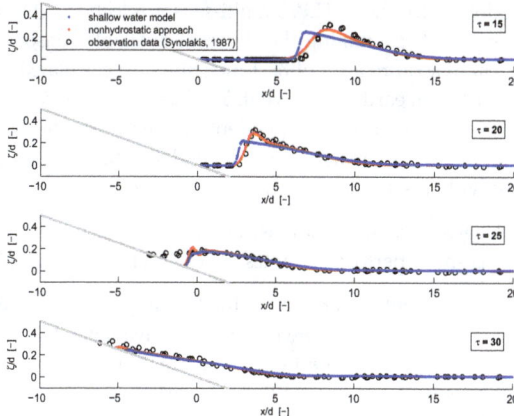

Figure 25. Snapshots at different times $\tau = t\sqrt{g/d}$ for a solitary wave with a maximum wave height of $a_s = a/d = 0.3$.

the hydrostatic shallow water model. As a consequence of boundary condition (13) the wave front experiences a vertical displacement. Since this dispersive process cannot be represented by the shallow water model, its wave front steepens in an unnatural way. Furthermore, the snapshot at $\tau = 30$ shows clearly that there are some differences with respect to inundation, which is very important as concerns tsunami warning.

7. Discussion

The combination of non-conforming velocity with linear elevation suggests a well-rounded choice for shallow-water modelling on unstructured triangular grids, with a particular focus on simulating tsunami wave propagation. Although our approach was initially inspired by the algorithm proposed by Hanert et al. [10] the resulting model is essentially different from it in a number of key directions. First, it is equipped with wetting and drying algorithms and can simulate inundation caused by tsunami. Second, it suggests a choice of stably working discretization of the momentum advection which all improve over the original method of Hanert et al. [10] and differ between themselves in a degree of smoothing applied. Third, it uses the Smagorinsky horizontal viscosity which is crucial for keeping the dissipation on the level that does not affect the quality of the solution. Finally, the explicit time stepping and use of the nodal quadrature (mass matrix lumping) of the time derivative term in the continuity equation ensure numerically efficient performance while providing a straightforward and easy to implement algorithm.

Nonhydrostatic effects become important when vertical acceleration is not negligible. For tsunami wave generation and propagation it can happen at the very initial stage and during run up. In this context, we present an algorithm of the nonhydrostatical pressure for vertical averaged equations.

We introduced the architecture of the German-Indonesian Tsunami Early Warning System (GITEWS) and methodology for the GITEWS multi-sensor selection. The selection algorithm uses a multi sensor approach combining the different available sensor types (SeisComP3, CGPS and Tide Gauges) to acquire a set of best matching pre calculated scenarios (Tsunami Scenario Repository, TSR) to an earthquake event. The basic principle is to reduce the number of possible tsunami scenarios by using independent measurements of the same event. Only a small number of scenarios can match the independent measurements, even with high uncertainty in each individual set of measurements, since the combination needs to fit.

One of the important stages of the successful use of the numerical model is its verification and validation. This work has been performed on test cases and published in [33].

We simulated the tsunami event generated by the Sumatra-Andaman (2004) and Tōhoku (2012) earthquakes. The model results were compared to available data (tide gauge, satellite altimetry, and field measurements in the inundation area). Given still only approximately known parameters of the tsunami source the coincidence between the model and observation is indeed good. Not only the arrival time of the first wave is reliably simulated, but the entire shape of the signal is reproduced reasonably well, and with correct amplitude. The model can be considered as an easy to use and reliable tool which not only serves the purposes of

GITEWS but can be employed for other tasks which can be described in the framework of shallow water equations (with exception of true shock waves for which continuous elevation is a suboptimal choice).

Simulations suggest strong nonlinear interaction between the tsunami and tidal waves. The major difference between tsunami simulations with and without tides occurs in the run-up region. Two mechanism of nonlinear interaction were found to be directly related to the morphometry of the object. In areas with high variability in morphometry (sharp bathymetry, complex coastline, etc.), the main role is played by the nonlinear interaction of tidal velocities and tsunamis velocity. Another mechanism of nonlinear interaction operates through the changes in the thickness of the water layer in the presence of tides, which is typical for shallow areas. In this case besides the amplitude of the incoming wave, the arrival time can vary due to the change in wave phase speed. These results lead us to conclude that the account of tidal dynamics may prove to be necessary for the faithful modelling of tsunami waves.

For comparison of the results delivered by the shallow water model in the original state and with the nonhydrostatic extension on the other two standard test cases are executed. This example shows dispersive effects that occur when a solution is running up a plain beach.

The comparison between the results of the original shallow water model and the model with the nonhydrostatic correction, two test cases were performed. They illustrate the importance of dispersive effects, which may have implication to inundation prediction. The inclusion of nonhydrostatic effects may therefore be necessary for successful modeling of tsunami wave propagation.

Brief overview

The wave propagation model TsunAWI based on finite elements was presented. We also introduced the architecture of the Tsunami Early Warnin System and methodology for the multi-sensor selection. We simulated the tsunami events generated by the Sumatra-Andaman (2004) and Tōhoku (2012) earthquakes. Part of our work addresses the influence of tidal dynamics on tsunami wave propagation in coastal areas. The impact of nonhydrostatic effects is investigated in this study.

Acknowledgements

TsunAWI is the result of joint developments of different groups within AWI namely the computing center and the climate dynamics section. We acknowledge Jörn Behrens, Dmitry Sein, and Dmitry Sidorenko for contributions to the model development. We acknowledge Ralf Kiefl from the German Aerospace Agency (DLR) for providung us with inundation areas estimated from satellite images and NOAA/PMEL/NOAA Center for Tsunami Research for providing satellite altimetry data. We acknowledge Widjo Kongko for providing us with field measurements in Aceh region. The tide gauge data were provided by the Survey of India and National Institute of Oceanography and by the University of Hawaii Sea level Center within the Global Sea Level Observing System (GLOSS-IOC). Matthias Mück from DLR provided us with the Intermap DTM and DSM topography data in Mataram.

Author details

Alexey Androsov, Sven Harig, Annika Fuchs, Antonia Immerz, Natalja Rakowsky, Wolfgang Hiller and Sergey Danilov
Alfred Wegener Institute for Polar and Marine Research, POB 120161, 27515 Bremerhaven, Germany

8. References

[1] Kienle J, Kowalik Z, Murty T. S (1987) Tsunamis generated by eruptions from Mount St.Augustine Volcano, Alaska. Science, 236, 1442–1447.

[2] Greenberg D. A, Murty T, Ruffman A (1987) A numerical model for Halifax Harbor tsunami due to the 1917 explosion. Marine Geodesy, 16, 153–167.

[3] Baptista A, Priest G, Murty T (1993) Field survey of the 1992 Nicaragua Tsunami. Marine Geodesy, 16, 1692–1703.

[4] Mofjeld H, Gonzales F, Titov V, Venturato A, Newman J (2007) Effects of tides on maximum tsunami wave heights: probability distributions. Journal of Atmospheric and Oceanic Technology, 24, 117–123.

[5] Weisz R, Winter C (2005) Tsunami, tides and run-up: a numerical study. Papadopoulos G. A Satake K (eds.), Proceedings of the International Tsunami Symposium, Chania, Greece, 27-29 June 2005.

[6] Kowalik Z, Proshutinsky T, Proshutinsky A (2006) Tide-tsunami interactions. Science of Tsunami Hazards, 24, 242–256.

[7] Titov V. V, Synolakis C. E (1995) Evolution and Runup of Breaking and Nonbreaking waves using VTSC-2. Journal of Waterway, Port, Coastal and Ocean Engineering, ASCE, 121, 308–325.

[8] Oliger J, Sundström A (1978) Theoretical and practical aspects of some initial boundary value problems in fluid dynamics. SIAM J. Appl. Math., 35, 419–446.

[9] Androsov A, Klevanny K, Salusti E, Voltzinger N (1995) Open boundary conditions for horizontal 2-d curvilinear-grid long-wave dynamics of a strait. Adv. Water Resour., 18, 267–276.

[10] Hanert E, Le Roux D, Legat V, Deleersnijder E (2005) An efficient Eulerian finite element method for the shallow water equations. Ocean Modelling, 10, 115–136.

[11] Maßmann S, Androsov A, Danilov S (2008) Intercomparison between finite element and finite volume approaches to model North Sea tides. Continental Shelf Research, 30, 680–691.

[12] Lynett P, Wu T.-R, Liu P.-F (2002) Modeling wave runup with depth-integrated equations. Coastal Engineering, 46, 89–107.

[13] Smagorinsky J (1963) General circulation experiments with the primitive equations. Mon. Wea. Rev., 91, 99–164.

[14] Shewchuk J (2006) Triangle: engineering a 2d quality mesh generator and Delaunay triangulator. manocha D. (ed.). In: Lin MC, Applied computational geometry: towards geometric engineering, pp. 203–222, Springer, Heidelberg.

[15] Frey W, Field D (1991) Mesh relaxation: a new technique for improving triangulations. Int. J. Numer. Methods Eng., 31, 1121–1133.

[16] Casulli V (1999) A semi-implicit finite difference method for non-hydrostatic, free-surface flows. International Journal for Numerical Methods in Fluids, 30, 425–440.

[17] Cheung K. F, Yamazaki Y, Kowalik Z (2009) Depth-integrated, non-hydrostatic model for wave breaking and run-up. International Journal for Numerical Methods in Fluids, 61, 473–497.

[18] Walters R. A (2005) A semi-implicit finite element model for non-hydrostatic (dispersive) surface waves. International Journal for Numerical Methods in Fluids, 49, 721–737.

[19] Fleischer J, Häner R, Herrnkind S, Kloth A, Kriegel U, Schwarting H, Wächter J (2010) An integration platform for heterogeneous sensor systems in GITEWS - Tsunami Service Bus. Natural Hazards and Earth System Science, 10, 1239–1252.

[20] Steinmetz T, Raape U, Teßmann S, Strobl C, Friedemann M, Kukofka T, Riedlinger T, Mikusch E, Dech S (2010) Tsunami early warning and decision support. Natural Hazards and Earth System Science, 10, 1839–1850.

[21] Behrens J, Androsov A, Babeyko A. Y, Harig S, Klaschka F, Mentrup L (2010) A new multi-sensor approach to simulation assisted tsunami early warning. Natural Hazards and Earth System Science, 10, 1085–1100.

[22] Babeyko A, Hoechner A, Sobolev S. V (2008) Modelling tsunami generation for local tsunami early warning in Indonesia. Geophysical Research Abstracts, 10, eGU2008-A-01523.

[23] Harig S, Chaeroni C, Pranowo W. S, Behrens J (2008) Tsunami simulations on several scales: Comparison of approaches with unstructured meshes and nested grids. Ocean Dynamics, 58, 429–440.

[24] The general bathymetric chart of the oceans (GEBCO). Webpage (http://www.gebco.net).

[25] Shuttle radar topography mission X-SAR / SRTM. Webpage (http://www.dlr.de/srtm/).

[26] Tanioka Y, Yudhicara, Kususose T, Kathiroli S, Nishimura Y, Iwasaki S.-I, Satake K (2006) Rupture process of the 2004 great Sumatra-Andaman earthquake estimated from tsunami waveforms. Earth Planets Space, 58, 203–209.

[27] Gayer G, Leschka S, Nöhren I, Larsen O, Günther H (2010) Tsunami inundation modelling based on detailed roughness maps of densely populated areas. Natural Hazards and Earth System Sciences, 10, 1679–1687.

[28] Egbert G S E (2002) Efficient inverse modeling of barotropic ocean tides. Journal of Atmospheric and Oceanic Technology, 19, 183–204.

[29] Androsov A, Voltzinger N (2005) The Straits of World ocean - the general approach to modeling. Nauka, 188 pages, in russian.

[30] Harbitz C (1992) Model simulations of tsunamis generated by the Storegga Slide. Marine Geology, 105, 1–21.

[31] Løvholt F, Harbitz C, Haugen K. B (2005) A parametric study of tsunamis generated by submarine slides in the Ormen Lange/Storegga area off western Norway. Marine and Petroleum Geology, 22, 219–231.

[32] Synolakis C. E, Bernard E. N, Titov V. V, Kânoğlu U, González F. I (2007) Standards, criteria, and procedures for NOAA evaluation of tsunami numerical models. NOAA

Tech. Memo. OAR PMEL-135, NOAA/Pacific Marine Environmental Laboratory, Seattle.

[33] Androsov A, Harig S, Behrens J, Schröter J, Danilov S (2008) Tsunami modelling on unstructured grids: Verification and validation. Proceedings of the International Conference on Tsunami Warning (ICTW), Bali, Indonesia.

Chapter 2

Shear Wave Propagation in Soft Tissue and Ultrasound Vibrometry

Yi Zheng, Xin Chen, Aiping Yao, Haoming Lin, Yuanyuan Shen, Ying Zhu, Minhua Lu, Tianfu Wang and Siping Chen

Additional information is available at the end of the chapter

1. Introduction

Studies have found that shear moduli, having the dynamic range of several orders of magnitude for various biological tissues [1], are highly correlated with the pathological statues of human tissue such as livers [2, 3]. The shear moduli can be investigated by measuring the attenuation and velocity of the shear wave propagation in a tissue region. Many efforts have been made to measure shear wave propagations induced by different types of force, which include the motion force of human organs, external applied force [4], and ultrasound radiation force [5].

In past 15 years, ultrasound radiation force has been successfully used to induce tissue motion for imaging tissue elasticity. Vibroacoustography (VA) uses bifocal beams to remotely induce vibration in a tissue region and detect the vibration using a hydrophone [5]. The vibration center is sequentially moved in the tissue region to form a two-dimensional image. Acoustic Radiation Force Imaging (ARFI) uses focused ultrasound to apply localized radiation force to small volumes of tissue for short durations and the resulting tissue displacements are mapped using ultrasonic correlation based methods [6]. Supersonic shear image remotely vibrates tissue and sequentially moves vibration center along the beam axis to create intense shear plan wave that is imaged at a high frame rate (5000 frames per second) [7]. These image methods provide measurements of tissue elasticity, but not the viscosity.

Because of the dispersive property of biological tissue, the induced tissue displacement and the shear wave propagation are frequency dependent. Tissue shear property can be modeled by several models including Kelvin-Voigt (Voigt) model, Maxwell model, and Zener model [8]. Voigt model effectively describes the creep behavior of tissue, Maxwell model effectively describes the relaxation process, and the Zener model effectively describes both creep and relaxation but it requires one extra parameter. Voigt model is often used by many researchers because of its simplicity and the effectiveness of modeling soft tissue. Voigt model consists of a purely viscous damper and a purely elastic spring connected in parallel. For Voigt tissue, the tissue motion at a very low frequency largely depends on the elasticity, while the motion at a very high frequency largely depends on the viscosity [8]. In general, the tissue motion depends on both elasticity and viscosity, and estimates of elasticity by ignoring viscosity are biased or erroneous.

Back to the year of 1951, Dr. Oestreicher published his work to solve the wave equation for the Voigt soft tissue with harmonic motions [9]. With assumptions of isotropic tissue and plane wave, he derived equations that relate the shear wave attenuation and speed to the elasticity and viscosity of soft tissue. However, Oestreicher's method was not realized for applications until the half century later.

In the past ten years, Oestreicher's method was utilized to quantitatively measure both tissue elasticity and viscosity. Ultrasound vibrometry has been developed to noninvasively and quantitatively measure tissue shear moduli [10-16]. It induces shear waves using ultrasound radiation force [5, 6] and estimates the shear moduli using shear wave phase velocities at several frequencies by measuring the phase shifts of the propagating shear wave over a short distance using pulse echo ultrasound [10-16]. Applications of the ultrasound vibrometry were conducted for viscoelasticities of liver [16], bovine and porcine striated muscles [17, 18], blood vessels [12, 19-21], and hearts [22]. A recent *in vivo* liver study shows that the ultrasound vibrometry can be implemented on a clinical ultrasound scanner of using an array transducer [23].

One of potential applications of the ultrasound vibrometry is to characterize shear moduli of livers. The shear moduli of liver are highly correlated with liver pathology status [24, 25]. Recently, the shear viscoelasticity of liver tissue has been investigated by several research groups [23, 26-28]. The most of these studies applied ultrasound radiation force in liver tissue regions, measured the phase velocities of shear wave in a limited frequency range, and inversely solved the Voigt model with an assumption that liver local tissue is isotropic without considering boundary conditions. Because of the boundary conditions, shear wave propagations are impacted by the limited physical dimensions of tissue. Studies shows that considerations of boundary conditions should be taken for characterizing tissue that have limited physical dimensions such as heart [22], blood vessels [19-21], and liver [8], when ultrasound vibrometry is used.

2. Shear wave propagation in soft tissue and shear viscoelasticity

The shear wave propagation in soft tissue is a complicated process. When the tissue is isotropic and modeled by the Voigt model, the phase velocity and attenuation of the shear wave propagation in the tissue are associated with tissue viscoelasticity. Oesteicher documented the detailed derivations of the solution of the sound wave equation for Voigt tissue [9]. We extended the solution to other models [8] for the applications of ultrasound vibrometry [8]. In this section, we provide the simplified descriptions of the shear wave propagation in tissue modeled by Voigt model, Maxwell model, and Zener model.

Assuming that a harmonic motion produces the shear wave that propagates in a tissue region, the phase velocity $c_s(\omega)$ of the wave can be estimated by measuring the phase shift $\Delta\phi$ over a distance Δz:

$$c_s(\omega) = \omega\Delta z / \Delta\phi \tag{1}$$

The phase velocity is associated with the tissue property, which can be found by solving the wave equation with a tissue viscoelasticity model. For a small local region, the wave is approximated as a uniform plane wave, which has a simple form in isotropic medium:

$$\frac{d^2\mathbf{S}}{dz^2} + k^2\mathbf{S} = 0 \tag{2}$$

where $\mathbf{S}$ is the phasor notation of the displacement of the time-harmonic field of the shear wave, z is the wave propagation distance which is perpendicular to the direction of the displacement of the shear wave, and the complex wave number is

$$k = k_r - ik_i \tag{3}$$

The solution of (2) is a standard solution of a homogeneous wave equation:

$$\mathbf{S} = \hat{x}S_0e^{-ikz} \tag{4}$$

where S_0 is the displacement at $z = 0$, $\hat{x}$ is an unit vector in x direction. The plane wave is independent in y direction. The real time time-harmonic shear wave is:

$$S(\omega, z, t) = \hat{x}\,\mathrm{Re}\left\{\mathbf{S}e^{i\omega t}\right\} = \hat{x}S_0e^{-k_iz}\cos(\omega t - k_rz) \tag{5}$$

Although attenuation coefficient $\alpha = -k_i$ carries information of the complex modulus of tissue, the phase measurement is often more reliable because it is relatively independent to transducers and measurement systems. The phase velocity is the speed of the wave propagating at a constant phase, which is a solution of $d(\omega t - k_rz)/dt = 0$:

$$c_s(\omega) = \frac{dz}{dt} = \omega/k_r \tag{6}$$

The complex wave number k of the plane shear wave is a function of the frequency and the complex modulus of the medium [9]:

$$k = \sqrt{\rho\omega^2/\mu} \tag{7}$$

where ρ is the density of the tissue and the complex modulus that connects stress σ and strain ε:

$$\mu = \sigma/\varepsilon = \mu_1 + i\omega\mu_2 \tag{8}$$

which describes the relationship between stress and strain in the Voigt tissue. The Voigt model consists of an elastic spring μ1 and a viscous damper μ2 connected in parallel, which represents the same strain in each component as shown in Figure 1.

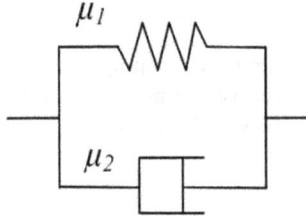

Figure 1. Voigt model consists of an elastic spring μ1 and a viscous damper μ2 connected in parallel.

The relation between stress σ and strain ε of the Maxwell tissue is:

$$\sigma = \mu_1 \varepsilon + \mu_2 \frac{d\varepsilon}{dt} \tag{9}$$

For a harmonic motion, (9) becomes:

$$\sigma = (\mu_1 + i\omega\mu_2)\varepsilon \tag{10}$$

which is the same as (8). Substituting (8) into (7) and finding the real part of the wave number, the phase velocity of the shear wave in Voigt tissue can be obtained from (6):

$$c_s(\omega) = \sqrt{\frac{2(\mu_1^2 + \omega^2 \mu_2^2)}{\rho(\mu_1 + \sqrt{\mu_1^2 + \omega^2 \mu_2^2})}} \tag{11}$$

The elasticity μ1 and viscosity μ2 are two constants and independent to the frequency.

A numerical example of phase velocity of Voigt tissue is shown in Figure 2. Equation (11) shows that $c_s(\omega)$ increases at the rate of square root of the frequency and there is no the upper limit for $c_s(\omega)$. As shown in the Figure 2, the phase velocity is determined by both elasticity and viscosity. Ignoring the viscosity introduces errors and biases for elasticity estimates. However, examining the velocities at the extreme frequencies is useful for understanding the model and obtaining initial values for numerical solutions of μ1 and μ2. In tissue characterization applications, μ1 is often in the order of a few thousands and μ2 is often less than 10. Thus, when the wave frequency is very low (less than a few Hz),

$$\mu_1 \approx \rho c_s^2(\omega) \qquad \forall \ \text{very low } \omega. \tag{12}$$

When the frequency is very high (higher than a few tens of kHz),

$$\mu_2 \approx \rho c_s^2(\omega) / 2\omega \qquad \forall \ \text{very high } \omega. \tag{13}$$

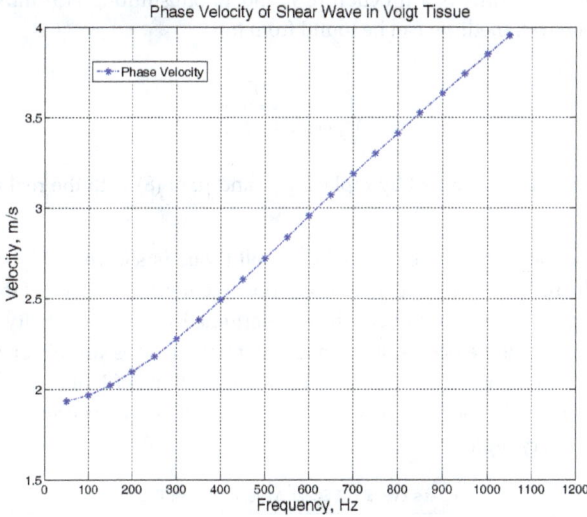

Figure 2. Plot of phase velocity of shear wave having μ_1=3 kpa and μ_2=1 pa.s in Voigt tissue

A broad frequency range is needed to accurately estimate both μ_1 and μ_2. (12) and (13) are only useful for estimating initial values for the numerical solutions of (11) with measured velocities, and they should not be used for final estimates.

Equation (7) can be used for other models for the plane shear wave having a single frequency. The Maxwell model consists of a viscous damper η and an elastic spring E connected in series, which represents the same stress in each component, as shown in Figure 3.

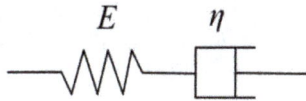

Figure 3. Maxwell model consists of a viscous damper η and an elastic spring E connected in series.

The relation between stress σ and strain ε of the Maxwell tissue is:

$$\frac{1}{E}\frac{d\sigma}{dt} + \frac{\sigma}{\eta} = \frac{d\varepsilon}{dt} \tag{14}$$

For a harmonic motion, (14) becomes:

$$\frac{\sigma}{\varepsilon} = \frac{i\omega\eta E}{E + i\omega\eta} = \mu = \frac{\omega^2\eta^2 E}{E^2 + \omega^2\eta^2} + i\frac{\omega\eta E^2}{E^2 + \omega^2\eta^2} \tag{15}$$

which is the complex shear modulus of the Maxwell model. Unlike the Voigt model, real and imaginary components of (15) are functions of the frequency. When the frequency is fixed, the complex modulus is a function of η and E. Substituting (15) into (7), the shear wave speed in Maxwell medium can be found from (6):

$$c_s(\omega) = \sqrt{\frac{2E}{\rho(1 + \sqrt{1 + E^2/\omega^2\eta^2})}} \qquad (16)$$

Equation (16) can be also obtained by replacing μ_1 and μ_2 of (8) with the real and imaginary terms of (15).

A numerical example of phase velocity of Maxwell tissue is shown in Figure 4. Note that $c_s(\omega)$ gradually increases to a limit that is proportional to the square root of the elasticity. As shown in the Figure 4, the phase velocity is determined by both elasticity and viscosity. However, examining the velocities at the extreme frequencies is useful for understanding the model and obtaining initial values for numerical solutions of E and η. $E \approx \rho C_s^2(\omega)$ for a very large ω, $\eta \approx \rho C_s^2(\omega)/2\omega$ for a very small ω, $c_s(\omega)$ is zero for ω=0, and $c_s(\omega)$ approaches $\sqrt{E/\rho}$ when ω is very high.

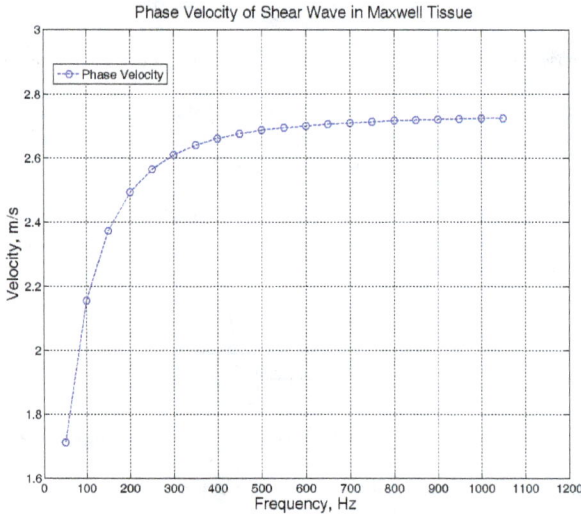

Figure 4. Plot of phase velocity of shear wave having E = 7.5 kpa and η = 6 pa.s in Voigt tissue

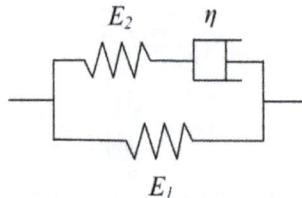

Figure 5. Zener model adds an elastic spring E_1 to the Maxwell model (η, E_2) in parallel.

The Zener model adds an additional elastic spring, having the elasticity of E_1, to the Maxwell model (η, E_2) in parallel. The Zener model combines the features of the Voigt model and the Maxwell models and describes both creep and relaxation. Based on the Maxwell model, the complex shear modulus of the Zener model can be readily obtained:

$$\mu = E_1 + \frac{i\omega\eta E_2}{E_2 + i\omega\eta} = E_1 + \frac{\omega^2\eta^2 E_2}{E_2^2 + \omega^2\eta^2} + i\frac{\omega\eta E_2^2}{E_2^2 + \omega^2\eta^2} \tag{17}$$

Substituting (17) into (7), the shear wave speed in Zener medium can be found from (6):

$$c_s(\omega) = \sqrt{\frac{2(\omega^2\eta^2(E_1 + E_2)^2 + E_1^2 E_2^2)}{\rho(\omega^2\eta^2(E_1 + E_2) + E_1 E_2^2 + \sqrt{(\omega^2\eta^2(E_1 + E_2)^2 + E_1^2 E_2^2)(\omega^2\eta^2 + E_2^2)}}} \tag{18}$$

Equation (18) shows that $E_1 + E_2 \approx \rho C_s^2(\omega)$ for a very large ω, $E_1 \approx \rho C_s^2(\omega)$ for a very small ω, η is proportional to the slop of the speed curve, and $c_s(\omega)$ approaches $\sqrt{(E_1 + E_2)/\rho}$ when ω is very high. A numerical example of phase velocity of Zener tissue is shown in Figure 6.

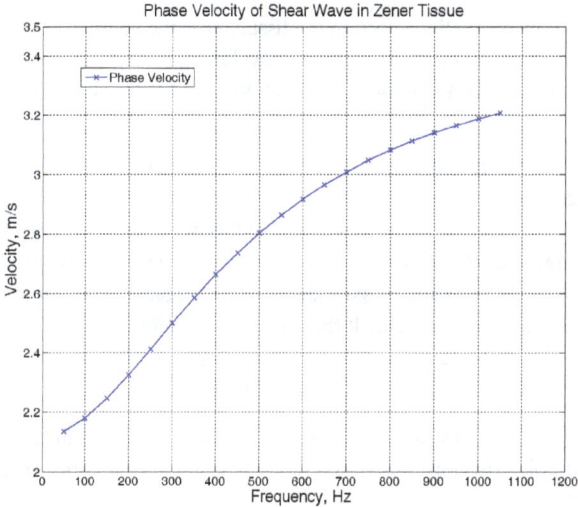

Figure 6. Plot of phase velocity of shear wave having E_1 = 4.5 kpa, η = 1.5 pa.s, and E_2 =7.5 ka in Zener tissue

3. Ultrasound vibrometry

Ultrasound vibrometry has been developed to induce shear wave in a tissue region, measure phase velocity of the shear wave, and calculate the tissue viscoelasticity based on (11), or (16), or (18). The basics of the ultrasound vibrometry are described in details in references [11-17, 32]. Ultrasound vibrometry induces tissue vibrations and shear waves using ultrasound radiation force and detects the phase velocity of the shear wave propagation using pulse-echo ultrasound.

From the solution of the wave equation, equation (5) can be represented by a harmonic motion at a location,

$$d(t) = D\sin(\omega_s t + \phi_s) \tag{19}$$

where $\omega_s = 2\pi f_s$ is the vibration angular frequency, the vibration displacement amplitude D and phase ϕ_s depend on the radiation force and tissue property. (19) is another representation of (5). Applying detection pulses to the motion that causes the travel time changes of detection pulses and phase shift changes of the return echoes, the received echo becomes [11]:

$$r(t,k) = |g(t,k)|\cos\left(\omega_0 t + \phi_0 + \beta\sin(\omega_s(t+kT) + \phi_s)\right) \tag{20}$$

where T is the period of the push pulses shown in Figure 9 and the modulation index is:

$$\beta = 2D\omega_0 \cos(\theta) / c \tag{21}$$

where c is the sound propagation speed in the tissue, ω_0 is the angular modulation frequency of detection tone bursts, $g(t,k)$ is the complex envelope of $r(t,k)$, ϕ_0 is a transmitting phase constant and θ is an angle between the ultrasound beam and the tissue vibration direction.

Received echo $r(t,k)$ is a two-dimensional signal. When one detection pulse is transmitted, its echo from the different depth of tissue is received as t changes. In medical ultrasound field, variable t is called fast time. When multiple detection pulses are transmitted, the multiple echo sequences are received as k changes. Variable k is called as slow time. $r(t,k)$ in fast-time t is called as fast-time signal to represent the echo signal in beam axial direction or the depth location in the tissue. Its variation in slow-time k called slow-time signal to represent the signals from one echo to another echo. If there is no tissue motion, $r(t,k)$ will be the same for different k values. The tissue motion information is carried by modulation index β and phase ϕ_s. A quadrature demodulator is used to obtain β and phase ϕ_s.

As shown in Figure 7, a quadrature demodulator is applied to extract the motion information from $r(t,k)$. The complex envelop consists of the in-phase and quadrature term [29]:

$$g(t,k) = I(t,k) + jQ(t,k) \tag{22}$$

Operating on the in-phase and quadrature components I and Q with input $r(t,k)$, we obtain the tissue motion in slow time [11]:

$$s_A(t,k) = \tan^{-1}(Q/I) - \text{mean of } \tan^{-1}(Q/I) = \beta\sin\left(\omega_s(t+kT) + \phi_s\right) \tag{23}$$

A phase constant can be added to the local oscillator of the demodulator [11] to avoid zeros in I. The signal extracted by (23) is proportional to the displacement of a harmonic motion induced by the push pulses.

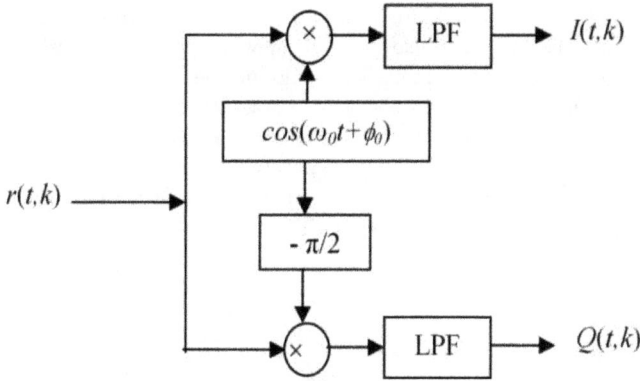

Figure 7. Block diagram of quadrature demodulator

Another motion detection method [14] uses a complex vector that is a multiplication between two successive complex envelops [29]

$$X(t,k) + iY(t,k) = g(t,k+1)g^*(t,k) \qquad (24)$$

Thus, the motion velocity in slow time can be obtained,

$$s_B(t,k) = \tan^{-1}\left(\frac{Y(t,k)}{X(t,k)}\right) = -2\beta \sin(\omega_s T / 2)\cos\left(\omega_s(t+kT) + \phi_s + \omega_s T / 2\right) \qquad (25)$$

which is proportional to the velocity of the tissue harmonic motion for $\omega_s T/2 \ll 1$. Thus, $\sin(\omega_s T/2) \approx \omega_s T/2$ and the velocity amplitude is $\omega_s T\beta$, which is also $D\omega_s = \beta\omega_s c / 2\omega_0 \cos(\theta)$ because of the derivative relation between (19) and (25)..

The slow-time signal s(t,k) represents the tissue motion at a particular location, its amplitudes and phases change over distances are described by (5). The measurements of amplitudes and phases at two locations are used to calculate attenuation and phase velocity. As shown in (1), the phase velocity is related to the frequency and inversely related to the phase difference $\Delta\phi$ over a short distance Δd. Thus, estimating the phase differences is the key step of the ultrasound vibrometry. The phase difference can be obtained by comparing phases ϕ_s of the slow-time signals $s(t,k)$ at two locations z and z+Δz:

$$\Delta\phi = \phi_s(z) - \phi_s(z + \Delta z) \qquad (26)$$

There are several methods to estimate the phases of slow-time signals: Fourier transform, correlation method, and Kalman filter [14]. The estimated phase of the slow-time signal at a location include some phase constants due to the tissue location t and different pulse k, and phase $\phi_s = -k_r z$. Given a tissue location (axial location) in fast time, all constant phases are removed by (26) except the phase shift $\Delta\phi$ in the lateral location.

Ultrasound vibrometry is developed to induce the shear wave described by (19) and detect the phase shift $\Delta\phi$ described by (26) for characterizing the tissue shear property using (1) and (11), (14), and (16). Ultrasound virbometry uses interleaved periodic pulses to induce shear wave and detects the phase velocity of shear wave propagation using pulse-echo ultrasound. Figure 8 shows an application setup of the ultrasound vibrometry. An ultrasound transducer transmits push beams to a tissue region to induce vibrations and shear waves. The push beams are periodic pulses that have a fundamental frequency f_v and harmonics nf_v. During the off period of the push pulses, the detection pulses are transmitted and echoes are received by the transducer at lateral locations that are away from the center of the radiation force applied, as shown in Figure 9. In some of our applications, fundamental frequency f_v of the push pulses is in the order of 100 Hz, and pulse repetition frequency f_{PRF} of the detection pulses is in the order of 2 kHz.

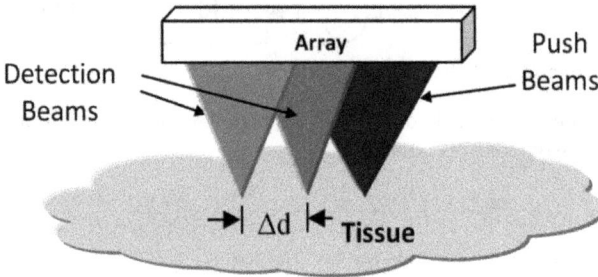

Figure 8. Array transducer for transmitting ultrasound radiation force and detecting shear wave propagation

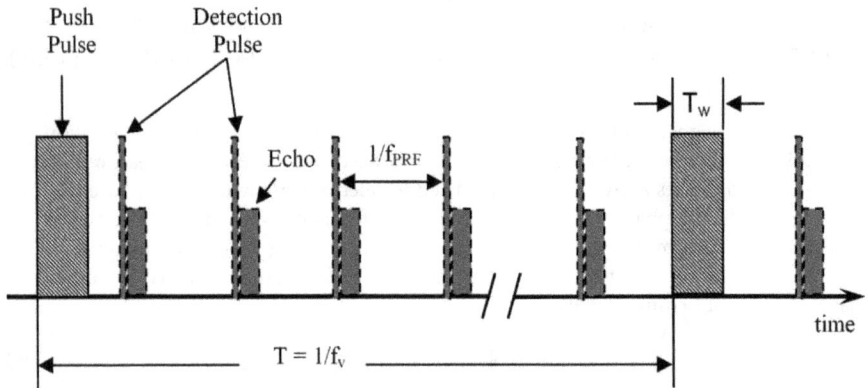

Figure 9. Interleaved push pulses for ultrasound radiation force and detection pulses

There are different variations of the excitation pulses beside the on-off binary pulses: continuous waves [11], non-uniform binary pulses [15], and composed pulses or Orthogonal Frequency Ultrasound Vibrometry (OFUV) pulses [30, 31]. The OFUV pulses can be designed to enhance higher harmonics to compensate the high attenuations of high harmonics. The OFUV pulses have multiple binary pulses in one period of the fundamental period [30, 31]. Other variations of the ultrasound vibrometry include consideration of background motion and boundary conditions that require more complicated models of tissue motions [13] and wave propagations [22].

4. Finite element simulation of shear wave propagation

Simulations using Finite Element Method (FEM) were conducted to understand the shear wave propagation in tissue. The simulation tool is COMSOL 4.2. The simulated tissue region is a two-dimensional axisymmetric finite element model of a viscoelastic solid with a dimension of 100 mm × 100 mm, as shown in Figure 10. The size of domain $\Omega 1$ is 100 mm × 80 mm. The domain is divided to 25,371 mesh elements and the average distance between adjacent nodes is 0.95 mm. The schematic diagram shown in Figure 10 includes simulation domains ($\Omega 1$, $\Omega 2$, $\Omega 3$) and boundaries (B1,B2). A line source (with a length of 60 mm) in the left of the solid represents as an excitation source of the shear wave.

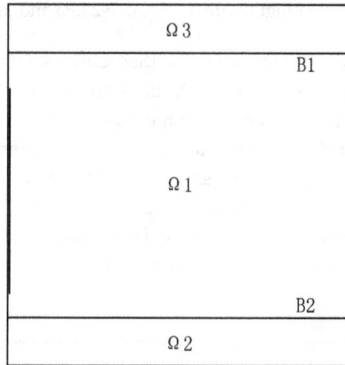

Figure 10. Schematic diagram of simulated tissue region (domain) and

All domains had the same material property of the Voigt tissue and all boundaries were set free to avoid reflections. The material parameters were: density of 1055 kg/m^3, Poisson's ratio of 0.499, and Voigt rheological model of the viscoelasticity model. The Voigt model was converted and represented in the form of Prony series. The store modulus and loss modulus were calculated using frequency response analysis for demonstrating the conversion of the Prony series. The complex shear modulus of the Voigt model is the same as (8):

$$\mu(\omega) = \mu_1 + i\omega\mu_2$$

where elasticity modulus μ_1 and viscosity modulus μ_2 were set to be 2 kPa and 2 Pa*s, respectively, in this simulation.

Transient analysis was used and the time step for solver was one eightieth of the time period of the shear wave. Uniform plane shear wave was produced by oscillating the line source with ten cycles of harmonic vibrations in the frequency range from 100 Hz to 400 Hz with a maximal displacement in the order of tens of micrometers. The displacements of the shear wave were recorded for post-processing at 8 locations, 1 mm apart, along a straight line that is normal to the line source. The phases of the wave were estimated by the Kalman filter and the average phase shifts were estimated using a linear fitting method [14]. The estimates of shear wave velocity and viscoelasticity are shown in Table 1.

	Shear Wave Velocity (m/s)				Viscoelasitcity Estimation	
	100Hz	200Hz	300Hz	400Hz	μ_1(kPa)	μ_2(Pa*s)
Reference value	1.5574	1.9372	2.3470	2.7362	2	2
Measurement 1	1.46238	1.91972	2.37439	2.66872	1.69	1.90
Measurement 2	1.50648	1.94791	2.42833	2.86637	1.63	2.10
Measurement 3	1.52748	1.92955	2.46275	2.81089	1.74	2.10
Average	1.49479	1.9216	2.44359	2.77296	1.69±0.056	2.03±0.11
Std	0.02828	0.02455	0.05672	0.08517		

Table 1. Estimated Viscoelasticity of Voigt tissue having μ_1 = 2 kPa and μ_2= 2 Pa*s

The shear wave velocities in red represent the theoretical values of wave speeds in Voigt tissue. The estimates of the speeds and viscoelasticity moduli of three simulations are shown by three sets of the measurement. Their average values are close to the theoretical values as shown in Figure 11, except the elasticity μ_1. Note that the differences between the average velocities and the reference velocities are less than 9% but the estimate error of μ_1 is 15.5%. It is due to the fact that viscoelasticity moduli are proportional to the square of the phase velocity. Any small estimation errors of phase introduce large biases in the estimates of viscoelasticity, which is an intrinsic weakness of the ultrasound vibrometry, demonstrated by this example.

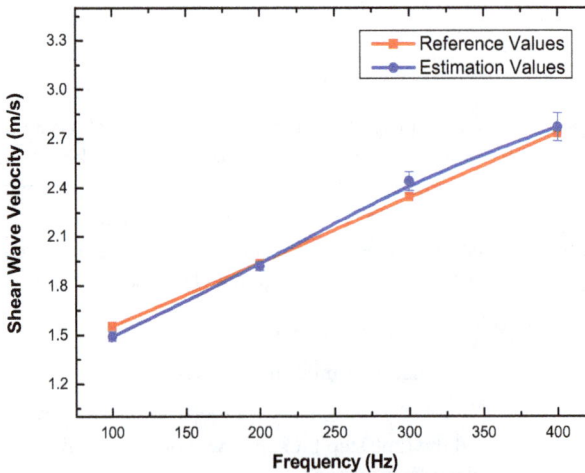

Figure 11. Estimated shear phase velocities and set reference values

5. Experiment system and results

Experiments were conducted for evaluating ultrasound vibrometry. The diagram of an experiment system is shown in Figure 12. This system mainly consists of a transmitter to produce the ultrasound radiation force and a receiver unit using a SonixRP system. Two arbitrary signal generators were utilized to generate the system timing and excitation waveform. The waveform was amplified by a power amplifier having a gain of 50 dB to drive an excitation transducer for inducing vibrations in a tissue region. The SonixRP system was applied to detect the vibration using pulse-echo mode with a linear array probe. The SonixRP is a diagnostic ultrasound system packaged with an Ultrasound Research Interface (URI). It has some special research tools which allow users to perform flexible tasks such as low-level ultrasound beam sequencing and control. The center frequency of the excitation transducer was 1 MHz. The center frequency of the linear array probe was 5 MHz and the sampling frequency of SonixRP was 40 MHz. The excitation transducer and detection transducer were fixed on multi-degree adjustable brackets and were controlled by three-axis motion stages.

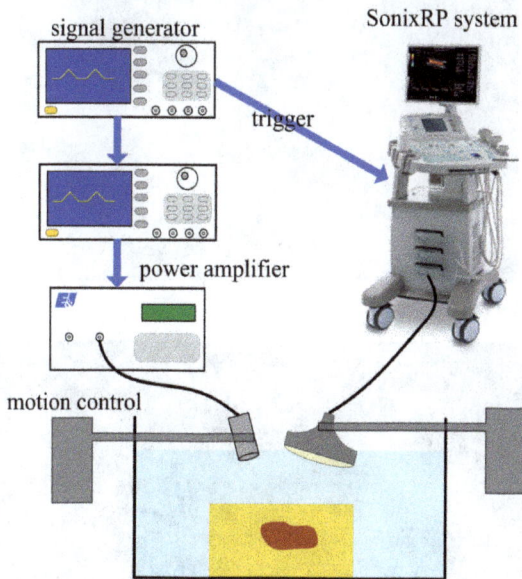

Figure 12. Block diagram of the experiment system

The picture of experiment system setup is shown in Figure 13. The left lobe of a SD rat liver was embedded in gel phantom and placed in water tank. Before experiment, the SonixRP URI was run first to preview the internal structure of the liver. In the interface shown in Figure 14, the B-mode image and RF signal of a selected scan line were displayed together to help users selecting test points inside the liver tissue. The positions of the excitation transducer and the detection probe were adjusted to focus on two locations in the liver at the same vertical depth.

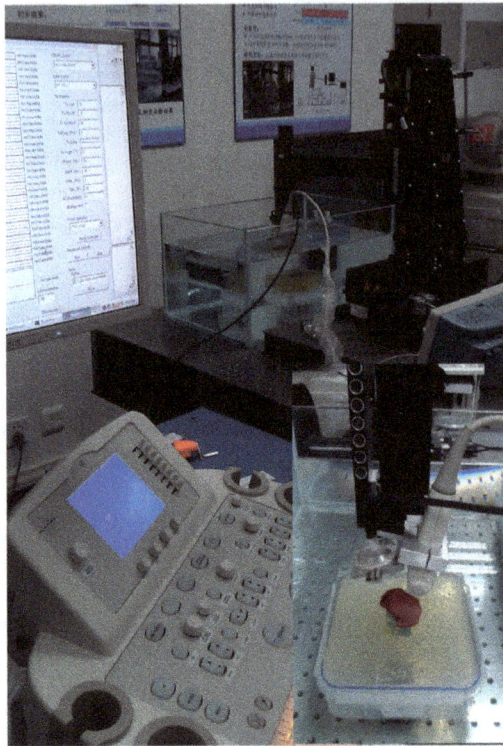

Figure 13. Experiment setup with SonixRP.

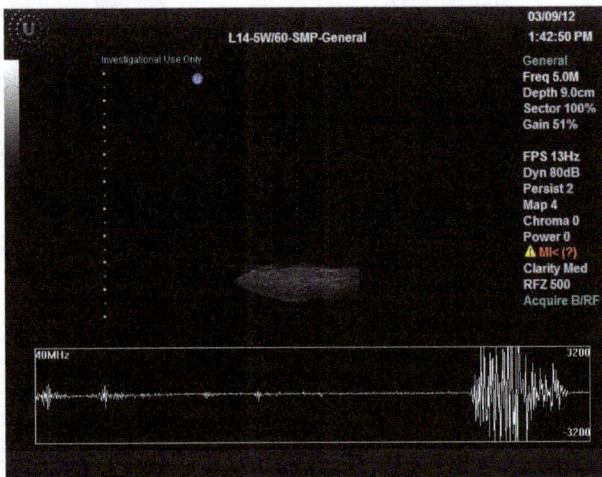

Figure 14. Ultrasound Research Interface (URI) of SonixRP

Computer programs based on the software development kit (SDK) of SonixRP were developed for detecting the vibrations and shear wave propagation. The programs defined a specific detection sequencing and timing that repeatedly transmit pulses to a single scan line and repeatedly receive the echoes with a PRF of 2 KHz. The timing of the excitation and detection pulses is shown in Figure 15. The pulse repetition frequency of the excitation pulses was 100 Hz.

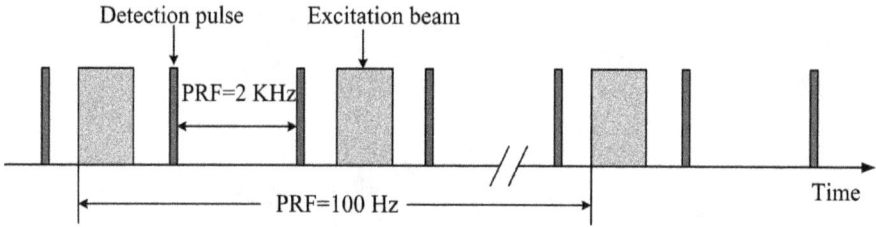

Figure 15. Timing sequence of the experiment system

An example of the typical fast-time RF ultrasound signal acquired by the SonixRP is shown in Figure 16. Figure 16a shows the echo through the entire liver tissue region, while Figure 16b shows the echo around the focus point (75 mm in depth) in the liver tissue.

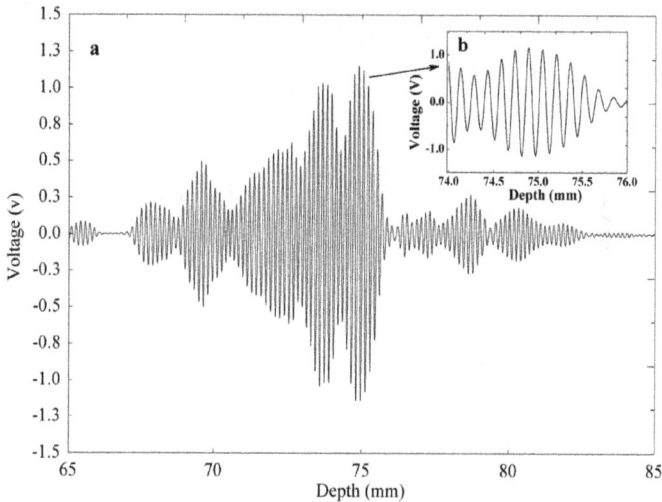

Figure 16. Ultrasound RF echo (a) through the entire liver and (b) around the focus point in the liver tissue

The vibration of shear wave at a location was extracted from I and Q channels using the I/Q estimation algorithm described by equation (23). Figure 17a shows the vibration displacement and Figure 17b shows the spectral amplitude of the vibration.

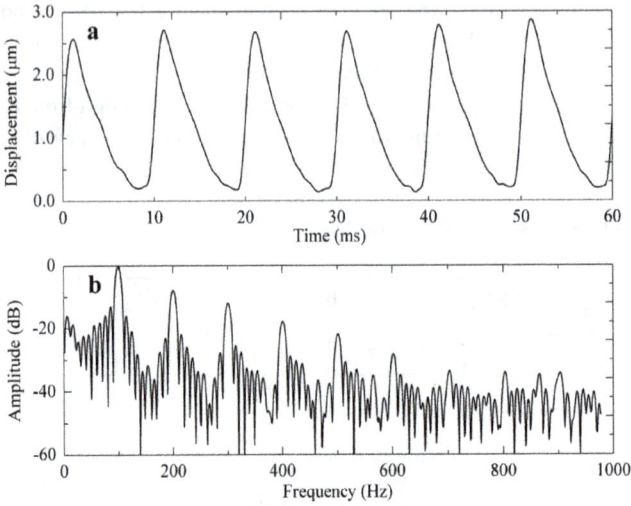

Figure 17. Displacements of the vibration and its frequency spectrum

The extracted displacement signal $s_B(t,k)$ was processed by the Kalman filter [14] that simultaneously estimates phases of the fundamental frequency and all harmonics. Figure 18 shows estimated vibration phase shifts of the first four harmonics over a distance up to 4 mm. Linear regression was conducted to calculate the shear wave propagation speed for each frequency.

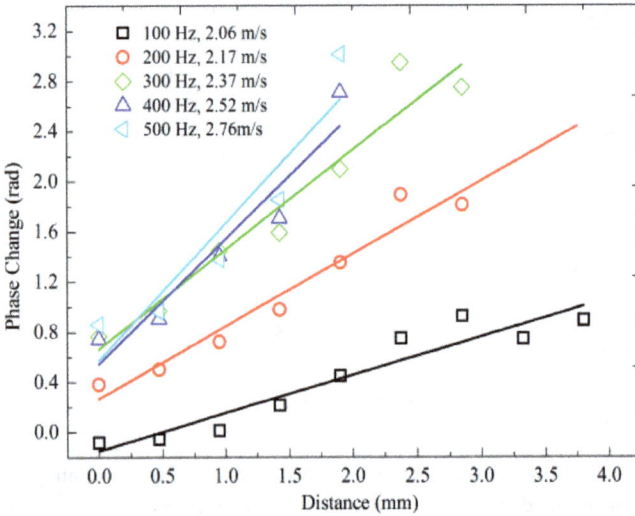

Figure 18. Estimates of phase shifts over distances using vibration displacements and Kalman filter

Figure 19 shows the phase velocities at different harmonics and the fitting curves of three models: Voigt, Maxwell, and Zener models. The fitting values are shown in Table 2. As shown by the figure and table, the Voigt model and Zener model fit the measurements of the phase velocity of the liver tissue better than the Maxwell model for this liver.

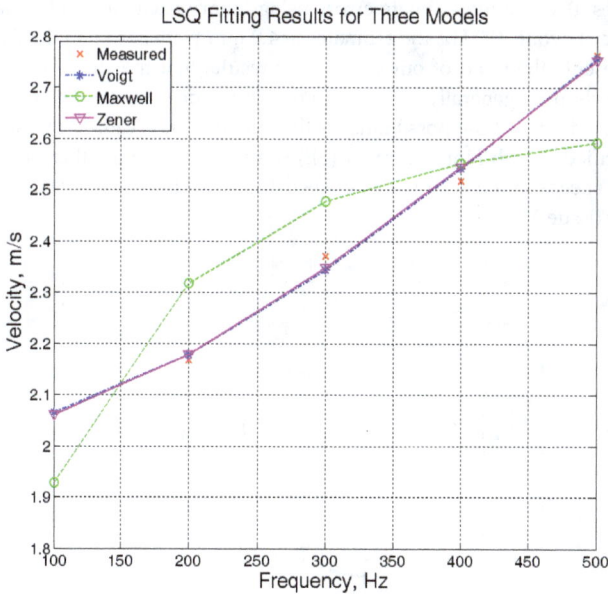

Figure 19. Curve fittings of three models with the estimates of the phase velocities of the liver tissue

Voigt Model, μ_1, μ_2, fitting error	4.10 kPa	1.51 Pa·s	0.019 m/s
Maxwell Model, E, η, fitting error	7.18 kPa	4.27 Pa·s	0.143 m/s
Zener Model, E_1, E_2, η, fitting error	4.07 kPa, 45.9 kPa	1.47 Pa·s	0.020 m/s

Table 2. Estimated viscoelasticity moduli of three models

The second experiment was conducted to demonstrate the impact of boundary conditions. Because boundary conditions play very important roles in wave propagation, *in vitro* experiments were also conducted to investigate shear moduli of the superficial tissue of livers (0.4 mm below the capsule) and the deep tissue of livers (4.9 mm below the capsule). The excitation pulses were tone bursts having a center frequency of 3.37 MHz and a width of 200 µs for the binary excitation pulses and 100 µs or less for the OFUV excitation pulses. The pulse repetition frequency of the excitation pulses was 100 Hz. The broadband detection pulses had a center frequency of 7.5 MHz and pulse repetition frequency (PRF) of 4 kHz. Liver phantoms using fresh swine livers were carefully prepared so that the interface between the gelatin and the liver was flat. The thicknesses of liver samples were more than 2 cm and the areas were about 4×4 cm². The phantom was immersed in a water tank.

The shear wave speeds were measured from 100 Hz to 800 Hz over a distance up to 5 mm away from the center of the radiation force application. Figure 20 shows the estimates of the shear wave speeds. Each error bar was the standard deviation of 30 estimates from five data sets of repeated measurements and six distances (1 to 4 mm, 1 to 5 mm, etc). The estimates from 100 Hz to 400 Hz were almost identical for the binary excitation pulses and the OFUV excitation pulses. Because the estimate errors using binary excitation pulses were too high for the frequency beyond 400 Hz, the estimates at 4.9 mm were based on the OFUV method. Figure 20 represents the trend of our experiment results that the shear wave speed in the superficial liver tissue is generally higher than that in the deep tissue. The results should be carefully examined. One of the possibilities is that we think it is caused by the liver capsule as we have verified it with Finite Element (FE) simulations, and another possibility is that the shear wave speeds of the gelatin are between 3 to 4 m/s from 100 to 800 Hz, higher than that in the liver tissue.

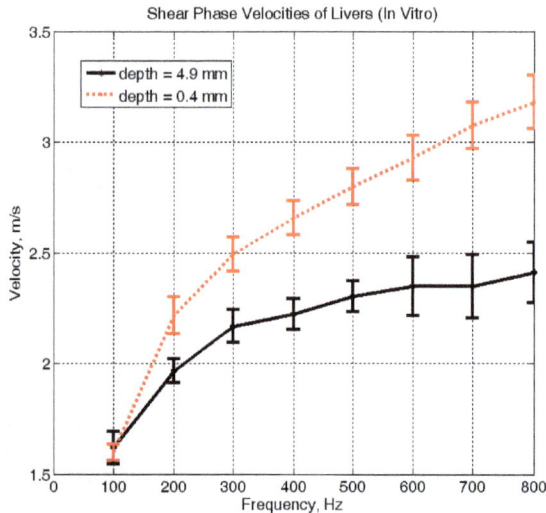

Figure 20. Shear wave speeds in superficial and deep liver tissues

The estimates of shear wave speeds at deep tissue of 4.9 mm and superfical tissue of 0.4 mm were used to numerically solve for the shear moduli of the three models. The curves generated by the models were compared with the measurements. As shown in Figures 21a and 21b, we find that the Voigt model may not always suitable for modeling liver shear viscoelasticity, at least for *in-vitro* applications with increased frequencies of shear waves in some of our studies. On the other hand, we find that the Zener models matches the measurements very well with very small fitting errors as shown in the Figure 21 and Table 3.

Table 3 shows the estimated shear moduli of different models with two different frequency ranges at two different depths in liver tissues based on our experiment data. Each modulus is an average of 30 estimates from 5 data sets and 6 distances. All elasticity has the unit of kPa and all viscosity has the unit of Pa·s. The fitting errors (m/s) are the deifferences between the measurements and calculated shear wave speeds using the models. The changes represent the variations of the estiamtes from one frequency range to another. The statistics are not conclusive because of the small number of samples. But this study indicates the variations of estimates and importance of the selection of tissue viscoelasticity models.

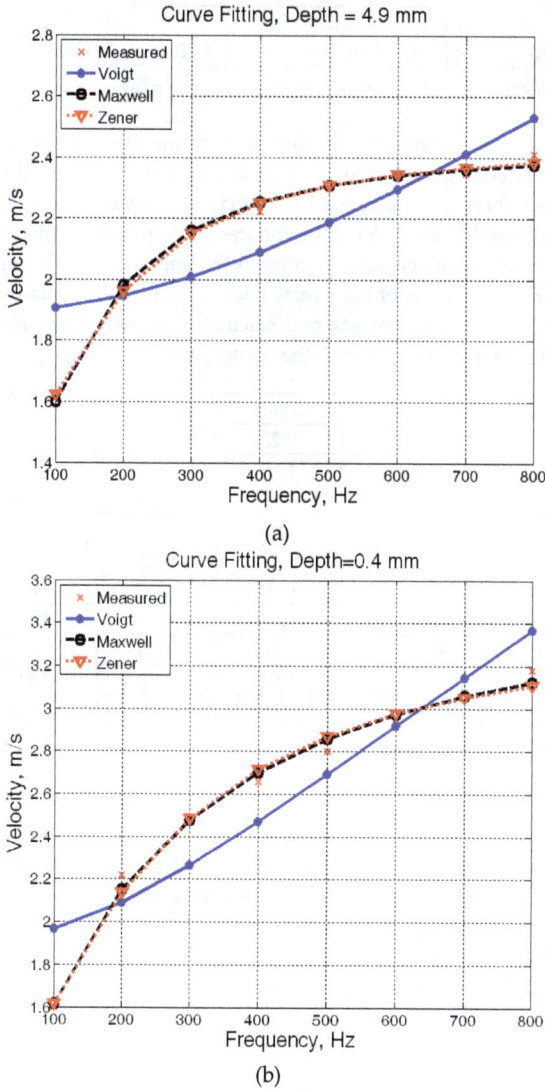

(a)

(b)

Figure 21. Model fittings for shear wave in the deep tissues (a), and superficial tissue (b)

Depth=0.4 mm	100 to 400 Hz	100 to 800 Hz	Changes
Voigt, μ_1, μ_2, fitting error	2.48, 2.00, 0.152	3.71, 1.46, 0.204	50%, 27%
Maxwell, E, η, fitting error	10.7, 2.50, 0.043	11.7, 2.36, 0.048	10%, 6%
Zener, E_1, E_2, η, fitting error	0.578, 9.033, 2.85, 0.028	1.34, 9.843, 2.56, 0.0569	132%, 9%, 10%
Depth=4.9 mm			
Voigt, μ_1, μ_2, fitting error	2.74, 1.35, 0.108	3.59, 0.791, 0.151	31%, 41%
Maxwell, E, η, fitting error	5.68, 2.82, 0.016	5.90, 2.70, 0.021	4%, 4%
Zener, E_1, E_2, η, fitting error	1.49, 4.20, 2.44, 0.015	1.70, 4.25, 2.19, 0.018	14%, 1%, 10%

Table 3. Estimates of Shear Moduli (elasticity in kPa, viscosity in Pa·s)

The third experiment was conducted to demonstrate the effectiveness of the ultrasound vibrometry to characterize the injury of liver tissue. Table 4 shows that the measured shear moduli of the livers thermally damaged by a microwave oven using different amount of cooking time (3, 6, 9, and 12 seconds). All estimates were from the superficial tissue region. It shows that the shear wave speeds estimated in the superficial tissue region are effective for indicating the damage levels of the livers. The errors are the standard deviations of the differences between the measurements and calculated speeds of the models. The Zener model provides the best curve fitting with the minimum fitting error.

		3 sec.	6 sec.	9 sec.	12 sec.
Voigt Model	μ_1	9.23	9.67	11.2	13.0
	μ_2	1.60	1.72	2.54	3.01
	Error, (m/s)	0.103	0.114	0.114	0.121
Maxwell Model	E	18.3	19.6	32.3	39.2
	η	3.60	3.73	3.93	4.48
	Error, (m/s)	0.117	0.173	0.172	0.231
Zener Model	E_1	7.68	8.40	9.60	11.9
	E_2	15.0	18.0	35.0	63.3
	η	1.90	1.91	2.81	3.10
	Error, (m/s)	0.029	0.034	0.0344	0.102

Table 4. Shear Moduli of thermally damaged livers

6. Discussion

Shear moduli have very high dynamic ranges and are highly correlated with the pathological statues of human tissue. The solutions of the wave equation with constitutional models of tissue viscoelasticity show that the shear moduli of tissue can be estimated by measuring the phase velocity and attenuation of shear wave propagation in the tissue. However, it is a challenge to effectively and remotely generate vibrations and shear waves in a tissue region. It is also a challenge to measure shear wave because shear wave attenuates very fast as the propagation distance increases.

In the past fifteen years, the use of pulsed and focused ultrasound beams has been demonstrated as an effective method to remotely induce localized vibrations and shear waves in a tissue region. Several useful technologies have been developed for characterizing tissue viscoelasticity: Vibroacoutography, ARFI, Supersonic imaging, and ultrasound vibrometry, etc.

The ultrasound vibrometry is only technique that quantitatively estimates both tissue elasticity and viscosity. We found that the estimates of tissue elasticity by ignoring the viscosity are erroneous. Shear phase velocity are frequency dependent because the dispersive property of the biological tissue. Therefore, regardless of the usefulness of the viscosity, accurate estimates of tissue elasticity require the inclusion of the viscosity in the tissue models, as indicated by the solutions of the wave equation with three viscoelasticity models.

The ultrasound vibrometry transmits periodic push pulses to induce vibrations and shear waves in a tissue region, and detects the shear wave propagation using the pulse-echo ultrasound. The push pulses and detection pulses are interleaved so that one array transducer can be used for the applications of both pulses. The application of the array transducer allows the detection over a distance so that the phase velocities of several harmonics can be measured for calculating shear moduli.

Accurate estimates of shear moduli require an extended frequency range over an extended distance. The current technology is only effective for a few hundred Hertz in the frequency and a few mm in the distance away from the center of the radiation force applied. Shear wave having a high frequency attenuates very quickly as distance increases. Other vibration methods such as OFUV may be worth to explore.

We found that the shear wave speeds of livers are location dependent or dispersive in locations. Our experiment results indicate that the shear moduli estimated from a superficial tissue region and from a deep tissue region can be significantly different. Boundary conditions play a very important role in shear wave propagation and its phase velocity. The solution of the wave equation with boundary conditions should be considered for a tissue region that has a limited physical size. Some studies in this area have been done for myocardium and blood vessel walls.

The measurements of the ultrasound vibrometry are based on the assumption that tissue under the test is isotropic, which is not true for most tissues. Nevertheless, the measurements may be useful in clinical practices, which need to be evaluated *in vivo* experiments and clinical studies. On the other hand, the solutions of the wave equation with anisotropic tissue are needed.

Limited by the extensive contents in this chapter, we do not discuss the application of the Kalman filter in this work. The Kalman filter has great potential to include more complicated tissue models and motion models that are not fully explored yet, at least are not publically reported yet. On the other hand, Fourier transform and correlation method are also effective tools to calculate phases of the slow-time signals, if the motion model is simply sinusoidal.

Our experiments demonstrate that the ultrasound vibrometry can be readily implemented by using commercial medical ultrasound scanners with minimum alterations. Our experiment results also demonstrate that the ultrasound vibrometry is effective to characterize the stiffness and injury levels of livers.

We find that the Zener model fit the shear wave speeds of the livers better than the Voigt model and Maxwell model in almost all cases that include different frequency ranges, different locations, and different tissue conditions. Our study also indicates that the Voigt model is sensitive to the change of the observation frequency. Measurements at higher frequencies should be included when the Voigt model is used. In this case, the OFUV is useful to enhance the higher frequency components of the shear waves. The Zener model and Maxwell model appear to be less impacted by the frequency changes with our experiment data.

7. Conclusion

Tissue pathological statues are related to tissue shear moduli, which can be estimated by measuring the phase velocity of shear wave propagation in a tissue region. Ultrasound vibrometry is an effective tool to quantitatively measure tissue elasticity and viscosity. Ultrasound vibrometry induces vibrations in a tissue region using pulsed and focused ultrasound radiation force and detects the shear wave propagation using pulse-echo ultrasound. Experiment results demonstrate the effectiveness of the ultrasound vibrometry for characterizing tissue stiffness and liver damages.

Author details

Yi Zheng and Aiping Yao
Department of Electrical and Computer Engineering,
St. Cloud State University, St. Cloud, Minnesota, USA

Xin Chen, Haoming Lin, Yuanyuan Shen,
Ying Zhu, Minhua Lu, Tianfu Wang and Siping Chen
Department of Biomedical Engineering, School of Medicine, ShenZhen Univeristy, ShenZhen, China

Xin Chen, Haoming Lin, Yuanyuan Shen,
Ying Zhu, Minhua Lu, Tianfu Wang and Siping Chen
National-Regional Key Technology Engineering Laboratory for Medical Ultrasound,
ShenZhen, China

Acknowledgement

This research was supported in part through a grant from National Institute of Health (NIH) of USA with a grant number of EB002167 and a grant from Natural Science Foundation of China (NSFC) with a grant number of 61031003.

8. References

[1]   A.P. Sarvazyan, O.V. Redenko, S. D. Swanson, J.B. Fowlkers, and S.Y. Emelianov, "Shear wave elasticity imaging: a new ultrasonic technology of medical diagnostics," *Ultrason. Med. Biol.* 24:1419-1435, 1998.

[2]   W. Yeh, P. Li, Y. Jeng , H. Hsu, P. Kuo, M. Li, P. Yang, and P. Lee, "Elastic modulus measurements of human and correlation with pathology," Ultrasound in Med. & Biol., Vol. 28, No. 4, pp. 467–474, 2002.

[3]   L. Huwart, F. Peeters, R. Sinkus, L. Annet, N. Salameh, L.C. ter Beek, Y. Horsmans, and B.E. Van Beers, "Liver fibrosis: non-invasive assessment with MR elastography," NMR Biomed, vol. 19, pp. 173-179, 2006.

[4]   J. Ophir, I. Cespedes, H. Ponnekanti, Y. Yazdi, and X. Li, "Elastography: A quantitative method for imaging the elasticity of biological tissues," Ultrason. Imag., vol. 13, pp. 111–134, 1991.

[5]   M. Fatemi and J. F. Greenleaf, "Ultrasound-Stimulated Vibro-Acoustic Spectrography," Vol. 280. no. 5360, pp. 82 – 85, Science 3 April 1998.

[6]   K. Nightingale, M. Scott Soo, R. Nightingale, and G. Trahey, "Acoustic Radiation Force Impulse Imaging:In Vivo Demonstration of Clinical Feasibility," Ultrasound Med. Biol., 28(2): 227-235, 2002.

[7]   J.Bercoff, M. Tanter, and M. Fink, "Supersonic Shear Imaging: A New Technique for Soft Tissue Elasticity Mapping," IEEE transactions on Ultrasonics, Ferroelectrics, and Frequency Control, 51(4), pp. 396-409, 2004.

[8]   Y. Zheng, A. Yao, K. Chen, E. Zheng, T. Wang, and S. Chen, "Impacts of the Capsule on Estimation of Shear Viscoelasticity of Livers," IEEE International Ultrasonics Symposium Proceedings, October, 2011.

[9]   H. L. Oestreicher, "Field and impendence of an oscillating sphere in a viscoelastic medium with an application to biophysics," Journal of Acoustical Society of America, vol. 23, pp. 707-714, June 1951.

[10] S. Chen, M. Fatemi, and J. F. Greenleaf, "Quantifying elasticity and viscosity from measurement of shear wave speed dispersion," Journal of Acoustical Society of America, vol. 115, no. 6, pp. 2781-2785, 2004.

[11] Y. Zheng, S. Chen, W. Tan, and J.F. Greenleaf, "Kalman filter motion detection for vibro-acoustography using pulse echo ultrasound," Proceedings of 2003 IEEE Ultrasunics Symposium, 2003, pp. 1812 – 1815.

[12] Y. Zheng, S. Chen, X. Zhang, and J. F. Greenleaf , "Detection of shear wave propagation in an artery using pulse echo ultrasound and Kalman filtering, "Proceedings of 2004 IEEE International Ultrasonic Symposium, 2004, pp. 1251-1253.

[13] Y. Zheng, A. Yao, S. Chen, and J. F. Greenleaf, "Measurement of shear wave using ultrasound and Kalman filter with large background motion for cardiovascular studies," Proceedings of the 2006 IEEE Ultrasonics Symposium, 2006, pp. 718-721.

[14] Y. Zheng, S. Chen, W. Tan, R. Kinter, and J.F. Greenleaf, "Detection of tissue harmonic motion induced by ultrasonic radiation force using pulse echo ultrasound and Kalman filter," IEEE Transaction on Ultrasound, Ferroelectrics, and Frequency Control, February, vol. 54, no. 2, pp. 290-300, 2007.

[15] Y. Zheng, A. Yao, S. Chen, and J. F. Greenleaf,"Rapid Shear Wave Measurement for SDUV with broadband excitation pulses and non-uniform Sampling, "Proceedings of 2008 IEEE International Ultrasonics Symposium, 2008, pp.217-220.

[16] S. Chen, R.R. Kinnick, C. Pislaru, Y. Zheng, A. Yao, and J.F. Greenleaf, "Shearwave dispersion ultrasound vibrometry (SDUV) for measuring tissue elasticity and viscosity," IEEE Transaction on Ultrasonics, Ferroelectric, and Frequency Control, 56(1): 55-62, 2009.

[17] M. W. Urban, S. Chen, and J.F. Greenleaf, "Error in estimates of tissue material properties from shear wave dispersion ultrasound vibrometry," *IEEE Trans. Ultrason. Ferroelectr. Freq. Control,* vol. 56. pp. 748-758, 2009.

[18] X. Zhang and Greenleaf J.F., "Measurement of wave velocity in arterial walls with ultrasound transducers," *Ultrasound Med. Biol,* vol. 32, pp. 1655-60, 2006.

[19] X. Zhang, R.R. Kinnick,M. Fatemi, and J.F. Greenleaf, "Noninvasive method for estimation of complex elastic modulus of arterial vessels," *IEEE Trans. Ultrason. Ferroelectr. Freq. Control,* vol. 52, pp. 642-52, 2005.

[20] X. Zhang and J.F. Greenleaf, "Noninvasive generation and measurement of propagating waves in arterial walls," *J. Acoust. Soc. Am,* vol. 119, pp. 1238-43, 2006.

[21] M. Bernal, M.W. Urban, and J.F. Greenleaf, "Estimation of mechanical properties of arteries and soft tubes using shear wave speeds," in *Ultrasonics Symposium (IUS), 2009 IEEE International,* 2009, pp. 177-180.

[22] C. Pislaru, Urban M.W., Nenadic I, and Greenleaf J.F., "Shearwave dispersion ultrasound vibrometry applied to in vivo myocardium," in *Engineering in Medicine and Biology Society, 2009. EMBC 2009. Annual International Conference of the IEEE,* 2009, pp. 2891-2894.

[23] H. Xie, V. Shamdasani, A.T. Fernandez1, R. Peterson, Mi. Lachman, Y. Shi, J. Robert, M. Urban, S. Chen, and J.F. Greenleaf, "Shear Wave Dispersion Ultrasound Vibrometry (SDUV) on an Ultrasound System: *in vivo* Measurement of Liver Viscoelasticity in Healthy Animals," proceedings of 2010 *IEEE Ultrasunics Symposium,* pp. 912-915.

[24] M.L. Palmeri, M.H. Wang, J.J. Dahl, K.D. Frinkley and K.R. Nightingale, "Quantifying hepatic shear modulus in vivo using acoustic radiation force," Ultrasound in Med. & Biol., vol. 34, no. 4, pp. 546-558, 2008.

[25] T. Deffieux, G. Montaldo, M. Tanter, and M. Fink, "Shear Wave Spectroscopy for *In Vivo* Quantification of Human Soft Tissues Visco-lasticity," *IEEE* Transaction on Medical Imaging, pp. 313 to 322,VOL. 28, NO. 3, 2009.

[26] N.C. Rouze, M.H. Wang, M.L. Palmeri, and K.R. Nightingale, "Robust Estimation of Time-of-Flight Shear Wave Speed Using a Radon Sum Transformation," Proceeding of IEEE IUS 2010, pp. 21 to 24.

[27] Y. Zheng, and J.F. Greenleaf, "Stable and Unbiased Flow Turbulence Estimation from Pulse Echo Ultrasound," IEEE Trans. on Ultrasonics, Ferroelectric. Frequency Control, vol. 46, pp. 1074-1087, 1999.

[28] Y. Zheng, A. Yao, S. Chen, M. W. Urban, R. Kinnick and J.F. Greenleaf,, Orthogonal Frequency Ultrasound Vibrometry," American Society of Mechanical Engineering Congress, November, 2010, IMECE2010-39095.

[29] Y. Zheng, A. Yao, S. Chen, M. W. Urban, Y. Liu, K. Chen and J.F. Greenleaf, "Composed Vibration Pulse for Ultrasound Vibrometry,", IEEE International Ultrasonics Symposium Proceedings, October, 2010, pp. 17-20.

[30] Y. Wang, S. Chen, T. Wang, T. Zhou, Q. Li, Y. Zheng, and X. Chen, "Development of a generic ultrasound vibro-acoustic imaging platform for tissue elasticity and viscosity". *Journal of Innovative Optical Health Sciences,* Vol. 5, No. 1, pp. 1250002 (1-7), 2011.

Chapter 3

Acoustic Wave Propagation in a Pulsed Electro Acoustic Cell

Mohamad Abed A. LRahman Arnaout

Additional information is available at the end of the chapter

1. Introduction

The pulsed electro-acoustic technique [1] is presented to the Electrical Engineering community where it can find many applications, from the development of improved materials for electrical insulation to the control of electrostatic surface discharge (ESD) phenomena [2]. This phenomenon could involve serious damage to the satellite structure. In order to get a better control on the discharge it is necessary to clarify the nature, the position and the quantity of stored charges with time and to understand the dynamics of the charge transport in solid dielectrics used in space environment. Since its first implementation, the PEA method has been improved and adapted to many configurations of measurement: in 2D and 3D resolution [3] [4], with remote excitation [5] [6], on cables [7] [8] and under alternative stress [9] [10].

Recently, based on the PEA method, two original setups to measure space charge distribution in electron beam irradiated samples have been developed, and are called 'open PEA' and 'Short-Circuit PEA'. One of the weaknesses of this current technique is spatial resolution, about 10 μm. Indeed, dielectrics materials used in satellite structure have a thickness around 50 μm. Our work aims at improving the spatial resolution of a cell measurement by analyzing: electrical component, signal treatment, electrode material and sensor. In this paper, we only focused on the study of acoustic wave generation and their propagation. An electro-acoustic model has been developed with commercial software COMSOL®. This model is one-dimensional, and system of equations with partial differential functions is solved using a finite element method in non-stationary situations. Results show the propagation of acoustic wave vs. time in each part of cell measurement: sample, electrodes, piezoelectric sensor, and absorber. Influence of sensor geometry on the quality of output signal is also analyzed.

2. PEA method

The PEA measurement principle is given in Figure 1. Let us consider a sample having a thickness d presenting a layer of negative charge ρ at a depth x. This layer induces on the electrodes the charges ρ_d and ρ_0 by total influence so that:

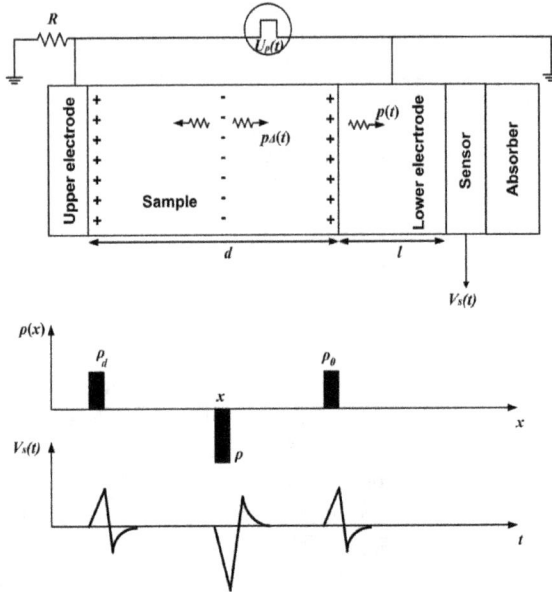

Figure 1. Schematic diagram of a PEA system.

$$\rho_d = \frac{-x}{d}\rho$$
$$\rho_0 = \frac{x-d}{d}\rho \tag{1}$$

Application of a pulsed voltage $U_p(t)$ induces a transient displacement of the space charges around their positions along the x-axis under Coulomb effect. Thus elementary pressure waves $p_\Delta(t)$, issued from each charged zone, with amplitude proportional to the local charge density propagates inside the sample with the speed of sound. Under the influence of these pressure variations, the piezoelectric sensor delivers a voltage $V_s(t)$ which is characteristic of the pressures encountered. The charge distribution inside the sample becomes accessible by acoustic signal treatment.

The expression of pressure waves reaches the piezoelectric detector is as follows:

$$p(t) = \int_{-\infty}^{+\infty} P_\Delta(x,t) = \int_{-\infty}^{+\infty} U_p(t - \frac{l}{v_e} - \frac{x}{v_s}).\rho(x).dx \tag{2}$$

With, v_e and v_s are the sound velocity of the electrode and the sample respectively.

Various parameters relevant to the spatial resolution of PEA method are clearly identified: as the thickness of the piezoelectric sensor, the bandwidth of system acquisition, etc... To quantify the influence of each parameter, a simplified electro acoustic model is proposed based on PEA cell.

3. Simulation set up

Our approach consists to establish an electro acoustic model from five sub-domains which represent the essential PEA cell element [11]. Each sub-domain is defined by the material and the thickness Table 1. As the samples are very thin compared to the lateral dimensions, we will consider a one-dimensional modelling. Each element is defined by a segment of length equal to the actually thickness.

Sub-domain	Material	Thickness (μm)
Upper electrode	Linear Low Density Poly Ethylene (LLDPE)	1000
Sample	Poly Tetra Fluoro Ethylene(PTFE)	300
Lower electrode	Aluminum	10 000
Piezoelectric sensor	PolyVinyliDene Fluoride(PVDF)	1→9
Absorber	Poly Methyl Meth Acrylate (PMMA)	2000

Table 1. Characteristics of each sub-domain in PEA model

Theoretically, the acoustic wave propagation is completely described by a partial differential equation (3).

$$\frac{1}{\rho_0 c^2}\frac{\partial^2 p}{\partial t^2} + \frac{\partial}{\partial x}(-\frac{1}{\rho_0}(\frac{\partial p}{\partial x} - q)) = Q_{int} \tag{3}$$

Where p represents the acoustic pressure (N.m^{-2}), c sound velocity (m.s^{-1}), ρ_0 density of material (kg.m^{-3}), q (N.m^{-3}) and Q_{int} (N.kg^{-1}.m^{-1}) reflect respectively the effect of external and the internal forces in the domain.

Acoustic pressure is obtained by resolving equation 3. In order to simplify our model, some assumptions are defined below:

- Attenuation and dispersion are not taken into account.
- Acoustic waves are generated by Coulomb forces created by the application of electric pulse on the electric charges present in sample.
- There is no acoustic source within the model: $Q_{int} = 0$.

After these assumptions (3) becomes:

$$\frac{1}{\rho_0 c^2}\frac{\partial^2 p}{\partial t^2} + \frac{\partial}{\partial x}(-\frac{1}{\rho_0}(\frac{\partial p}{\partial x} - q)) = 0 \tag{4}$$

Given the different assumptions, we have grouped the five sub-domains into three categories.

- Sub-domain which contains an acoustic source: sample, upper and lower electrode.
- Sub-domain that excludes acoustic source: absorber (PMMA).
- Piezoelectric sub-domain: piezoelectric sensor (PVDF).

Acoustic wave behavior in PEA simplified model depends on the acoustic impedance of each sub-domain. This impedance is equal to the product of sound velocity and density of material Table 2.

Sub-domain	Upper electrode	Lower electrode	Sample	Piezoelectric sensor	Absorber
ρ_0 (kg.m^{-3})	940	2700	2200	1780	1190
c (m.s^{-1})	2200	6400	1300	1270	2750
Z (kg.m^{-2}.s^{-1})	2068 x10^3	17280 x10^3	286 x10^3	2260.6 x10^3	3272.5 x10^3

Table 2. Acoustic parameters of each sub-domain in the model.

The application of an electric field on a sample (which contains electric charges) induces a mechanical force. This force consists of four terms [12] (5).

$$q_k = -\frac{1}{2}E_i.E_j\frac{\partial \varepsilon_{ij}}{\delta x_k} - \frac{1}{2}.\frac{\partial(\alpha_{ijkl} \cdot E_i \cdot E_j)}{\delta x_l} + \rho.E_k + kp_i.\frac{\partial E_k}{\partial x_i} \tag{5}$$

With:

- $\frac{1}{2}E_i.E_j\frac{\partial \varepsilon_{ij}}{\delta x_k}$: force produced by the variation of the electric permittivity.

- $\frac{1}{2}.\frac{\partial(\alpha_{ijkl} \cdot E_i \cdot E_j)}{\delta x_l}$: force provided by electrostriction effect.

- $\rho.E_k$: force created by the presence of electric charge in the sample.

- $kp_i.\frac{\partial E_k}{\partial x_i}$: force produced by the presence of electric dipoles in the sample.

- E_i, E_j, E_k: electric field components.
- ε_{ij}: electric permittivity.
- α_{ijkl}: electrostriction tensor.
- ρ: electric charge in the sample.
- kp_i: electric dipoles coefficient.

In 1D and without electric dipole equation (5) is:

$$q = -\frac{1}{2}E^2\frac{\partial \varepsilon}{\partial x} - \frac{1}{2}\frac{\partial(\alpha E^2)}{\partial x} + \rho E \tag{6}$$

Knowing that: $div(E) = \dfrac{\rho}{\varepsilon}$

$$q = -\frac{1}{2}E^2\frac{\partial \varepsilon}{\partial x} - \frac{1}{2}\frac{\partial(\alpha E^2)}{\partial x} + E\frac{\partial \varepsilon E}{\partial x} \tag{7}$$

Considering that: $\dfrac{\partial \varepsilon E^2}{\partial x} = \varepsilon E\dfrac{\partial E}{\partial x} + E\dfrac{\partial \varepsilon E}{\partial x}$ so

$$q = -\frac{1}{2}E^2\frac{\partial \varepsilon}{\partial x} - \frac{1}{2}\frac{\partial(\alpha E^2)}{\partial x} + \frac{\partial \varepsilon E^2}{\partial x} - \varepsilon E\frac{\partial E}{\partial x} = -\frac{1}{2}E^2\frac{\partial \varepsilon}{\partial x} - \frac{1}{2}\frac{\partial(\alpha E^2)}{\partial x} + \frac{\partial \varepsilon E^2}{\partial x} - \frac{1}{2}\varepsilon\frac{\partial E^2}{\partial x} \tag{8}$$

Recognizing that: $-\dfrac{1}{2}E^2\dfrac{\partial \varepsilon}{\partial x} - \dfrac{1}{2}\varepsilon\dfrac{\partial E^2}{\partial x} = -\dfrac{1}{2}\dfrac{\partial \varepsilon E^2}{\partial x}$ therefore,

$$q = -\frac{1}{2}\frac{\partial \varepsilon E^2}{\partial x} - \frac{1}{2}\frac{\partial(\alpha E^2)}{\partial x} = \frac{1}{2}\frac{\partial((\varepsilon - \alpha)E^2)}{\partial x} \tag{9}$$

Due to the application of electric pulse, the electric field varies from E to E+ΔE(t), so (9) becomes:

$$q = \frac{1}{2}\frac{\partial((\varepsilon - \alpha)E^2)}{\partial x} + \frac{\partial((\varepsilon - \alpha)(E\Delta E(t)))}{\partial x} + \frac{1}{2}\frac{\partial((\varepsilon - \alpha)(\Delta E(t)^2)}{\partial x} \tag{10}$$

In order to simplify our model we consider that α=0 and ΔE, ε are uniforms. After simplification of (10), the mechanical force in our model is the product of electric charges present in sample by the applied electric pulse.

$$q = \rho.\Delta E(t) \tag{11}$$

In our model we consider that:

- Electric charge profile ϱ is established by three Gaussian shapes of width = 3μm. A normalized negative one at sample center and two positives at sample interfaces Figure 2-a.
- A normalized square electric pulse which has a 5ns pulse width that has almost the same value of the experimental pulse Figure 2-b.

Piezoelectric transducer is used to convert electrical energy into mechanical energy and vice versa. The active element is basically a piece of polarized material, PVDF in our case. When the acoustic wave propagates in the PVDF, an electric voltage will appear at its interface, which it related to the direct piezoelectric effect [13].

The piezoelectric relations are given in equation (12).

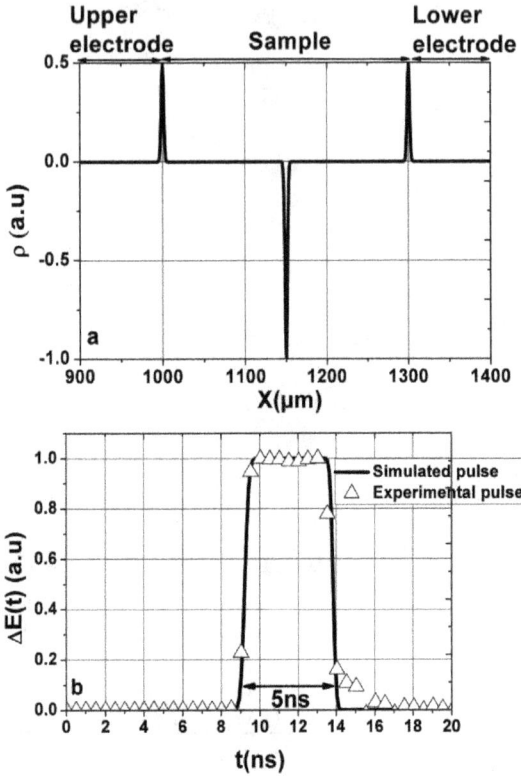

Figure 2. Modeling of mechanical force which is the product of electric charge distribution represented in (a) and the electric pulse represented in (b).

$$\begin{cases} D_i = d_{ikl}T_{kl} + \varepsilon_{ij}^T E_j \\ S_{ij} = s_{ijkl}^E T_{kl} + d_{ijk}E_k \end{cases} \tag{12}$$

With, D_i is the electric charge density displacement, E_i the electric field, S_{ij} the strain and T_{kl} the stress. $s^E{}_{ijkl}$ is the compliance, $\varepsilon^T{}_{ij}$ the permittivity. d_{ikl} is the matrix for the direct piezoelectric effect and d_{ijk} is the matrix for the converse piezoelectric effect.

In one dimension and referring to PEA case (D=0 because we have an open circuit configuration and $T_{kl}=-p$), the electric field vs. pressure waves is written as follow:

$$E_{PVDF} = \frac{d_{33}}{\varepsilon_{33}^T}p = g_{33}p \tag{13}$$

Output voltage signal can be obtained by integrating the electric field i.e. pressure wave along the PVDF thickness:

$$V = - \int_0^{d_{PVDF}} E_{PVDF}(x)dx = \int_0^{d_{PVDF}} -g_{33}p(x)dx \tag{14}$$

3.1. Sub-domain without acoustic source

This case represents the PMMA absorber. In principle, this material (PMMA) is used to avoid reflections of waves at the interface with the piezoelectric sensor. Unfortunately due to the difference between their acoustic impedance, we always have acoustic wave reflection at its interfaces. We estimate acoustic pressure in this sub-domain by neglecting the acoustic source, q = 0 in (4).

3.2. Boundary conditions

In our model, two kinds of boundary conditions are considered:

- *Dirichlet boundary condition*: pressure at external interfaces is considered as null in our model (upper electrode and PMMA):

$$p = 0 \tag{15}$$

- *Continuity condition*: The internal interfaces of the geometry are specified by the continuity of velocity vibration and acoustic pressure:

$$\frac{1}{\rho_i} \times \left(\nabla p_i - q_i \right) = \frac{1}{\rho_{i+1}} \times \left(\nabla p_{i+1} - q_{i+1} \right) \tag{16}$$

4. Simulation results

Simulation of acoustic wave in the model is realized by the commercial software Comsol. This software is based on the finite element method [14]. The domain of calculation is divided into several uniform elements of width Δx. Time step, Δt, in the computation is chosen with respect to Friedrich-Levy [15] condition:

$$\Delta t < \frac{\Delta x}{c} \tag{17}$$

In our model $\Delta x = 1\mu m$ and $\Delta t = 0.1$ ns.

Figure 3 shows the pressure for T=20 ns, for 300µm sample thick and 9µm sensor thick. This figure is divided into three regions. The sample is presented in region 2, however regions 1 and 3 represent respectively, and the adjacent upper and lower electrode. An electrical pulse is applied to probe the space charge distribution. Under the effect of the electric field pulse, the charges are shifted and come back to their original position, creating an acoustic wave (Coulomb forces effect).

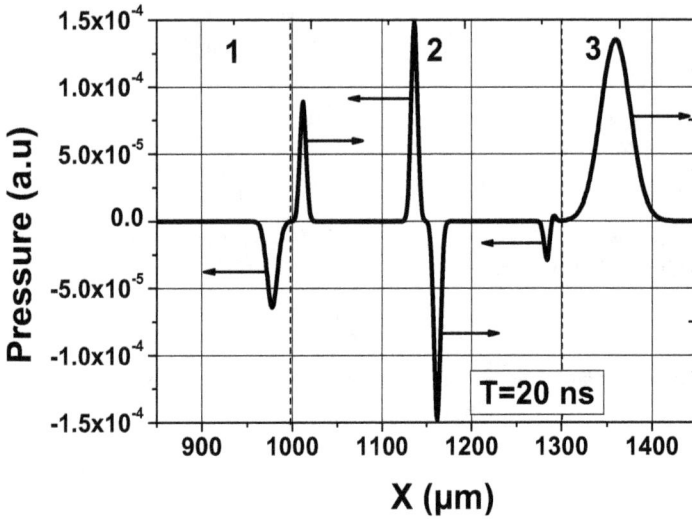

Figure 3. Acoustic wave generation in the sample.

When acoustic wave is completely generated for T=20 ns, we observe two pressure waves move at opposite way Figure 3: the first (at the left of Figure 3) moves towards the upper electrode and the second (at the right of Figure 3) moves towards the lower electrode. We can also observe a spreading more important for the pressure wave inside the lower electrode than the wave inside sample. This feature goes with the different values of the sound velocity between PTFE (sample) and Aluminum (lower electrode) Table 2. The same conclusion can be done for the upper electrode; value of sound velocity for the upper electrode is about 2 times larger than the sample.

Figure 4 shows acoustic wave for T=0 across the five sub-domains of PEA model i.e. upper electrode, sample, lower electrode, piezoelectric sensor and the absorber.

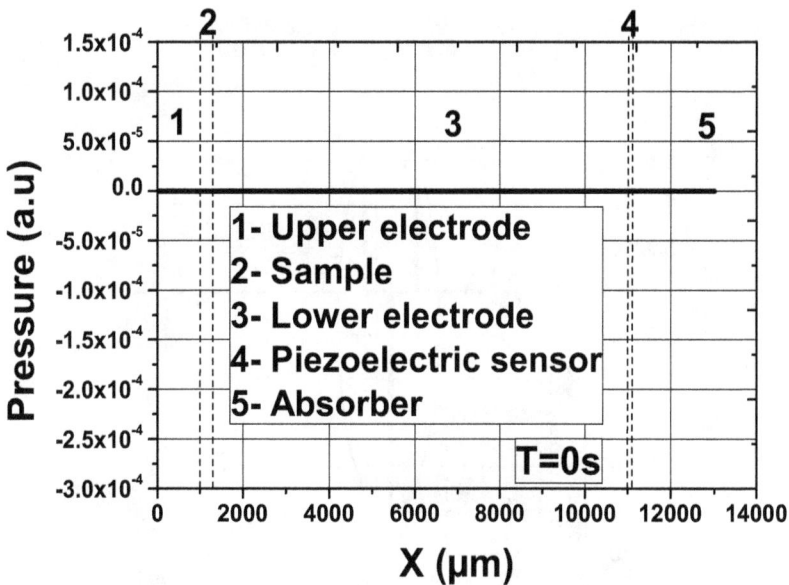

Figure 4. PEA model for T=0s.

For t= 800 ns we observe that the useful signal moves through lower electrode, Figure 5-a, and reaches the piezoelectric transducer at $t = 1.5$ µs, Figure 5-b. These results show that reflexion phenomena induce a lot of unwanted signals able to disturb the quality of the useful signal.

Figure 6 shows the output voltage signal for the five values of sensor thickness.

Only the useful signal (that its shape corresponds to the charge shape) has been integrated along the PVDF thickness referred to equation (14). We can observe when the sensor thickness decreases and reaches 1 µm, the output voltage signal leads to the same shape of space charge distribution inside the sample, Figure 2-a. For a larger thickness than 1 µm, the shape leads to a very different shape of the space charge distribution. The reflexion of acoustic waves on PVDF and PMMA interfaces plays an essential role in this case and involves some interference between the incident wave and reflected wave. This interference leads to a degradation of the output signal as shown in Figure 8. In order to improve the quality of this signal for thicknesses larger than 1 µm, it will be necessary to adapt acoustic impedances between sensor and absorber.

PMMA absorber is a related component with the piezoelectric sensor. Due to the difference between the acoustic impedance of these both materials, the transmission of acoustic waves is not optimally performed. Only a part of the wave is transmitted to the absorber, the other part is reflected on the interface and interferes with the incident wave. To improve the transmission coefficient wave, an acoustic impedance matching must be done i.e. the related materials should have the same acoustic impedance which is the product of material density and material sound velocity. Therefore, in our study, the PMMA absorber is replaced by a PVDF.

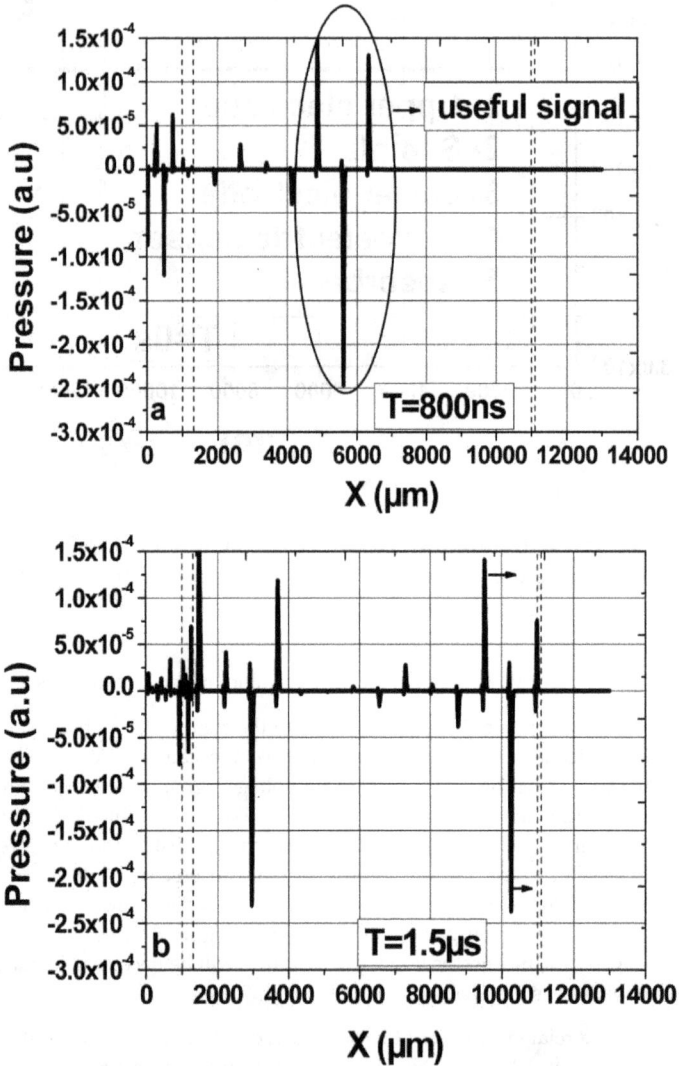

This simulation has been realized with a piezoelectric transducer thickness equal to 9 μm. With different sensors thicknesses, the same pressure wave evolution has been observed between sample and sensor (sensor not included). Only the shape of the output voltage signal is affected by the transducer thickness.

Figure 5. Acoustic wave propagation in the model . a) for T=800ns.b) for T=1.5μs

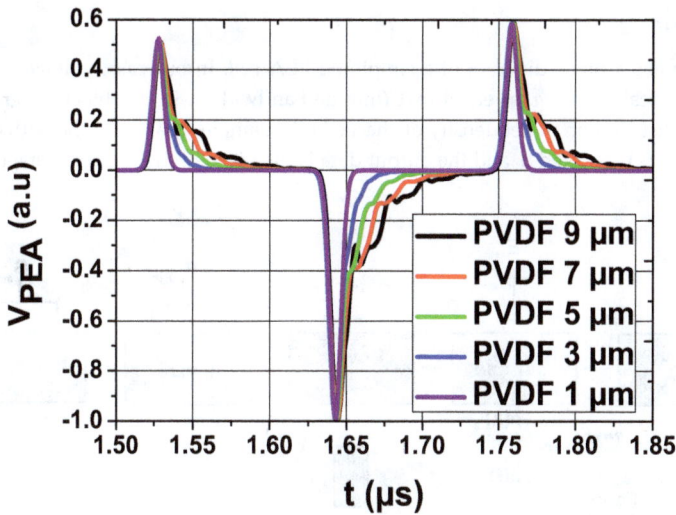

Figure 6. Potential signal for different PVDF thicknesses

Figure 7 shows the output voltage of piezoelectric sensor with and without a matched interface. The thickness of the sensor is fixed at 9 µm. This figure shows clearly the influence of the acoustic impedance on the quality of the output signal. Indeed, the reflected wave at the interface of PVDF / PMMA induces a strong distortion of the voltage signal that appears at each part of the useful signal. This distortion may affect the data processing, so a matched interface must be realized during the design of the new optimized cell.

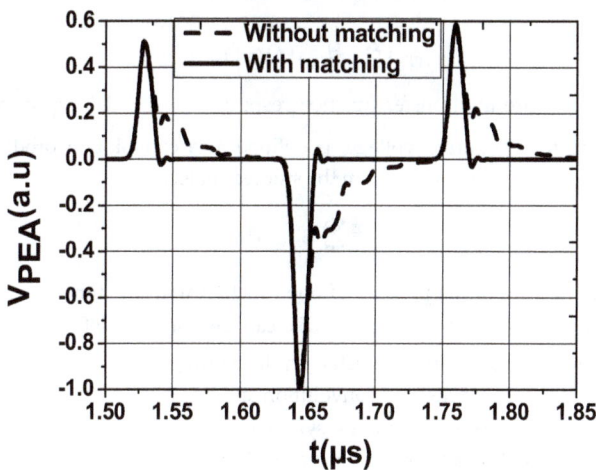

Figure 7. PVDF output signal with and without a matched interface between the sensor and the absorber

5. Analysis

Figure 7 shows a block diagram of a simplified PEA cell. In our case, voltage amplifier is considered ideal with a gain equal to 1 (infinite bandwidth). In this block diagram, input data is the distribution of net density of charge in the sample (charges at the both electrodes and in the bulk), denoted ϱ, and the output data is the piezoelectric sensor voltage, denoted V_{PEA}.

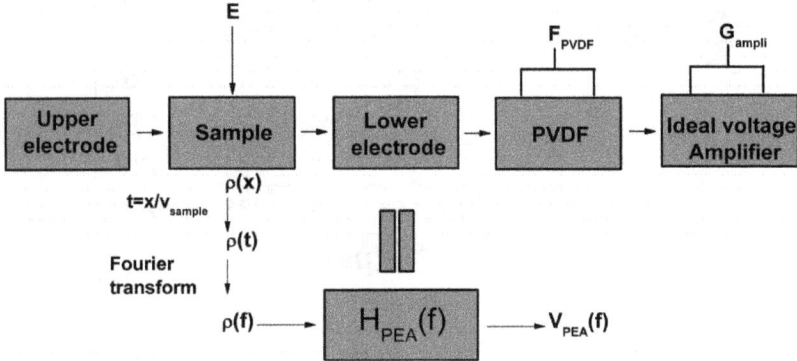

Figure 8. PEA bloc diagram

According to this figure, equations (18) and (19), written in frequency domain, will allow defining the transfer function matrix:

$$H_{PEA}(f) = E(f) \times F_{PVDF}(f) \times G_{ampli}(f) \qquad (18)$$

$$V_{PEA}(f) = H_{PEA}(f)\rho(f) \qquad (19)$$

Where F_{PVDF} and G_{ampli} are the transfer function respectively of PVDF sensor and voltage amplifier. According to (19) output voltage, is defined as a convolution product between transfer function and net density of charge in the studied dielectric:

$$v_{PEA_i} = \sum h_{PEA_{i,j}} \rho_j \qquad (20)$$

Convolution being the sum of the product of one function with the time reversed copy of the other function, a symmetric Toeplitz matrix can be used to define H_{PEA} [16]. This diagonal-constant matrix is a matrix in which each descending diagonal from left to right is constant. It is especially used for discrete convolution and it is completely determined by the first row. The following matrix A illustrates a symmetric Toeplitz matrix of order n and the following vector v represents exactly the same matrix A shown above where:

$$A_{ij} = a_{i-1,j-1} \text{ for } i = 2, n \text{ and } j = 2, n$$

$$A = \begin{pmatrix} a_{11} & a_{21} & \cdot & a_{n1} \\ a_{21} & a_{11} & \cdot & \cdot \\ \cdot & \cdot & \cdot & a_{21} \\ a_{n1} & \cdot & a_{21} & a_{11} \end{pmatrix} \quad v = \begin{pmatrix} a_{11} \\ a_{21} \\ \cdot \\ a_{n1} \end{pmatrix} \tag{21}$$

In our case, vector v is the impulse response i.e. the values of the first peak of output voltage for a polarized material, (denoted and usually named "calibration signal") and hence, equation (19) can be re-written as a linear function:

$$\underbrace{\begin{pmatrix} v_{PEA1} \\ v_{PEA2} \\ \cdot \\ v_{PEAn} \end{pmatrix}}_{V_{PEA}} = \underbrace{\begin{pmatrix} v_{ref1} & v_{ref2} & \cdot & v_{refn} \\ v_{ref2} & v_{ref1} & & \cdot \\ \cdot & \cdot & \cdot & v_{ref2} \\ v_{refn} & \cdot & v_{ref2} & v_{ref1} \end{pmatrix}}_{H_{PEA}} \underbrace{\begin{pmatrix} \rho_1 \\ \rho_2 \\ \cdot \\ \rho_n \end{pmatrix}}_{\rho} \tag{22}$$

Using the simulated, it is relatively easy to define the transfer function matrix. Knowing V_{PEA} and H_{PEA} the purpose of the next section is to analyze and to improve the condition number of H_{PEA} matrix.

In numerical analysis, the condition number of a matrix, denoted C in equation (23), measures the dependence of the solution compared to the data problem [17], in order to check the validity of a computed solution with respect to its data. Indeed, data from a numerical problem depends on experimental measurements and they are marred with error. We can say the condition number associated with a problem is a measure of the difficulty of numerical problem calculation. A problem with a condition number close to 1 is said to be well-conditioned problem, while a problem with a high condition number is said to be ill-conditioned problem. The condition number of the PEA Toeplitz matrix is equal to 400,000 and hence this very high value shows that our system is very ill-conditioned.

$$C = \|H_{PEA}\|_2 \|H_{PEA}^{-1}\|_2 \tag{23}$$

Where;

$$\|H_{PEA}\|_2 = \sqrt{\max(\det(H_{PEA}{}^t \cdot H_{PEA} - \lambda I))}$$

λ : Eigenvalues of H_{PEA}.

i.   Identity matrix

The identification method of the matrix is preserved; our goal is to improve the matrix condition number by studying the influence of intrinsic parameters for a PEA cell. Three parameters are studied: the thickness of piezoelectric sensor, the shape of pulse voltage, and the matching interfaces.

Table 3 shows condition number values for different PEA configuration. As we can see on this table, a "theorical" optimized cell measurement can be defined using a piezo-electric of 1μm, an impedance matching for all materials interfaces, and applying a Gaussian shape for electrical pulse. The condition number for this optimized cell is about 200 times smaller than the one from classical measurement cell! However, it should be noted that the choice of a Gaussian pulse decreases the expected resolution of the measure, it will be necessary to establish eventually a compromise between high spatial resolution and a well-posed system.

Thickness of PVDF	Pulse Shape	Matching interfaces	Condition number
9μm	Square	without matching	4.13 x 105
1μm	Square	without matching	4.2 x 104
9μm	Gaussian	without matching	9.2 x 104
9μm	Square	(PVDF/PMMA)	5.4 x 104
9μm	Square	(Electrode/sample)	1.3 x 105
1um	Gaussian	All interfaces	2 x 103

Table 3. Impact of PEA intrinsic parameters on transfer matrix condition number

In the next section, different deconvolution techniques are going to be used in order to recover the repartition of the net density of charge imposed on the Comsol model.

Wiener filtering is commonly used to restore degraded signals or images by minimizing mean square error. It is based on a statistical approach i.e. this filter is assumed to have knowledge of the spectral properties of the original signal and noise. Wiener filter must be physically realizable and causal and it is frequently used in the process of deconvolution. In frequency domain, its equation can be written as following:

$$H_{Wiener}(f) = \frac{1}{H_{PEA}(f)} \left[\frac{|H_{PEA}(f)|^2}{|H_{PEA}(f)|^2 + \frac{1}{SNR(f)}} \right] \tag{24}$$

and equation (19) becomes:

$$\rho(f) = H_{Wiener}(f) V_{PEA}(f) \tag{25}$$

Where ; H_{PEA} is the transfer function matrix and $SNR(f)$ the signal-to-noise ratio. When there is zero noise (i.e. infinite signal-to-noise), the term inside the brackets equals 1, which means that the Wiener filter is simply the inverse of the system. However, as the noise at certain frequencies increases, the signal-to-noise ratio drops, so the term inside the square bracket also drops. This means that the Wiener filter attenuates frequencies dependent on their signal-to-noise ratio. As explain previously, the condition number for H_{PEA} is relatively high and the coefficient $a = 1 / SNR(f)$ will be estimated using L-Curve method.

The L-Curve method consists of the analysis of the piecewise linear curve whose break-point are:

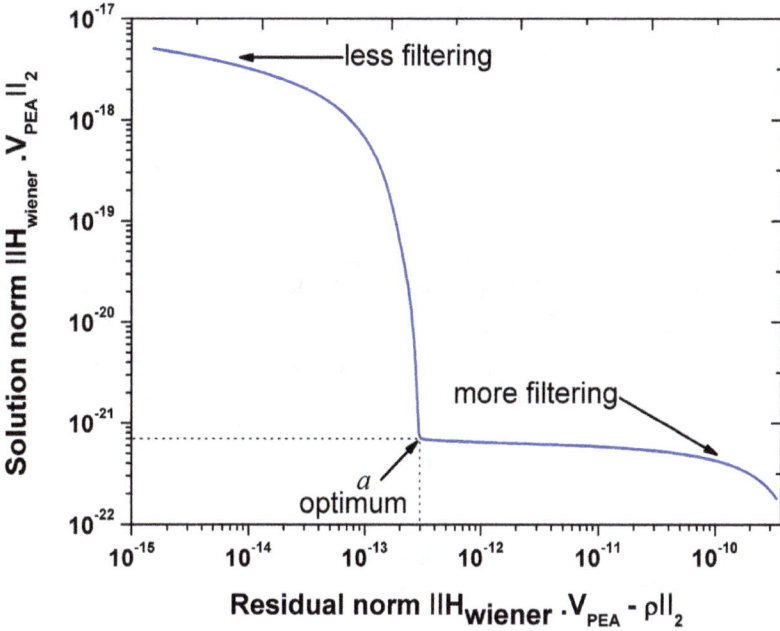

Figure 9. L-curve shape for unoptimized PEA cell

This curve, in most cases, exhibits a typical "L" shape, and the optimal value of the regularization parameter a is considered to be the one corresponding to the corner of the "L", Figure 10 [18-19]. The corner represents a compromise between the minimization of the norm of the residual and the semi-norm of the solution. This is particular evident in Fig.8, the horizontal branch of the "L" is dominated by the regularization error, while the vertical branch shows the sharp increase in the semi-norm caused by propagation errors.

Our approach consists to establish the deconvolved charge by using the transfer matrix H_{PEA} which was established. Gaussian filter is not accounted. The main objective of this section is to analyze the shape of the recovered charges using current PEA cell and the shape of the recovered charges using an optimized cell.

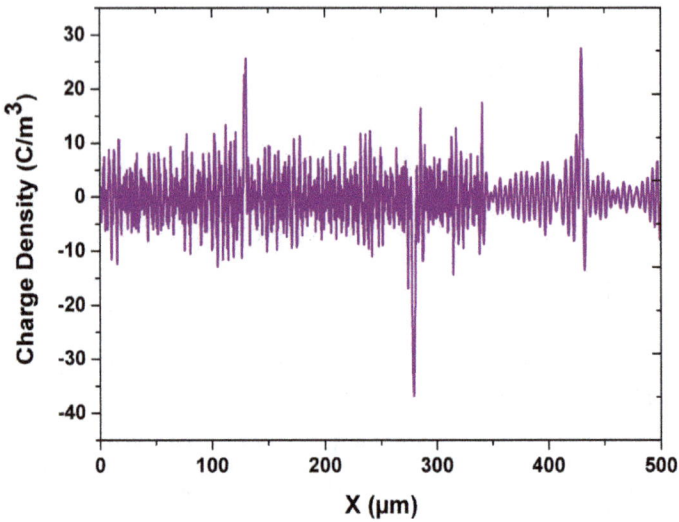

Figure 10. Net density of charges estimated using Wiener filter and a =8x10[-21] Current PEA cell with C = 400,000 for H_{PEA}

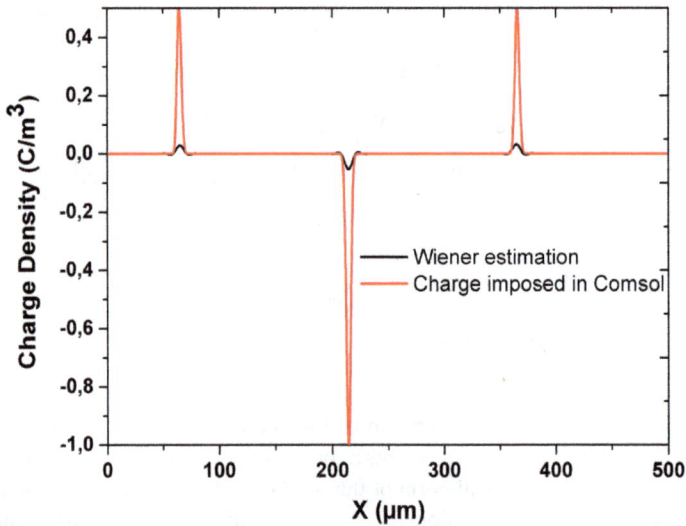

Figure 11. Net density of charges estimated using Wiener filter and a =3x10[-12] Optimized PEA cell with C = 2,000 for H_{PEA}

Results presented in Figure 10 and Figure 11 show the net density of charges estimated by Wiener method using the current PEA cell, Figure 10, and using an optimized cell, Figure 11.

Figure 10 shows that the recuperated charges by Wiener deconvolution are completely wrong, only noise is recovered and amplified! The inversion of this matrix is equivalent to applying a high pass filter, thus amplifying any high frequency noise. That is mainly for this reason that a Gaussian filter is usually applied for the signal treatment of PEA cell.

Using an optimized cell, result shown in Figure 11 has the similar shape, but unfortunately its amplitude is lower by 10 times than the net density of electric charge, and presents oscillations on both sides of the useful signal. These observations are typical of the adjustment when the system is ill conditioned. It is a compromise between filtering, and precision. This compromise is achieved by the settlement of regularized parameters determined by L-Curve method. Based on the previous results, the further work aims to redefine the method used for the calibration signal in order to have a condition number much more less than one obtained previously.

6. Conclusion

A one-dimensional numerical model based on acoustic wave propagation and established using COMSOL® was developed for a PEA cell with the objective to understand how sensor thickness is involved in the output voltage signal. In this model, transmission and reflection are taken into account, only attenuation and diffusion are neglected. Partial differential equation has been resolved using finite element method. In this paper, simulation results have been presented for different thicknesses of PVDF sensor from 1 μm to 9 μm. Results show the interest to use an ultra-thin piezoelectric sensor for improving the spatial resolution of the PEA cell. This model also permits to analyze acoustic wave behavior from its generation in sample to its conversion to an electric signal by piezoelectric transducer. Referred to this model, a PEA transfer function has been developed using a Toeplitz matrix, kind of matrix based on convolution principle. In order to improve the condition number of the transfer matrix, an optimized PEA cell has been defined based on a piezo-electric of 1μm, an impedance matching for all materials interfaces, and applying a Gaussian shape for electrical pulse. Deconvolution results show that a high condition number involves a strong deformation of estimated charge compare to a condition number much lower. Moreover, this study has highlighted the limits of regularization when the matrices are ill-conditioned: presence of oscillations, loss of information, etc.

Author details

Mohamad Abed A. LRahman Arnaout
Toulouse University, Plasma and Energy Conversion Laboratory, Toulouse, France

7. References

[1] Takada T Maeno T and Kushibe H (1987) An electric stress-pulse technique for the measurement of charges in plastic plate irradiated by an electron beam IEEE Trans. Electr. Insul. 4 pp 497-501.

[2]   Blaiman Keith G (1996) Space experiment design for electrostatic charging and discharging Proc. Int Conference on Astronautics pp115-122.

[3]   Maeno T (2001) Three-dimensional PEA Charge Measurement System IEEE Trans. Dielectr. Electr. Insul. 8 pp 845-848.

[4]   Fukuma M Maeno T and Fukunaga K (2005) High Repetition Rate Two-dimensional Space Charge Measurement System Proc. Int Symposium on Electrical Insulating Materials -Kitakyushu pp 584-587.

[5]   Griseri V Fukunaga K Maeno T Laurent C Levy L and Payan D (2004) Pulsed Electro-acoustic Technique Applied to In-situ Measurement of Charge Distribution in Electron irradiated Polymer IEEE Trans. Dielectr. Electr. Insul.11 pp 891-898.

[6]   Perrin C (2007) Apport de la technique électro-acoustique pulsée à la mesure et à l'analyse du transport de charges dans les diélectriques sous faisceau d'électron (Toulouse III University) pp 1-162.

[7]   Bodega R 2006 Space Charge Accumulation in Polymeric High Voltage DC Cable Systems -Delft University pp 1-183.

[8]   Muronaka T Tanaka Y Takada T Maruyama S and Mutou H (1996) Measurement of Space Charge Distribution in XLPE Cable using PEA system with Flat Electrode Annual Report Conference on Electrical Insulation and Dielectric Phenomena pp 266-269.

[9]   Alison JM Dissado LA and Fothergill JC (1998) The pulsed-electro-acoustic method for the measurement of the dynamic space charge profile within insulators The dielectrics society pp 93-121.

[10] Thomas C Teyssedre G and Laurent C(2008) A New Method for Space Charge Measurements under Periodic Stress of Arbitrary Waveform by the Pulsed Electro-acoustic Method IEEE Trans. Dielectr Electr Insul. 15 pp 554-559.

[11] Bernstein J.B (1991) Analysis of the electrically stimulated acoustic wave method for observing space charge in semi-insulating films Phvs. Rev. B. 44 pp 10804-10814.

[12] Holé S and Ditchi T (2000) Influence of divergent electric fields on space-charge distribution measurements by elastic methods Phys. Rev. B. 61. pp.13528-13539

[13] Berquez. L and Franceschi J.L (1998) Modeling and characterization of transducers for therrnoacoustic signal Sensors and Actuators A: Physical , pp. 115-120.

[14] Cook R.D Malkus D.S and Plesha M.E (1989) Concepts and Applications of Finite Element Analysis, 3rd Ed., JohnWiley and Sons.

[15] Kesserwani G Ghostine R Vazquez J Mosé R and Ghenaim A (2007) Simulation unidimensionnelle de l'écoulement à surface libre avec un schéma numérique TVD en discrétisation implicite et explicite La Houille Blanche 5 pp 101-106.

[16] Gray R.M, Toeplitz and circulant Matrices (2006) AReview Foundations and Trends® in communications and Information theory 2 pp155-239,.

[17] Tikhonov A.N Arsenin V.Y (1997) Solution of ill-posed problems wiley New York.

[18] Hansen P.C 1994 Regularization Tools: A MATLAB package for analysis and solution of discrete ill-posed problems Num. Algoritms 6 pp1-35.

[19] Hansen P.C O'Leary D.P (1993) The use of the L-Curve in the regularization of discrete ill-posed problems SIAMJ.Comput 14 N.6 pp 1487-1503.

Chapter 4

Radio Wave Propagation Phenomena from GPS Occultation Data Analysis

Alexey Pavelyev, Alexander Pavelyev, Stanislav Matyugov,
Oleg Yakovlev, Yuei-An Liou, Kefei Zhang and Jens Wickert

Additional information is available at the end of the chapter

1. Introduction

The aim of this book chapter is to reconsider the fundamental principle of radio-occultation (RO) remote sensing and to find new applications of RO method.

The RO remote sensing can be performed with any two cooperating satellites located on opposite sides with respect to the Earth's limb and moving to radio shadow. Several RO missions are working now aboard the Low Earth Orbit satellites. These missions provide global monitoring of the atmosphere and ionosphere of the Earth at different altitudes with high spatial resolution and accuracy. Their data are very important for meteorology, weather prediction. The RO data can be used to detect the climate changes, connections between the ionospheric, atmospheric processes, and solar activity, and to estimate conditions for radio navigation and radio location.

Up to now the RO inverse problem solution was based on the assumptions that the atmosphere and ionosphere are spherically symmetric and that the influence of turbulent and irregular structures on the retrieved vertical profiles of refractive index is insignificant [1,2]. The vertical profiles of refractive index are usually determined by measurement of the Doppler shift of radio wave frequency [1–3]. Information contained in the amplitude part of the radio-holograms was almost not addressed earlier, and this fact impeded separation of the contributions from layers and turbulent (small-scale) structures.

A new important relationship between the second-order time derivative (acceleration) of the phase path (eikonal), Doppler frequency, and intensity variations of the radio occultation (RO) signal was revealed by theoretical considerations and experimental analysis of the radio-holograms recorded onboard of the CHAMP and FORMOSAT-3 satellites [4–6]. Using the detected relationship, a possibility of determining the altitude, position, and inclination of plasma layers in the ionosphere from the RO data has been revealed [4–8]. The proposed calculation technique is simpler than the phase-screen [9] and back-propagation methods [10, 11]. The detected relationship is also important for estimation of the total attenuation of radio waves in a satellite communication link by combining the analysis of information contained in the amplitude and phase channels of the radio-holograms. The mentioned relationship makes it possible to convert the eikonal acceleration (or the time derivative of the Doppler frequency shift) into the refractive attenuation.

The total absorption of radio waves in the decimeter wavelength range at a frequency of 930 MHz was earlier determined experimentally [12, 13] in the "MIR" orbital station–geostationary satellites communication link. In those papers, attenuation was removed from the amplitude data with the use of the time dependence of the derivatives of the phase and Doppler frequency shifts. Measurements of the total absorption for determining the water content in the stratosphere and troposphere will be performed in the future radio-occultation missions [14] at three frequencies near the water-vapor absorption line at the wavelength of 1.35 cm. For analysis and processing of these measurement data, a technique [15, 16] using the integral Fourier operators (Canonical Transform (CT) and Full Spectrum Inversion Fourier analysis (FSI)) is proposed. In [16, 17] a radio-holographic technique of the total absorption measurements has been previously proposed. Following this technique, the refractive attenuation effect on the amplitude of the field transformed by an integral Fourier operator is ruled out by using the relationship between the refractive attenuation and the second-order time derivative of the phase difference of the recorded and reference signals.

Unlike the methods used in [15–17], the eikonal acceleration/intensity technique does not use any integral transform and can be directly employed for determining the total absorption of radio waves in the case of significant refractive attenuation under the condition of single path propagation. Moreover, the combined analysis of the eikonal acceleration and radio wave intensity makes radio vision of the atmospheric and ionospheric layers possible, i.e., allows the layers to be detected by observing correlated variations in the eikonal acceleration and intensity against the background of an uncorrelated contribution of turbulent inhomogeneities and small-scale structures and permits one to measure the layer parameters.

The book chapter is organized as follows. In section 2 the basic rules are given for describing radio waves propagation in a spherically symmetric medium including a new relationship for the refractive attenuation. In section 3 an advanced eikonal /intensity technique is introduced and applied to find the total absorption from analysis of the RO data. In section 4 a locality principle and its applications to determining the location, slope, and height of plasma layers in the ionosphere are described. Comparison with the back-propagation radio-holographic method is carried out. In section 5 the seasonal changes of the bending angle during four years of observations in Moscow and Kamchatka areas are described. Conclusions and references are given in section 6 and section 7, respectively.

2. Basic rules for radio waves propagation in a spherically symmetric medium

To obtain basic relationships describing the radio wave propagation in a spherically symmetric medium it is necessary to use a formula [18] for electromagnetic field $\mathbf{E}$ in an inhomogeneous medium that follows from Maxwell equations:

$$\nabla^2 \mathbf{E} + k_0^2 n^2 \mathbf{E} = -\text{grad}\left(\frac{\mathbf{E}\,\text{grad}\,\varepsilon}{\varepsilon}\right) \tag{1}$$

where k_0 is the wave number of radio wave in free space, n, ε are the refractive index and dielectric permittivity of medium, respectively. If vector $\mathbf{E}$ is perpendicular to grad ε, then (1) can be transformed to a homogeneous wave equation:

$$\nabla^2 E + k_0^2 n^2 E = 0 \tag{2}$$

where $E(\mathbf{r})$ is a component of the field $\mathbf{E}$. Solution of the eqn. (2) can be presented in the form [18,19]:

$$E(\mathbf{r}) = E_a(\mathbf{r})\,\exp\left(ik_0\psi(r)\right) \tag{3}$$

where $E_a(\mathbf{r})$, $\psi(r)$ – are the complex amplitude and phase path (eikonal) of radio wave. The eikonal $\psi(r)$ can be described by relationship:

$$\psi(r) = \int n(\mathbf{r})\,dl \tag{4}$$

where λ_0 is the wavelength in free space, and integration in (4) is fulfilled along a ray trajectory of radio wave. After substitution (3) into (2) one can obtain:

$$\nabla^2 E + k_0^2 n^2 E = \left\{ \begin{array}{l} \nabla^2 E_a(\mathbf{r}) - \kappa_0^2 E_a(\mathbf{r})(\nabla\psi)^2 + k_0^2 n^2 E_a(\mathbf{r}) \\ +i\kappa_0\left[2(\nabla E_a(\mathbf{r})\nabla\psi) + E_a(\mathbf{r})\nabla^2\psi\right] \end{array} \right\} \exp\left(ik_0\psi\right) = 0 \tag{5}$$

Two terms in the curly brackets of equation (5) formally differ by a factor – the imaginary unit i. Therefore to fulfill equation (5) these terms must be zero separately [18]:

$$\nabla^2 E_a(\mathbf{r}) - \kappa_0^2 E_a(\mathbf{r})\left(\nabla\,\psi\right)^2 + n^2(\mathbf{r})\kappa_0^2 E_a(\mathbf{r}) = 0 \tag{6}$$

$$2\nabla E_a(\mathbf{r})\nabla\,\psi + E_a(\mathbf{r})\nabla^2\psi = 0 \tag{7}$$

Under the geometric optics assumptions [18,19]:

$$\frac{\left|\nabla^2 E_a(\mathbf{r})\right|}{\left|\kappa_0^2\,E_a(\mathbf{r})\right|} \ll \left(\nabla\psi\right)^2,\ n^2(\mathbf{r}) \tag{8}$$

the next equations are valid [18,19]:

$$(\nabla \psi)^2 = n^2(\mathbf{r}); \quad |\nabla \psi| = n(\mathbf{r}); \nabla \psi = \mathbf{l}_0 n(\mathbf{r}) \tag{9}$$

$$E_a(\mathbf{r})\nabla^2\psi + 2(\nabla E_a(\mathbf{r})\nabla \psi) = 0 \tag{10}$$

where $\mathbf{l}_0$ is a unit vector oriented along the radio ray. The first and second relationships (9) are the eikonal equations. Formula (10) connects variations of the eikonal ψ and gradient of amplitude $E_a(\mathbf{r})$. Relationship (10) is known as a transfer equation for the field amplitude [19].

It follows from the relationships (9) that the ray equation has a form [18,19]:

$$\frac{\partial}{\partial l}(\mathbf{l}_0 n) = \operatorname{grad} n \tag{11}$$

where ∂l is an element of the length of the radio ray. In the case of spherical symmetry with a center located at the center of the Earth the gradient of the refractivity $\operatorname{grad} n$ and vector $\mathbf{r}$ have the same directions, and the impact parameter p is constant along the radio ray [18, 19]:

$$rn(r)\sin \gamma = p = \operatorname{const} \tag{12}$$

where γ is the angle between directions to the center of spherical symmetry and tangent to the radio ray.

One can obtain a relationship for the refractive attenuation of radio wave by multiplying equation (10) by $E_a(\mathbf{r})$:

$$E_a^2(\mathbf{r})\nabla^2\psi + 2E_a(\mathbf{r})\left(\nabla E_a(\mathbf{r})\nabla \psi\right) = 0 \tag{13}$$

It follows from (13)

$$\operatorname{div}\left(E_a^2(\mathbf{r})n(\mathbf{r})\,\mathbf{l}_0\right) = 0 \tag{14}$$

According to the Gauss theorem the next relationship is valid along a ray tube:

$$n(\mathbf{r})E_a^2(\mathbf{r})\Delta A = \operatorname{const} \tag{15}$$

where ΔA is the cross section square of a ray tube.

The relationships (11), (12), and (15) present basic rules describing the ray direction and power conservation laws in the spherically symmetric medium. From (15) one can obtain important formula for the refractive attenuation when the transmitter and receiver are located in a medium with arbitrary values of the refraction index.

In the case of spherical symmetric medium one can consider according to [18,20] a ray tube having at point G in the plane of Figure 1 the angular size $d\gamma$. This tube in the figure plane is bounded by lines GL and GL_1 (dotted line in Figure 1). The circle of radius R, with center at point O, intersects with the dotted line GL_1 at L_1, so that the arc $LL_1 = Rd\theta$. From the geometry (Figure 1) it follows that vector $\mathbf{LL}_2$ whose magnitude is $LL_2 = Rd\theta \cos \gamma_1$ perpendicular to the line GL_1. The side walls of the ray tube are located in the planes which are intersecting with straight line OG. The dihedral angle between these planes is equal to $d\gamma$. The size of the ray tube in the plane perpendicular to the figure plane at point L_1 is equal to $R \sin \theta d\gamma$. Therefore the cross section of the ray tube may be described by the relationship:

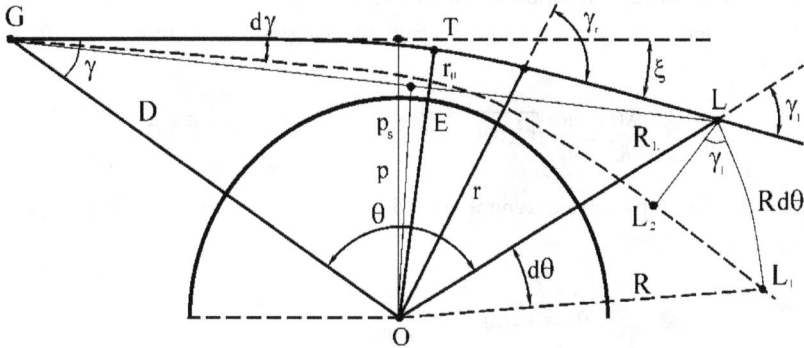

Figure 1. Ray tube in the radio occultation scheme. The center of spherical symmetry coincides with Earth's center O.

$$\Delta A = R^2 \left(d\gamma \right)^2 \frac{d\theta}{dp} \frac{dp}{d\gamma} \cos \gamma_1 \sin \theta \tag{16}$$

where $R = R_L$ is the distance from point L_1 to the center of spherical symmetry, γ_1, θ are the angles between the direction to the center of spherical symmetry, tangent to the radio ray, and the direction to the transmitter of radio wave, respectively. $d\gamma$ is the angular size of the ray tube. The value $\frac{dp}{d\gamma}$ can be obtained from (12):

$$\frac{dp}{d\gamma} = n(D)D\cos\gamma = \sqrt{n^2(D)D^2 - p^2} \tag{17}$$

where D is the distance from transmitter to the center of spherical symmetry. After substitution of (16), (17) in (15) with accounting for (12) one can obtain:

$$E_a^2(\mathbf{r})R\left(d\gamma\right)^2 \frac{d\theta}{dp} \sqrt{n^2(D)D^2 - p^2} \sqrt{n^2(R)R^2 - p^2} \sin\theta = const = C \tag{18}$$

The refractive attenuation $X(\mathbf{r})$ can be defined as a ratio of intensities of radio wave in the medium and in free space:

$$X(\mathbf{r}) = \frac{E_a^2(\mathbf{r})}{E_0^2(\mathbf{r})} \qquad (19)$$

where $E_0(\mathbf{r})$ is the radio field emitted by the same transmitter in free space. The radio field $E_0(\mathbf{r})$ can be described by a relationship:

$$E_0^2(\mathbf{r}) = \frac{C_T}{R_0^2} \qquad (20)$$

where R_0 is the distance from the transmitter to the current point $\mathbf{r}$, and C_T is constant which can include the transmitter's power and antenna gain. Substitution (19), (20) in (18) gives:

$$C_T \frac{X(\mathbf{r})}{R_0^2} R \frac{d\theta}{dp} \sqrt{n^2(D)D^2 - p^2} \sqrt{n^2(R)R^2 - p^2} \sin\theta (d\gamma)^2 = C \qquad (21)$$

The next relationships connect central angle θ, impact parameter p, bending angle $\xi(p)$, and distance R_0:

$$\theta = \pi + \xi(p) - \sin^{-1} \frac{p}{n(R)R} - \sin^{-1} \frac{p}{n(D)D} \qquad (22)$$

$$\theta = \xi(p) + \sin^{-1} \frac{p}{n(R)R} - \sin^{-1} \frac{p}{n(D)D} \qquad (23)$$

$$\frac{R}{R_0} \sin\theta = \sin\gamma, \text{ if } R_0 \to 0 \qquad (24)$$

Formula (23) is valid when the tangent point on the ray trajectory, where the ray is perpendicular to the gradient of refractivity, is absent [20]. Equations (22)-(24) allow transforming the formula (21):

$$X(\mathbf{r}) \left| \frac{d\xi}{dp} \frac{\sqrt{n^2(D)D^2 - p^2} \sqrt{n^2(R)R^2 - p^2}}{R_0} - \frac{\sqrt{n^2(R)R^2 - p^2} \pm \sqrt{n^2(D)D^2 - p^2}}{R_0} \right| C_T d\Omega = C \qquad (25)$$

$$d\Omega = \sin\gamma (d\gamma)^2$$

where $d\Omega$ is the solid angle occupied by the ray tube.

Constant C can be determined from (25) by estimating the refractive attenuation near the transmitter, when $R_0 \to 0$ (in equation (25) one should choose the lower sign). When $R_0 \to 0$ $X(\mathbf{r})$ is assumed to be equal to unity, i.e. $E_a^2(\mathbf{r}) = E_0^2(\mathbf{r})$, $R_0 \to 0$. This requires that the production of the antenna gain and emitted power along the ray $GTL\ C_T$ (20) does not change after the installation of the transmitter from free space in a medium with the refractive index $n(D)$. Under these conditions one can obtain from (25) when $R_0 \to 0$:

$$C = C_T n(D) d\Omega \tag{26}$$

Thus the refractive attenuation can be evaluated from (25) and (26) as:

$$X(\mathbf{r}) = \frac{R_0^2 n(D) D \sin \gamma}{RD \sin \theta \left| \frac{d\xi}{dp} \right| \sqrt{n^2(D)D^2 - p^2} \sqrt{n^2(R)R^2 - p^2} - \sqrt{n^2(R)R^2 - p^2} \pm \sqrt{n^2(D)D^2 - p^2}} \tag{27}$$

$$X(\mathbf{r}) = \frac{pR_0}{p_s \sqrt{n^2(D)D^2 - p^2} \sqrt{n^2(R)R^2 - p^2} \left| \frac{d\theta}{dp} \right|}; \ p_s = \frac{RD \sin \theta}{R_0}; \ p = n(D) D \sin \gamma \tag{28}$$

The refractive attenuation $X(\mathbf{r})$ represented by (27) and (28) satisfies the mutuality principle and does not depend on changing locations of the transmitter and receiver.

Previously refractive attenuation $X_p(\mathbf{r})$ has been defined as a ratio of the power flows in the medium and in free space. The magnitude of $X_p(\mathbf{r})$ has been obtained in the form [20]:

$$X_p(\mathbf{r}) = \frac{pR_0}{p_s \sqrt{D^2 - p^2} \sqrt{R^2 - p^2} \left| \frac{d\theta}{dp} \right|} \tag{29}$$

where p_s is the impact parameter corresponding to the line of sight GL. The difference between $X_p(\mathbf{r})$ (29) and $X(\mathbf{r})$ (28) consists in accounting for the refractivity near the transmitter and receiver.

Equations (28) for the refractive attenuation generalize the relationship (29) for the case when the transmitter and receiver are located in a spherically symmetric inhomogeneous medium. This relationship can be appropriate for RO data analysis during experiments provided in the planetary and Earth's atmospheres and ionospheres.

3. Total absorption

A new important relationship between the second-order time derivative (acceleration) of the phase path (eikonal), Doppler frequency, and intensity variations of the radio occultation (RO) signal has been established by theoretical considerations and experimental analysis of the radio-holograms recorded onboard of the CHAMP and FORMOSAT-3 satellites [4, 6-8]. The detected relationship makes it possible to convert the eikonal acceleration (or the time derivative of the Doppler frequency shift) measured using the RO phase data into the refractive attenuation and then exclude it from the RO amplitude data to obtain the total absorption. The method of measuring of the total absorption from joint analysis of the RO amplitude and phase variations is described below.

Layout of a RO experiment in the transionospheric link using the high-stability, synchronized by atomic-clock, radio signals of GPS navigation system is shown in Figure 2. Point O is the center of spherical symmetry of the Earth's atmosphere. The radio waves emitted by a GPS satellite located at point G enter the receiver input onboard a low-orbit satellite (point L) upon passage along the GTL ray, where T is the perigee of the ray. At point T, the distance h from the ray to the Earth's surface is minimal and the gradient of the refractive index $N(h)$ is perpendicular to the trajectory GTL (Figure 2). Projection of the point T on the Earth's surface determines the geographic coordinates of the studied region. Records of signals along the trajectory of a low-orbit satellite at two frequencies, $f_1 = 1575.42$ MHz and $f_2 = 1227.6$ MHz, are one-dimensional radio-holograms, which contain the amplitudes $A_1(t)$ and $A_2(t)$,

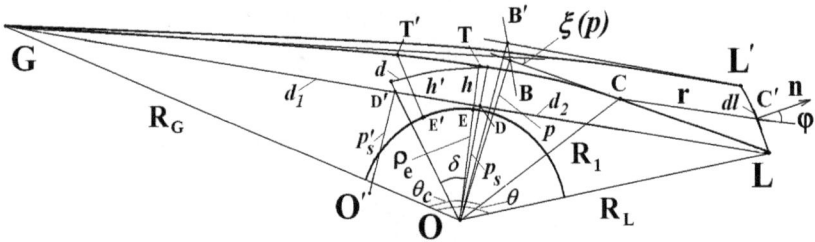

Figure 2. Main geometrical parameters describing the RO experiment conditions.

and the eikonal increments $\Phi_1(t)$ and $\Phi_2(t)$ of a radio field. The vertical velocity of the radio occultation ray in the perigee amounts to about 2 km/s, which is significantly greater than the velocities of motion of layers in the ionosphere and the atmosphere. Thus, the RO data are the instantaneous one-dimensional radio-holograms of the ionosphere and atmosphere. In the case of global spherical symmetry of the ionosphere and atmosphere, the following relations between the phase-path increments $\Phi(t)$ and the refractive attenuation $X(t)$ of radio waves [4-8] are fulfilled:

$$\lambda \frac{dF_d(t)}{dt} = m\frac{d^2\Phi(t)}{dt^2} = ma = 1 - X \tag{30}$$

where λ is the wavelength, $F_d(t)$ is the Doppler frequency of RO signal. In the RO experiments, parameter m can be determined from the orbital data.

Equations (30) relate the refractive attenuation $X(t)$ and the eikonal acceleration a in a form similar to the classical-dynamics equation. Equations (30) determine equivalence of the acceleration of the eikonal a to the time derivative of the Doppler shift $F_d(t)$ and the refractive attenuation $X(t)$. Thus, Eq. (30) make it possible to convert the eikonal acceleration and/or time derivative of the Doppler shift $F_d(t)$ into the refractive attenuation $X(t)$. This is important for estimation of the total absorption in the atmosphere.

The attenuation of intensity of radio waves $X_a(t)$ can be determined from the amplitude data in the form of a ratio of the intensity $I_a(t)$ of a radio signal propagating across the atmosphere to its intensity $I_s(t)$ in free space:

$$X_a(t) = \frac{I_a(t)}{I_s(t)} \tag{31}$$

The experimental quantity $X_a(t)$ is the product of the refractive attenuation $X_p(t)$ and the total absorption coefficient $\Gamma(t)$ and depends on the gain calibration errors of transmitter and receiver. However, the eikonal acceleration depends only on the refractive attenuation $X_p(t)$. This makes it possible to determine the atmospheric absorption $\Gamma(t)$ for a sufficiently stable level of the amplitude from the relations where $X(t)$ is the refractive attenuation of radio waves which was converted from the eikonal data. Parameter m can be found from the data describing the relative motion of the GPS satellite and the low-orbit satellite with respect to the spherical-symmetry center - point O (Figure 2) in the GOL plane. The quantity $X(t)$ calculated from Eq. (1) can be used to remove the refractive attenuation effect from the amplitude data:

$$\Gamma = 1 - \frac{X_a(t)}{X(t)} \tag{32}$$

Eqs. (30) - (32) permit one to estimate the total absorption of radio waves Γ along the ray GTL by amplitude and phase measurements of the radio-holograms.

The results of estimation of the total absorption Γ as a function of the altitude in the atmosphere for the experiment onboard the CHAMP satellite (No. 0159 at 14:54 UT) are shown in Figure 3. The experiment has been performed in June 16, 2003. Experiment corresponds to a polar region with geographical coordinates 83.0 N 258.6 W. The refractive attenuations $X_a(h)$ and $X(h)$, which were calculated from the RO amplitude and phase data using Eqs. (1) and (2) are shown in Figure 3 on the left (curves 1 and 2, respectively). Smooth curves 3 on the left in Figure 3 show the approximation obtained by a least-squares method. Slow trends in the refractive attenuations $X_a(h)$ are practically coinciding and vary from 0 dB at altitudes greater than 34 km to $-(10–15)$ dB at altitudes of about 5 km as seen in Figure 3 (left panel). This is experimental proof of the fulfillment of the relations described by Eqs. (1) that have been used for calculation of the refractive attenuation $X(h)$ from the RO phase-channel data of the satellite radio-holograms. Significant correlation between the high-frequency part of the variations in $X(h)$ and $X_a(h)$ takes place (Figure 3). Good correspondence between the variations in $X(h)$ and $X_a(h)$ exists at altitudes of 5 to 32 km

(Figure 3, left panel). At altitudes greater than 30 km, variations in $X_a(h)$ are greater than those in $X(h)$. This difference can be related to possible variations in the receiver (transmitter) gain or, which is more probable, to the ionospheric scintillation effect. The scintillation index S4 was equal to 2.7% in the experiment, and such a value corresponds to moderately disturbed conditions in the ionosphere. Relationships between the refractive attenuations retrieved from the amplitude and phase variations are important for estimation of the altitude dependence of the total absorption in the atmosphere. This dependence is shown by curve 1 in Figure 3 (right panels). Smooth curve 2 corresponds to the total absorption Γ found by a least-squares method, its value nearly corresponds to the absorption in atmospheric oxygen in accordance with [21]. Calculations show that the influence of the absorption in atmospheric oxygen can be tangible at altitudes less than 15 km. Experimental data agree, on the average, with this conclusion.

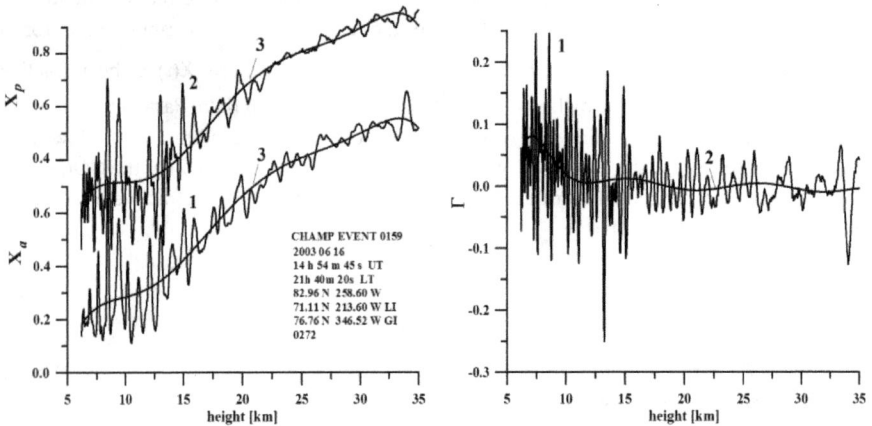

Figure 3. Left. Comparison of the refractive attenuation $X(h)$ retrieved from the eikonal data and attenuation $X_a(h)$ found from the intensity variations of the RO signal. Right. Difference of the refractive attenuations $X - X_a$ (curve 1) and absorption coefficient Γ found by a least-squares method (curve 2).

Obtaining more exact information on total absorption requires averaging of significant variations in the experimental values of Γ. For obtaining the dependence $\Gamma(h)$, the vertical profiles $X(h)$ and $X_a(h)$ were approximated by polynomials using a least-squares method. In Figure 4, left panel, curves 1–5 correspond to the resulting vertical profiles $X(h)$ and $X_a(h)$, and curves 6–10, Figure 4, middle panel, are related to the dependences $\Gamma(h)$ for five radio-occultation sessions performed using the CHAMP satellite in June 16, 2003. For convenience, curves 1–4, 6, 7, 9, 10 were displaced for comparison along the vertical axis. All sessions (No. 122, 02:27 LT, 77.6 N 141.0 W; No. 173, 17:35 LT, 80.9 N 337.1 W; No. 0030, 20:59 LT, 77.9 N 83.5 W; No. 0159, 21:40 LT, 83.0 N 258.6 W; and No. 0203, 16:56 LT, 76.3 N

37.9 W) correspond to the north polar regions. At altitudes between 12 and 30 km, the profiles $X(h)$ and $X_a(h)$ almost coincide. At altitudes less than 12 km, splitting of the curves $X_a(h)$ and $X(h)$ (e.g., the curves a and p, Figure 4, left panel), which begins at different heights, is observed. This effect is notable for all curves 6–10 in Figure 4 (middle panel). The differences at the initial height of splitting can be related to slow variations in the radio signal amplitude. The existence of splitting is probably a signature of the presence of small, but tangible integral atmospheric absorption, whose magnitude is, on the average, close to the values mentioned in [21] and to the magnitude 0.0096 ± 0.0024 dB/km of absorption per unit length measured in [12, 13, 22] (curves 6–10). The total absorption is varied within the limits 0.034 – 0.081 in the altitude range 12–5 km (Figure 4, middle panel, curves 6–10). The total absorption is near zero at the altitudes greater 12-15 km. The estimated value of the absolute statistical and systematic errors in the total absorption Γ is ± 0.01.

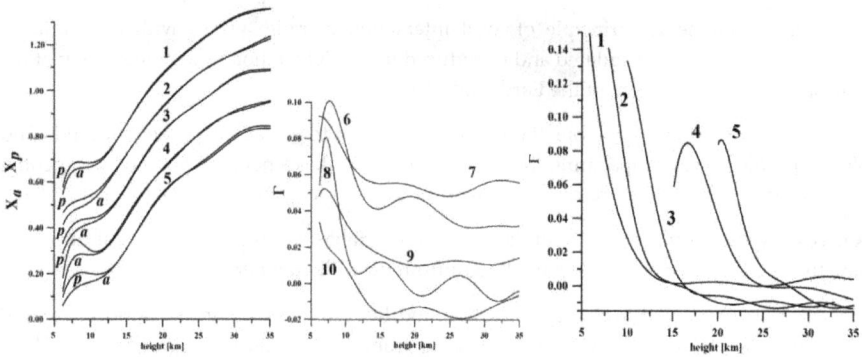

Figure 4. Left and middle panels. Comparison of the refractive attenuation $X(h)$ (index "p") retrieved from the eikonal data and attenuation $X_a(h)$ (index "a") found from the intensity variations of the RO signal (curves 1-5). Curves 6-10 in the middle panel describe the total absorption Γ calculated using Eq. (3). Curves 6,7, and 9,10 are displaced for convenience of comparison. The minimal and maximal values of parameter are: curve 6 -0.01, 0.062; curve 7 -0.001, 0.043; curve 8 -0.01, 0.081; curve 9 -0.0064, 0.0336; curve 10 -0.002, 0.046. Right panel. Averaged values of the total absorption in the Earth's atmosphere found for five days (Curves 1 – 5 correspond to found from CHAMP RO data averaged during February 24; June 16; May 03; November 30; July 07, 2003, respectively).

In Figure 4 (right panel) the averaged values of the total absorption Γ in the Earth's atmosphere are presented for five different days. These values have been found by the introduced eikonal/intensity method from CHAMP RO data for five days: February 24; June 16; May 03; November 30; July 07, 2003, respectively. According to Figure 4 (right panel), the maximal values of the total absorption Γ are containing for different days in the range 0.06 – 0.14 and correspond to the altitudes interval 6 km – 8 km. Averaging significantly reduces the statistical error of measurements, however the systematic errors remain. The systematic errors can be estimated from Figure 4 using negative values of the total absorption Γ as ±

0.01. Analysis of experimental data shows that additional attenuation as compared with the theoretical dependence can exist at altitudes smaller than 8 km. This can be related to the absorption effect in clouds and water vapor.

The introduced method is a perspective tool for investigation of seasonal and annual variations and geographical distributions of the total absorption from RO data. Also this method is, possibly, may be applied to study the influence of the tropical hurricanes and typhoons on the altitude profiles of water vapor in the stratosphere and tropopause.

4. Locality principle and RO remote sensing

A possibility to find the total absorption from joint analysis the RO amplitude and phase data described in section 3 is an important consequence of general locality principle valid in the of RO remote sensing of spherical symmetric atmospheres and ionospheres of the Earth and planets. Up to now this principle is implicit unformulated property of the RO method.

Below the fundamental principle of local interaction of radio waves with a spherically symmetric medium is formulated and introduced in the RO method of remote sensing of the atmosphere and ionosphere of the Earth and planets.

In accordance with this principle, the main contribution to variations of the amplitude and phase of radio waves propagating through a medium makes a neighborhood of a tangential point where gradient of the refractive index is perpendicular to the radio ray.

A necessary and sufficient condition (a criterion) is established to detect from analysis of RO data the displacement of the tangential point from the radio ray perigee.

This criterion is applied to the identification and location of layers in the atmosphere and ionosphere by use of GPS RO data. RO data from the CHAllenge Minisatellite Payload (CHAMP) are used to validate the criterion introduced when significant variations of the amplitude and phase of the RO signals are observed at RO ray perigee altitudes below 80 km.

The detected criterion opens a new avenue in terms of measuring the altitude and slope of the atmospheric and ionospheric layers. This is very important for the location determination of the wind shear and the direction of internal wave propagation in the lower ionosphere, and possibly in the atmosphere.

The new criterion provides an improved estimation of the altitude and location of the ionospheric plasma layers compared with the back-propagation radio-holographic method previously used.

4.1 Application of GPS RO method to study the atmosphere and ionosphere

The radio occultation (RO) method employs the highly-stable radio waves transmitted at two GPS frequencies $f_1 = 1575.42$ MHz and $f_2 = 1227.60$ MHz by the GPS satellites and recorded at a GPS receiver onboard low Earth orbiting (LEO) satellite to remote sense the Earth's ionosphere and neutral atmosphere [4,5,10,11,17,23-42,46,47,52-56,59-61]. When applied to ionospheric investigations the RO method may be considered as a global tool and can be compared with the global Earth- and space-based radio tomography [42,43]. The RO method delivers a great amount of data on the electron density distribution in the upper and lower ionosphere that are important sources for modernizing the current information over the morphology of the ionosphere and ionospheric processes [44,45]. The RO method has been actively used to study the global distribution of sporadic E-layers in dependence of latitude, longitude, altitude and local time [5, 27-32,39-41,45-47]. These investigations have produced useful data on climatology and the formation process of sporadic E-layers which depend mainly on the Earth's magnetic field and meteor impact according to the theory of the wind shear mechanism of plasma concentration [48-51]. The thermospheric wind and atmospheric tides seem to be the main energy sources for this mechanism [39].

Therefore the spatial distributions of sporadic E layers are important for investigating the connections of natural processes in the neutral and ionized components of the ionosphere. The location and intensity of sporadic E-layers plays a critical role for the quality of radio communications in the HF frequency band. The RO measurements in the atmosphere can be affected significantly by ionospheric contributions since the RO signals propagate through two different parts of the ionosphere.

Usually the ionospheric influence in the RO measurements may be described through a relatively slow change in the excess phase without noticeable variations in the amplitude of RO signals. This effect can be effectively reduced by a number of different methods of ionospheric correction [10,52,53].

However disturbed ionosphere may significantly change not only the phase but also the amplitude of the RO signals. Strong amplitude and phase frequency dependent variations in the RO signals are often surprisingly observed within the altitudes of the RO ray perigee $h(T)$ between 30 and 80 km above the main part of the neutral atmosphere and below the E-layer of the ionosphere. The effects of strong phase and amplitude variations of the RO signals at a low altitude provide a good source of information for the remote sensing of the atmosphere and ionosphere including detecting and studying the internal gravity waves propagating in the atmosphere and ionosphere [54]. Accurate knowledge of spatial location, height and inclination of the sporadic E-layers is important for the estimation of the off-equatorial height-integrated conductivity [44,45]. The RO low altitude amplitude variations have been interpreted as a contribution from the inclined ionospheric layers displaced relative to the RO ray perigee, and equations for the determination of the height and slope of inclined plasma layers from the known displacement of layers have been developed [27].

The altitudes of sporadic E-layers have been evaluated as the height of the RO radio ray perigee in recent times [28,39-41]. A relationship between the eikonal (phase path) and amplitude variations in the GPS/MET RO data has been analyzed in [53] and conclusions have been made that (i) the amplitude variations in distinction to the phase of RO signal have a strong dependency on the distance from observation point to the location of an ionospheric irregularity and (ii) the location of the irregularities in the low ionosphere may be determined by measuring the distance between the observation point up to a phase screen which should be located perpendicularly to the RO ray trajectory at its perigee.

A radio-holographic back-propagation method has been suggested and applied for location of the irregularities in E- and F-layers of the ionosphere [10,11]. A relationship between the derivatives of the phase, eikonal, Doppler frequency on time and intensity of radio waves propagating through the near Earth's space has been detected from both theoretical considerations and experimental analysis of the RO radio-holograms [4,5,31,36-38,47]. The introduced eikonal acceleration technique can de used for locating layers in the ionosphere and atmosphere.

The aim of this section is to demonstrate the possibility of identifying the contributions and measuring parameters of the inclined plasma layers by means of an analytical criterion. A test of a suggested method is provided by use of CHAMP RO data.

4.2. Criterion for layer locating

The scheme of RO experiments is shown in Figure 2. A navigational satellite G emitted highly-stable radio waves which after propagation through the ionosphere and atmosphere along the radio ray GTL arrived to a receiver onboard the Low Earth Orbital (LEO) satellite L. The amplitudes and phase variations of the RO signals are recorded as a function of time, sent to the ground stations with orbital data and analyzed with an aim to find the physical parameters of the neutral atmosphere and ionosphere along the trajectory of the RO radio ray perigee – point T (Figure 2). The receiver onboard LEO records the amplitude $A_1(t)$, $A_2(t)$ and the excess phase path $\Phi_1(t)$, $\Phi_2(t)$ of the GPS transmitted radio wave signals as a function of time t at two GPS frequencies.

The global spherical symmetry of the ionosphere and atmosphere with a common centre of symmetry is the cornerstone assumption of the RO method. Under this assumption a small area centered at tangent point T (Figure 2) where the RO ray is perpendicular to the gradient of refractivity, makes a significant contribution to the amplitude and phase variations of RO signals despite the prolonged path GTL (Figure 2). Under the global spherical symmetry condition the tangent point coincides with the RO ray perigee T. The size of this area along the ray GTL is equal to the horizontal resolution of the RO method $\Delta_h = 2\left(2l_f\rho_e\right)^{1/2}$, where $l_f = \left(\lambda d_2\right)^{1/2}$ is the size of the Fresnel zone, λ is the wavelength, ρ_e is the distance TO, d_2 is the distance TL which is nearly equal to DL (Figure 2). The magnitude of Δ_h corresponds to the minimal horizontal length of a layer estimated by the RO method.

The quiet ionosphere introduces regular trends in the excess phases at two GPS frequencies which can be removed by the ionospheric correction procedure [25,53]. The contributions in the phase and amplitude variations of RO signals of the intensive sporadic E-layers at the altitude interval 90-120 km is significantly greater than the impact of the F-layer turbulent structures [25]. Impact of a regular layer on the RO signal depends on position relative to the RO ray perigee. The length, l_{ce}, of coherent interaction of the RO signal with a layer having the vertical width l depends on the elevation angle ε between the local horizon direction and ray trajectory: $l_{c\varepsilon} \approx \dfrac{l}{\sin \varepsilon}$. For the RO ray perigee the elevation angle, ε, is zero, and the corresponding value l_c is described by relationship:

$$l_c = 2(2l\rho_e)^{1/2} \tag{33}$$

The ratio G of the lengths l_c and $l_{c\varepsilon}$ is equal to:

$$G = \frac{l_c}{l_{c\varepsilon}} = 2\sqrt{\frac{2\rho_e}{l}} \sin \varepsilon \tag{34}$$

Under spherical symmetry condition $\sin \varepsilon$ is about 0.25 at the altitude of ionospheric F-layer 250 km, and one can obtain from (34):

$$G \approx 0.5\sqrt{\frac{2\rho_e}{l}} \approx \frac{0.57 \cdot 10^2}{l^{1/2}} \tag{35}$$

If the vertical width l is about one kilometer, the contribution to the phase variations of a layer disposed in the RO ray perigee differs by about a hundred times on the impact of the similar layer located in the F-region. Therefore as a rule the RO method is effective tool for layers detection and measurements of their parameters with high vertical resolution and accuracy along of the trajectory of the RO ray perigee.

The next connection between the excess phase path (eikonal) $\Phi(t)$ acceleration a and the refractive attenuation of electromagnetic waves $X_p(t)$ have been detected and validated [4,5,31,36-38]:

$$1 - X_p(t) = ma, \; a = \frac{d^2\Phi(t)}{dt^2}, \; m = d_2(1 - d_2 / R_0) / (dp_s / dt)^2 \tag{36}$$

where d_2, R_0 are the distances along the straights lines DL and GL, respectively, p, p_s are the impact parameters corresponding to the ray GTL and the straight line GL (Figure 1). Note, that the distance d_2 is nearly equal to distance TL within an accuracy corresponding to the horizontal resolution of the RO method (about 100-300 km). Parameters m and dp_s/dt may be evaluated from the orbital data. The first formula (36) has been derived under condition [37]:

$$\left| (p - p_s)\frac{dR_{1,2}}{dt} \right| \ll \left| p_s \frac{dp_s}{dt} \right| \tag{37}$$

where R_1, R_2 are the distances OG, OL, respectively, (Figure 1). Condition (37) holds for RO studies of the atmospheres and ionospheres of the Earth and planets because the module of difference $p - p_s$ is always well below the magnitudes of p, p_s. If absorption is absent the magnitude $X_p(t)$ describes the refractive attenuation determined from the amplitude data:

$$X_p(t) \equiv X_a(t) \tag{38}$$

$$X_a(t) = I / I_0 \tag{39}$$

where I_0, I are the intensities of the RO signals measured before and after the immersion of the RO ray in the atmosphere, respectively. It should be noted that the total absorption in the atmosphere can be determined by excluding the refractive attenuation found from measurements of the eikonal acceleration at the same frequency by use of the first Eq. (36).

$$\Gamma = 1 - X_a(t) / X_p(t) \tag{40}$$

Eqn. (36) and (40) are the basis of the proposed method for determining the total absorption by measuring the time dependence of the intensity and eikonal of the RO signal at one frequency [31]. This method is much simpler than the previously used method based on estimation of the refractive attenuation on the first derivative of the bending angle on the impact parameter. When the total absorption is absent, it follows from (36) and (38), if the center of symmetry is located at point O :

$$1 - X_p(t) \equiv 1 - X_a(t) = ma \tag{41}$$

Relationship (41) establishes equivalence of the values $X_p(t), X_a(t)$ in the case of the spherical symmetry with centre O . Criterion (38) is a necessary and sufficient condition to ensure that the tangential point coincides with the radio ray perigee. This criterion is valid when the total absorption is absent and the requirement of the global spherical symmetry is fulfilled. In this case variations of the refractive attenuations found from the phase and amplitude variations of the RO signal should be the same at any time and can be attributed to the influence of the medium near the ray perigee (the locality principle). Therefore the RO method is based on an implicit locality principle and the RO method results correspond to the trajectory of motion of the RO ray perigee in the case of a spherically symmetric medium.

However the locality principle has more general meaning. Therefore it is necessary to extend the theory of the RO method to develop an appropriate technique to find the locations of the tangent points on the RO ray. This is an aim of the last part of this section.

In some cases the centers of spherical symmetry in the two parts of the ionosphere located on the path GTL (Figure 2) do not coincide with that of the neutral atmosphere [4,31,32,46,47]. In particular, this effect can be caused by the displacement of the centre of spherical symmetry O' of an ionospheric part of the ray GTL from the point O (Figure 2). In this case according to the derivation made previously [37] the inequality (37) is also valid after changing the distances $R_{1,2}$ to $R'_{1,2}$ and impact parameters p, p_s to p', p'_s because smallness of the difference $p' - p'_s$ as compared with any of the values p', p'_s . Therefore the identity (38) is valid also in the new coordinate system with centre at point O' (Figure 2):

$$X_p'(t) \equiv X_a(t) \qquad (42)$$

where $X_p'(t)$ - is a new value of the refractive attenuation relevant to a new center of spherical symmetry:

$$1 - X_p'(t) = m'a, \ a = \frac{d^2\Phi(t)}{dt^2}, \ m' = d_2'(1 - d_2' / R_0) / (dp_s' / dt)^2 \qquad (43)$$

where m' - is a new value of the parameter m relevant to a new center of spherical symmetry O', d_2' is the distance $D'L$, respectively (Figure 2). As compared with formula (36) the first equation (43) is different with new values of the refractive attenuation $X_p'(t)$ and parameter m'. The refractive attenuation $X_a(t)$ found from the amplitude data (39) and the eikonal acceleration a do not depend on location of the spherical symmetry centre. Identity (42) extends the criterion (38) to general case in which the centre of spherical symmetry is shifted to an arbitrary point.

This allows one to formulate the locality principle for remote sensing of layered spherically symmetric medium in the absence of absorption. A certain point of the radio ray is tangential if and only if the refractive attenuations found from the second derivative of the eikonal on time and intensity variations of the radio waves passed through the medium are equal. In this case both the intensity and the second derivative of the eikonal variations are mainly influenced by a small neighbourhood of the tangential point.

The principle of locality allows one to determine the location of a tangential point and to find the altitude, slope and displacement of a layer from the radio ray perigee. According to Eqn. (36), (43) it follows:

$$1 - X_a(t) \equiv \frac{m'}{m}(1 - X_p) \qquad (44)$$

where the refractive attenuation X_p is determined from Eq. (33) using measured value a; coefficients m', m - correspond to the centres of spherical symmetry O and O'. It follows from (36), (43), (44):

$$X_p - X_a(t) = \left(\frac{m'}{m} - 1\right)(1 - X_p) = \left[\frac{d_2'(1 - d_2' / R_0)(dp_s / dt)^2}{d_2(1 - d_2 / R_0)(dp_s' / dt)^2} - 1\right](1 - X_p) \qquad (45)$$

If the displacement of the center of spherical symmetry satisfies the following conditions:

$$d_2 / R_0, \ d_2' / R_0 \ll 1; \quad \frac{dp_s}{dt} \approx \frac{dp_s'}{dt} \qquad (46)$$

then one can find from (45):

$$X_p - X_a(t) = \frac{d_2' - d_2}{d_2}(1 - X_p) = \frac{d}{d_2}(1 - X_p) \qquad (47)$$

where d is the distance DD' (Figure 2). In the case of small refraction effect the distance d is approximately equal to the length of arc TT'. Relationship (47) establishes a natural connection between the displacement of the tangential point from the radio ray perigee d and variations of the refractive attenuations $X_a(t)$ and X_p.

Let us consider the refractive attenuation variations as the analytical signals in the form:

$$1 - X_p(t) = ma = A_p(t)\mathrm{Re}\left[\exp j\chi_p(t)\right]; \ 1 - X_a(t) = A_a(t)\mathrm{Re}\left[\exp j\chi_a(t)\right] \qquad (48)$$

where $A_p(t)$, $A_a(t)$; $\chi_p(t), \chi_a(t)$ are, correspondingly, the amplitudes and phases of the analytical signals, relevant to the functions $1 - X_p(t)$ and $1 - X_a(t)$. The amplitudes and phases $A_p(t)$, $A_a(t)$; $\chi_p(t), \chi_a(t)$ describe atmospheric (ionospheric) modulations of the refractive attenuation variations $1 - X_p(t)$ and $1 - X_a(t)$. The phases $\chi_p(t), \chi_a(t)$ differ from the excess phase path (eikonal) $\Phi(t)$. In the case when the variations $1 - X_p(t)$ and $1 - X_a(t)$ can be described by a narrowband process the functions $A_p(t)$, $A_a(t)$; and $\chi_p(t), \chi_a(t)$ can be found by the numerical Hilbert transform or by other methods of the digital data analysis.

After substitution (48) in (44) one can obtain:

$$A_a(t)\mathrm{Re}\left[\exp j\chi_a(t)\right] \equiv \frac{m'}{m} A_p(t)\mathrm{Re}\left[\exp j\chi_p(t)\right] \qquad (49)$$

The ratio $\frac{m'}{m}$ is supposed to be nearly constant during the RO measurement event. For fulfilling (49) the phases $\chi_p(t)$ and $\chi_a(t)$ should be equal, but the amplitudes $A_a(t)$ and $A_p(t)$ are different. In this case one can obtain from (49) under the conditions (46) an alternative relationship for the displacement d in the form:

$$d = d_2' - d_2 = d_2 \frac{A_a - A_p}{A_p}; \ d_2 = \sqrt{R_2^2 - p_s^2}; \ m' = \frac{A_a}{A_p}m \qquad (50)$$

Equation (50) establishes a rule: location of a tangent point on the ray trajectory can be fulfilled using the analytical amplitudes of the refractive attenuation variations $A_{a,p}$; the displacement d is positive or negative depending on the sign of difference $A_a - A_p$, the tangent point T' is located on the parts GT or TL, respectively. The phases $\chi_p(t)$ and $\chi_a(t)$ should be equal within some accuracy determined by a quality of measurements. From the last equation (50) one can find the coefficient m' if the magnitude m is known.

Note, that equation (50) is valid when the distance of one of the satellites from the ray perigee T is many times greater than the corresponding value for the second one. This condition is fulfilled for the planetary RO experiments provided by use of the communication radio link spacecraft-Earth and GPS occultations [31].

Correction to the layer height Δh and its inclination δ with respect to the local horizontal direction can be obtained from the displacement d [27]:

$$\delta = d / \rho_e, \quad \Delta h = 0.5 d \delta \tag{51}$$

where ρ_e is the distance TO (Figure 1).

Condition of the spherical symmetry with new center O' justifies application of the Abel's transform for solution of the inverse problem. For the Abel's transform the next formula is used [55]:

$$N(p_0) = -\frac{1}{\pi} \int_{p_0}^{\infty} \ln\left[\frac{p}{p_0} + \sqrt{\left(\frac{p}{p_0}\right)^2 - 1}\right] \frac{d\xi(p)}{dp} dp; \quad \frac{dN(p_0)}{dh} = \frac{1 + N(p_0)}{\left(1 - \frac{dN(p_0)}{dp_0}(r_e + h)\right)} \frac{dN(p_0)}{dp_0} \tag{52}$$

where p_0 is the magnitude of the impact parameter p corresponding to ray GTL in the initial instant of time t_0, $N(p_0)$ and $\frac{dN(p_0)}{dh}$ are the refractivity and its vertical gradient. the derivative of the bending angle $\xi(p)$ on the impact parameter p $\frac{d\xi(p)}{dp}$ can be found from the refractive attenuation X by use of equation obtained previously [20]:

$$\frac{d\xi}{dp} \approx \left(1 - \frac{1}{X}\right) \frac{R_0}{\sqrt{R_1^2 - p^2} \sqrt{R_2^2 - p^2}} \tag{53}$$

where R_0 is the distance GL (Figure 1). from (36), (52), (53) one can obtain the modernized formula for the Abel inversion:

$$N(p_0) = -\frac{1}{\pi} \int_{t_0}^{t_x} \ln\left[\frac{p(t)}{p_0} + \sqrt{\left(\frac{p(t)}{p_0}\right)^2 - 1}\right] \frac{m'a}{\sqrt{R_2^2 - p(t)^2}} \frac{dp_s}{dt} dt \tag{54}$$

Factor m' in (54) can be estimated from the last equation (50). Magnitude $m'a$ in (54) may be changed by the value $1 - X_a$ to use directly the RO amplitude data for the Abel inversion.

Note, that equation (52) provides the Abel's transform in the time domain t_0, t_x where a layer contribution does exist. The linear part of the regular trend due to influence of the upper ionosphere is removed because the eikonal acceleration a in (54) contains the second derivative on time. However the influence of the upper ionosphere is existing because it contributes in the impact parameter $p(t)$. Also nonlinear contribution of the upper ionosphere remains in the eikonal acceleration a. Equation (54) gives approximately that part of the refractivity altitude distribution which is connected with influence of a sharp plasma layer. The electron density vertical distribution in the earth's ionosphere $N_e(h)$ is connected at GPS frequencies with the refractivity $N(h)$ via relationship:

$$N_e(h) = -\frac{N(h)f^2}{40.3} \tag{55}$$

where f is carrier frequency [Hz], $N_e(h)$ is the electron content $\left[\dfrac{el}{m^3}\right]$.

4.3. Analysis of CHAMP experimental data

To consider a possibility to locate the plasma layers we will use a CHAMP RO event 005 (November 19, 2003, 0 h 50 m UT, 17.3 S, 197.3 W) with strong quasi-regular amplitude and phase variations. The refractive attenuations of the CHAMP RO signals X_a, X_p found from the intensity and eikonal data are shown in Figure 5 (left panel) as functions of the RO ray perigee altitude h. The eikonal acceleration a has been estimated by double differentiation of a second power least square sliding polynomial over a sliding time interval $\Delta t = 0.5\ s$. This time interval corresponds approximately to the vertical size of the Fresnel's zone of ~1 km since the vertical component of the radio ray was ~2.1 km/s. The refractive attenuation X_p is derived from the evaluated magnitude a using equation (36); m value is obtained from the

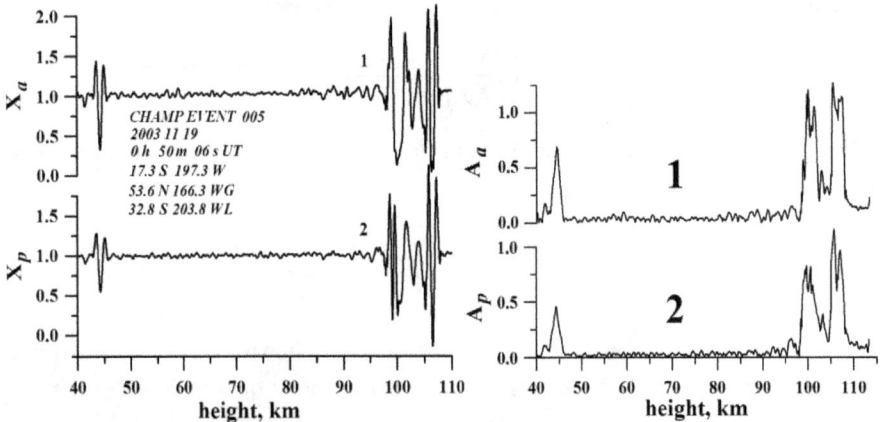

Figure 5. Left plot: the refractive attenuations X_a, X_p found from the intensity and eikonal ro data at frequency f_1 (curves 1 and 2, respectively). Right plot: the amplitudes A_a, A_p of analytical signals corresponding to the variations of the refractive attenuations X_a, X_p (curves 1 and 2).

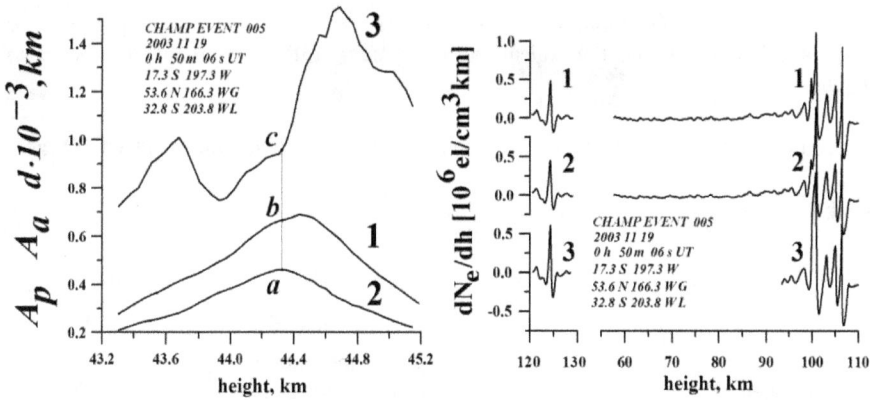

Figure 6. Left: evaluation of the plasma layer displacement from the RO perigee d. Right: results of restoration of the vertical gradients of the electron density.

orbital data. The refractive attenuation X_a is derived from the ro amplitude data by a least square method with averaging in the same time interval of 0.5 s. in the altitude ranges of 42-46 km and 98-106 km, the refractive attenuations variations X_a, X_p are strongly connected and may be considered as coherent oscillations caused by sporadic layers (Figure 5, left panel). Using the Hilbert numerical transform, the amplitudes A_a, A_p of analytical signals related to $X_a - 1$ and $X_p - 1$ have been computed and are shown in Figure 5 (right panel). In the altitude range of 42-46 km, the amplitudes A_a and A_p are nearly identical, but the magnitude of A_a is about 1.5 times greater than that of A_p. Accordingly, a plasma layer is displaced from the RO ray perigee T in the direction to the satellite G (Figure 2). A similar form of variations of the refractive attenuations $X_a - 1$ and $X_p - 1$ allows locating the detected ionospheric layer. The displacement d corresponding to a plasma layer recorded at the 44 km altitude of the RO ray perigee is shown in Figure 6 (left). The curves 1 and 2 in Figure 6 (left) correspond to the amplitudes A_a, A_p. Curve 3 describes the displacement d found from the amplitudes A_a, A_p using equation (50). The changes of d are concentrated in the altitude range of 720-1500 km when the functions A_a, A_p vary near their maximal values of 0.46 and 0.69 in the ranges of $0.2 \leq A_p \leq 0.46$ and $0.2 \leq A_a \leq 0.69$ respectively. The statistical error in the determination of the ratio $\dfrac{A_a - A_p}{A_p}$ in equation (50) is minimal when A_p is maximal. Point a in Figure 6 (left panel) marks the maximum value of A_p, and the points b and c denote the corresponding values $A_a = 0.67$ and $d = 940\ km$ respectively, the plasma is displaced from the RO ray perigee T in direction to the navigational satellite G (Figure 2). If the relative error in the measurements of A_p is 5%, then, according to Figure 6 (left) the accuracy in the estimation of d is about $\pm 120\ km$. The inclination of a plasma layer to a local horizontal direction calculated using eqns. (51) is approximately equal to $\delta = 10.4° \pm 0.2°$.

The vertical gradient $\dfrac{dN_e}{dh}$ of the electron density distribution $N_e(h)$ for the given RO event is shown in Figure 6 (right). Curves 1 and 2 correspond to the vertical gradient $\dfrac{dN_e}{dh}$ retrieved using eq. (52) and (54) respectively. Curve 3 is related to the vertical gradient $\dfrac{dN_e}{dh}$ retrieved using the refractive attenuation X_a and formula (54). The real altitude of the ionospheric layers is indicated on the horizontal axis in Figure 6 (right). Two ionospheric layers are seen (curves 1, 2, and 3 in Figure 6, right). The first layer is located on the line GT at the 120-130 km altitudes at a distance ~ 950 km from point T. The second layer is located near the RO perigee at the 98-108 km altitudes (Figure 5 and Figure 6, right). From the comparison of the refractive variations X_a, X_p (Figure 5, left) and the vertical gradients of the electron content (Figure 6, right) the width of the sporadic E-layers is nearly equal to the altitude interval of the amplitude variations of the RO signals. From Figure 6 (right), the variations of the vertical gradient of the electron density are concentrated in the interval $-\dfrac{1.1 \cdot 10^6 el}{cm^3 km} < \dfrac{dN(h)}{dh} < \dfrac{1.1 \cdot 10^6 el}{cm^3 km}$. These magnitudes of $N_e(h)$ are typical for intensified sporadic E-layers [45]. The height interval of the amplitude variations is nearly equal to the height interval of the variations in the electron density and its gradient.

The second example of the identification and location of sporadic plasma layer in the lower ionosphere is shown in Figure 7 for CHAMP RO event 211 (July 04 2003, 10 h 54 m LT, 2.1 N, 145.6 W) with intensive sporadic e layers. The refractive attenuations X_a and X_p of the CHAMP RO signals at f_1 obtained from the intensity and eikonal data are shown in Figure 7 (a) as functions of the RO ray perigee altitude h. The refractive attenuations variations X_a, X_p are strongly correlated and can be considered as coherent oscillations caused by a single sporadic e-layer. as shown in Figure 7 (b), the amplitudes A_a, A_p corresponding to $X_a - 1$, $X_p - 1$ are attained from the Hilbert numerical transform and the magnitude A_a is about 1.3 times greater than A_p. This means that a corresponding plasma layer is displaced from the RO ray perigee T in the direction to the satellite g (Figure 1). The displacement d of the tangent point can be determined from the amplitude variations A_a, A_p. The displacement d, the correction of the altitude Δh, the corrected height h' of the plasma layer maximum, and the slope of the plasma layer relative to the horizontal direction δ are shown in Figure 7, c, d. Curves 1, 2 and 3 in Figure 7 (d) are the amplitudes A_a, A_p and the corrected height h' of the plasma layer maximum on the RO ray perigee altitude h respectively. Curves 1, 2 and 3 in Figure 4 (c) are the displacement d (its values are marked

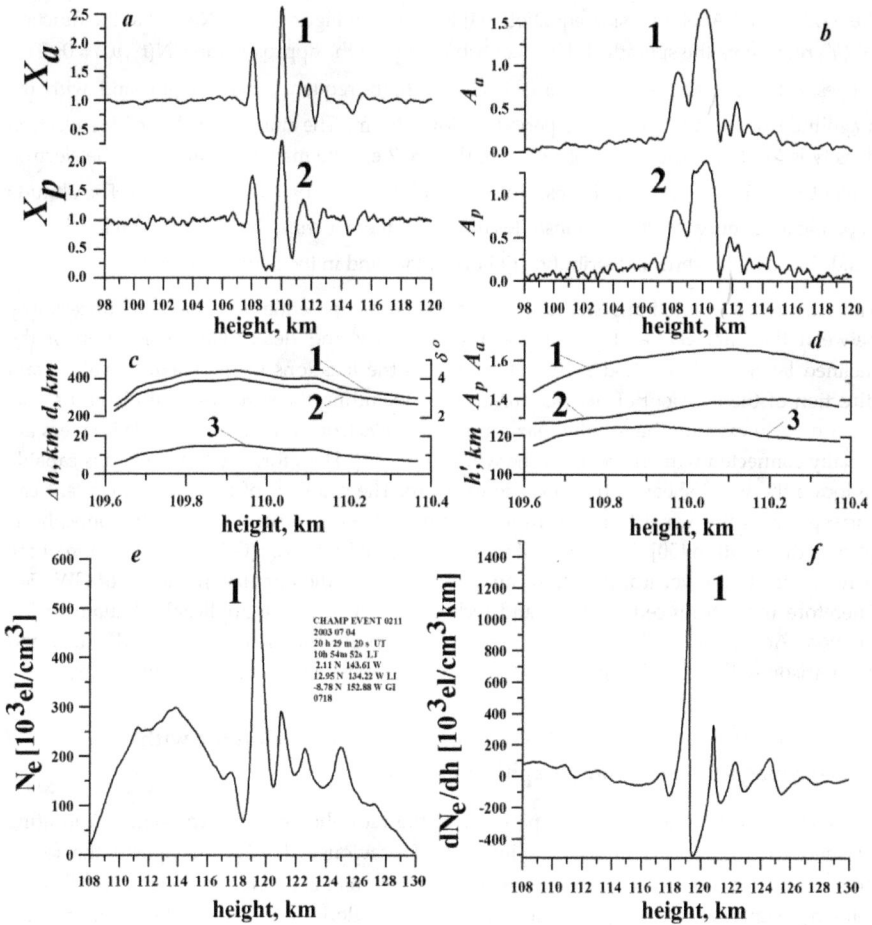

Figure 7. Panels a-d: identification and location of a layer in the lower ionosphere. Panel e: distribution of the electron density in the identified sporadic Es layer. Panel f: distribution of the gradient of electron density.

at the left vertical axis), the layer slope δ [degrees] (right vertical axis), and the correction Δh respectively. the changes of d, Δh, and δ are concentrated in the ranges of 240-400 km, 5-15 km, and 2.2°...3.2° when the altitude of the ro ray perigee changes in the range of 109.6-110.4 km. From these changes the average values of d, Δh, and δ are determined, i.e. $d = 350 \ km \pm 50 \ km$; $\Delta h = 10 \ km \pm 5 \ km$, and $\delta = 3.1°\pm0.3°$. It is concluded that the detected sporadic layer is displaced from the ro ray perigee by 350 km in the direction to the gps satellite and the altitude of which is 10 km greater than the height of the point T. The height distribution of the electron density $N_e(h')$ and its altitude gradient $\dfrac{dN_e(h')}{dh'}$ recalculated from the modernized Abel inversion equation (54) is shown in Figure 7, e, f. Note, that the function $N_e(h')$ represents the sporadic E-layer contribution with the approximation $N(t_x, p_0) = 0$. This suggests that the above calculation reflects the high-frequency part $N_e(h')$ and with the magnitude of the vertical spatial periods below 10 km. The maximal value of the electron density is located at the height of 119.2 km (Figure 7 e). The maximal gradient of the electron content $\sim 1.4 \cdot 10^6$ [$el / cm^3 km$] is observed at the altitude of 119.0 km (Figure 7 f). The altitude dependent quantity $N_e(h')$ demonstrates the wave-like structure that is possibly related to the wind shears in the vertical distribution of horizontal wind in the neutral gas [50].

The introduced method appears to have a considerable potential to resolve the uncertainty between the part GT and LT of the ray trajectory and determine the location of the inclined layers. This method accurately indicates the locations of the maximal values and direction of the gradient of the electron density including the distance, altitude and slope. According to existing theory, the maximum of the electron content in sporadic E-layers are usually connected with influence of the wind shear [45]. Therefore the RO method is capable to locate the wind shear in the lower ionosphere. The gradient of the electron content can correspond to the wave fronts of different kinds of wave influencing on the ionospheric plasma distribution [50]. In the case of the internal gravity waves (GW) the inclination of the wave vector to the vertical direction can be used to find the angular frequency of GW [54]. Therefore the introduced criterion and technique extended the applicable domain of RO method. Additional validation of this method through analyzing the CHAMP data and comparison with ground-based ionosonde information is the task for the future work.

4.4. Comparison of the eikonal acceleration/intensity technique with back-propagation radio-holographic methods

The analytic technique can be compared with the radio-holographic approach for locating plasma structure in the ionosphere introduced previously [10,11]. In general the radio-holographic back-propagation may be carried out using a Green function $G(|\mathbf{r}|)$ as a reference signal and a complex field $\phi(l)$ measured along a part of orbital trajectory of a LEO satellite ($L'L$) (Figure 2) [30]:

$$\phi(C) = \left[ik_0 / (2\pi) \right]^{1/2} \int_{LL'} r^{-1/2} G(|\mathbf{r}|) \, \phi(l) \, \cos\varphi \, dl \tag{56}$$

where $|\mathbf{r}|$ is the distance CC' (Figure 2), $\phi(C)$ is the radio fields restored by a back-propagation method at point C, φ is the angle between the vector $\mathbf{r}$, connecting the observation point C and current integration element dl with center C', and normal $\mathbf{n}$ to the curve $L'L$ (Fig.1).

The Green function $G(|\mathbf{r}|)$ is a solution of the scalar wave equation:

$$(\Delta + k_0^2 n^2(r)) G(\mathbf{r}, \mathbf{r}') = -4\pi\delta(\mathbf{r} - \mathbf{r}') \tag{57}$$

where, k_0 is the module of the wave vector in the free space, $\mathbf{r}, \mathbf{r}'$ are the vectors indicating the coordinates of a point in a medium and a source of the field, $n(r)$ is the spatial distribution of the refraction index. It is supposed below that $n(r)$ depends only on a radial coordinate r of layered structures relative to a centre of spherical symmetry.

To obtain the optimal values of the vertical resolution and accuracy in measuring physical parameters in the atmosphere and ionosphere, usually the Green function $G(|\mathbf{r}|)$ in (56) may be chosen in the form depending on the model of a layered medium $n_m(r)$ [30]. As the simplest case of a reference signal, the Green function $G(|\mathbf{r}|)$ describing spherical waves in the free space can be selected [56]:

$$G(|\mathbf{r}|) = \exp(i\pi / 4)\exp(-ik_0|\mathbf{r}|) / |\mathbf{r}|^{1/2} \tag{58}$$

The Green function $G(|\mathbf{r}|)$ corresponding to radio waves emitted by a point source in a spherical symmetric layered medium has been suggested [30]:

$$G(|\mathbf{r}|) = A_G(|\mathbf{r}|, \varphi)\exp(-ik_0\Phi_G(|\mathbf{r}|)) \tag{59}$$

where, $\Phi_G(|\mathbf{r}|)$, $A_G(|\mathbf{r}|, \varphi)$ are the eikonal and amplitude of the Green function $G(|\mathbf{r}|)$.

The complex wave field $\phi(l)$ at the orbital trajectory $L'L$ in the wave-optics approximation is given by relationship [30]:

$$\phi(l) = X^{1/2}(p)\exp\left[ik_0\Phi(p)\right] \tag{60}$$

where, p is the impact parameter depending on the location of the element dl, $X(p)$ is the refractive attenuation along ray GTL. The eikonal $\Phi(p)$ and distance CC' $|\mathbf{r}|$ are presented by the following relationships:

$$\Phi(p) = \sqrt{R_L^2 - p^2} + \sqrt{R_G^2 - p^2} + p\xi(p) + \kappa(p) \tag{61}$$

$$|\mathbf{r}| = \sqrt{R^2 + R_1^2 - 2RR_1\cos(\theta - \theta_c)} \tag{62}$$

where, $\xi(p)$ is the bending angle, $\kappa(p)$ is the main refractivity part depending on the distribution of the vertical gradient of refractivity along the radio ray GTL, R_G, R, and R_1 are the distances OG, OC, and OC', respectively, O is the spherical symmetry center, θ_c is the central angle with the vertex at O between directions OG and OC (Figure 2). The bending angle $\xi(p)$ is the negative derivative of the main refractivity part $\kappa(p)$ with respect to p:

$$\xi(p) = -\frac{d\kappa(p)}{dp} \tag{63}$$

The central angle θ (Figure 2) is connected with the impact parameter p of the ray GTL by the following equation:

$$\theta = \pi + \xi(p) - \sin^{-1}\frac{p}{R_1} - \sin^{-1}\frac{p}{R_g} \tag{64}$$

The back-propagated field $\phi(C)$ may be evaluated by stationary phases. In the case of circular orbits of the transmitting and receiving satellites the integration along the curve $L'L$ for obtaining the field $\phi(C)$ may be provided on the central angle θ. The phase of the back-propagated field in point C is equal to:

$$\psi(p) = \sqrt{R_L^2 - p^2} + \sqrt{R_G^2 - p^2} + p\xi(p) + \kappa(p) - \Phi_G(|\mathbf{r}|) \tag{65}$$

The form of the eikonal $\Phi_G(|\mathbf{r}|)$ depends on the Green functions eqns (58) and (59) used for the back propagation:

$$\Phi_G(|\mathbf{r}|) = |\mathbf{r}| \tag{66}$$

$$\Phi_G(|\mathbf{r}|) = \sqrt{R_1^2 - p_b^2} - \sqrt{n_m^2(R)R^2 - p_b^2} + p_b\xi_m(p_b) + \kappa(p_b) \tag{67}$$

where, the refraction index $n_m(R)$ in point C and the bending angle $\xi_m(p_b)$ are corresponding to the refractivity distribution in a medium for which Green function $G(|\mathbf{r}|)$ (i.e. equation (59)) is known. For Green function (58) $n_m(R) \equiv 1$, $\xi_m \equiv 0$. The central angle $\theta - \theta_c$ is connected with the impact parameter p_b of the back-propagated ray CC' (Figure 2) by:

$$\theta - \theta_c = \xi_m(p_b) + \sin^{-1}\frac{p_b}{n_m(R)R} - \sin^{-1}\frac{p_b}{R_1} \tag{68}$$

The stationary phase method can be applied to evaluate the back-propagating field. For the stationary point, the following equation holds:

$$\frac{\partial \psi}{\partial \theta} = 0 \tag{69}$$

After substitution (65)-(68) into (69) one obtains:

$$\frac{\partial \psi}{\partial \theta} = \frac{\partial \Phi}{\partial p}\frac{\partial p}{\partial \theta} - \frac{\partial \Phi_G(|\mathbf{r}|)}{\partial p_b}\frac{\partial p_b}{\partial \theta} = p\frac{\partial \theta}{\partial p}\frac{\partial p}{\partial \theta} - p_b\frac{\partial(\theta - \theta_c)}{\partial p_b}\frac{\partial p_b}{\partial \theta} = p - p_b = 0 \tag{70}$$

From equation (70) the impact parameters p, p_b related to ray GTL and the back propagated ray CC' are identical in the stationary point C'. Therefore the field $\phi(C)$ is back-propagating in the occultation plane along the tangents to the occultation rays at any point of the orbital trajectory $L'L$ (Figure 2). The back-propagated rays are the straight lines or curves depending on the form of Green functions (58) or (59) used. The stationary phase method gives the next expression for the back-propagated field $\phi(C)$:

$$\phi(C)=A_G(|\mathbf{r}|,\varphi)X^{1/2}(p)\cos\varphi\ \exp\left[ik_0(\Phi(p)-\Phi_G(|\mathbf{r}|))\right]\left|\frac{\partial^2\Psi}{\partial\theta^2}\right|^{-1/2} \tag{71}$$

$$X^{1/2}(p)=\sqrt{\frac{pR_0}{p_s\sqrt{R_G^2-p^2}\sqrt{R_1^2-p^2}}\left|\frac{\partial\theta}{\partial p}\right|^{-1}} \tag{72}$$

$$\frac{\partial^2\Psi}{\partial\theta^2}=\frac{\dfrac{\partial\theta}{\partial p}-\dfrac{\partial(\theta-\theta_c)}{\partial p_b}}{\dfrac{\partial\theta}{\partial p}\dfrac{\partial(\theta-\theta_c)}{\partial p_b}} \tag{73}$$

The derivatives $\dfrac{\partial\theta}{\partial p},\dfrac{\partial(\theta-\theta_c)}{\partial p_b}$ can be found from the relationships (64), (68) as follows:

$$\frac{\partial\theta}{\partial p}=\frac{\partial\xi}{\partial p}-\frac{1}{\sqrt{R_G^2-p^2}}-\frac{1}{\sqrt{R_1^2-p^2}} \tag{74}$$

$$\frac{\partial(\theta-\theta_c)}{\partial p_b}=\frac{\partial\xi_m}{\partial p_b}+\frac{1}{\sqrt{n_m^2(R)R^2-p_b^2}}-\frac{1}{\sqrt{R_1^2-p_b^2}} \tag{75}$$

From (64)-(75) the following formula for $\chi=\left|\dfrac{\partial^2\Psi}{\partial\theta^2}\right|^{-1/2}$ can be derived:

$$\chi=\sqrt{\frac{\left[\dfrac{\partial\xi_m}{\partial p_b}\sqrt{n_m^2(R)R^2-p^2}\sqrt{R_1^2-p^2}-|\mathbf{r}|'\right]\left(\dfrac{\partial\xi}{\partial p}\sqrt{R_G^2-p^2}\sqrt{R_1^2-p^2}-R_0'\right)}{\left(R_L^2-p^2\right)\left|\dfrac{\partial\xi_m}{\partial p_b}-\dfrac{\partial\xi}{\partial p}\right|\sqrt{n_m^2(R)R^2-p^2}\sqrt{R_G^2-p^2}+R_x\right|}} \tag{76}$$

$$R_x=\sqrt{n_m^2(R)R^2-p^2}+\sqrt{R_G^2-p^2}$$
$$R_0'=\sqrt{R_G^2-p^2}+\sqrt{R_1^2-p^2}\approx R_0;\ |\mathbf{r}|'=\sqrt{R_1^2-p^2}-\sqrt{n_m^2(R)R^2-p^2}\approx|\mathbf{r}| \tag{77}$$

Under condition:

$$n_m(R)R=p \tag{78}$$

the next relationship follows from eqns (72), (76), (77):

$$\chi X^{1/2} = \sqrt{\frac{R_0'\,|\mathbf{r}'|\,\left| X \left(\frac{\partial \xi}{\partial p} \frac{\sqrt{R_G^2 - p^2}\sqrt{R_1^2 - p^2}}{R_0} - 1 \right) \right|}{\sqrt{R_G^2 - p^2}\left(R_L^2 - p^2\right)}} = \sqrt{\frac{pR_0'\,|\mathbf{r}'|\left(R_1^2 - p^2\right)^{-1}}{p_s\sqrt{R_G^2 - p^2}}} \approx const \qquad (79)$$

where, p_s is the impact parameter corresponding to the straight line GL (Figure 2).) Factor $\chi X^{1/2}$ in (79) is independent of the refractive angles ξ and ξ_m. Therefore the amplitude variations are minimal at the geometric places in space determined by condition (78). According to [11] this corresponds to the position of a layer which can be estimated by finding the location of the minimum of the amplitude modulation of the 2-D back-propagating electromagnetic field. This property can be used as a main condition for locating layered structures in atmosphere and ionosphere. The accuracy of the location determination depends on: (i) the form of Green function used for the back-propagation; (ii) the structure and form of the (ionospheric) irregularities.

The simplest form of the Green function (58) has been used [10,11] to locate plasma layers in the E- and F- regions in the ionosphere. In this case $n_m(R) \equiv 1$, the back-propagated rays are straight lines, and condition (78) has the following form:

$$R = p \qquad (80)$$

From condition (80) the curve BB' in Figure 2 indicates the place, where the amplitude of the back-propagated field is constant. The curve BB' may be approximated by a straight line because the bending angle is small in the RO case. The inaccuracy in the determination of distance $T'L$ by back-propagation may be evaluated as the distance of the curve BB' to a new ray perigee T' $T'B \approx p'\xi/2$ (Figure 2). The proposed technique gets the length $T'L$ as the sum $TL + d$. The systematic inaccuracy of this technique is equal to the difference $T'L - TL - d$ which usually is smaller than that of the considered back-propagation method. For a more complex form of the Green function (59) the back-propagated rays are curved. If the Green function (59) corresponds to a real refractivity distribution in layered structures, then condition (78) gives an accurate location of the ray perigee T (Figure 2).

4.5. Locality principle and its importance for RO remote sensing

Locality principle allowed designing new analytic technique for locating the inclined layered structures (including sporadic Es layers) in the ionosphere. The location of the ionospheric layers including their altitude, displacement from the RO ray perigee and slope relative to the horizontal direction can be determined using the introduced criterion that compares the refractive attenuations found from the RO amplitude and phase data. Depending on the sign of the refractive attenuations the displacement of a plasma layer from the RO ray perigee should be positive (in the direction to a GPS satellite and vise versa). The magnitude of the displacement can be found from a ratio of the refractive attenuation's difference to the magnitude of the refractive attenuation from the RO phase data. The altitude and slope of a plasma layer can be found from the known value of its displacement.

Therefore the standard estimation of a layer's altitude as a height of RO ray perigee should be revised due to underestimation of the altitude of inclined plasma structures in the lower ionosphere.

The current radio-holographic back-propagation method implicitly uses the relationship between the eikonal acceleration and intensity variations of RO signals to locate irregularities in the ionosphere. The accuracy of this method depends on the form of the Green function used for the back-propagation. If the Green function corresponding to the propagation in the free space is used, then the inaccuracy of back-propagation method is proportional to the bending angle. The analytic technique is simpler and more precise than the previously published back-propagation method.

By use of the introduced criterion the RO method is capable to locate and determine the direction and magnitude of the gradient of electron density in the lower ionosphere. The gradient of the electron content indicates the direction of the different kinds of wave fronts in the ionosphere. In the particular case of the internal gravity waves (GW) the inclination of the wave vector to the vertical direction can be used to find the angular frequency and the parameters of GW.

The introduced criterion and technique extended the applicable domain of RO method to remote sense the waves in the lower ionosphere. This conclusion has a general importance for the planetary and terrestrial radio occultation experiments in a broad range of frequencies.

5. Bending angle: Seasonal changes

The RO method has important radio meteorological application. Previously the radio meteorological parameters (refractive angle, refractive attenuation, phase path excess, total absorption, and other) have been recalculated from the temperature, humidity and pressure delivered from the current meteorological observations. Nowadays the RO method directly measured the bending angle, refractive attenuation, phase path excess, total absorption, etc.) from the amplitude and phase delay of RO signal. Thus the RO radio meteorological observation are very important for estimation of condition for radio wave propagation, radio navigation, and radio climate in the near Earth space.

In this section the seasonal change of the bending angles as an important radio meteorological parameter will be considered.

Atmospheric refraction caused by gradients of the refractive index of air leads to a deviation of the direction of radio wave propagation from straight line connecting transmitter and receiver. Practical problems require to study variations of the bending angle, refractive attenuation and other radio parameters as functions of the coordinates of transmitter and receiver. When the altitude of a radio link is low, changes in the vertical profiles of temperature, pressure and humidity introduce main contribution in the refraction effects. Meteorological parameters depend on the climate and weather in different geographical positions, which was the cause of origin of radio meteorology – a branch of radio science which used the weather information for analysis of the electromagnetic waves propagation conditions in radio communication and radar applications [38,57,58]. The vertical and horizontal distributions of the pressure, temperature, and humidity found from meteorological measurements are approximated by use of different models to find the altitude and spatial dependences of the refractive index, bending angle, refractive attenuation and absorption of radio waves. However the meteorological measurements are local, and relevant parameters are variable, which inevitably leads to discrepancy between the measured and calculated values of the bending angles.

The innovative RO method is a new important tool for direct measurements of the radio meteorological parameters and for investigation of radio climate of the Earth at different altitudes in the atmosphere with a global coverage. In contrast to previously used goniometric methods with a narrow antenna pattern or interferometers for measuring refraction effects and their variations in radio links, the RO method directly determines with high accuracy the bending angle from measurements of the Doppler frequency of radio wave. The measured bending angle does not depend on the wavelength, orbits of satellites, and characteristics of the transmitting and receiving devices. The measured bending angle is delivered with high accuracy without any assumptions concerning the structure of the atmosphere, and can be regarded as an independent quantitative radio meteorological parameter in different regions of the Earth. It is essential that the spatial and temporal distributions of refractive properties can be obtained over a long period of time, which will contain daily, seasonal, and long-term radio climatic changes in the atmosphere. This information can be applied for detailed analysis of radio wave propagation conditions along the Earth's surface.

The aim of this section is to establish the applicability of the bending angle as an indicator of the global state of the atmosphere. The annual and seasonal variations of the refractive parameters above Russia and some territories are analyzed and discussed.

5.1. Method of measurement

In determining the angle of refraction by the radio occultation method, the measured parameters of coherent radio waves with the frequencies $f_1 = 1575.42$ MHz and $f_2 = 1227.60$ MHz radiated by GPS satellites and received after transmission through the atmosphere were used. The radio waves were received by low earth orbit satellites FORMOSAT-3. A constellation of ~30 GPS satellites orbiting the Earth at a height of 20 000 km and of 6 low-earth-orbit satellites orbiting the Earth at a height of 800 km provided from 1400 to 1800 atmospheric soundings in various regions of the Earth. The measurements of the atmospheric component of the phase path increment and, respectively, of the Doppler shift of the signal frequency, determined by the atmospheric refraction, were performed with a sampling frequency of 50 Hz. The mean time of sensing the atmosphere in the altitude 50 – 0 km interval was 90 s. Owing to refraction, the rate of changing of the height of the ray perigee decreases as the ray descends into denser layers of the atmosphere; therefore, the step of the measurement of the Doppler shift of the signal was ~50 m for stratosphere and ~5 m for the lower troposphere.

In analyzing the space–time variations of the bending angle, the results of 4252 occultation atmospheric soundings performed from June 2006 to July 2010 in the region of European Russia with coordinates of 50°N to 60°N and 30°E to 40°E were used. The extent of this region is 1100 km along the meridian and ~600 km along the latitude circle. In this region, three to seven measurement sessions were conducted every day.

The method for determining the bending angle $\xi(h)$ as a function of the minimum ray path height is based on the relation between the atmospheric component of Doppler frequency shift Δf_a and $\xi(h)$. This relation is most simple for the occultation sensing, when one of the satellites is at a large distance from the ray path perigee. In this case $\Delta f_a \approx \lambda^{-1} V_1 \xi(h)$, where λ is the wavelength, V_1 is the projection of the vector of the satellite velocity in the occultation plane on the perpendicular to the straight line connecting the satellites. The accuracy of determining the angle of refraction depends on the error of the frequency shift measurement, which is affected by the errors of measuring the atmospheric phase path increment, coordinates, velocities of the navigation and low earth orbit satellites, and also by the influence of the ionosphere and multipath propagation. The contribution of these errors is analyzed in detail in study [59]. This contribution can be minimized by improving the receiving equipment installed on the low earth orbit satellite and the procedures for measuring and processing the raw data. At present, the instrumental error of the bending angle measurements is no more than $5 \cdot 10^{-7}$ rad.

The sources of systematic error are related to the effect of the ionosphere and multipath propagation. The influence of the ionosphere is eliminated using two frequency ionospheric correction [53]. However, the ionospheric correction cannot completely remove the bending angle fluctuations caused by small scale electron concentration irregularities. In sensing the upper stratosphere, the bending angle errors caused by this factor can be as high as $3 \cdot 10^{-6}$ rad, but they rapidly decrease as the height decreases [55]. Below 30 km, this error component can be ignored. Of greater importance to meteorological applications are the refractive angle errors caused by multipath in sensing the lower troposphere. To solve this problem, several radio-holography methods for processing occultation data have been developed. The number of publications on this subject is very large, the descriptions of these methods and references to the original publications are given in [38,60].

The refractive angle is determined from the measurements of the signal frequency, a quantity that can be measured with a maximum accuracy. The results of the analysis made in [61] show that, in the middle latitude atmosphere, at heights of 5 to 30 km, the discrepancy between the measured and calculated (with the use of various models of the atmosphere) bending angles does not exceed ±1%. The height profile of the bending angle may contain inaccuracies related to the errors in height measurements. At the initial stage of data processing, the dependence of the refractive angle $\xi(p)$ on the impact parameter $p = (\rho_e + h)n(\rho_e + h)$ is determined where ρ_e is the Earth's radius. The height of the atmospheric layer is determined as the height above the surface of the geoid described by reference ellipsoid WGS84 (World Geodetic System 1984). The error of the height evaluation in our data does not exceed ±100 m. It is necessary to study the variability of the refractive angle height profiles at different time periods and in various regions of the Earth. As an example such an analysis is performed using a vast region of Russia.

5.2. Mean bending angle vertical profile

In analyzing space–time refraction variations, one should eliminate the influence of the regular component. To this end, a model of the bending angle vertical profile $\xi(h)$ derived, e.g., from long term observations is required. It is clear that regional models of $\xi(h)$ are in better agreement with the measurement results than the global model. The variability of the parameters of these models for different regions is of interest for radio-meteorology. As an example, the vertical profiles $\xi(h)$ for the middle latitudes of Russia are described in this section. The mean vertical profile of $\xi(h)$ and ranges of variations of the refractive angle at different heights were obtained by averaging the data of 8711 measurement sessions performed during a four year period. In addition to the data obtained in the region with coordinates of 50°N to 60°N and 30°E to 40°E, the results of 4459 atmospheric soundings performed in the same latitude belt of 50°N to 60°N but at a longitude of 160°E to 170°E were used. A second region includes Kamchatka's eastern coast and the adjacent water area of the Bering Sea and is characterized by marine climate. The use of the measurement data obtained in the two regions allowed deriving analytical dependence $\xi(h)$ suitable for analyzing the space–time variations of the refractive angle observed in the regions that are in different climatic conditions. The performed analysis showed that, in the altitude range from 0 to 50 km, the mean vertical profile of the refractive angle $\xi_a(h)$ is described by

$$\xi_a(h) = \exp\left(a + bh + ch^2 + d_1^3\right) \tag{81}$$

If the bending angle is expressed in milliradians and the height is expressed in kilometers, the coefficients in the exponent (81) have the values given in Table 1.

h , km	a	b km⁻¹	c km⁻²	d_1 km⁻³
≤ 12.4	3.226	−0.154	3.765×10^{-3}	-1.487×10^{-3}
≥ 12.4	3.611	−0.166	4.128×10^{-4}	-6.374×10^{-6}

Table 1. Coefficients in equation (1)

The height profiles of the refractive angles measured in the two regions and calculated with model (1) are compared in Table 2. Root mean square values of σ_m significantly exceed the measurement errors and characterize essential refraction variations caused by the difference of the meteorological conditions.

h, km	ξ_m, mrad	σ_m, mrad	ξ_a, mrad	$\xi_m - \xi_a$, mrad
0.2	23.96	2.19	24.41	−0.45
0.4	23.57	2.25	23.68	-0.11
0.6	23.03	2.23	22.98	0.06
0.8	22.50	2.19	22.30	0.20
1	21.95	2.15	21.65	0.29
2	19.94	1.91	18.75	0.19
3	16.16	1.43	16.32	-0.16
4	13.99	0.98	14.28	-0.29
5	12.26	0.64	12.25	-0.29
6	10.88	0.45	11.06	-0.18
7	9.76	0.38	9.76	-0.01
8	8.82	0.35	8.63	0.18
9	7.99	0.32	7.64	0.35
10	7.17	0.34	6.75	0.41
12	5.36	0.42	5.25	0.11
14	3.83	0.21	3.87	-0.04
16	2.81	0.15	2.82	-0.01
18	2.07	0.11	2.06	0.01
20	1.51	0.08	1.50	0.01
25	0.69	0.04	0.68	0.002
30	0.31	0.02	0.31	-0.003

Note: h is the height of the ray perigee relative to the Earth's surface, ξ_m, ξ_a are the bending angles determined by averaging from the measured data and calculated with model (1), respectively, σ_m is the rms value of the bending angle for the four year period.

Table 2. Comparison of the averaged height profiles and variations of the refractive angle

Of interest is the distribution of these variations. The refractive angle disturbances at the corresponding heights are distributed, with respect to the mean value of ξ_m in accordance with a near normal law. In 90% of the measurements, for heights h = 0.2, 1, 4, and 8 km, the mean refractive angles are 23.88 ± 3.63, 22.12 ± 3.72, 13.98 ± 1.71, and 8.76 ± 0.51 mrad, respectively. It follows from Table 2 that the largest absolute and relative variations σ_m, σ_m / ξ_m of the refractive angle are observed in the troposphere at altitudes below 8 km. These variations are due to weather changes in the regions being sounded. In the stratosphere above 14 km, relative refractive angle variations σ_m / ξ_m are approximately half the same one in the troposphere. Model (81) of the refractive angle altitude profile is in good agreement with the experimental data, since its deviation from the mean bending angle, ξ_m, is several times smaller than the observed rms variations σ_m. At heights above 14 km, the deviation of the experimental data from the model data is no more than 1%. Note that analytical model (1) with the coefficients listed in Table 1 does not describe the individual features of the refraction that are, e.g., related to the influence of the tropopause. This factor is responsible for the marked difference between ξ_m and ξ_a over the altitude interval from 9 to 12 km, heights that are typical for the tropopause in the midlatitudes. It should also be noted that this model is not intended for deriving meteorological parameters, e.g., a temperature–height profile. The model describes a significant decrease in refraction over the height interval from 0 to 30 km. The model can be used in detecting small regional and seasonal variations of the vertical profile.

5.3. Seasonal and diurnal bending angle variations

The bending angle rapidly decreases from ~24 mrad to ~0.3 mrad as the ray perigee height h increases from $h = 0.2$ km to $h = 30$ km. In analyzing small bending angle disturbances, caused by various factors, it is necessary to eliminate the component related to a rapid decrease in the atmospheric density occurring with an increase in height. To this end, it is possible to use obtained approximation of mean height profile of the bending angle ξ_a (81) and consider the refractive angle disturbances observed in individual measurement sessions in reference to this approximation. The most significant refractive angle disturbances are observed in the troposphere and lower stratosphere. Therefore, in analyzing the disturbances in angle ξ, we will restrict our consideration to a height interval of 0 to 14 km and compare the observational results obtained in summer and in winter.

As an illustration of slow disturbances in the height profile of the bending angle, let us consider the vertical profiles of the deviations of the refractive angles, observed in individual measurement sessions, from the values calculated with model (81), i.e., $\Delta\xi(h) = \xi - \xi_a$). Profile $\Delta\xi(h)$ allows detecting influence of regular layered structures of various nature on the refraction.

Shown in Figure 8 are four profiles $\Delta\xi(h)$ obtained in morning, in January 2007. The conditions for measurements are determined by the day of the year (DOY) counted from January 1, by the Universal Time Coordinated (UTC) expressed in hours and minutes, and by the geographic latitude and longitude of the region (Table 3). The time and the coordinates of the region are usually given for the moment when the straight satellite to satellite line touches the Earth's surface. This moment corresponds to the ray perigee height $h \sim 13$ km. The local time (LT) in the region being sounded is, on the average, three hours later than the UTC. The winter night is characterized by a stationary state of the troposphere with layered structures and pronounced maximums of $\Delta\xi(h)$ at heights of 1.5–2.7 km and 3.5–4 km, which correspond to significant vertical gradients of refractivity. The maximum values of $\Delta\xi(h)$ are as high as 4 mrad, and the thickness of the layers is 0.7 to 1.5 km. Stronger disturbances were observed in July 2010. In summer daytime, the deviations of the refractive angle from the mean values are 4–10 mrad at heights of 0.7 to 5 km. It seems likely that this phenomenon is due to a higher content of water vapor and significant variations in the vertical distribution of water vapor in summer as compared to winter. For the height interval of 9 to 14 km, depending on the season and time of day, smooth refractive angle deviations from mean values $\Delta\xi(h)$, caused by the temperature inversion in the tropopause, are recorded. Unlike the refractive angle disturbances in the lower troposphere, which are, as a rule, related to the variations in humidity, the temperature inversion in the tropopause results in an insignificant increase of angle ξ in a range of 0.6 to 1.3 mrad. With a narrow tropopause, the position of the local maximum of $\Delta\xi(h)$ is close to the height of the minimum temperature, which is determined from the occultation sounding data and meteorological sounder measurements.

Curves in Figure 4	Date January 2007	UTC, h, min	N, deg	E, degr.	Curves in Figure 4	Date January 2007	UTC, h, min	N, deg	E, deg
1	01	04:34	57.34	39.9	5	10	04:17	58.29	41.71
2	01	04:44	53.99	39.31	6	20	16:19	56.51	42.56
3	06	05:10	54.03	34.66	7	19	06:26	53.41	35.62
4	10	04:05	58.03	33.55	8	20	06:41	57.54	41.76

Table 3. Coordinates and time of experiments

In addition to slow (seasonal and diurnal) variations in the refractive angle, significant rapid fluctuations caused by atmospheric irregularities are observed. Using the results of 200 measurement sessions, the fluctuations of the bending angle observed in winter and in summer without separating them in time of day were analyzed. It is necessary to eliminate the regular and slow variations in order to estimate the rapid fluctuations. To this end, a filtering procedure was used that involves subtraction of function $\xi_S(h)$ obtained by smoothing the refractive angle in the sliding window $\Delta h = 2$ km.

Changes in the conditions of refraction in regions with different climates and on different time scales will have both individual and general laws. To detect these patterns let us look at the changes of refraction in a homogeneous area of the climatic conditions at different time intervals and analyze the seasonal and annual changes. To reduce the influence of spatial factors let us to limit the area in which the seasonal variations in the angle of refraction are investigated, and to select the cell size of approximately 400x400 km², extending in latitude from 54.0 N to 58.0 N and longitude from 35.0 E to 41.0 E. The center of this area is located in the vicinity of Moscow. In the period from January 1, 2007 and November 30, 2009 in this area was carried out 1232 RO soundings of the atmosphere. In each month of the year from 25 to 29 soundings at different times of day were held. Let us consider the seasonal changes of the bending angle at two altitudes: in the middle stratosphere at 15 km, and in the upper troposphere at 9 km (Figure 8). Note that the most accurate radio occultation measurements in the stratosphere and upper troposphere have been provided at these altitudes. At the altitude 17 km in the middle stratosphere, there is a positive trend, i.e. strengthening of refraction with time. This increase is nearly equal to 0.07. mrad during four years. In contrast to the stratospheric region in the upper troposphere at an altitude of 9 km in the period under review there was a negative trend in refraction, whose value is amounted to 0.11 mrad. When reducing the height, this trend is weakening and at the altitude 4 km the long-term trend of refraction is practically not observed.

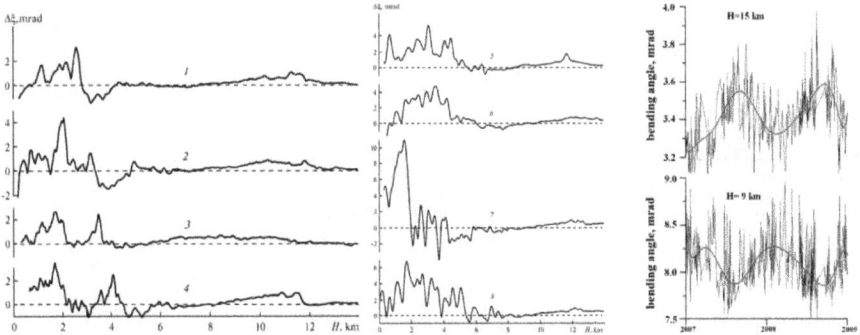

Figure 8. Left panel. Examples of the refractive angle variations obtained in January 2007. Middle panel. Examples of the refractive angle variations obtained in July 2010. Right panel. Annual changes of the bending angle at the 9 km and 15 km altitudes from FORMOSAT-3 RO data in Moscow region.

The future task is to investigate the trends as functions of time and geographical position in different climatic zones for longer periods. However it is clear is that the angle of refraction is a sensitive indicator of the state of the troposphere – stratosphere system. Seasonal changes in the refractive properties observed in the stratosphere and the troposphere are evident, but they manifest themselves in different ways. In the stratosphere, there are quasi-harmonic changes with a period of 12 ± 0.5 months and the amplitude of about 0.12 mrad relative to the average trend. Maximum values of the bending angle occur in late July - early August, and the minimal during February - March. Seasonal changes in the upper troposphere also contain a component with a period of 12 ± 0.5 months, but they are opposite to the phase variations in the stratosphere. Their amplitude is in average 0.23 mrad. The maximum refraction occurs in March near the vernal equinox, and the minimum - in August. The influence of a weak quasi-monochromatic component is seen in the middle troposphere at an altitude of 4 km in the bending angle variations. Maximum values of the bending angle is ~ 15 mrad are observed, usually in late summer - early autumn, and in the rest of the year they are 3-5 mrad. This behavior corresponds to refraction in the middle and lower troposphere due to weather changes, which essentially smoothes the effect of changing seasons of the year.

6. Conclusions

The fundamental principle of local interaction of radio waves with a spherically symmetric medium is formulated and introduced in the RO method of remote sensing of the atmosphere and ionosphere of the Earth and planets.

In accordance with this principle, the main contribution to variations of the amplitude and phase of radio waves propagating through a medium makes a neighborhood of a tangential point where gradient of the refractive index is perpendicular to the radio ray.

A necessary and sufficient condition (a criterion) is established to detect from analysis of RO data the displacement of the tangential point from the radio ray perigee.

This criterion is applied to the identification and location of layers in the atmosphere and ionosphere by use of GPS RO data. RO data from the CHAllenge Minisatellite Payload (CHAMP) are used to validate the criterion introduced when significant variations of the amplitude and phase of the RO signals are observed at RO ray perigee altitudes below 80 km.

The new criterion provides an improved estimation of the altitude and location of the ionospheric plasma layers compared with the back-propagation radio-holographic method previously used.

The detected criterion opens a new avenue in terms of measuring the altitude and slope of the atmospheric and ionospheric layers. This is important for the location determination of the wind shear and the direction of internal wave propagation in the lower ionosphere, and possibly in the atmosphere.

The locality principle makes it possible to convert the eikonal acceleration (or the time derivative of the Doppler shift) into refractive attenuation. This is important for estimation of the total absorption of radio waves on the satellite-to-satellite transionospheric communication paths. This dependence is also important for measuring the water vapor content and atmospheric gas minorities in the future radio-occultation missions in view of the possibility to remove the refractive attenuation effect from the amplitude data. The advantages of the proposed method were tested by analysis of the CHAMP satellite radio-occultation data.

The obtained results indicate that measurements of the total absorption on radio occultation paths can potentially be used for monitoring of the atmospheric-oxygen content provided that the transmitter and receiver gain calibration is substantially improved. It follows from the above analysis that the comparison of the refractive attenuations retrieved from the amplitude and phase variations of a radio-occultation signal is necessary for the detection of layered structures in the atmosphere.

The total absorption, refractive attenuation, bending angle, bending angle, and index of refraction are important radio meteorological parameters which can be measured directly with a high accuracy by the radio occultation method. The prolonged radio occultation data base is very important for determination of the radio climate changes at different altitudes in the atmosphere with a global coverage.

Author details

Alexey Pavelyev, Alexander Pavelyev, Stanislav Matyugov and Oleg Yakovlev
FIRE RAS, Fryazino, Russia

Yuei-An Liou
CSRSR, NCU, Taiwan

Kefei Zhang
RMIT University School of Mathemathical & Geospatial Sciences, Melbourne, Australia

Jens Wickert
GeoForschungsZentrum Potsdam (GFZ-Potsdam), Telegrafenberg, Potsdam, Germany

Acknowledgement

This work was supported in part by Russian Fund of Basic Research (grant No. 10-02-01015-a), Program of Presidium RAS No. 22, Program PSD- IY.13 of the Branch of Physical Sciences of the Russian Academy of Sciences, by Grants of the National Science Council of Taiwan NSC 101-2111-M-008-018; 101-2221-E-008-0, by Australian Research Council Project ARC-LP0883288 of the Department of Industry, Innovation, Science, and Research of Australia International Science Linkage under Project DIISR/ISL-CG130127, and by the Australia Space Research Program Project endorsed to research consortiums led by the Royal Melbourne Institute of Technology University. The authors are grateful to the GeoForschungsZentrum (Potsdam, Germany) for the data obtained during the CHAMP satellite mission.

7. References

[1]   Yakovlev O. Space Radio Science. London: Taylor and Francis; 2003.

[2]   Gurvich A, Krasilnikova T. Application of the navigational satellites for radio occultation of the Earth's atmosphere. Cosmic Res.1987 6: 89-95.

[3]   Melbourne W. Radio Occultations Using Earth Satellites: A Wave Theory Treatment, Jet Propulsion Laboratory, California Institute of Technology, Boulder, 2004.

[4]   Liou Y, Pavelyev A. Simultaneous observations of radio wave phase and intensity variations for locating the plasma layers in the ionosphere. Geophys. Res. Lett. 2006; 33: L23102.

[5]   Pavelyev A, Liou Y, Wickert J, et al. Effects of the ionosphere and solar activity on radio occultation signals: Application to CHAllenging Minisatellite Payload satellite observations. J. Geophys. Res. 2007; 112: A06326, doi: 10.1029/2006JA011625.

[6]   Liou Y, Pavelyev A, Liu S, et al. FORMOSAT-3/COSMIC GPS Radio Occultation Mission: Preliminary Results. IEEE Trans. Geosci. Remote Sensing 2007; 45: 3813-3823.

[7]   Pavelyev A, Wickert J, Liou Y. Localization of plasma layers in the ionosphere based on observing variations in the phase and amplitude of radio waves along the satellite-to-satellite path. Radiophys. Quantum Electron 2008; 51(1): 1-7

[8]   Pavelyev A, Liou Y, Wickert J, et al. Location of layered structures in the ionosphere and atmosphere by use of GPS occultation data. Adv. Space Res. 2008; 42: 224-228.

[9]   Sokolovskiy S. Inversion of radio occultation amplitude data. Radio Sci. 2000; 35: 97-105.

[10] Gorbunov M, Gurvich A, Shmakov A. Back-propagation and radioholographic methods for investigation of sporadic ionospheric E-layers from Microlab-1 data. Int. J. Remote Sensing 2002; 23: 675-682.

[11] Sokolovskiy S, Schreiner W, Rocken C, Hunt D. Detection of high-altitude ionospheric irregularities with GPS/MET. Geophys. Res. Lett. 2002; 29: 1033-1037.

[12] Pavelyev A, Zakharov A, Kucheryavenkov A, et al. The features of propagation of radio wave reflected from terrestrial surface at small elevation angles on radio telecommunication link low orbital satellite-GEO. J. Commun. Tech. Electron 1997; 42: 45-51.

[13] Pavelyev A, Volkov A, Zakharov A, et al. Bistatic radar as a tool for the Earth's observation using small satellites. Acta Astronautica 1996; 39: 721-730.

[14] Kirchengast G, Hoeg P in: Steiner A, Kirchengast G, Foelsche U Eds. Occultations for Probing Atmosphere and Climate, Springer, New York 2004; 201-212.

[15] Gorbunov M, Kirchengast G. Processing X/K Band Radio Occultation Data in Presence of Turbulence. Radio Sci. 2005; 40: RS6001.

[16] Lohman M, Jensen A, Benzon H, Nielsen A. Radio Occultation Retrieval of Atmospheric Absorption based on FSI. Report 03-20, Danish Meteorological Institute: Copenhagen; 2003.

[17] Jensen A, Lohmann M, Nielsen A,. Benzon H. Full Spectrum Inversion of Radio Occultation Signals. Radio Sci.2003; 39: 1040-1052.

[18] Yakovlev O, Yuakubov V, Uruadov V, Pavelyev A. Radio waves propagation. Moscow Lenand Ed; 2009.

[19] Kravtsov Yu., Orlov Y. Geometrical Optics of Inhomogeneous Media. Springer: Berlin; 1990.

[20] Pavelyev A, Kucheryavenkov A. On a possibility of investigation of the Earth's atmosphere by radio occultation method. J. Commun. Tech. Electron. 1978; 7: 1345-1353.

[21] Kislyakov A, Stankevich K. Absorption of radio waves in the atmosphere. Radiophys. Quantum Electron. 1967; 10: 1244-1268.

[22] Yakovlev O, Matyugov S, Vilkov I. Attenuation and scintillation of radio waves in the atmosphere from data of radio occultation experiments in the satellite-to-satellite link. Radio Sci. 1995; 30: 591-599.

[23] Rius A, Ruffini G, Romeo A. Analysis of ionospheric electron density distribution from GPS/MET occultations. IEEE Trans. Geosci. Remote Sens. 1998; 36(2): 383–394.

[24] Hajj G, Romans L. Ionospheric electron density profiles obtained with the Global Positioning System: Results from GPS/MET experiment. Radio Sci. 1998; 33(1) 175-190.

[25] Igarashi K, Pavelyev A, Hocke K, Pavelyev D, Wickert J. Observation of wave structures in the upper atmosphere by means of radio holographic analysis of the radio occultation data. Advances in Space Research 2001; 27: 1321-1326.

[26] Wickert J, Reigber C, Beyerle G, Konig R, Marquardt C, Schmidt T, Grunwaldt L, Galas R, Meehan T,. Melbourne W, Hocke K. Atmosphere sounding by GPS radio occultation: First results from CHAMP. Geophys. Res. Lett. 2001; 28(9): 3263–3266.

[27] Wickert J, Pavelyev A, Liou Y, Schmidt T, Reigber C, Igarashi K, Pavelyev A, Matyugov S. Amplitude scintillations in GPS signals as a possible indicator of ionospheric structures. Geophys. Res. Lett. 2004; 31(15): L24801, 1-4.

[28] Wu D, Ao C, Hajj G, Juarez de la Torre, J. Mannucci A. Sporadic E morphology from GPS-CHAMP radio occultation. J. Geophys. Res. 2005; 110: A01306, doi:10.1029/2004JA010701.

[29] Pavelyev A, Liou Y, Huang C, Reigber C, Wickert J, Igarashi K, Hocke K. Radio holographic method for the study of the ionosphere, atmosphere and terrestrial surface using GPS occultation signals. GPS Solut., 2002; 6: 101-108.

[30] Pavelyev A, Liou Y, Wickert J. Diffractive vector and scalar integrals for bistatic radio-holographic remote sensing. Radio Sci. 2004; 39(4): RS4011, 1-16, doi:10.1029/2003RS002935.

[31] Pavelyev A, Liou Y, Wickert J, Gavrik A, Lee C. Eikonal acceleration technique for studying of the Earth and planetary atmospheres by radio occultation method. Geophys. Res. Lett. 2009; 36. doi:10.1029/2009GL040979, L21807, 1-5.

[32] Pavelyev A, Zhang K, Wang C, Kuleshov Y, Liou Y, Wickert J. Identification of Inclined Ionospheric Layers Using Analysis of GPS Occultation Data. IEEE Trans. Geosci. Rem. Sens. 2011; 49(6): part 2, 2374-2384. doi:10.1109/TGRS.2010.2091138.

[33] Liou Y,. Pavelyev A, Huang C, Igarashi K, Hocke K. Simultaneous observation of the vertical gradients of refractivity in the atmosphere and electron density in the lower ionosphere by radio occultation amplitude method. Geophys. Res. Lett. 2002; 29(19), 43-1-43-4. doi:10.1029/2002GL015155.

[34] Liou Y, Pavelyev A, Huang C, Igarashi K, Hocke K, Yan S. Analytic method for observation of the GW using RO data. Geophys. Res. Lett. 2003; 30(20): ASC 1-1 – 1-5., doi:10.1029/2003GL017818.

[35] Liou Y, Pavelyev A, Pavelyev A, Wickert J, Schmidt T. Analysis of atmospheric and ionospheric structures using the GPS/MET and CHAMP radio occultation data base: A methodological review. GPS Solut. 2005; 9 (2): 122–143.

[36] Pavelyev A, Liou Y, Zhang K et al. Identification and localization of layers in the ionosphere using the eikonal and amplitude of radio occultation signals Atmos. Meas. Tech., 2012; 5(1): 1–16, doi:10.5194/amt-5-1-2012.

[37] Pavelyev A, Zhang K, Wang C et al. Analytical method for determining the location of ionospheric and atmospheric layers from radio occultation data Radiophys. Quantum Electron 2012; 55(3): 168-175.

[38] Liou Y, Pavelyev A, Matyugov S, Yakovlev O, Wickert J. 2010 Radio Occultation Method for Remote Sensing of the Atmosphere and Ionosphere. Edited by Liou Y. INTECH Published by In-The Olajnica 19/2, 32000 Vukovar, Croatia, 170, 45, ISBN 978-953-7619-60-2.

[39] Arras C, Wickert J, Jacobi C, Heise S, Beyerle G. Schmidt T. A global climatology of ionospheric irregularities derived from GPS radio occultation. Geophys. Res. Lett. 2008; 35, L14809, doi: 10.1029/2008GL034158.

[40] Arras C, Jacobi C, Wickert J. Semidiurnal tidal signature in sporadic E occurrence rates derived from GPS radio occultation measurements at higher midlatitudes. Ann. Geophys., 2009; 27: 2555–2563.

[41] Arras C, Jacobi C, Wickert J, Heise S, Schmidt T. (2010), Sporadic E signatures revealed from multi-satellite radio occultation measurements. Adv. Radio Sci. 2010; 8: 225–230. doi:10.5194/ars-8-225-2010.

[42] Jakowski N, Leitinger R, Angling M. (2004), Radio occultation techniques for probing the ionosphere. Ann. Geophys. 2004; supplement to 47(2/3): 1049-1066.

[43] Kunitsyn V, Tereshchenko E. Ionospheric Tomography. Springer-Verlag: Berlin: 2003.

[44] Kelley M, Wong V, Hajj G, Mannucci A. On measuring the off-equatorial conductivity before and during convective ionospheric storms. Geophys. Res. Lett. 2004; 31: L17805, doi:10.1029/2004GL020423.

[45] Kelley M, Heelis A. The Earth's ionosphere: plasma physics and electrodynamics. Elsevier Science.2009; 96: 556, International geophysics Second Edition Cornell University Engineering School of Electrical Engineering Academic Press, ISBN 10: 0120884259/0-12-088425-9, ISBN 13: 9780120884254

[46] Pavelyev A, Liou Y, Wickert J, Zhang K, Wang C, Kuleshov Y. Analytical model of electromagnetic waves propagation and location of inclined plasma layers using occultation data. Prog. Electromagn. Res. 2010; 106: 177-202, doi: 10.2528/PIER10042707.

[47] Pavelyev A, Liou Y, Wickert J, Schmidt T, Pavelyev A, Matyugov S. Phase acceleration: a new important parameter in GPS occultation technology. GPS Solut. 2010; 14(1): 3-14. doi:10.1007/s10291-009-0128-1.

[48] Whitehead J. The formation of the sporadic-E layer in the temperate zones. J. Atmos. Terr. Phys. 1961; 20 49–58.

[49] Plane J. (2003), Atmospheric Chemistry of Meteoric Metals. Chem. Rev. 2003; 103. 4963–4984.

[50] Pavelyev A, Tsuda T, Igarashi K, Liou Y, Hocke K. Wave structures in the electron density profile in the ionospheric D and E-layers observed by radio holography analysis of the GPS/MET radio occultation data. J. Atmos. Solar-Terr. Phys. 2003; 65(1): 59-70.

[51] Haldoupis C, Pancheva D, Singer W. Meek C, Mac-Dougall J. An explanation for the seasonal dependence of midlatitude sporadic E layers. J. Geophys. Res. 2007; 112: A06315, doi:10.1029/2007JA012322.

[52] Melbourne W, Davis E, Duncan C, Hajj G, Hardy K, Kursinski E, Meehan T, Young L, Yunck T. The Application of Spaceborne GPS to Atmospheric Limb Sounding and Global Change Monitoring. Jet Propul. Lab., Pasadena, Calif. 1994; 94-18: 147.

[53] Vorob'ev V, Gurvich A, Kan V, Sokolovskiy S, Fedorova O, Shmakov A. (1997), The structure of the ionosphere from the GPS-"Microlab-1" radio occultation data: Preliminary results. Cosmic Research 4, 74-83. (In Russian)

[54] Gubenko V, Pavelyev A, Andreev V. Identification of wave origin of temperature fluctuations and determination of the intrinsic frequency of internal gravity waves in the Earth's stratosphere derived from radio occultation data. J. Geophys. Res. 2008; 113: D08109, doi:10.1029/2007JD008920.

[55] Hocke K. Inversion of GPS meteorology data, Ann. Geophys. 1997;15. 143-152.

[56] Gorbunov M, Sokolovskiy S, Bengtson L. Advanced algorithms of inversion of GPS/MET satellite data and their application to reconstruction of temperature and humidity Tech. Rep Report No. 211. Max Planck Institute for Meteorology. Hamburg. 1996.

[57] Bean B, Dutton E. Radio Meteorology. Central Bureau of Standards. Boulder. Colorado. 435 p., 1966.

[58] Kolosov M, Shabel'nikov A. Refraction of Electromagnetic Waves in the Atmosphere of the Earth, Venus and Mars (Sovetskoe Radio, Moscow, 1976) [in Russian].

[59] Kursinski E, Hajj G, Hardy K, Linfield R, Hardy K. Observing Earth's atmosphere with radio occultation measurements using the Global Positioning System. J. Geophys. Res. 1997; 102 (D19), 23429.

[60] Yakovlev O, Pavelyev A, Matyugov S. Satellite Monitoring of the Earth: Radio Occultation Monitoring of the Atmosphere and Ionosphere (Librokom, Moscow, 2010) [in Russian].

[61] Loescher A, Lauritsen K, Sørensen M. The GRAS SAF Radio Occultation Processing Inter comparison Project ROPIC New Horizons in Occultation Research (Springer Verlag, Heidelberg, 2009), pp. 49–62.

Chapter 5

Radio Wave Propagation Through Vegetation

Mir Ghoraishi, Jun-ichi Takada and Tetsuro Imai

Additional information is available at the end of the chapter

1. Introduction

Vegetation is an indispensable feature of most outdoor wireless channel environments. The interaction between radio waves and vegetation has been researched for several decades. Because of the complex structure of the foliage, composed of randomly oriented trunks, branches, twigs and leaves, the involved physical process in the propagation of the radio wave through vegetation is complex. Accurate modeling of the propagation of radio waves through tree foliage, generally requires accurate electromagnetic description of the tree geometry, including its branches and leaves, valid over a wide range of frequencies. Originally empirical models were developed to describe the propagation of radio waves in the vegetation. In other approaches this interaction is analyzed using ray-based techniques. More recently theoretical –statistical and analytical– approaches became favorable. As it is discussed in the next section, major disadvantage of all these models is that the final outcome is basically presented by providing an excess attenuation to that caused by free space propagation.

1.1. Existing approaches

Empirical models have been developed to characterize the the propagation of radio waves in the vegetation for years. Their significant advantage is their simplicity. The drawback is that, like any other empirical model, the formulated model is strictly related to the specific measured data set and fails to give any indication as to the physical processes involved in the propagation within the channel. These models usually provide either the mean attenuation of the propagation signal caused by vegetation or calculate the link budget. Parameters in these models, e.g. frequency, incident angles, direct-path length through vegetation and other parameters associated with the specific environment under which measurements were performed, are usually computed through regression curves fitted to measurement data. Among many, the modified exponential decay model suggested in [1], and the COST 235 model [2] can be mentioned. These models are expressed as equations in exponential forms to give the specific attenuation as a function of path length and operating frequency. The attenuation of trees as a function of vegetation depth has been shown to be more accurately represented by dual slope attenuation functions [3]-[7]. To accommodate this dual slope, an empirical nonzero gradient model was developed to follow the dual gradient of the measured attenuation curve [8]. The initial slope describes the loss experienced by the coherent component, whereas the second slope describes the dominance of the incoherent component, which occurs at a much reduced rate. An important disadvantage of the semi-empirical vegetated radio channel models, common to other approaches such as radiative transfer theory, is that their little account of the dynamic effects in the channel and no account of the wideband effects of the vegetation medium.

Another approach in the the analysis and prediction of the vegetated radio channel is ray-tracing [9]-[16]. These have to be carefully designed and used, taking into account the frequency of the radio wave, the dimension of interacting objects and their distance to the sources to fulfill the far field condition. Therefore in different frequencies the mechanisms by which the wave propagates can vary dramatically. The scattering has been modeled deterministically in many different ways depending on the electrical density of the vegetative medium. At lower frequencies, where individual components of the vegetation (trunks, branches, twigs and leaves or needles) and their separations are small by comparison with the radio wavelength, considering the vegetation as a homogeneous dielectric slab, the propagation has been modeled in terms of a lateral wave at treetop heights [9]. At frequencies above 200 [MHz] or so, a single slab becomes inadequate. As the scale of the changes in density and structure of the vegetation become greater than the order of a wavelength, and layered representations of the vegetation should be used [10]. A full-wave analysis of the radiowave propagating along mixed paths inside a four-layered forest model applicable to frequencies up to 3 [GHz] was proposed in [11], [12], which consists of four isotropic and homogeneous dielectric. The first layer is the semi-infinite free-space, whereas the second layer represents the forest canopy. The third and fourth layers model the trunk and the semi-infinite ground plane, respectively. As the distance between the transmitter and the receiver is very long, the radio wave propagation through the stratified forest is characterized by the lateral wave that mainly propagates on top of the canopy along the air-canopy interface. For short distances, however, such a propagation is denominated by the direct or coherent component. When the receiver is at a clearing distance from the dense vegetated area the edge of the forest is treated as a source of diffraction [14], and the uniform theory of diffraction (UTD) associating a double-diffracted component over the canopy and a transmission component which includes the exact calculation of refraction angles is also used [15].

To model the incoherent component which is the dominant propagation mechanism for long distances inside vegetation, theoretical models, which are more complicated but more generic and applicable to any arbitrary foliage wave propagation scenario, are used. Two major approaches, namely the radiative transfer theory and the analytical theory, have been pursued to develop these models [17],[18]. These two methods are closely related as they are addressing the same problem of the wave propagation in randomly distributed particles. In fact, the radiative transfer theory can be derived from analytical approach by applying some approximations [17], and they have proven equivalent for the application of radar in the forest canopies [19].

In the method of radiative transfer theory, the vegetation medium is modeled as a statistically homogeneous random medium of scatterers which is characterized by parameters such as the absorption cross-section per unit volume, the scatterer cross-section per unit volume and the scattering function of the medium [20]. The scattering function (phase function) is characterized by a narrow forward lobe and an isotropic background. The model considers a plane wave incident from an air half space upon the planar interface of a vegetation half space. The basic equation of the radiative energy transfer theory is expressed in terms of the specific intensity for which the radiative transfer theory gives the specific value at a given point within the vegetation medium as a sum of a coherent component and an incoherent component. The coherent component is reduced in intensity due to absorption and scattering of the incident wave, and the incoherent component due to the scattered wave. Each scatterer is assumed to have a directional scattering profile, or phase function. As the constituents of the tree are relatively large compared to the wavelength at micro- and millimeter wave frequencies, the scattering function is assumed to consist of a strongly scattering forward lobe, which can be assumed to be Gaussian with an isotropic background level. The radiative transfer theory predicts the dual slope nature of the measured attenuation versus vegetation depth curves and provides a physical interpretation. The equation based on the radiative transfer theory allows the prediction of the attenuation curves in which the received signal is reduced linearly by scattering and absorption of the incident signal. As the receiver is moved deeper into the vegetation, and the direct coherent component is reduced further still, the isotropically scattered component becomes significant. Due to the increase of scattering volume as we move deeper into the medium, the scattering signal level tends to be maintained, leading in turn to an attenuation rate which is significantly reduced at these depths. The model however requires four input parameters namely the ratio of the forward scattered power to the total scattered power, the beam-width of the phase function, the combined absorption and scattering coefficient, and the albedo. These are extracted from path-loss measurement data so that the approach makes itself a semi-empirical model in essence. Direct computation of these parameters, such as the albedo and the phase function, is very difficult, because the vertical profile of the foliage is inhomogeneous, i.e. the distinction exists between the trunk layer and the crown layer, whereas the radiative transfer approach is generally applied to a homogeneous medium. In order to overcome this limitation, an improved version of the discrete radiative transfer is proposed for isolated vegetation specimens [21],[22]. However this requires discretization of the foliage into small cells which is numerically intractable for large propagation distances.

The alternative approach in the problem of wave propagation in randomly distributed particles is the analytical approach [17]. This is usually in the format of Foldy-Lax solution for point scatterers [18],[23], which has been widely used to estimate the signal attenuation in the the foliage [24]-[32]. In this approach the Born approximation is applied to account only the first term in the equation as considering higher terms can complicate the computations prohibitively [18]. It predicts the exponential decay of the radio field corresponding to the linear foliage path-loss (in dB) versus the wave propagation distance. In [24]-[30], the inhomogeneous forest structure was represented by using a realistic-looking fractal tree model. The statistics of the received field are then obtained through a Monte Carlo technique which generates random forest structures according to prescribed statistical botanical features, such as tree height, branch and leaf orientation angles, and tree locations. Another approach is to model leaves as flat circular lossy dielectric discs, and branches as finitely long circular lossy dielectric cylinders [31],[32]. The disadvantage of the analytical approach stems from the fact that Born approximation accounts only for single scattering, which has been shown to overestimate the foliage path-loss at high frequencies or over long distance propagation where the multiple-scattering effects become important [28],[29]. Another concern with this method is the required computation time. Computing foliage path-loss over long distances in a forested environment can be prohibitively time-consuming even with the single-scattering model. This difficulty can be circumvented by treating the forest as a statistically homogeneous medium along the direction of wave propagation and only analyzing the wave propagation behaviors in a typical block of forest, which can then be reused for all forest blocks [30]. Furthermore a main concern in Born approximation is its validity is restricted to scatterers with a dielectric permittivity close to unity and that the effect of multiple scattering from the discrete scatterers are not negligible [33]. To overcome these limitations the Feynman diagrams are converted to the set of expanded green functions presented in integral operator form is suggested as an alternative to Born approximation [33],[34].

To benefit from the ray-tracing and theoretical approaches at the same time, the model proposed in [35] combines the effects of three individual propagation modes, i.e. diffraction from the side and top of the foliage, ground reflection and direct (through vegetation) propagation. In this approach the extent of the vegetation is modeled as rectangular hexahedrons (boxes). The loss experienced by the diffracted waves over the vegetation as well as those around the vegetation are treated as double isolated knife-edge diffraction. If the ground reflection is passed through vegetation, the path loss due to propagation through vegetation is added to the reflection loss. The values for the permittivity and conductance of the ground are obtained from ITU-R recommendations [36]. For the direct through vegetation propagation component the radiative transfer approach is adopted and the necessary parameters for specific geometries, species and frequencies are measured and provided in tables [35]. This model was adopted by ITU-R and later works published as recommendations of ITU-R have improved the tables of parameters for some kind of trees [37].

While the above mentioned models mostly ignore the channel dynamics over time, narrowband analysis of the wind effect on an isolated tree in the anechoic chamber is reported in [38]-[40].

1.2. Directional analysis of the vegetation radio channel

Either if the interaction between radio signal and foliage is modeled based on diffraction theories and ray-tracing, radiative transfer theory, analytical theory, statistical, or empirical approaches, it is usually aimed to provide an excess attenuation to that caused by free space propagation. On the other hand, as these models are strongly dependent on measurements for their evaluation and modifications, a considerable number of experiments have been accomplished to analyze the foliage influence on the propagation channel as well as to evaluate the proposed models. The ultimate target in most of these experiments however is to measure the attenuation of the radio wave caused by the vegetation. The problem is that even though such a result proves useful for specific purposes, it provides only limited knowledge about the interactions in the channel. On the other hand, the assumption of homogeneous media of randomly distributed scattering points is widely used in analytical and statistical approaches, an assumption to be examined yet. Obviously with a high-resolution spatial analysis of the vegetation radio channel, the existing methods and their assumptions can be further evaluated and if necessary modified. Moreover, such an analysis can be used in design and performance analysis of modern wireless systems equipped with multiple-antenna technologies to improve the capacity. Never the less, and in spite of such significant benefits, such measurements and analyses are rarely reported due to practical limitations.

Few recent reports address –indirectly in most cases– the spatial characteristics of the radio channel in vegetation. The delay power spectrum of the received signal through a single isolated tree for different angle-of-incidents is evaluated for a wideband signal in [41]. The spatial correlation of a multiple antenna system operated in the forest is evaluated in [42]. The scattering effect of the foliage in suburban scenarios is examined by directive antennas [43], and delay spread of the received signal as a result of scattering by trees was observed [44]. In a recent publication, the delay and angular spread of the wireless channel for the application in positioning of mobile users is reported [45].

In this chapter the directional radio channel in dense vegetation is investigated. An experimental approach is necessary because of the complexity of the underlying phenomena. The methodology used for this purpose, is the directional analysis of a carefully captured measurement data. First the measurement campaign and setup parameters as well as the signal model are introduced in section 2, and then two analysis methods for the measured data are discussed. In the first approach, the experimental data is analyzed by calculating the received signal dispersion in delay and azimuth-of-arrival of the propagated waves which is done using a Capon beam-forming technique [46]. One advantage of such analysis is that no presumption on the distribution of the dispersed signal was necessary. In the second approach, a high-resolution parameter estimation technique was adopted to acquire a more accurate knowledge of the channel, particularly to identify the involving propagation mechanisms. Results of the analysis by each method, including the multipath cluster identification and propagation mechanism determination are presented in section 3. Section 4 provides discussions on the findings of the chapter, the analysis and results, and argues how the directional analysis, aiming at the identification of dominant propagation mechanisms in the channel, can meaningfully improve the insight toward the problem. Section 5 sums up important conclusions of the chapter.

2. Method

2.1. Experimental investigation

As it was expressed, to clarify the influence of the vegetation on the radio channel, and specifically to re-examine those fundamental assumptions usually presumed in such studies we had to gather a set of experimental data with necessary resolution in delay and spatial domains. For this propose a dense vegetated area was chosen in the southern Kanagawa, Japan. A schematic of the measurement scenario is illustrated in Fig. 1 where the height of the base-station antenna could be altered and the mobile-station antenna height is fixed. The employed channel sounder is a double-directional sounder at the center frequency of 2.22 [GHz]. The sounder specifications and the measurement set-up parameters are found in Table 1. At the transmitter a sleeve and a slot antenna are used to send the vertically and horizontally polarized signal in different time slots. The cylindrical array antenna at the receiver switches the vertical and horizontal patch antenna feeds, 96 elements for each polarizations. The transmitter antennas are mounted on a measurement bucket capable to elevate up to 15 [m]. Measurements were performed on the 4, 6, 9, 12 and 15 [m] of the transmitter antenna heights. The receiver antenna array is installed on the roof-top of a

f_c	2.22 [GHz]
Tx signal	OFDM, 2048 FFT points
Number of subcarrier	897
Tx signal bandwidth	50 MHz
Tx power	30 [dBm]
Tx antenna	sleeve, slot (different time slots)
Tx antenna height	15, 12, 9, 6, 4 [m]
Rx antenna	Cylindrical array (4 ring stacked) 96 elements V/H
Rx antenna height	3.5 [m] (installed on the van's roof-top)
Tx-Rx separation	100 [m]
Tallest trees height	10 [m]

Table 1. Specifications of Experiment

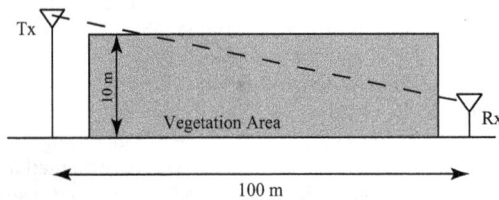

Figure 1. Measurement scenario.

measurement van which carries the receiver system. The van is parked deep in the vegetation area with a transmitter to receiver horizontal distance of about 100 [m]. Here we only report the static channel that is the measurement van has been parked during the measurement. The tallest trees are mostly Japanese cedar whereas in lower layers of the vegetation several kinds of trees are found as it can be seen in Fig. 2. No significant scattering other than trees is observable around the receiver.

With such a measurement setup we are able to achieve a more accurate analysis of the radio channel. Two different analysis methods are introduced in next sections and then the interpretations are compared.

2.2. Analysis approaches

A major part of the reported experiments regarding the interaction of radio with vegetations targeting the spatial analysis use directional antennas or an array antenna with beam-forming scheme. A good reason for this is the simplicity of such measurements, compared to more complicated ones concluding in high-resolution analysis. An important but less obvious reason is that since in case of propagation through vegetation we know little about the distribution of effective scatterers, beam-forming as a robust approach seems the appropriate candidate for the analysis. This is expressed in contrast to the model based estimation methods although the resolution of the beam-forming is inferior. In this section two methods

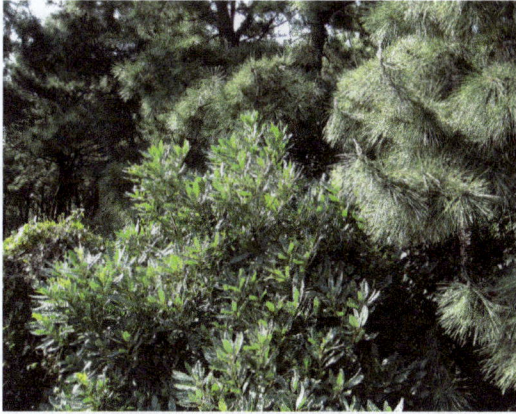

Figure 2. Different types of trees in the foliage.

for the analysis of the measurement data are presented. First we introduce a Capon beam-former to analyze and provide the spread of the received radio signals. In the next step a high-resolution SAGE algorithm is introduced for the analysis of the multidimensional measurement data.

2.2.1. Signal model

Assuming multipath in the channel, the signal observed at the output of the receiver array can be represented in vector notation as [48]

$$\mathbf{Y}(t) \doteq [Y_1(t), \cdots, Y_{N_r}(t)]^{\mathrm{T}}$$
$$= \sum_{l=1}^{L} \mathbf{s}(t; \boldsymbol{\theta}_l) + \sqrt{\frac{N_0}{2}} \mathbf{w}(t), \tag{1}$$

where $\mathbf{Y}(t)$ is the N_r dimensional vector of received signals at the receiver array with N_r antenna elements and $\mathbf{s}(t; \boldsymbol{\theta}_l)$ is the contribution of the lth multipath to the received signal, again a vector with N_r elements and it is assumed that L multipath are successfully received by the receiver. The N_r dimensional complex, temporally and spatially white noise is denoted by $\mathbf{w}(t)$ and N_0 is a positive constant.

The parametric characteristic of the lth propagation wave is described by vector $\boldsymbol{\theta}_l$ which is represented for the current directional and polarimetric measurement as

$$\boldsymbol{\theta}_l \doteq [\boldsymbol{\Omega}_{rl}, \tau_l, \mathbf{A}_l]. \tag{2}$$

Here the subscript 'l' indicates the correspondence to the lth path and the parameters are angle-of-arrival $\boldsymbol{\Omega}_{rl} = [\varphi_{rl}, \vartheta_{rl}]$ with φ_{rl} indicating the azimuth-of-arrival for the propagation wave calculated in degrees, counter clockwise with its origin along the direction of receiver toward the transmitter, and ϑ_{rl} being the elevation-of-arrival of the propagation wave computed in degrees from the horizon with increasing values upward, so called coelevation.

Other parameters are excess-delay τ_l and complex magnitude $\mathbf{A}_l$ is the vector to represent the weights for the co- and cross-polarization components according to

$$\mathbf{A}_l \doteq \begin{bmatrix} \alpha_{vl} \\ \alpha_{hl} \end{bmatrix}. \tag{3}$$

It is observed that due to the static measurement scenario Doppler-frequency of the propagation multipath is not considered, otherwise its value is expected to be zero or negligible. The contribution of the lth multipath to the received signal can therefore be expressed in vector notation as

$$\mathbf{s}(t; \boldsymbol{\theta}_l) \doteq [s_1(t; \boldsymbol{\theta}_l), \cdots, s_{N_r}(t; \boldsymbol{\theta}_l)]^{\mathrm{T}}$$
$$= \mathbf{C}_r(\boldsymbol{\Omega}_{rl}) \mathbf{A}_l u(t - \tau_l). \tag{4}$$

The matrix $\mathbf{C}_r(\boldsymbol{\Omega})$ in (4) is defined as $\mathbf{C}_r(\boldsymbol{\Omega}) \doteq [\mathbf{c}_{rv}(\boldsymbol{\Omega}), \mathbf{c}_{rh}(\boldsymbol{\Omega})]$ (v and h subscripts indicate V- and H-polarizations) where the N_r dimensional-vector receiver array radiation pattern for the propagation wave impinging from direction $\boldsymbol{\Omega}_{rl} = [\varphi_{rl}, \vartheta_{rl}]$ at the receiver is denoted as $\mathbf{c}_{rp_r}(\boldsymbol{\Omega}_{rl})$ and is defined as

$$\mathbf{c}_{rp_r}(\boldsymbol{\Omega}) \doteq [f_{n_r,p_r}(\boldsymbol{\Omega}) \exp\{j2\pi\lambda^{-1}\boldsymbol{\Omega} \cdot \mathbf{r}_{n_r}\}]; \quad n_r = 1, \cdots, N_r]^{\mathrm{T}}, \tag{5}$$

for the receiver polarization $p_r \in \{v, h\}$ and $f_{n_r,p_r}(\boldsymbol{\Omega})$ denotes the radiation pattern for the n_rth element of the array. The transmitted signal at any arbitrary time instance t is an impulse train in the frequency domain which is obtained as

$$u(t) = \sum_{b=-B}^{B} \delta(f_c + bf_s), \tag{6}$$

where f_c is the center frequency of the sounding signal, f_s is the DFT frequency shift and the signal bandwidth is therefore equal to $(2B + 1)f_s$. The energy of the signal is assumed as P. A delayed version of such a signal is represented as

$$u(t - \tau) = e^{-j2\pi f_s \tau} u(t). \tag{7}$$

Thus the signal model (4) can be represented as

$$\mathbf{s}(t; \boldsymbol{\theta}_l) = e^{-j2\pi f_s \tau_l} \mathbf{C}_r(\boldsymbol{\Omega}_{rl}) \mathbf{A}_l u(t). \tag{8}$$

It is assumed that each snapshot of the sounding signal is transmitted during T_s (sounding signal duration), in I consecutive intervals each longer than transmitted signal duration.

2.2.2. Beam-forming

Capon beam-former, or minimum variance distortionless response (MVDR) beam-former, is proved optimum for estimating an unknown random plane-wave signal, being received in the presence of noise, to provide a minimum variance unbiased estimate [49]. This is equivalent to passing the signal through a distortionless filter which minimizes the output variance. A significant advantage of this algorithm when used in the parameter estimation context is that the provided spatial spectral estimate does not rely on any underlying signal model [49]. The approach employed in this section was previously used to estimate the azimuth of the distributed source components in [50]. Following the same method, the first and second moments of the azimuth-of-arrival and excess-delay spread of the received radio waves are calculated.

We consider the directional but stationary scenario where the L multipath are received in D_m received signal clusters at each measurement m (at each specific transmitter antenna height) as

$$\mathbf{Y}_m(t) = \sum_{d=1}^{D_m} \sum_{l \in C_{m,d}} \mathbf{s}(t; \boldsymbol{\theta}_l) + \sqrt{\frac{N_0}{2}} \mathbf{w}(t), \qquad m = 1, \cdots, M, \qquad (9)$$

where $C_{m,d}$ is the set of multipath in dth cluster of mth measurement. It is observed that the definition of cluster is dependent to the resolution of the analysis scheme. Hence in the next section where high-resolution algorithm is used, a distinct definition for the multipath cluster is employed. Furthermore, in the beam-forming analysis of the signal we neglect the elevation-of-arrival ϑ_r of the received signal because of low resolution of the analysis in this dimension and therefore the propagation in azimuth plane only is considered. In the current analysis although the resolution is not enough to separate multipath, but the delay spread is already available by performing the inverse discrete Fourier transform on the measured data samples with the resolution $\tau_{res} = ((2B+1)f_s)^{-1}$ and the azimuth-power spectrum at each delay is computed by a Capon beam-forming due to its better resolution in comparison to the conventional approach. The Capon spectrum is obtained as [49]:

$$P_m(\varphi_r, \tau_k)\Big|_{p_t p_r} = \frac{1}{\mathbf{c}_{rp_r}^H(\varphi_r, 0) \, \mathbf{R}_{\mathbf{Y}_{mp_t}}^{-1}(\tau_k) \, \mathbf{c}_{rp_r}(\varphi_r, 0)}, \qquad (10)$$

where $\mathbf{R}_{\mathbf{Y}_{mp_t}}$ is the received signal covariance matrix for the measurement m and transmitter antenna polarization of p_t, $\mathbf{c}_{rp_r}(\varphi_r, 0)$ is the receiver array response in the azimuth plane for the polarization p_r and $\tau_k = (k-1)\tau_{res}$ is a specific delay within the range. Figure 3 presents a sample of computed delay-azimuth-power spectrum for the transmitter antenna height of 15 [m] ($m = 1$), where the transmitted signal and the receiving antenna feeds are in vertical polarization ($p_t p_r = vv$). The figure indicates –and it is the case for most measured data in this campaign– that the radio signal is received as a cluster of multipath with probably a Gaussian or von-Mises spread in the azimuth, and an exponential spread along the delay. We therefore modify the signal model to $D_m = 1$ for all values of m and drop the index d in the beam-forming data analysis. In the analysis the normalized spectrum $\dot{P}_m(\varphi_r, \tau)$ to the minimum value of the spectrum within a support around the considered multipath cluster is used which is calculated for each measurement m at polarization combination $p_t p_r$. The size of the support shall be selected large enough to include the diffuse components but not so large to have any multipath from other clusters in. It is noticed however that in a beam-forming approach the array antenna response could not be de-embedded from the measured data. Another disadvantage is low spatial resolution resulting in the single cluster presumption.

2.2.3. High-resolution parameter estimation

A space-alternative generalized EM algorithm (SAGE) was employed to estimate the vector of parameters $\boldsymbol{\theta}_l$ for each propagation wave within the resolution of the system [51]. Observation of the estimated paths indicates a large cross polarization ratio in most cases and therefore the results presented here does not discuss the cross polarization of the multipath. Thus

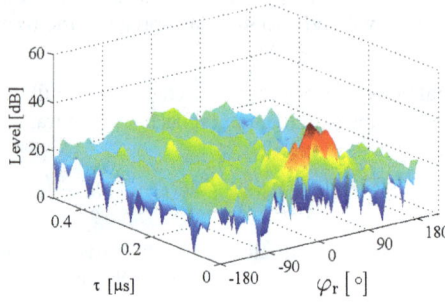

Figure 3. delay-azimuth-power spectrum of the received waves for transmitter antenna height of 15 m, VV case.

the subscript indicating the polarization of the complex magnitude $\alpha_{p,l}$, i.e. $p_r \in \{v, h\}$, is dropped in further discussions.

Several tens of multipath are detected at each measurement points, while this number could reach well beyond 100 in a couple of measurements. The array calibration indicates that array response for extreme elevation-of-arrival values is highly unreliable. Hence we have neglected those estimated multipath impinging with too high or too low elevation-of-arrival, i.e. we consider the acceptable range of elevation-of-arrival as $\vartheta_{rl} \in (-75°, 75°)$.

3. Analysis

3.1. Moments of radio wave spread –beam-forming

Having the delay-azimuth-power spectrum of the measured data computed we can obtain the mean values $\bar{\varphi}_m$ and $\bar{\tau}_m$, and the standard deviations σ_{φ_m} and σ_{τ_m} for each measurement m simply as:

$$\bar{\varphi}_m(\tau_k) = \frac{\sum_{\varphi_r \in \Phi} \varphi_r \dot{P}_m(\varphi_r, \tau_k)}{\sum_{\varphi_r \in \Phi} \dot{P}_m(\varphi_r, \tau_k)}, \tag{11}$$

$$\sigma_{\varphi_m}(\tau_k) = \left(\frac{\sum_{\varphi_r \in \Phi} (\varphi_r - \bar{\varphi}_r)^2 \dot{P}_m(\varphi_r, \tau_k)}{\sum_{\varphi_r \in \Phi} \dot{P}_m(\varphi_r, \tau_k)} \right)^{1/2}, \tag{12}$$

$$\bar{\tau}_m(\varphi_{rq}) = \frac{\sum_{\tau \in \Xi} \tau_k \dot{P}_m(\varphi_{rq}, \tau)}{\sum_{\tau \in \Xi} \dot{P}_m(\varphi_{rq}, \tau)}, \tag{13}$$

$$\sigma_{\tau_m}(\varphi_{rq}) = \left(\frac{\sum_{\tau \in \Xi} (\tau - \bar{\tau})^2 \dot{P}_m(\varphi_{rq}, \tau)}{\sum_{\tau \in \Xi} \dot{P}_m(\varphi_{rq}, \tau)} \right)^{1/2}, \tag{14}$$

where $\Xi = \{0, \tau_{res}, \cdots, T_s\}$, $\Phi = \{-\varphi_{max}, \cdots, -\varphi_{res}, 0, \varphi_{res}, \cdots, \varphi_{max}\}$, with φ_{res} being the sampling resolution of the beam-forming.

The mean and standard deviation values of the excess-delay and azimuth-of-arrival for the direct path, $\bar{\varphi}_m|_{\tau_k=0}$, $\bar{\tau}_m|_{\varphi_q=0}$, $\sigma_{\tau_m}|_{\varphi_q=0}$, $\sigma_{\varphi_m}|_{\tau_k=0}$, are obtained for all combinations of

m	h_{tm}	$p_t p_r$	$\bar{\varphi}_m\,[\,^\circ]$	$\sigma_{\varphi m}\,[\,^\circ]$	$\bar{\tau}_m\,[\text{ns}]$	$\sigma_{\tau_m}\,[\text{ns}]$
1	15	VV	0.6	16.6	1.8	3.6
1	15	HV	5.6	17.6	3.7	7.9
1	15	VH	1.2	10.3	1.1	0.7
1	15	HH	1.3	9.0	1.0	0.4
2	12	VV	13.8	13.8	8.7	13.2
2	12	HV	13.6	22.3	19.9	15.7
2	12	VH	17.8	13.8	8.6	13.8
2	12	HH	1.8	12.4	1.7	3.3
3	9	VV	-0.3	10.8	1.1	0.5
3	9	HV	10.6	6.7	1.2	2.0
3	9	VH	8.7	13.1	2.3	5.9
3	9	HH	5.0	8.4	1.0	0.3
3	6	VV	7.0	9.0	1.0	0.3
4	6	HV	-0.8	5.7	1.0	0.2
4	6	VH	11.2	9.6	1.4	3.5
4	6	HH	-5.7	6.6	1.3	0.6
5	4	VV	0.4	10.4	1.0	0.3
5	4	HV	3.0	17.5	1.6	1.5
5	4	VH	5.7	22.1	1.2	1.4
5	4	HH	-0.7	5.1	1.0	0.1

Table 2. Radio Signal Spread Estimated Moments

transmitter antenna height h_{tm} and transmitter-receiver polarizations $p_t p_r$ are presented in Table 2. Considering the delay resolution of the measurement, 20 [ns], the small values for the standard deviation of excess-delay σ_τ agrees with the previously reported results. This is because the dispersion is caused due to scatterings by leaves and branches along the propagation path.

To get a sense of azimuth standard deviations it is necessary to obtain the azimuth resolution for the current analysis. Figure 4 shows the Capon spectrum for a single path measured in the anechoic chamber. While the Rayleigh resolution is slightly larger than the theoretical value 11.5° it shall not be confused with the resolution of the standard deviation. The computed standard deviation for the Capon spectrum of Fig. 4 is $\sigma_\varphi = 4.8°$. This means any deviation larger than this value is caused by the dispersion of the radio wave while propagating through foliage.

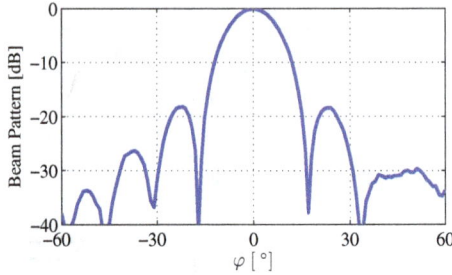

Figure 4. Capon spectrum of a single path in the azimuth.

3.2. Multipath cluster identification –high-resolution analysis

A simple manual cluster identification algorithm was applied to the estimated multipath. As the purpose is to identify dominant propagation mechanisms, only those clusters conveying significant power are discussed. Each cluster carries a total power calculated as

$$g_{m,c} = \sum_{l \in C_{m,c}} g_l, \qquad \begin{matrix} m = 1, \cdots, M \\ c = 1, \cdots, C \end{matrix} \qquad (15)$$

where $g_l = |\alpha_l|^2$ is the estimated path-gain, M indicates the number of measurements, C is the number of identified clusters and $C_{m,c}$ represents the set of estimated mutilpath in cth cluster of measurement m. The mean and standard deviation of the estimated parameters for each cluster can be obtained as

$$\bar{x}_{m,c} = \frac{\sum_{l \in C_{m,c}} g_l x_l}{g_{m,c}}, \qquad \begin{matrix} m = 1, \cdots, M \\ c = 1, \cdots, C \end{matrix} \qquad (16)$$

$$\sigma_{x_{m,c}} = \left(\frac{\sum_{l \in C_{m,c}} (x_l - \bar{x}_{m,c}) g_l}{g_{m,c}} \right)^{1/2}, \qquad \begin{matrix} m = 1, \cdots, M \\ c = 1, \cdots, C \end{matrix} \qquad (17)$$

Here x is a substitute variable for the estimated path parameters except for the path magnitude, i.e. $x \in \{\tau, \varphi_r, \vartheta_r\}$.

Table 3 displays most significant clusters identified for each measurement. In this table different number of clusters are presented for each measurement with the specifications of the measurement such as the measurement index m, transmitter antenna height h_{tm}, polarizations of transmit and received signal $p_t p_r$ and receive signal noise level σ_n^2; as well the specifications of the clusters are exhibited as the index of the cluster for each measurement c, number of multipath in the cluster $L_{m,c}$, cluster's associated power $g_{m,c}$ followed the average and standard deviation of excess-delay, azimuth- and elevation-of arrival of the cluster.

One observation is that significant clusters are of mixed polarization combinations, that is the channel acts random interactions with the radio wave in terms of polarization. Also notable is that there could be found almost no significant cluster associated with back scattering. Moreover it is observed that for the measurement number 2 with the transmit antenna height of $h_{t2} = 12$ [m] the received signal is weaker compared to other measurements. Hence the number of estimated multipath as well as identified clusters are smaller. In the next subsection these strictly identified clusters are grouped to recognize what we call propagation mechanisms.

3.3. Propagation mechanisms –high-resolution analysis

With the strict classification of multipath in the previous section we have derived clusters in the Table 3 which can not be so informative when it comes to the propagation mechanism analysis. In this section those clusters are further arranged in groups of related clusters to provide a better knowledge of the channel. Here we are specifically interested in clusters with close mean angle-of-arrival ($\bar{\varphi}_{m,c}$, $\bar{\vartheta}_{m,c}$) to identify propagation mechanisms in the channel. It is observed that clusters with different excess-delays could exist in the same group, hence those clusters belonging to the same group may not represent identical propagation mechanism in the strict sense, e.g. both multiple-scattering and single-scattering multipath could be included in the same group, but as long as we are interested in the final interaction of the propagation wave in the channel the current approach suits. As an example consider clusters $C_{1,1}$, $C_{1,4}$, $C_{1,5}$, clearly their close mean angle-of-arrival values hint their last interaction to the channel coming from the same source. By rearrangement of the clusters in such groups within each measurement, we further examined any existing link between these groups among all measurements and found the following classes of received multipath:

A. The foliage close to the receiver in the direction of transmitter scatters a great amount of radio energy. The clusters with a mean azimuth-of-arrival value in the interval $\bar{\varphi}_{m,c} \in (-6°, 10°)$ and with a slight positive elevation-of-arrival belong to this class. This class includes clusters $C_{1,1}$, $C_{1,4}$, $C_{1,5}$, $C_{2,2}$, $C_{3,1}$, $C_{3,2}$, $C_{3,4}$, $C_{3,6}$, $C_{4,1}$, $C_{4,2}$, $C_{4,4}$, $C_{4,7}$, $C_{4,8}$, $C_{4,11}$, $C_{5,1}$, $C_{5,3}$, $C_{5,5}$, $C_{5,15}$.

B. Scattering from the foliage close to the receiver and with azimuth-of-arrival values in the interval $\bar{\varphi}_{m,c} \in (17°, 30°)$ and with slight positive values of elevation-of-arrival can be associated to this class. Clusters in this class are $C_{1,2}$, $C_{1,3}$, $C_{1,6}$, $C_{1,8}$, $C_{1,9}$, $C_{2,1}$, $C_{2,3}$, $C_{3,3}$, $C_{3,5}$, $C_{3,7}$, $C_{4,3}$, $C_{4,9}$, $C_{4,10}$, $C_{5,6}$, $C_{5,9}$, $C_{5,10}$.

C. Third class of identified clusters represents the foliage located in the azimuth-of-arrival in the interval $\bar{\varphi}_{m,c} \in (-40°, -16°)$ and again with slight positive values of elevation-of-arrival. This class of clusters include $C_{1,11}$, $C_{3,9}$, $C_{4,5}$, $C_{4,6}$, $C_{5,2}$, $C_{5,7}$, $C_{5,14}$.

Few clusters remain unclassified conveying minor radio energy compared to these three classes of clusters. It is important to remember that this classification was done among different measurements particularly in terms of transmitter antenna heights. Figure 5 schematically illustrates top view of the radio channel and three major classes of identified clusters. It can be said that major propagation mechanisms in this channel are associated with one of these classes. Class A represents the forward scattering component. Even though there is no line-of-sight from the foliage to the transmitter in any of the measurements, this class introduces the most powerful propagation mechanism in the channel. The radio signal finds its way through the leaves and in between tree trunks to reach to this foliage and be rescattered toward the receiver. It is observed that strongest clusters in all measurements, except measurement 2, are put in this class.

To describe the scatterings from classes B and C it is noticed that even though the measurement area is densely vegetated, the density of foliage is not homogeneous

m	h_{tm}	$p_t p_r$	σ_n^2 [dB]	c	$L_{m,c}$	$g_{m,c}$ [dB]	$\bar\tau_{m,c}$ [ns]	$\sigma_{\tau m,c}$ [ns]	$\bar\phi_{m,c}$ [°]	$\sigma_{\phi m,c}$ [°]	$\bar\vartheta_{m,c}$ [°]	$\sigma_{\vartheta m,c}$ [°]
1	15	VV	3.6	1	12	73.8	20	13.1	-3	3.9	26	12.2
1	15	VV	3.6	2	6	49.3	20	10.4	27	6.1	9	9.9
1	15	HH	3.9	3	2	48.7	20	4.9	23	0.0	1	0.0
1	15	HV	3.8	4	1	48.1	0	0.0	-6	0.0	26	0.0
1	15	VH	3.6	5	3	45.6	20	12.7	-1	1.4	12	2.5
1	15	VV	3.6	6	1	38.3	60	0.0	28	0.0	3	0.0
1	15	HH	3.9	7	1	37.4	40	0.0	-139	0.0	1	0.0
1	15	VH	3.6	8	3	34.4	40	10.0	27	5.0	6	10.3
1	15	HV	3.8	9	2	34.3	0	9.8	24	0.5	13	7.8
2	12	VV	2.6	1	5	50.4	40	10.0	17	7.9	2	3.8
2	12	VV	2.6	2	1	32.4	0	0.0	-1	0.0	19	0.0
2	12	HV	2.7	3	2	27.5	20	4.5	19	6.3	1	0.0
3	9	HV	2.4	1	11	65.0	20	9.6	-3	6.8	6	9.0
3	9	VH	2.5	2	2	57.8	20	0.0	5	6.7	11	0.0
3	9	HV	2.4	3	4	55.9	40	11.8	23	5.6	3	0.8
3	9	VH	2.5	4	3	45.5	40	8.7	10	0.9	11	0.0
3	9	HV	2.4	5	2	43.7	0	4.6	23	2.3	7	9.6
3	9	HV	2.4	6	5	45.5	60	7.8	1	3.8	2	5.2
3	9	HV	2.4	7	2	34.6	80	5.0	21	1.0	1	0.0
3	9	HV	2.4	8	2	34.5	60	8.5	-60	5.1	3	0.0
4	6	HH	2.6	1	7	70.6	20	8.3	-2	4.6	8	6.5
4	6	VV	2.4	2	4	64.5	0	8.8	8	7.5	3	1.5
4	6	HH	2.6	3	4	60.2	40	7.5	19	5.5	2	3.1
4	6	HV	2.4	4	3	54.2	20	8.5	-6	3.4	1	0.5
4	6	VV	2.4	5	2	49.4	0	7.6	-16	5.3	10	9.8
4	6	VV	2.4	6	5	48.0	40	14.2	-40	7.7	1	5.2
4	6	HH	2.6	7	3	42.9	100	5.0	-2	1.3	11	0.0
4	6	VV	2.4	8	3	42.5	60	5.0	5	2.6	1	0.0
4	6	VH	2.4	9	2	42.5	0	4.6	19	0.5	5	0.0
4	6	HH	2.6	10	1	41.0	20	0.0	30	0.0	5	0.0
4	6	HH	2.6	11	4	40.8	60	9.1	0	5.6	11	0.0
5	4	VV	2.8	1	10	77.3	20	8.0	0	5.1	9	3.6
5	4	VV	2.8	2	4	57.6	20	4.8	-29	2.0	30	7.4
5	4	HV	2.6	3	6	57.5	20	4.6	1	8.6	29	5.9
5	4	VV	2.8	4	3	54.2	20	0.0	-101	2.7	32	1.7

Table 3. Identified Clusters

m	h_{tm}	$p_t p_r$	σ_n^2 [dB]	c	$L_{m,c}$ [dB]	$g_{m,c}$ [dB]	$\tau_{m,c}$ [ns]	$\sigma_{\tau m,c}$ [ns]	$\bar{\varphi}_{m,c}$ [°]	$\sigma_{\varphi m,c}$ [°]	$\bar{\vartheta}_{m,c}$ [°]	$\sigma_{\vartheta m,c}$ [°]
5	4	HH	2.7	5	2	53.8	40	0.0	1	3.7	13	4.5
5	4	VV	2.8	6	3	50.4	20	8.7	27	3.2	9	2.5
5	4	HH	2.7	7	3	46.5	40	9.5	-28	5.6	20	4.2
5	4	VV	2.8	8	3	43.8	20	0.0	145	7.4	5	2.0
5	4	HH	2.7	9	3	40.2	40	9.3	34	5.8	9	3.4
5	4	VV	2.8	10	2	38.8	80	9.3	20	0.9	5	1.4
5	4	VV	2.8	11	2	36.8	20	0.0	98	5.9	7	0.5
5	4	VV	2.8	12	3	36.6	20	8.7	46	3.1	33	9.8
5	4	HH	2.7	13	1	35.4	60	0.0	3	0.0	-40	0.0
5	4	HV	2.6	14	2	35.2	20	7.7	-18	0.8	28	8.4
5	4	VH	2.8	15	3	34.3	20	9.7	0	4.6	17	4.5
5	4	VV	2.8	16	2	33.9	40	0.0	-98	4.4	29	4.4

Table 3. Identified Clusters *(Continued)*.

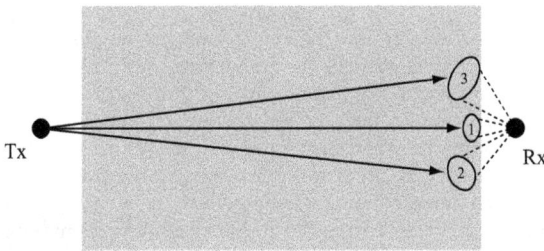

Figure 5. Schematic top view of the radio channel and three major classes of received multipath. No line-of-sight exists between transmitter and antenna and these classes.

everywhere. This produces airy spaces within the dense vegetation which act similar to canyons to conduct radio signals. Figure 6 displays such spaces in the vegetation viewed from the measurement 4 transmitter antenna position, $h_{t4} = 6$ [m]. When these guided signals arrive to one or more foliage with line-of-sight toward the receiver they make strong clusters of received propagation waves. It seems this is what happens in case of multipath classes B and C. Notable is that strong clusters from measurements with higher transmit antenna contribute to class B whereas those from measurements with lower antenna height, e.g. measurement 5, appear strongly in class C.

Again it is reminded that there is not any line-of-sight from the transmitter to the foliages associated with any of classes, however the configuration of foliage within the vegetation area causes the radio signal to be nonuniformly received in format of clusters. The significant point in the classification process is that clusters from every transmitter height, and from a variety of polarization combinations are taking part in each multipath class. The interpretation can be that even if the interactions of multipath to the vegetation is of random nature and number, when it comes to the angle-of-arrival the receiving radio signal can be from a deterministic selected number of directions.

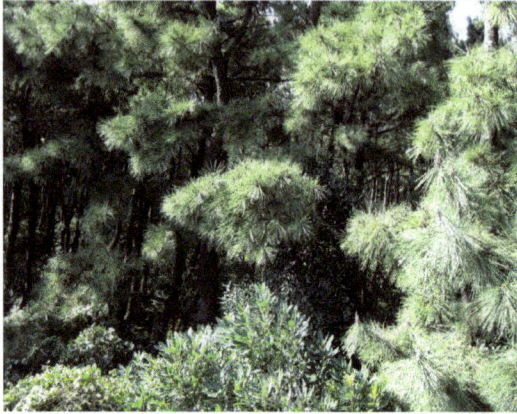

Figure 6. Airy spaces within dense vegetated area. Viewed from transmitter antenna location at $h_{t4} = 6$ [m].

4. Discussion

Typical measurements for the analysis of the interaction between radio waves and vegetation, as well to evaluate the existing models, generally target only the field attenuation along distance. Moreover, measurement schemes and facilities usually do not acquire enough resolution to support an accurate analysis of the active mechanisms and basically provide limited insight into the physical phenomena which in reality is complex. There are nearly universal assumptions, e.g. vegetation as a homogeneous media of randomly distributed scatterers, used in most corresponding studies which have never been evaluated. In this chapter the directional radio channel in dense vegetation is investigated through the analysis of appropriate measurement data. Two different analysis methods are used for the comparison of results.

First the dispersions in delay and azimuth-of-arrival of the received waves are derived by using a beam-forming. Investigating the mean delay values in the Table 2 shows that in some cases components other than the direct path have been involved in the analysis. The interpretation of some large spreads in azimuth and the distance of the mean azimuth values from zero shall be related to this fact. The main conclusion of the analysis however is to confirm the received signal spatial spread due to interaction with foliage, and to provide rough estimates of this spread. It is observed that even though the beam-forming approach is robust, easy to implement and does not require any presumption regarding the received signal, its spatial resolution proves limited. Hence the received signal energy through different propagation mechanisms can not be accurately separated. Moreover the array antenna response is included in the channel and can not be de-embedded.

For the propagation mechanisms to be distinguished, high resolution parameter estimation of the measured data is necessary. The well known SAGE algorithm is used to identify received propagation multipath, which are then grouped in clusters using a simple clustering algorithm. Specifically looking at the measured receiving radio signal with high resolution in angular domain, three major classes of received multipath associated with separate final scattering foliage are identified. Thus contrary to the widely assumed homogeneous random scattering media it was observed that the radio waves in the vegetated channel are received from distinct directions in clusters of multipath. The identified classes of multipath correspond to two significant propagation mechanisms other than forward scattering by the foliage obstructing the line-of-sight –normally used as the main mechanism in different approaches. It seems airy spaces in the vegetated area can have crucial influence in directing the radio signal toward specific directions, to be redirected to the receiver by a foliage with the line-of-sight toward the receiver.

The significant point is that by beam-forming analysis of the measurement data it is not possible to identify those propagation mechanisms, although limited insight regarding the spatial spread of the direct received path achieved. Thus appropriate measurements assisted with high-resolution data analysis can meaningfully change our understanding of the vegetation wireless channel and substantially modify corresponding models.

It is now well known that in most practical wireless channels the diffuse scattering plays a considerable role in terms of carrying the radio energy [52],[53]. In comparison to the specular component, the estimation and modeling of which is rather straightforward, the estimation and modeling of the diffuse component can be rather complicated. Nevertheless, due to the random shape and direction of trees and branches in any foliage, it is expected that a major part of energy is transferred by means of diffuse component. Further high-resolution investigation and modeling of the radio channel including the diffuse component is left as a future task.

5. Conclusion

In this chapter the radio channel in dense vegetation is investigated through the directional analysis of carefully gathered measurement data. By looking at the measured receiving radio signal with high resolution in angular domain, three major classes of received multipath associated with separate final scattering foliage are identified. The identified classes of multipath correspond to two significant propagation mechanisms other than forward scattering by the foliage obstructing the line-of-sight which is normally considered as the main mechanism in such channels. The airy spaces in the vegetated area is probably accounted for directing the radio signal toward specific directions. These are redirected toward the receiver by a foliage in its line-of-sight. Thus contrary to the widely assumed homogeneous random scattering media it was observed that the radio waves in the vegetated channel are received from distinct directions in clusters of multipath. Results of the high-resolution analysis are in contrast to the beam-forming analysis of the measured data which is also presented in the chapter for comparison. This proves that state-of-the-art measurements assisted with high-resolution analysis can modify our understanding and modelings of the phenomena. Moreover to fully analyze and understand these channels, the diffuse component also has to be estimated and modeled which is left as a future task.

Author details

Mir Ghoraishi
Tokyo Institute of Technology, Japan,
University of Surrey, United Kingdom

Jun-ichi Takada
Tokyo Institute of Technology, Japan

Tetsuro Imai
NTT DOCOMO Inc., Kanagawa, Japan

6. References

[1] M. Weissberger, "An initial critical summary of models for predicting the attenuation of radio waves by trees," ESD-TR-81-101, EMC Analysis Center, Annapolis MD, USA, 1982.

[2] COST 235, Radiowave propagation effects on next generation fixed services terrestrial telecommunications systems, Final Report, ISBN 92-827-8023-6, Commission of the European Union, 1996.

[3] F. Schwering, E. Violette, R. Espeland, "Millimeter-wave propagation in vegetation: Experiments and theory," *IEEE Trans. Geoscience and Remote Sensing*, Vol. 26, pp. 355Ü367, May 1984.

[4] R. Tewari, S. Swarup, M. Roy, "Radio wave propagation through rain forests of india," *IEEE Trans. Antennas and Propagation*, Vol. 38, No. 4, pp. 433Ü449, Apr. 1990.

[5] M. Al-Nuaimi, A. Hammoudeh, "Attenuation functions of microwave signals propagated through trees," *Electronics Letters*, Vol. 29, No. 14, pp. 1307-1308, July 1993.

[6] A. Seville, "Vegetation attenuation: modelling and measurements at millimetric frequencies," *Int. Conf. Antennas and Propagation*, Vol. 2, pp.5-8, April 1997.

[7] M. Al-Nuaimi, A. Hammoudeh, "Measurements and predictions of attenuation and scatter of microwave signals by trees," *IEE Proc. Microwaves, Antennas and Propagation*, Vol. 141, No. 2, pp. 70-76, April 1994.

[8] R. Stephens, M. Al-Nuaimi, R. Calderinha, "Characterisation of depolarization of radio signals by single trees at 20 GHz," *Fifteenth National Radio Science Conference (URSI)*, B12, 1-7, Feb. 1998.

[9] T. Tamir, "On radio-wave propagation in forest environments," *IEEE Trans. Antennas and Propagation*, Vol. 15, No. 6, pp. 806-817, Nov. 1967.

[10] L. Li, J. Koh, T. Yeo, M. Leong, P. Kooi, "Analysis of radiowave propagation in a four-layered anisotropic forest environment," *IEEE Trans. Geoscience and Remote Sensing*, Vol. 37, No. 4, pp. 1967-1979, July 1999.

[11] G. Cavalcante, D. Rogers, A. Giardola, "Radio loss in forests using a model with four layered media," *Radio Science*, Vol. 18, 1983.

[12] L. Li, T. Yeo, P. Kooi, M. Leong, "Radio wave propagation along mixed paths through a four-layered model of rain forest: an analytic approach," *IEEE Trans. Antennas and Propagation*, Vol. 46, No. 7, pp. 1098-1111, July 1998.

[13] Y. de Jong, M. Herben, "A tree-scattering model for improved propagation prediction in urban microcells," *IEEE Trans. Vehicular Technology*, vol. 53, pp. 503.513, Mar. 2004.

[14] A. H. Lagrone, "Propagation of VHF and UHF electromagnetic waves over a grove of trees in full leaf," *IEEE Trans. Antenna and Propagation*, Vol. 25, pp. 866-869, Nov. 1977.

[15] R. Matschek, B. Linot, H. Sizun, "Model for wave propagation in presence of vegetation based on the UTD associating transmitted and lateral waves," *Proc. IEE National Conf. Antennas and Propagation*, pp. 120-123, April 1999.

[16] F. Schubert, B. Fleury, P. Robertson, R. Prieto-Cerdeira, A. Steingass, A. Lehner, "Modeling of Multipath Propagation Caused by Trees and Its Effect on GNSS Receiver Performance," *European Conf. Antennas and Propagation (EUCAP 2010)*, 2010.

[17] A. Ishimaru, *Wave Propagation and Scattering in Random Media*, Academic Press, 1978.

[18] L. Tsang, J. Kong, K. Ding, C. Ao, *Scattering of Electromagnetic Waves*, Wiley, 2001.

[19] S. Saatchi, K. McDonald, "Coherent effects in microwave backscattering models for forest canopies," *IEEE Trans. Geoscience and Remote Sensing*, Vol. 35, No. 4, July 1997.

[20] R. Johnson, F. Schwering, "A Transport Theory of Millimeter Wave Propagation in Woods and Forests," Research and Development Technical Report, CECOM-TR-85-1, Feb. 1985

[21] D. Didascalou, M. Younis, W. Wiesbeck, "Millimeter-wave scattering and penetration in isolated vegetation structures," *IEEE Trans. Geoscience and Remote Sensing*, Vol. 38, pp. 2106Ű2113, Sep. 2000.

[22] T. Fernandes, R. Caldeirinha, M. Al-Nuaimi, J. Richter, "A discrete RET model for millimeter-wave propagation in isolate tree formations," *IEICE Trans. Communications*, Vol. E88-B, No. 6, pp. 2411Ű2418, Jun. 2005.

[23] L. Foldy, "The multiple scattering of waves," *Physical Review*, Vol. 67, No. 3, pp. 107-119, 1945.

[24] L. Li, J. Koh, T. Yeo, M. Leong, P. Kooi, "Analysis of radiowave propagation in a four-layered anisotropic forest environment," *IEEE Trans. Geoscience and Remote Sensing*, Vol. 37, No. 4, pp. 1967-1979, July 1999.

[25] Y. Lin, K. Sarabandi, "A Monte Carlo coherent scattering model for forest canopies using fractal-generated trees," *IEEE Trans. Geoscience and Remote Sensing*, Vol. 37, pp. 440Ű451, 1999.

[26] Y. Lin, K. Sarabandi, "Retrieval of forest parameters using a fractal-based coherent scattering model and a genetic algorithm," *IEEE Trans. Geoscience and Remote Sensing*, Vol. 37, pp. 1415Ű1424, May 1999.

[27] I. Koh, K. Sarabandi, "Polarimetric channel characterization of foliage for performance assessment of GPS receivers under tree canopies," *IEEE Trans. Antennas and Propagation*, Vol. 50, No. 5, pp. 713Ű72, 2002.

[28] I. Koh, F. Wang, K. Sarabandi, "Estimation of coherent field attenuation through dense foliage including multiple-scattering, *IEE Trans. Geoscience and Remote Sensing*, Vol. 41, pp. 1132Ű1135, 2003.

[29] F. Wang, K. Sarabandi, "An enhanced millimeter-wave foliage propagation model," *IEEE Trans. Antennas and Propagation*, Vol. 53, No. 7, pp. 2138Ű2145, Jul. 2005.

[30] F. Wang, K. Sarabandi, "A Physics-Based Statistical Model for Wave Propagation Through Foliage," *IEEE Trans. Antennas and Propagation*, Vol. 55, No. 3, Part 2, pp. 958-968, March 2007.

[31] S. Torrico, H. Bertoni, R. Lang, "Modeling tree effects on path loss in a residential environment," *IEEE Trans. Antennas and Propagation*, Vol. 46, No. 6, pp. 872-880, June 1998.

[32] S. Torrico, R. Lang, "A Simplified Analytical Model to Predict the Specific Attenuation of a Tree Canopy," *IEEE Trans. Vehicular Technology*, Vol. 56, No. 2, pp. 696-703, March 2007.

[33] N. Blaunstein, I. Kovacs, Y. Ben-Shimol, J. Andersen, D. Katz, P. Eggers, R. Giladi, K. Olesen, "Prediction of UHF path loss for forest environments," *Radio Science*, Vol. 38, No. 3, 2003.

[34] N. Blaunstein, D. Censor, D. Katz, "Radio propagation in rural residential areas with vegetation," *Progress in Electromagnetic Research*, PIER 40, pp. 131-153, 2003.
[35] N. Rogers, A. Seville, J. Richter, D. Ndzi, N. Savage, R. Caldeirinha, A. Shukla, M. Al-Nuaimi, K. Craig, E. Vilar, J. Austin, A generic model of 1-62 Ghz radio attenuation in vegetation - Final report, Available On-line.
[36] ITU-R Recommendation P.527-3, 1978-1982-1990-1992-2000.
[37] ITU-R Recommendation P.833-4, Oct. 11, 2004.
[38] R. Lewenz, "Path loss variation due to vegetation movement," *Proc. IEE National Conf. Antennas and Propagation*, pp. 97-100, April 1999.
[39] M. Hashim, S. Stavrou, "Measurements and modeling of wind influence on radio wave propagation through vegetation," *IEEE Trans. Wireless Communications*, vol. 5, pp. 1055-1064, May 2006.
[40] Y. Lee, Y. Meng, "Tropical weather effects on foliage propagation," *European Conf. Antennas and Propagation (EuCAP2007)*, Nov. 2007.
[41] D. Ndzi, N. Savage, K. Stuart, "Wideband Signal propagation through Vegetation," XVII GA of URSI, Oct. 2005.
[42] Y. Meng, Y. Lee, B. Ng, "Study of the diversity reception in a forested environment," IEEE Trans. Wireless Communications, Vol. 8, No. 5, pp. 2302-2305, May 2009.
[43] M. Gans, N. Amitay, Y. Yeh, T. Damen, R. Valenzuela, C. Cheon, J. Lee, "Propagation measurements for fixed wireless loop (FWL) in a suburban region with foliage and terrain blockages," *IEEE Trans. Wireless Communications*, Vol. 1, pp. 302-310, Apr. 2002.
[44] G. Joshi, C. Dietrich, C. Anderson, W. Newhall, W. Davis, J. Isaacs, G. Barnett, "Near-ground channel measurements over line of sight and forested paths," *IEE Proc. Antennas and Propagation*, Vol. 152, pp. 589-596, Dec. 2005.
[45] C. Oestges, B. Villacieros, D. Vanhoenacker-Janvier, "Radio Channel Characterization for Moderate Antenna Heights in Forest Areas ," *IEEE Trans. Vehicular Technology*, Vol. 58, No. 8, pp. 4031-4035, Oct. 2009.
[46] M. Ghoraishi, J. Takada, C. Phakasoum, T. Imai, K. Kitao, "Azimuth and Delay Dispersion of Mobile Radio Wave Propagation through Vegetation," *European Conf. Antennas and Propagation (EuCAP 2010)*, April 2010.
[47] Mir Ghoraishi, Jun-ichi Takada, Tetsuro Imai, "Analysis of Radio Wave Dispersion Through Vegetation," *European Conf. Antennas and Propagation (EuCAP 2012)*, March 2012.
[48] X. Yin, *High-Resolution Parameter Estimation for MIMO Channel Sounding*, PhD Thesis, Aalborg University, July 2006.
[49] H. Van Trees, *Optimum Array Processing*, Wiley-Interscience, 2002.
[50] M. Tapio, "On the use of beamforming for estimation of spatially distributed signals," *IEEE Conf. Acoustics, Speech, Signal Processing (ICASSP)*, Vol. V, pp. 369-372, 2003.
[51] B. Fleury, M. Tschudin, R. Heddergott, D. Dahlhaus, K. Pedersen, "Channel Parameter Estimation in Mobile Radio Environments Using the SAGE Algorithm," *IEEE J. Selected Areas in Communications*, Vol. 17, No. 3, pp. 434-450, March 1999.
[52] A. Richter, R. Thoma, "Joint Maximum Likelihood Estimation of Specular Paths and Distributed Diffuse Scattering," *IEEE Vehicular Technology Conf. (VTC05-Spring)*, Vol. 1, pp. 11-15, June 2005.
[53] F. Mani, C. Oestges, "Evaluation of Diffuse Scattering Contribution for Delay Spread and Crosspolarization Ratio Prediction in an Indoor Scenario," *European Conf. Antennas and Propagation (EuCAP 2010)*, April 2009.

Chapter 6

Electromagnetic Waves and Their Application to Charged Particle Acceleration

Hitendra K. Malik

Additional information is available at the end of the chapter

1. Introduction

A wave is a disturbance that propagates through space and time, usually with the transference of energy from one point to another without permanent displacement of particles of the medium. The particles under this situation only oscillate about their equilibrium positions. If the particles oscillate in the direction of wave propagation, then the wave is called longitudinal wave. However, if these oscillations take place in perpendicular direction with the direction of wave propagation, the wave is said to be transverse in nature. Electromagnetic (EM) waves are transverse in nature. In electromagnetic waves such as light waves, it is the changes in electric field and magnetic field that represent the wave disturbance. The propagation of the wave is described by the passage of a waveform through the medium with a certain velocity called the phase (or wave) velocity. However, the energy is transferred at the group velocity of the waves making the waveform. Electromagnetic radiation is a form of energy exhibiting wave like behavior as it travels through the space. The electromagnetic radiation is classified based on the frequency of its wave. Figure 1 shows the electromagnetic spectrum that consists of radio waves, microwaves, infrared (IR) radiation, visible light, ultraviolet (UV) radiation, X-rays and gamma rays. T-rays shown in the spectrum represent the terahertz (THz) radiations. This region of frequency (10^{11}Hz to ~10^{13} Hz) had remained the last unexplored region between long wavelength and visible electromagnetic radiation for a long time due to the lack of efficient emitters and receptors. Interestingly this region of the THz rays demarcates the regions of most fascinating subjects of *electronics* and *photonics*.

2. Propagation of electromagnetic waves

In order to study the propagation of wave, we first let the sinusoidal variation of oscillating quantities as $\sim e^{i(\vec{k}\cdot\vec{r}-\omega t)}$ that are associated with the wave. Here $\vec{k}$ is the wave vector that tells about the direction of wave propagation and also gives the wavelength of oscillations as $\lambda = 2\pi / k$ and ω is the angular frequency of the oscillations that gives rise to the time period $T = 2\pi / \omega$. The relation between ω and k is called the dispersion relation, based on which the wave propagation is investigated in a medium. The propagation of electromagnetic waves in any medium can be understood based on the fundamental equations of electromagnetic wave theory, i.e. the Maxwell's equations, which were established by James Clerk Maxwell in 1873 and experimentally verified by Heinrich Hertz in 1888. These Maxwell's equations are

$$\vec{\nabla} \cdot \vec{D} = \rho \tag{1}$$

$$\vec{\nabla} \cdot \vec{B} = 0 \tag{2}$$

$$\vec{\nabla} \times \vec{E} = -\frac{\partial \vec{B}}{\partial t} \tag{3}$$

$$\vec{\nabla} \times \vec{H} = \vec{J} + \frac{\partial \vec{D}}{\partial t} \tag{4}$$

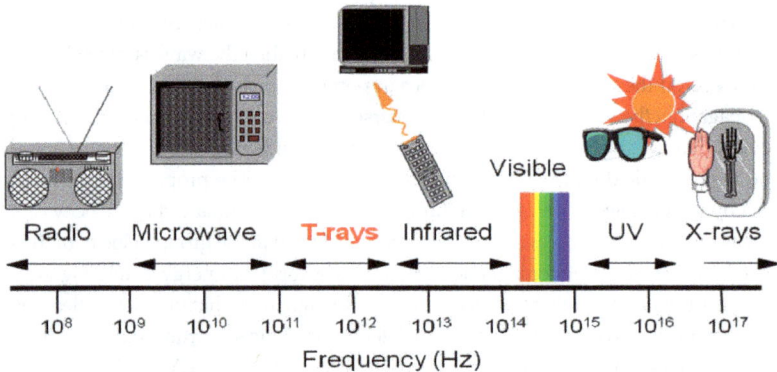

Figure 1. Electromagnetic spectrum.

Here $\vec{D}$, $\vec{B}$, $\vec{E}$, $\vec{H}$, $\vec{J}$ and ρ are respectively the electric displacement (C/m²), magnetic flux density (Wb/m²), electric field strength (V/m), magnetic field strength (A/m), electric current density (A/m²) and electric charge density (C/m³), which are the real functions of position and time. For an isotropic medium $\vec{D} = \varepsilon \vec{E}$, i.e. the vector $\vec{E}$ is parallel to $\vec{D}$, and $\vec{B} = \mu \vec{H}$, i.e. the vector $\vec{H}$ is parallel to $\vec{B}$. Here ε is the electric permittivity of the medium that tells about the polarization of the medium and is determined by the electrical properties of the medium. μ is the permeability of the medium that tells about the magnetization of the medium and is determined by the magnetic properties of the medium. Hence, the properties of the medium associated with the Maxwell's equations affect the electromagnetic wave propagation. Below we discuss the propagation of electromagnetic waves in different media.

2.1. EM wave propagation in free space / vacuum

For free space or vacuum $\varepsilon = \varepsilon_0$, $\mu = \mu_0$, $\vec{J} = 0$ and $\rho = 0$. If we put these values in Eqs.(1) – (4) and take the curl of Eq.(3), we get

$$\vec{\nabla}(\vec{\nabla} \cdot \vec{E}) - \nabla^2 \vec{E} = -\mu_0 \frac{\partial}{\partial t}(\vec{\nabla} \times \vec{H}) \tag{5}$$

The use of Eqs.(4) and (1) in Eq.(5) yields the following de-coupled equation in $\vec{E}$

$$\nabla^2 \vec{E} - \mu_0 \varepsilon_0 \frac{\partial^2 \vec{E}}{\partial t^2} = 0 \tag{6}$$

Similarly the wave equation for the field $\vec{H}$ is obtained as

$$\nabla^2 \vec{H} - \mu_0 \varepsilon_0 \frac{\partial^2 \vec{H}}{\partial t^2} = 0 \tag{7}$$

Equations (6) and (7) describe respectively the propagation of the fields $\vec{E}$ and $\vec{H}$ in free space. For the plane uniform wave, following are the solutions of these second order homogeneous differential equations

$$\vec{E}(\vec{r},t) = \vec{E}_0 e^{i(\vec{k} \cdot \vec{r} - \omega t)} \text{ and } \vec{H}(\vec{r},t) = \vec{H}_0 e^{i(\vec{k} \cdot \vec{r} - \omega t)} \tag{8}$$

The above solutions should satisfy the respective wave equations. For example, when we put the solution for $\vec{E}$ in Eq.(6) and replace $\vec{\nabla}$ by $i\vec{k}$ and $\frac{\partial}{\partial t}$ by $-i\omega$, we get for harmonic wave with single frequency

$$(k^2 - \omega^2 \mu_0 \varepsilon_0)\vec{E} = 0 \tag{9}$$

Since $\vec{E}$ cannot be zero for the existence of wave, the wave equation will be satisfied only if

$$k^2 - \omega^2 \mu_0 \varepsilon_0 = 0 \tag{10}$$

This is the dispersion relation of the electromagnetic wave in free space or vacuum. The ratio of ω and k gives rise to the phase velocity (say v) of the wave, i.e.

$$v = \frac{\omega}{k} = \frac{1}{\sqrt{\mu_0 \varepsilon_0}} = 3 \times 10^8 \text{ m/sec} = c, \text{the speed of light.}$$

Hence, it is clear that the electromagnetic wave propagates with the speed of light in free space.

In addition, we can examine the nature of the electromagnetic wave based on the directions of the fields $\vec{E}$, $\vec{H}$ and the wave vector $\vec{k}$. The use of solution (8) in Eq.(3) yields

$$\vec{k} \times \vec{E} = \omega \mu_0 \vec{H} \tag{11}$$

Similarly we obtain from Eq.(4)

$$\vec{k} \times \vec{H} = \omega \varepsilon_0 \vec{E} \tag{12}$$

Equation (11) says that the field vector $\vec{H}$ is perpendicular to both $\vec{k}$ and $\vec{E}$ vectors. Also the vector $\vec{E}$ is perpendicular to both $\vec{k}$ and $\vec{H}$ vectors [see Eq.(12)]. When we combine both the equations (11) and (12), it is inferred that the vectors $\vec{E}$, $\vec{H}$ and $\vec{k}$ form a set of orthogonal vectors such that the cross product of $\vec{E}$ and $\vec{H}$ is always in the direction of $\vec{k}$. For this reason, the energy associated with the electromagnetic waves is carried in the direction of wave propagation. On the other hand, Eq.(1) reveals that $\vec{k} \cdot \vec{E} = 0$ whereas Eq.(2) yields $\vec{k} \cdot \vec{H} = 0$. It means the oscillations of the electric field $\vec{E}$ are perpendicular to the direction of wave propagation; the same is the case with the magnetic field. Hence, it is evident that the electromagnetic waves are transverse in nature.

2.2. EM wave propagation in a dielectric

In an isotropic dielectric medium, the current density $\vec{J}$ and volume charge density ρ are zero. Also the vectors $\vec{D}$ and $\vec{B}$ are defined as $\vec{D} = \varepsilon_0 \vec{E} + \vec{P} \equiv \varepsilon \vec{E}$ and $\vec{B} = \mu_0 \vec{H} + \mu_0 \vec{M} \equiv \mu \vec{H}$ for the isotropic linear dielectric medium, which is polarizable and magnetic. Here the vectors $\vec{P}$ and $\vec{M}$ give respectively the measure of polarization and magnetization of the medium. Nonetheless, for the dielectric medium it would be sufficient to remember that ε_0 and μ_0 of free space have been replaced with ε and μ. Hence, for the dielectric medium the Maxwell's equations (1) – (4) take the form

$$\vec{\nabla} \cdot \vec{E} = 0 \tag{13}$$

$$\vec{\nabla} \cdot \vec{H} = 0 \tag{14}$$

$$\vec{\nabla} \times \vec{E} = -\mu \frac{\partial \vec{H}}{\partial t} \tag{15}$$

$$\vec{\nabla} \times \vec{H} = -\varepsilon \frac{\partial \vec{E}}{\partial t} \tag{16}$$

Following the similar procedure as done in the case of free space, the wave equations for the fields $\vec{E}$ and $\vec{H}$ are obtained as

$$\nabla^2 \vec{E} = \mu\varepsilon \frac{\partial^2 \vec{E}}{\partial t^2}, \quad \nabla^2 \vec{H} = \mu\varepsilon \frac{\partial^2 \vec{H}}{\partial t^2} \tag{17}$$

A comparison of these wave equations with Eqs.(6) and (7) reveals that the phase velocity v of the wave in a linear dielectric medium is

$$v = \frac{1}{\sqrt{\mu\varepsilon}} = \frac{1}{\sqrt{\mu_0 \mu_r \varepsilon_0 \varepsilon_r}} = \frac{c}{\sqrt{\mu_r \varepsilon_r}} \tag{18}$$

From the above equation, it is clear that the propagation velocity of an electromagnetic wave in a dielectric medium is less than that in free space. Also the refractive index, say n, can be evaluated as $n = \frac{c}{v} = \sqrt{\mu_r \varepsilon_r}$. Since for a non-magnetic dielectric medium $\mu_r \approx 1$, the refractive index is simply given by square root of the relative permittivity, i.e. $n \approx \sqrt{\varepsilon_r}$. This is also true for most materials as for them $\mu \approx \mu_0$ and hence $\mu_r \approx 1$.

2.3. EM wave propagation in a conductor

We consider a conducting linear and isotropic medium whose permeability is μ, permittivity is ε and the conductivity is σ. In the cases of vacuum and dielectrics or insulators, the conductivity is zero and hence the current density $\vec{J}$ was neglected in the Maxwell's equations. Moreover, the free charge density ρ was taken to be zero in these cases. In the case of conductors, the flow of charge however is not independently controlled and the current density in general cannot be neglected. Since any free charge supplied to a conductor gets dissipated, we can rather take $\rho = 0$. This can be seen based on the continuity equation $\frac{\partial \rho}{\partial t} + \vec{\nabla} \cdot \vec{J} = 0$. The use of Ohm's law $\vec{J} = \sigma\vec{E}$ and Gauss law of electricity $\vec{\nabla} \cdot \vec{D} = \rho$ in this equation leads $\frac{1}{\rho}\frac{\partial \rho}{\partial t} = -\frac{\sigma}{\varepsilon}$, the integration of which gives $\rho(t) = \rho(0)e^{\frac{-\sigma}{\varepsilon}t}$ together with $\rho(0)$ as the initial free charge density. This relation shows that if we put some free charge on a conductor, it will flow out to the edges in a characteristic time $\tau_f \equiv \frac{\varepsilon}{\sigma}$. For a perfect conductor this characteristic time $\tau_f = 0$ as $\sigma = \infty$, and for a good conductor τ_f will be much less than the other relevant times, for example $\frac{1}{\omega}$ in an oscillatory system, i.e. $\tau_f << \frac{1}{\omega}$. Under this situation, we can write the Maxwell's equations as

$$\vec{\nabla} \cdot \vec{E} = 0 \tag{19}$$

$$\vec{\nabla} \cdot \vec{H} = 0 \tag{20}$$

$$\vec{\nabla} \times \vec{E} = -\frac{\partial \vec{B}}{\partial t} = -\mu \frac{\partial \vec{H}}{\partial t} \tag{21}$$

$$\vec{\nabla} \times \vec{H} = \sigma \vec{E} + \varepsilon \frac{\partial \vec{E}}{\partial t} \tag{22}$$

Taking the curl of Eq.(21) and then making use of Eqs. (19) and (22), we get the following electromagnetic wave equation for the field $\vec{E}$ in a conductor

$$\nabla^2 \vec{E} = \mu\sigma \frac{\partial \vec{E}}{\partial t} + \mu\varepsilon \frac{\partial^2 \vec{E}}{\partial t^2} \tag{23}$$

Similarly the wave equation for the field $\vec{H}$ is obtained as

$$\nabla^2 \vec{H} = \mu\sigma \frac{\partial \vec{H}}{\partial t} + \mu\varepsilon \frac{\partial^2 \vec{H}}{\partial t^2} \tag{24}$$

In one-dimension (along z-axis) the wave equations are written as

$$\frac{\partial^2 \vec{E}}{\partial z^2} = \mu\sigma \frac{\partial \vec{E}}{\partial t} + \mu\varepsilon \frac{\partial^2 \vec{E}}{\partial t^2} \tag{25}$$

$$\frac{\partial^2 \vec{H}}{\partial z^2} = \mu\sigma \frac{\partial \vec{H}}{\partial t} + \mu\varepsilon \frac{\partial^2 \vec{H}}{\partial t^2} \tag{26}$$

If we compare Eq.(23) with Eq.(6), we notice that an additional term $\mu\sigma \frac{\partial \vec{E}}{\partial t}$ appears in the wave equation for the $\vec{E}$ field; the same is the case with Eq.(24) and an additional term $\mu\sigma \frac{\partial \vec{H}}{\partial t}$ appears. Hence, these wave equations are called modified wave equations for the electromagnetic field in a conductor. Owing to the inclusion of conductivity σ, both the additional terms are called the dissipative terms as these allow the current to flow through the medium.

We can assume the following plane wave solution (in one-dimension) to the wave equations (25) and (26)

$$\vec{E}(z,t) = \vec{E}_0 e^{i(kz - \omega t)} \text{ and } \vec{H}(z,t) = \vec{H}_0 e^{i(kz - \omega t)} \tag{27}$$

Putting the above solution of $\vec{E}$ in Eq.(25) or of $\vec{H}$ in Eq.(26) we get

$$k^2 = \mu\varepsilon\omega^2 + i\mu\sigma\omega \tag{28}$$

This relation shows that the wave vector is a complex quantity, say $k \equiv k_r + ik_i$. With this the fields $\vec{E}$ and $\vec{H}$ become

$$\vec{E}(z,t) = \vec{E}_0 e^{-k_i z} e^{i(k_r z - \omega t)} \text{ and } \vec{H}(z,t) = \vec{H}_0 e^{-k_i z} e^{i(k_r z - \omega t)} \qquad (29)$$

It is evident from the above expressions that when the electromagnetic wave propagates through a conductor, its amplitude decreases and hence the attenuation of the wave takes place. The distance through which the amplitude is reduced by a factor of $1/e$ is called skin depth (say δ). The skind depth is decided by the imaginary part of the wave vector, i.e. k_i, as it can be seen that

$$\delta \equiv \frac{1}{|k_i|} \qquad (30)$$

The real part of the wave vector determines the wavelength, phase velocity, and the refractive index in the usual manner, i.e. $\lambda = \frac{2\pi}{k_r}$, $v = \frac{\omega}{k_r}$ and $n = \frac{c}{v} = \frac{ck_r}{\omega}$. Putting $k = k_r + ik_i$ in Eq.(28) we obtain

$$k_r = \omega\sqrt{\frac{\mu\varepsilon}{2}}\left\{\sqrt{1+\left(\frac{\sigma}{\omega\varepsilon}\right)^2}+1\right\} \text{ and } k_i = \omega\sqrt{\frac{\mu\varepsilon}{2}}\left\{\sqrt{1+\left(\frac{\sigma}{\omega\varepsilon}\right)^2}-1\right\} \qquad (31)$$

It is evident that the propagation of the wave and the skin depth depend on the properties of the conductor and the frequency of the wave. Based on the expression of k_i, this can be seen that the skin depth for the electromagnetic waves having high frequencies is smaller. Since the skin depth is a measure of how far the wave penetrates into the conductor, the high frequency waves are found to penetrate less into the conductor. For example, in the case of copper, the skin depth of approximately 6 cm is obtained at the frequency of 1 Hz and it decreases to about 2 mm if the frequency is increased to 1 KHz. The skin depth causes the effective resistance of the conductor to increase at higher frequencies where the skin depth is smaller, thus reducing the effective cross-section of the conductor. If we talk in general about the skin depth, it is the tendency of an alternating electric current to distribute itself within a conductor with the largest current density near the surface of the conductor and decreased density at greater depths. Under this situation, the electric current flows mainly at the skin of the conductor. Hence, the word *skin* comes into picture.

During the wave propagation in conductors, unlike the cases of vacuum and dielectrics, the electric field and magnetic field vectors do not remain in phase. This can be seen as follows. Taking the direction of electric field $\vec{E}$ along the x-axis, we write it as $\vec{E}(z,t) = \hat{x}E_0 e^{-k_i z} e^{i(k_r z - \omega t)}$. Using this in Eq.(21), we get $\vec{H}(z,t) = \hat{y}\frac{kE_0}{\omega\mu} e^{-k_i z} e^{i(k_r z - \omega t)}$. Clearly the amplitude of field $\vec{H}$ contains k, which is a complex quantity and can be expressed in terms of its magnitude (say k') and phase (say θ_k) as $k = k' e^{i\theta_k}$. Here

$$k' = \sqrt{k_r^2 + k_i^2} = \omega \sqrt{\mu\varepsilon \left\{ 1 + \left(\frac{\sigma}{\varepsilon\omega} \right)^2 \right\}} \quad \text{and} \quad \theta_k = \tan^{-1}\left(\frac{k_i}{k_r} \right).$$ With this the field can be written as

$$\vec{H}(z,t) = \hat{y}\frac{k'E_0}{\omega\mu}e^{-k_i z}e^{i(k_r z - \omega t + \theta_k)}.$$ A   comparison   of   this   expression   with

$\vec{E}(z,t) = \hat{x}E_0 e^{-k_i z}e^{i(k_r z - \omega t)}$ clearly infers that $\theta_H - \theta_E = \theta_k$, where θ_H is the phase of the magnetic field and θ_E is the phase of the electric field. Hence, the magnetic field lags behind the electric field during the electromagnetic wave propagation in a conductor.

3. EM waves and plasma interaction

Our aim is to disucss the electromagnetic waves and plasma interaction in view of the particle acceleration. Hence, now we introduce the plasma as a new medium, which is sometimes referred to as the fourth state of the matter.

3.1. Plasma: Fourth state of matter

Everybody is well aware of three states of the matter, i.e. the solid, liquid and gas. In solids, the atoms are packed very close to each other and are fixed at definite positions. These are connected with each other by the interatomic forces. The atoms of solids start oscillating about their equilibrium positions when we supply energy to them, and as a result the interatomic forces become weaker and the atoms are separated significantly. This way the solid takes the form of liquid, where the atoms or molecules override. The liquid has a specific volume but does not have precise shape. So it changes shape according to the shape of the container in which it is kept. If we further supply energy to the atoms, the interatomic forces become insignificant, the atoms get separated and start moving freely. Under this situation, the liquid takes the form of gas. In gas, the atoms are not connected with each other and hence can move in any direction. The gas neither has precise shape nor the fixed volume. It takes the shape and volume of the container in which it is kept. If more energy is supplied to the atoms (or molecules) of a gas, the electrons from the outermost level of the atoms get easily detached and hence the atoms become ionized. As a result, we are left with the collection of ions, electrons and some neutrals (unionized atoms). This collection of charged and neutral particles is referred to as plasma. This is sometimes called the fourth state of matter, as it is found in natural conditions. For example, the gases near the sun are always in ionized state that qualifies for plasma. The species of the plasma being charged are connected with each other by the electromagnetic forces. This can be understood as follows. Since the charges separated with each other set up the electric field, the plasma species produce the electric field. However, the separation of charges of plasma is not fixed (as the species do not remain stationary). So this electric field is time varying field, which will generate magnetic field according to the Maxwell's fourth equation. On the other hand, the motion of charges generates current and hence the magnetic field. In view of this, the plasma species produce time varying magnetic field that will induce electric field according to the Maxwell's third equation. Thus, it can be said that the plasma species are connected with each other by the electromagnetic fields. In view of almost equal number of ions and electrons in the plasma, the plasma as a whole is neutral. However, the plasma is quasineutral, as we cannot neglect the internal forces at the same time. Moreover, if we attempt to disturb a part of the plasma, the whole body of the plasma gets perturbed due to the connection of all the species with each other. This property is known as collective behaviour of the plasma. Therefore, an ionized gas can qualify for plasma state, if it is quasineutral and it shows collective behaviour.

Another interesting property of the plasma is its ability to shield out the field that is applied on it. This can be better understood, for example, when we insert the electrodes of a battery into the plasma. Then the positive (negative) electrode attracts the electrons (ions) whose number is decided by the charge carried by the electrode. So an electron cloud is developed around the electrode that shields / cancels the external field. The thickness of this electron cloud is known as Debye length (say, λ_{De}). Since the electrons are light species compared with ions, the shielding is generally accomplished by the electrons only. It is clear that the field exists within the cloud or the Debye sphere (sphere with the radius λ_{De}). Now imagine if the Debye length is much less than the dimension (L) of the plasma. Then the bulk of the plasma will remain neutral. Therefore, $\lambda_{De} \ll L$ is the required condition for the quasineutrality. In aadtion, if the number of electrons in the Debye sphere (say, N_{De}) is much larger than unity, i.e. $N_{De} \gg 1$, then the condition of collective behaviour will be fulfilled. Any distance in the plasma system is measured in terms of Debye length λ_{De} and the time is measured in terms of reciprocal of plasma frequency (say f_{pe}). The plasma frequency is nothing but the natural frequency of the plasma, the same as all the materials have their natural frequencies. Actually this is the frequency of oscillations made by the electrons about their equilibrium positions. The plasma frequency f_{pe} and the Debye length λ_{De} in SI system of units are given by

$$f_{pe} = \frac{1}{2\pi}\left(\frac{n_0 e^2}{\varepsilon_0 m_e}\right)^{1/2} \text{ and } \lambda_{De} = \left(\frac{\varepsilon_0 k T_e}{n_0 e^2}\right)^{1/2}$$

In these expressions, n_0 is the plasma density, which is the common density of ions (n_i) and electrons (n_e), i.e. $n_0 = n_i = n_e$, T_e is the electron temperature, k is the Boltzmann constant ($k = 1.38 \times 10^{-23}$ J/K), e is the electronic charge and m_e is the electron mass. In plasmas, generally we do not talk about the temperature of the ions and electrons, but we specifically focus on their energies. It means the temperature is written in terms of the energy. For example, 1eV energy of the electron would be equal to its thermal energy kT_e (for two-dimensional system, and in general in plasma physics). So

$$1eV = kT_e$$

or $$1.6 \times 10^{-19} (J) = 1.38 \times 10^{-23} T_e (J/K)$$

or $$T_e = 11,600 \text{ K}.$$

It means 1eV energy is equivalent to 11,600 K temperature. The electron temperature in laboratory plasmas generally varies from 1 eV to 5 eV. For a plasma with number density of $10^{18}/m^3$ and temperature 2 eV, the Debye length comes out to be of the order of μm and the plasma frequency of the order of GHz (10^9 Hz).

Hence, it is clear that only the electrons would be able to respond to the high frequency field of the electromagnetic waves, for example, microwaves or lasers. As mentioned, our aim is to develop an understanding for the electromagnetic waves and plasma interaction for their possible applications to the particle acceleration. Below we discuss about this topic in greater detail and summarize the research conducted in this direction. At first we talk about the phenomena that may be realized during the interaction of electromagnetic waves and plasmas.

3.2. Some basic phenomena

According to linear theory, only the electromagnetic wave of frequency ω higher than plasma frequency ω_{pe} can propagate through the plasma. The wave whose frequency ω is below ω_{pe} gets reflected and the one with $\omega = \omega_{pe}$ gets absorbed resonantly in collisionless plasma. The plasma itself can support several types of electrostatic and electromagnetic waves such as electron plasma wave, ion acoustic wave, electron electromagnetic waves, etc. The interaction of electromagnetic wave with plasma can take place through the exciation of such waves and in this process the exchange of energy can be possible between the electromagnetic wave and plasma species. If the amplitude of wave is much higher than its nonlinear interaction with other collective modes in plasma, plasma instabilities are dominant. On the other hand, the wave can also decay by Landau damping and if plasma is underdense ($\omega > \omega_{pe}$) then the wave can decay in electrostatic wave and some other electromagnetic waves, resulting in parametric instabilities including Raman scattering, Brillouin scattering, etc. In case of large amplitude wave, the effect of ponderomotive force also comes into picture. This is very important phenomenon in view of harmonic generation, beat wave excitation, wakefield excitation for particle acceleration, self-focusing of laser beam, filamentation of laser beam, etc.

In the theory for resonance absorption, wave propagation in the resonance layer is described either by electron-ion collisions and thermal dispersion or by nonlinear effects like wave breaking, etc. [1 – 3]. The anomalous absorption of electromagnetic waves on a surface of an inhomogeneous unmagnetized plasma was theoretically predicted by Gildenburg [4]. Later this phenomenon was confirmed experimentally, out of which some experiments have shown that large amount of power can be absorbed by magnetized plasma at the electron and ion cyclotron frequencies. A usual way of coupling transverse waves into a plasma for the purpose of such resonant absorption has been to use a magnetic beach as suggested by Stix [5]. Breizman et al. [6] have presented a self-consistent theory of the rf-wave propagation and ion motion through the resonance. An important ingredient of the problem is the ion flow along the magnetic field. The flow velocity limits the time the ions spend at the resonance, which in turn limits the ion energy gain. A feature that makes the problem nonlinear is that the flow accelerates under the effect of $\vec{\nabla}B$ force and rf-pressure. This acceleration can produce a steep reduction in the plasma density at the resonance, resulting in partial reflection of the incident wave. The propagation and collisionless absorption of electromagnetic waves propagating in nonuniformly magnetized plasmas with regions of cyclotron resonance were computed by Kuckes [7]. He considered the particle dynamics associated with motion in a nonuniform magnetic field near cyclotron resonance explicitly and predicted the complete wave absorption above a critical plasma density.

The nonlinear behaviour of the large-amplitude plasma wave and the effect of an inhomogeneous plasma on its growth and saturation in a collisionless plasma due to the beating of two laser beams with frequencies much above the plasma frequency ω_{pe} has been considered taking into account the modulation of the Lorentz force by the large-amplitude plasma wave as well as the temporal variation of its phase [8]. In this case, a novel parametric instability as a result of the modulation of the Lorentz force by the large-amplitude plasma wave is found when the beat frequency is twice the plasma frequency. The high phase velocity electron plasma wave excited by collinear optical mixing has been detected directly [9], where the frequency, wave number, spatial extent, saturation time, and peak amplitude were all measured experimentally and found to be in reasonable agreement with the theoretical expectations. The resonant excitation of an electron plasma wave and its effects on the density profile steepening have been theoretically studied by using a modified, warm-capacitor model [10], where the scaling laws characterizing the process were established and the wave structure and density profile were self-consistently determined.

Chang et al. [11] have observed experimentally the parametric excitation of ion acoustic waves and cyclotron harmonic waves by a high frequency electric field with frequencies near the harmonics of the cyclotron frequency. They have verified both the wave vector and the frequency selection rules. Parametric excitation of longitudinal oscillations of plasma was studied by Kitsenko et al. [12] in a weak alternating electric field with frequency ω_0 close to that of electron-ion hydrodynamic longitudinal oscillations of cold plasma, $\omega_1 = \omega_{LH}\sqrt{\left[1+\left(m_i/m_e\right)\cos^2\theta\right]}$, where ω_{LH} is the lower hybrid resonance frequency. In their study, the angle θ between the direction of propagation of the oscillations and the magnetic field was close to $\pi/2$ and it was shown that oscillations can be excited in the plasma with frequencies much less than ω_0, if the drift velocity of the particles in the steady external magnetic field and the alternating electric field is less than the thermal velocity of the ions.

Optical investigations have been reported of the interaction of 0.3 TW, 250 fs Ti: sapphire laser pulses with underdense plasmas created from high density gas jet targets [13]. Time resolved shadowgraphy using a 2ω probe pulse, images of the transmitted radiation and images of 1ω and 2ω side radiations were presented for various gases. Their experimental results and analysis based on a simple numerical Gaussian beam model showed that ionization-induced refraction dominates the interaction process for all gases except hydrogen. The numerical modeling has also shown that for a given laser power there exists only a narrow density range in which self-focusing can be expected to occur. On the other hand, it has been observed that the nonlinear frequency shift of a strong electromagnetic wave in a plasma due to weak relativistic effects and the $\vec{v} \times \vec{B}$ force can cause modulation and self-focusing instabilities [14]. Kaw et al. [15] have shown that an electromagnetic wave interacting with a plasma is subject to instabilities that leads to light filamentation. Numerical studies of beam filamentation in laser produced plasma have been presented by Nickolas et al. [16] based on a parabolic wave equation, known as the Schroedinger equation, coupled with thermal transport equations for both the ions and electrons in two-dimensions. Also the results of a numerical code have been described which models the relativistic self-focusing of high intensity laser beams in plasmas by the nonlinear relativistic dependence of the optical constants on laser intensity [17]. Here rapid relativistic self-focussing down to a beam diameter of one micron in a distance of the order of the original beam diameter was observed. They also observed the production of GeV ions moving against the laser light.

3.3. Particle acceleration

Particle accelerators are among the largest machines built by humans. In the conventional linear accelerators (LINACs), the acceleration gradients are however limited to some tens of MeV/m. Since the energy gain of particles is the product of such acceleration gradient and the acceleration distance, we need to extend only the acceleration distances in order to reach high energies. That's why these tools for high energy physics are becoming larger and larger, and increasingly more expensive. For the first time, it was realized by Tajima and Dawson [18] that a laser beam propagating in a plasma can excite electron plasma wave, which being longitudinal can be used to accelerate electrons. To understand the underline principle for plasma based acceleration, consider the limits of conventional particle accelerators based on rf-waves propagating in corrugated metallic cavities. They are limited first by the availability of high peak power drivers and ultimately by electrical breakdown of the metal structure. These factors correspond to linear accelerating gradient of 20 – 100 MeV/m. Plasmas though are not limited by breakdown as they are already ionized and indeed can support electric fields of the order 10 – 100 GeV/m. Consequently, with regard to the energy gain of particles in accelerators, a plasma accelerator can cut down significantly the acceleration distance to boost particles from rest to several MeV over a short distance (less than the millimeter range) and still provide high quality electron beam. Thus, plasma based particle accelerators opened a new and exciting field of extreme gradient (beyond 1TV/m). There has been a tremendous progress in recent years, due to the advances in technology, particularly by the development of compact terawatt laser systems based on the technique known as chirped-pulse amplification (CPA).

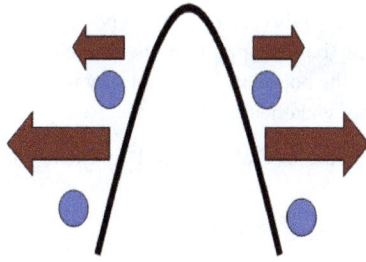

Figure 2. Laser intensity profile: Ponderomotive force.

3.3.1. Excitation of Langmuir waves: Wakefield generation

Although electric fields of the order of 1 TeV/m are readily achievable these days at the focus of a laser beam, these fields in vacuum cannot be used directly for the purpose of particle acceleration. This is because they are transverse and oscillatory in nature. However, if laser light can be used to excite Langmuir waves in plasma, these waves being longitudinal can be used to accelerate charged particles.

The motion of the electron in the presence of electric field is governed by the Lorentz force. In the case of high, nonuniform electromagnetic (or purely electric) field, the expression for Lorentz force has a second order term, which is proportional to the laser intensity gradient. This second order force term is known as the ponderomotive force, given by

$$\vec{F}_{pm} = \left\{ e^2 \left(1 + \alpha^2\right) / 4 m_e \omega^2 \right\} \vec{\nabla} E^2 \tag{32}$$

Here ω is the frequency of laser having the electric field E and α is the ellipticity of the laser light, which is equal to zero for the linearly polarized light and is unity for the circularly polarized light. The above expression is for ponderomotive force on a single electron. However, for the plasma the ponderomotive force on the electrons is defined for unit volume as per the following relation

$$\vec{F}_{pm} = -\left(\omega_{pe} / \omega\right)^2 \left\{ \left(1 + \alpha^2\right) / 2c \right\} \vec{\nabla} I \tag{33}$$

Thus, any spatial variation of the laser intensity I will act to push the electrons / ions from the region of higher intensity to the region of lower intensity through the ponderomotive force (Fig.2). This displacement of electrons creates large amplitude plasma wave, which is called the wake. The field corresponding to this wake, i.e. the wakefield, can reach up to 100 GeV/m provided there is a resonance between the plasma frequency ω_{pe} and the ponderomotive force. The concept of wakefield acceleration can be understood based on the following example. When a speed-boat travels in water, it produces two types of waves viz. bow waves and wakefield waves. The bow waves are conical waves having tip at the front end of the boat. These are produced because the velocity of the boat exceeds that of the water waves. The wakefield waves are waves set up at the back (or wake) of the boat, which travels with the velocity equal to the velocity of the boat. According to the principle of the Landau damping, a floating ball dropped in the wakefield wave of the boat will get accelerated to the velocity of the boat if its initial velocity is slightly less than that of the boat. This is exactly the principle of wakefield acceleration.

Below we discuss a few methods that are used to excite plasma wave and hence the wakefield.

3.3.1.1. Laser beat wave accelerator (LBWA)

In the LBWA method, the plasma wave is excited by beating two optical waves of slightly different frequencies. Two laser waves of frequencies ω_1 and ω_2, having polarization in the same direction, traveling in preformed plasma of uniform density n_0 (corresponding plasma frequency ω_{pe}) will beat at a frequency $\Delta\omega = \omega_1 - \omega_2$. If this frequency difference is exactly equal to the plasma frequency (i.e. $\Delta\omega = \omega_{pe}$), then strong Langmuir wave will be excited in the plasma by the longitudinal ponderomotive force of the beat wave. Since the beat wave moves with the laser pulse, the plasma wave will also move with a phase velocity equal to the group velocity (near light velocity) of the laser pulses. Then a properly placed bunch of electrons with a velocity slightly lesser than the laser group velocity will get accelerated by wave-to-particle energy transfer. However, in this process there is a problem of detuning of resonance condition, which is attributed to the modified plasma frequency $\left(\omega_{pe} \propto \sqrt{1/m_e}\right)$ due to the change in electron mass because of their reltivistic speeds in very large amplitude of the wakefield.

3.3.1.2. Laser wakefield accelerator (LWFA)

In beat wave acceleration scheme, it is necessary to have plasma of uniform density along with strict requirement on plasma density to exactly match with the beat wave frequency and clamping of field due to relativistic effects. Hence, laser wakefield acceleration shceme was proposed in which all the above problems are absent. For LWFA, one uses a short pulse of very high intensity. When such a high intensity laser pulse is incident on a gas, it ionizes the gas. The laser light propagates in this plasma with a velocity equal to the group velocity $\left(v_g\right)$ in plasma, which is nearly equal to the velocity of the light. The short laser pulse duration τ has a strong intensity variation in time and correspondingly in space. This leads to a strong longitudinal pondermotive force. The wavelength of this pondermotive force, and that of the density perturbation caused by it, is of the order of $2c\tau$. If this is made equal to the plasma wavelength (defined as $\lambda_p = 2\pi c / \omega_{pe}$), then high amplitude wakefields are produced due to resonance (Fig.3). Similar to the case of the boat, the laser wakefield moves with the pulse at a velocity equal to the group velocity of the laser pulse. Under this situation, a correctly injected bunch of electrons can be accelerated by the longitudinal field of the plasma waves (Fig.3), where an electron bunch is injected in the plasma wave midway between every two alternate plasma wave peaks. If the plasma wave itself moves with a phase velocity v_p and the electron beam moves with a velocity v_b, then the beam will be forced by the plasma wave to travel with a velocity equal to that of the plasma wave. This is because, if $v_b < v_p$, the electron bunch at point C (Fig.4) will start trailing from the midpoint and will experience a positive force due to electron bunch at point A. This will accelerate it in the +z direction till it attains a velocity equal to v_p. If $v_b > v_p$, then it will start drifting towards point B and the bunch at B will repel it backward till it slows down to a velocity equal to v_p. If the electron beam has a velocity much different from that of the plasma wave, it will cross the repulsive barriers at point A or B and its velocity will keep oscillating about its mean velocity. In other words, such a beam of electrons will not have a net exchange of energy with the plasma wave. Hence, a beam of electrons traveling with a velocity slightly less than that of the plasma wave will get accelerated. Moreover, if the phase velocity of the plasma wave is relativistic, then the slight gain in velocity corresponds to a large gain in the energy.

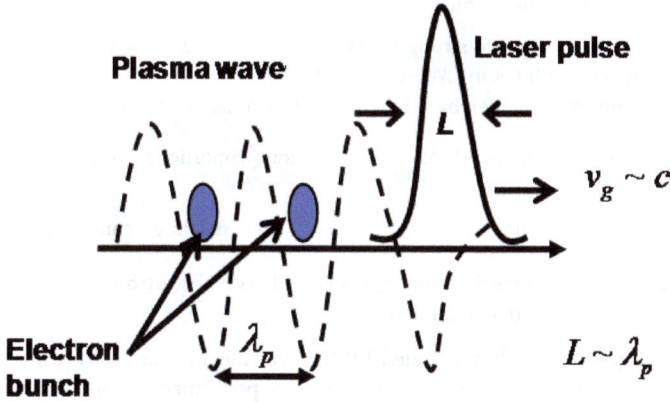

Figure 3. Schematic of LWFA.

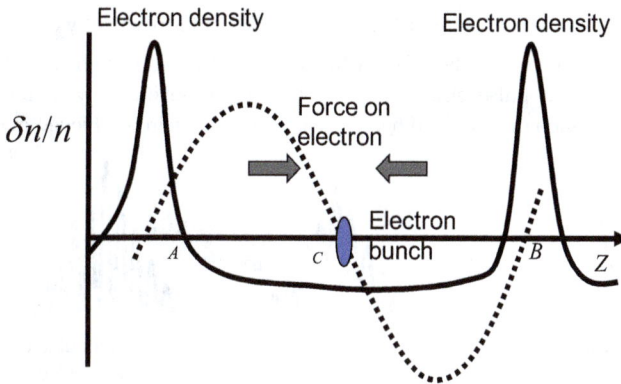

Figure 4. Force on an electron bunch trapped in an electron plasma wave.

3.3.1.3. Self modulated LWFA (SM-LWFA)

In this scheme, the electron plasma wave is excited resonantly by the modulation of the laser pulse envelope. This occurs for a laser pulse having length (L) few times longer than the plasma wavelength (λ_p) and pulse power larger than the power (critical power P_c) required to self-focus the laser beam. Owing to the finite pulse shape, a small plasma wave is excited non-resonantly, which results in growth of forward Raman scattering (FRS) instability. The FRS wave and the laser wave beat at the plasma frequency, which gives rise to an enhancement of the electron plasma wave. Thus, there exists an oscillating density perturbation within the pulse envelope. The laser pulse therefore sees a refractive index that is alternately peaked and dented at interval of $\lambda_p / 2$. As the phase velocity of the laser wave depends on the density, the modulation in density gives rise to redistribution of the photon flux within the laser pulse, which leads to modulations in the envelope with a period of λ_p.

This modulation gives rise to strong ponderomotive force with wavelength exactly equal to the plasma wavelength (as in LWFA). This strongly enhances the plasma wave amplitude. This effect grows in time, thereby transforming the initial laser pulse envelope into a train of shorter laser pulses with width of λ_p or duration proportional to $1/\omega_{pe}$. Since $\lambda_p \sim \sqrt{\dfrac{1}{n_0}}$ and $P_c \sim \dfrac{1}{n_0}$, the conditions $L > \lambda_p$ and $P > P_c$ for fixed laser parameters can usually be satisfied by operating at sufficiently high plasma density. Figure 5 shows the self-modulated scheme of laser wakefield acceleration.

The advantages of the self-modulated LWFA over the standard LWFA are the simplicity and enhanced acceleration. Simplicity is that a preformed density channel and pulse tailoring are not required for the matching condition of $L \sim \lambda_p$. Enhanced acceleration is achieved for the following reasons. First, the SM-LWFA operates at a higher density, which leads to a larger wakefield ($E \sim \sqrt{n_0}$). Second, the wakefield is resonantly excited by a series of pulses as opposed to a single pulse in the standard LWFA, relativistic optical guiding allows the modulated pulse structure to propagate for several Rayleigh lengths. This extends the acceleration distance and hence the large energy gain is achieved in this scheme.

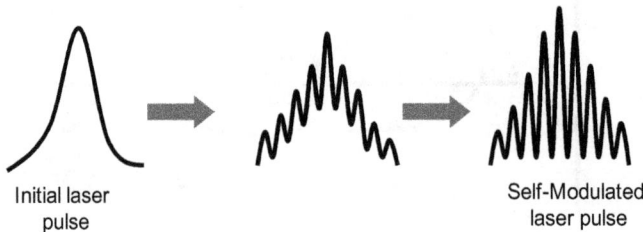

Initial laser
pulse

Self-Modulated
laser pulse

Figure 5. The self-modulated laser wakefield acceleration scheme.

So far we have seen that when a short laser pulse propagates through underdense plasma, a large amplitude plasma wave is excited in the wake of the laser pulse by the ponderomotive force associated with the temporal profile of the pulse. For tightly focused pulses ($k_p \times w_0 \leq 1$, where k_p and w_0 are the plasma wave vector and the beam size at the waist, respectively), both longitudinal and radial components of the ponderomotive force generate a density perturbation, whereas in loosely focusing geometry ($k_p \times w_0 \gg 1$), only a longitudinal electron plasma wave is generated. The amplitude of the wave is maximum when $\omega_{pe} \times \tau \sim 1$, where τ is the pulse duration and ω_{pe} is the plasma frequency.

3.3.1.4. Plasma wakefield accelerator (PWFA)

In a plasma wakefield accelerator (PWFA), the electron plasma wave is driven by one or more electron beams. Effectively the wakefield can be excited by a relativistic electron beam. This can be achieved if the electron beam terminates in a time shorter than the plasma period $1/\omega_{pe}$. In such a scheme, the ratio of energy gain to the drive beam energy (called transformation ratio) is limited to ≤ 2 for a symmetric driving beam in the linear regime. However, it can be increased by using an asymmetric drive beam.

3.3.2. Studies on particle acceleration

The researchers all over the world have made various attempts to accelerate the charged particles using wakefield and other mechanisms. Below we summarize the work done using lasers, microwaves and electron bunches.

3.3.2.1. Acceleration by wakefield

The investigations on the excitation of wakefield began with the pioneering work of Chen et al. [19], and the first experimental evidence was reported by Rosenzweig and coworkers [20, 21] followed by Nakajima et al. [22]. The wakefield generation has been widely studied experimentally, analytically and using simulations [23 – 29]. Nishida et al. [30] have successfully excited wakefield in the ion wave regime with long pulse duration by employing a variety of driving bunch shapes. Later, Aossey et al. [31] observed such type of wakefield in three-component plasma also. On the other hand, efforts have been made related to wakefield excitation by relativistic electron bunch [29], [32], and coupling of longitudinal and transverse motion of accelerated electrons in laser wakefield [25]. Lotov [24] has analytically studied the laser wakefield acceleration in narrow plasma filled channels. Analytical investigations on wakefield acceleration using a dielectric lined waveguide structure showed the acceleration gradient for electrons or positrons in the range of 50 – 100 MV/m for a few nC driving bunches [33]. In another wakefield accelerator, a peak acceleration gradient of 155 MeV/m was predicted for a 2 nC rectangular drive bunch [34]. Jing et al. [35] have found transverse wakefield of about 0.13 MeV/mnC (0.2 MeV/mnC) due to X-dipole modes (Y-dipole modes) in an X-band structure generated by an electron bunch in dipole-mode wakefield in a waveguide accelerating structure. Short microwave pulses have also been used in some experiments to excite a nonlinear large amplitude ion wave at resonance absorption region [36]. This has also been suggested that the wakefield of an ultra short laser pulse can be amplified by a second laser pulse copropagating behind with duration of a few plasma wavelengths or longer [37]. Malik [38] has analytically investigated the wakefield in waveguide generated by the different types of microwave pulses with moderate intensities.

For the purpose of efficient acceleration, it is necessary to excite the wakefield of a large amplitude along with its speed nearly equal to the speed of light. The wakefield is reported to be enhanced by the nonlinearities in response of plasma to ponderomotive force of a long smooth laser pulse of relativistic intensity whose pulse length is much larger than the half of the plasma wavelength [39]. The amplitude of the laser wakefield has also been found to increase by the ionization processes of the gases at comparatively higher laser peak intensities [40]. A capillary tube can be used as a waveguide in order to enhance the interaction length [41]. Tapered plasma channels have been proposed for the enhancement of interaction length to achieve greater acceleration [42]. However, in such interactions, when the plasma wave acquires sufficiently large amplitude it becomes susceptible to instability, which is also an important issue in nonlinear plasma physics [43 – 53] in addition to other types of waves, structures and instabilities [54 – 63] including the laser produced plasmas [64] that may support different types of growing waves under the effect of high magnetic field [65].

3.3.2.2. Acceleration using lasers

McKinstrie and Startsev [66] have proposed that a laser field can accelerate the pre-accelerated electron significantly. However, they neglected the effect of longitudinal field of the laser pulse. On the basis of 3-D particle-in-cell simulations for the ion acceleration from a foil irradiated by a laser pulse, Pukhov [67] has shown that at the front side the laser ponderomotive force pushes electrons inward and creates the electric field by charge separation, which drags the ions. Yu et al. [68] considered the electron acceleration from the interaction of an intense short pulse laser with low density plasma and the optimum condition for the acceleration in the wake was obtained. They showed that the electron acceleration within the pulse dominates as the pulse becomes sufficiently short. By using 2-D particle-in-cell simulation, Suk [69] has studied the electron acceleration based on self-trapping by plasma wake. Sentoku et al. [70] examined experimentally the interaction of short laser pulse with dense plasma target for the proton acceleration and found that the peak proton energy increases in inverse proportion to the target thickness. Singh and Tripathi [71] have studied the laser induced electron acceleration in a tapered magnetic wiggler where the IFEL resonance condition was maintained for longer duration. With regard to the importance of polarization effects, Kado et al. [72] have observed strongly collimated proton beam from Tantalum targets when irradiated with circularly polarized laser pulses. With the help of radially polarized ultra relativistic laser pulses, Karmakar and Pukhov [73] have shown that collimated attosecond GeV electron bunches can be produced by ionization of high-Z material. They also compared the results with the case of Gaussian laser pulses and found that the radially polarized laser pulses are superior both in the maximum energy gain and in the quality of the produced electron beams. Xu et al. [74]

made a comparison between circularly polarized (CP) and linearly polarized (LP) fields with regard to the laser driven electron acceleration in vacuum and found that the CP field can give rise to greater acceleration efficiency.

3.3.2.3. Acceleration using microwaves

The researchers have made efforts to use microwave field for the particle acceleration [35, 36, 75 – 85]. In microwave plasma interaction experiments, electron acceleration has been realized via the $\vec{v}_p \times \vec{B}$ process [75, 76] and that of resonance absorption during wave particle interaction [77]. In the $\vec{v}_p \times \vec{B}$ process, where $\vec{v}_p$ is the phase velocity of the wave, an electrostatic wave (e.g. electron plasma wave) propagates in a direction perpendicular to a magnetic field $\vec{B}$. Here an electron that is trapped in the wave trough gets accelerated in the $\vec{v}_p \times \vec{B}$ direction. In these experiments, the electrons could be accelerated up to 400 eV. In another experiment, a nonlinear large amplitude ion wave was excited by using short microwave pulses at the resonance absorption region [36], where a strong electron wave was found to be excited after shut-off of the incident microwave pulse and high energy electrons got emitted and accelerated by the electron wave wakefield. Hirshfield et al. [80] have proposed a cyclotron autoresonance accelerator using rf gyroresonant acceleration, where the resonance for a TE₁₁ mode was maintained along a waveguide by the applied magnetic field and group velocity axial tapers, and the maximum energy achieved by the electron beam in this process was up to 2.82 MeV. Yoder et al. [86] have measured the energy gain from a microwave inverse free electron laser accelerator including the energy change as a function of relative injection phase of the electron bunches. In this accelerator, the effective accelerating gradient was achieved as 0.43 MV/m and the gain for a 6 MeV electron bunch was observed about 360 keV. Carlsten [81] has done modal analysis and gain calculation for a sheet electron beam in a ridged waveguide slow wave structure. Kumar and Malik [87] have discussed the importance of obliquely applied magnetic field to an electron acceleration and obtained that the larger acceleration is possible when the condition $\omega_{pe} > \omega_c$ (ω_{pe} is the electron plasma frequency and ω_c is the electron cyclotron frequency) is achieved in the plasma filled waveguide. Also it was proposed to use the field of superposed mode in waveguide for the effective electron acceleration [88].

4. Case study: Wakefield by lasers and microwaves

Here we take an example of wakefield excitation by short pulse lasers in an infinite plasma [38] and by the microwave pulses in a rectangular waveguide [89, 90].

4.1. Wakefield by different types of laser pulses

A laser pulse with frequency ω (= $2\pi f$), intensity I_0 (corresponding field E_0) and pulse duration τ (= f_p^{-1} = $2\pi/\omega_{pe}$) is considered to propagate in a homogeneous plasma of density n_0

and excite wakefield E_x (corresponding potential ϕ) behind it. The ions (density n_i) in the plasma are taken to be immobile on the time scale of the interest and the plasma response to the electromagnetic field is given by the following cold and collisionless electron fluid equations

$$\partial n_e / \partial t + \vec{\nabla} \cdot \left(n_e \vec{v} \right) = 0 \tag{34}$$

$$d\vec{p}/dt = -e\left(\vec{E} + \vec{v} \times \vec{B} \right) \tag{35}$$

$$\vec{\nabla} \times \vec{E} = -\partial \vec{B}/\partial t \tag{36}$$

$$\vec{\nabla} \times \vec{B} = \mu_0 \vec{j} + \left(1/c^2 \right) \partial \vec{E}/\partial t \text{ with } \vec{j} = -n_e e \vec{v} \tag{37}$$

$$\vec{\nabla} \cdot \vec{B} = 0 \tag{38}$$

$$\varepsilon_0 \vec{\nabla} \cdot \vec{E} = e\left(n_i - n_e \right) \tag{39}$$

With the help of above equations one can easily obtain the following dispersion relation for the laser propagation in the plasma $\omega^2 = c^2 k^2 + \omega_{pe}^2$, from which the group velocity of the laser is found as $v_g = c\sqrt{\left(1 - \omega_{pe}^2 / \omega^2 \right)}$. Clearly the group velocity depends on the plasma density and it can be adjusted as per the requirement.

We consider one-dimensional weakly relativistic case for the nonevolving system, i.e. when all the quantities depend only on $\xi = x - v_g t$, and take the electron density $n_e = n_0 + n_e'$ together with n_e' as the density perturbation due to the laser pulse and n_0 as the unperturbed density in a homogeneous plasma where $\partial n_0 / \partial \xi = 0$. Then the fluid equations are integrated under the condition that the oscillating quantities vanish as $|\xi| \to \infty$ and also when the perturbations are not so great $\left(n_e'/n_0 \ll 1 \right)$. This yields

$$\frac{\partial^2 \phi}{\partial \xi^2} + \left(\frac{\omega_{pe}}{v_g} \right)^2 \phi - \left(\frac{c^2 e}{2 m_e v_g^4} \right) E^2 = 0 \tag{40}$$

This is the general equation for the wake potential ϕ which can directly use different envelopes of E, i.e. different shapes of the laser pulses. Here we concentrate on three types of the shapes, namely Gaussian-like (GL) pulse, rectangular-triangular (RT) pulse and rectangular-Gaussian (RG) pulse, as shown in Fig.6.

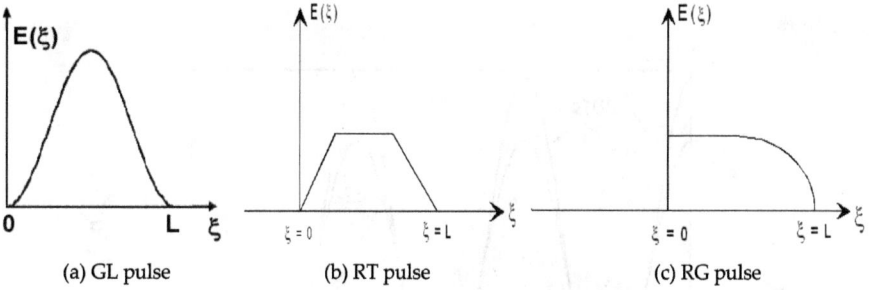

(a) GL pulse   (b) RT pulse   (c) RG pulse

Figure 6. Different shapes of the laser pulses with pulse length L (duration τ).

The wakefield E_{GL} for the case of GL pulse is obtained as $E_{GL}(\xi) = a_1 \cos(k_p L/2) \cos k_p (\xi - L/2)$, where $a_1 = \left(ec^2 E_0^2 \pi L / m_e v_g^4 \right)\left[1/\left(L^2 \alpha^2 - \pi^2 \right)\right]$ and $k_p = \omega_{pe}/v_g = 2\pi/\lambda_p$ together with λ_p as the plasma wavelength, which is described by the group velocity of the laser pulse in the plasma. The density perturbations behind the pulse are obtained as $(n_e'/n_0)_{GL} = \left(ea_1/2m_e v_g^2 \alpha \right)\left\{ 2\cos(k_p L/2)\sin k_p (\xi - L/2) + \left(L^2 \alpha^2 - \pi^2 \right)/\alpha \pi L \right\}$.

The wakefield E_{RT} for the case of RT pulse is obtained as $E_{RT}(\xi) = a_2 a_3 \cos k_p (\xi - L/2)$, where $a_2 = \left(ec^2 E_0^2 / 2m_e v_g^4 \alpha \right)$ and

$$a_3 = \left\{ \begin{array}{l} \left[\left(c_1 - 1 - \left(\dfrac{2\pi c_1}{L\alpha} \right) \sin\left(k_p L/4 \right) + \dfrac{(1-c_2)}{2}\cos\left(k_p L \right) + \dfrac{(1-c_2)}{2}\cos\left(2k_p L/3 \right) \right) \sin\left(k_p L/2 \right) \right] \\ -\left[\dfrac{2\pi c_1}{L\alpha}\cos\left(k_p L/4 \right) + \dfrac{(1-c_2)}{2}\sin\left(k_p L \right) + \dfrac{(1-c_2)}{2}\sin\left(2k_p L/3 \right) \right]\cos\left(k_p L/2 \right) \end{array} \right\}$$

together with $c_1 = L^2 \alpha^2 /\left(L^2 \alpha^2 - 4\pi^2 \right)$ and $c_2 = L^2 \alpha^2 /\left(L^2 \alpha^2 - 9\pi^2 \right)$. The density perturbations behind the pulse are obtained as $(n_e'/n_0)_{RT} = \left(ea_2/m_e v_g^2 \alpha \right)\left[1 + a_3 \sin k_p (\xi - L/2) \right]$.

For the case of RG pulse the wakefield E_{RG} and density perturbations $(n_e'/n_0)_{RG}$ are calculated as $E_{RG}(\xi) = -a_4 a_5 \cos k_p (\xi - L/2)$,

$(n_e'/n_0)_{RG} = -\left(ea_4 a_5 / m_e v_g^2 \alpha \right)\left\{ a_5 \sin k_p (\xi - L/2) - \left(L^2 \alpha^2 - \pi^2 \right) \right\}$. The constants a_4 and a_5 in these expressions are given by

$$a_4 = ec^2 E_0^2 /\left\{ 2m_e v_g^4 \alpha \left(L^2 \alpha^2 - \pi^2 \right) \right\}$$

$$a_5 = \left[\begin{array}{l} \left\{ 2\left(L^2 \alpha^2 - \pi^2 \right)\sin^2\left(k_p L/4 \right) + L^2 \alpha^2 \cos\left(k_p L/2 \right) - \pi\alpha L\sin\left(k_p L \right) \right\}\sin\left(k_p L/2 \right) \\ -\left\{ \pi^2 \sin\left(k_p L/2 \right) + \pi\alpha L\cos\left(k_p L \right) \right\}\cos\left(k_p L/2 \right) \end{array} \right]$$

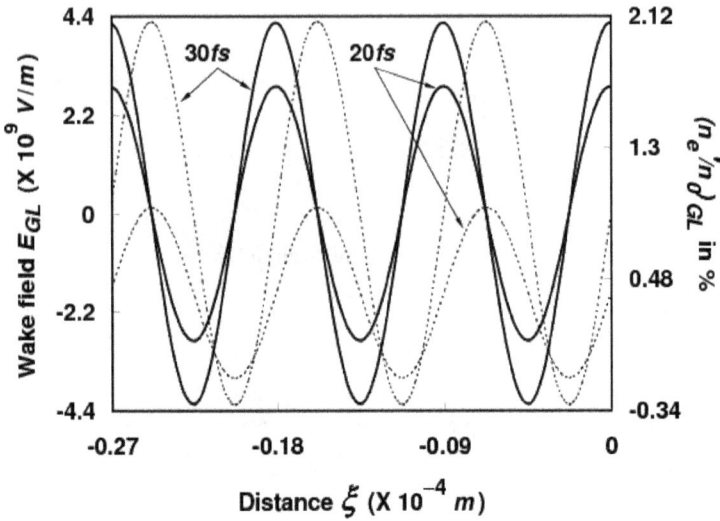

Figure 7. Wakefield E_{GL} and density perturbations $\left(n'_e/n_0\right)_{GL}$ behind the Gaussian-like pulse up to distance 3L for different pulse durations (τ = 20 ps, 30 ps) when the laser intensity is 3×10^{18} W/m^2 and laser frequency is 1.6 PHz.

In Fig. 7 we show the variation of wakefield E_{GL} by solid line graphs and of density perturbations $\left(n'_e/n_0\right)_{GL}$ by dotted line graphs behind the laser pulse ($\xi < 0$) up to the distance ξ = 3L when f =1.6 PHz and I_0 = 3×10^{18} W/m^2. The wakefield of the strength of 4.24×10^9 V/m is excited by the GL pulse and the density perturbations n_e' = 0.0208 times of the unperturbed density n_0 (= $1.37\times10^{25}/m^3$) for the pulse duration of 30 fs. A comparison of the graphs marked with *30 fs* and *20 fs* reveals that the effect of pulse duration is to increase the wakefield as well as the density perturbations. This may be attributed to the decreased plasma density n_0 for the larger pulse duration as we considered $\tau = f_{pe}^{-1} = 2\pi/\omega_{pe}$ as a condition for the wakefield excitation. For the fixed laser intensity, larger perturbations are realized in relatively lower density plasma and hence the enhanced field is obtained. Similar effects are observed for the cases of RT (Fig.8) and RG (Fig.9) pulses. Here the wakefield of the strength of 4.98×10^9 V/m ($4.28\times10^9 V/m$) and density perturbations of 0.023 (0.0209) times of the unperturbed density are obtained in case of RT (RG) pulse of the same duration of *30 fs*. This can also be seen that the pulses of higher intensity produce relatively larger wakefields and the density perturbations. However, a very weak effect of the laser frequency is noticed on the wakefields. In the present study of three pulses infers that the rectangular-triangular (RT) pulse is more suitable for the purpose of wakefield excitation in a homogeneous plasma.

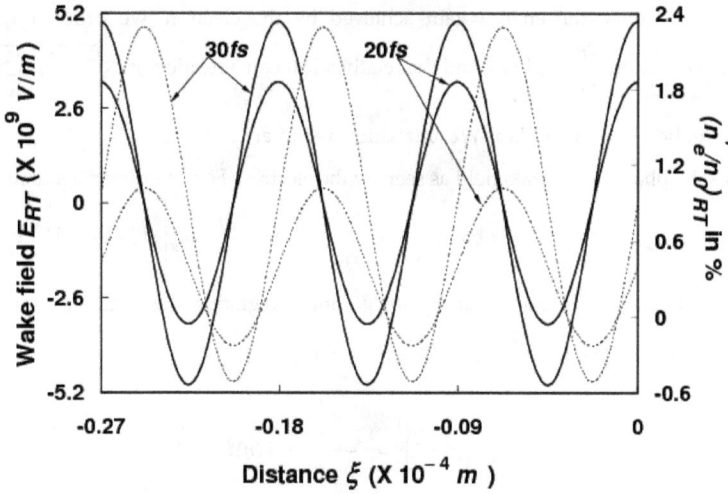

Figure 8. Wakefield E_{RT} and density perturbations ($n'_e/n_0)_{RT}$ behind the rectangular-triangular pulse up to distance $3L$ for different pulse durations (τ = 20 ps, 30 ps) and other parameters the same as in Fig.7.

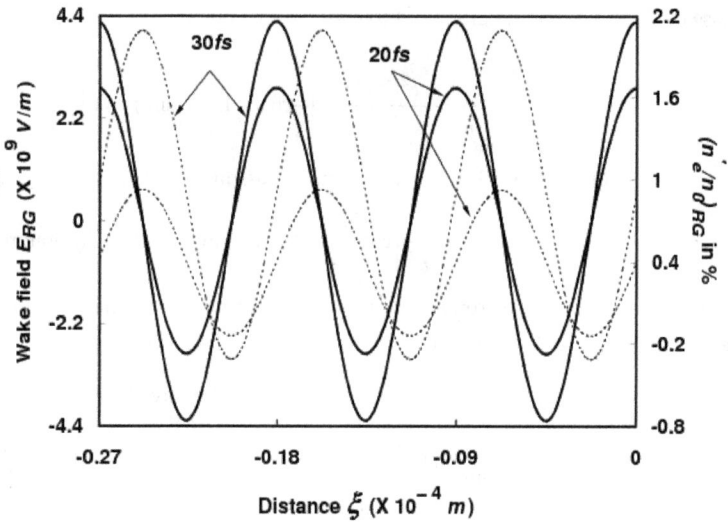

Figure 9. Wakefield E_{RG} and density perturbation $\left(n'_e/n_0\right)_{RG}$ behind the rectangular-Gaussian pulse up to distance $3L$ for different pulse durations (τ = 20 ps, 30 ps) and other parameters the same as in Fig.7.

4.1.1. Calculation of electron energy gain

In order to calculate the energy gain achieved by the electron, we proceed with the momentum equation $\dfrac{dp}{dt} = -eE(\xi)$ and the relativistic factor γ relation $\gamma = \sqrt{1 + p^2/m_e^2 c^2}$.

Here E is either E_{GL}, E_{RT} or E_{RG}. We introduce a similarity variable $\eta = k_p\left(\xi - L/2\right)$ that represents the phase of the wakefield as seen by the electron. For one-dimensional motion of the electron and $\xi = x - v_g t$, we obtain $\dfrac{d\gamma}{dt} = -\dfrac{peE(\eta)}{\gamma m_e^2 c^2}$ and $\dfrac{d\eta}{dt} = k_p\left[c\left(1 - 1/\gamma^2\right)^{\frac{1}{2}} - v_g\right]$ from the above relations. Dividing $d\gamma/dt$ by $d\eta/dt$ and integrating the resultant equation by taking $dx/dt = v_x = c\left(1 - 1/\gamma^2\right)^{1/2}$ and $\beta_r = v_g/c$ we get

$$\gamma - \beta_r\left(\gamma^2 - 1\right)^{\frac{1}{2}} = -\frac{e}{m_e c^2 k_p}\int E(\eta)d\eta \tag{41}$$

This is the general equation that describes the electron acceleration in the wakefield $E(\eta)$. By using the expressions of wakefield E for different shapes of the laser pulses we can determine the corresponding relativistic factor (or the energy gain).

For the case of GL pulse, the integration of the resultant equation with the initial value of γ as γ_0 at $\xi = 0$ yields

$$\gamma - \gamma_0 - \beta_r\left[\left(\gamma^2 - 1\right)^{\frac{1}{2}} - \left(\gamma_0^2 - 1\right)^{\frac{1}{2}}\right] = -\frac{e^2 E_0^2 \pi L}{m_e^2 v_g^4 k_p\left(L^2\alpha^2 - \pi^2\right)}\cos\left(k_p L/2\right)\left\{\sin\eta + \sin\left(\frac{k_p L}{2}\right)\right\}.$$

Without loss of generality we can assume $\gamma^2, \gamma_0^2 \gg 1$. Hence, $\Delta\gamma_{GL} = \gamma - \gamma_0 = -\dfrac{e^2 E_0^2 \pi L}{m_e^2 v_g^4 k_p\left(1 - \beta_r\right)\left[L^2\alpha^2 - \pi^2\right]}\cos\left(\dfrac{k_p L}{2}\right)\left\{\sin\eta + \sin\left(\dfrac{k_p L}{2}\right)\right\}.$ Therefore, the energy gain obtained by the electron during its acceleration in the wakefield excited by the GL pulse can be given by $\Delta W_{GL} = m_e c^2 \Delta\gamma_{GL}$.

Similarly the electron energy gain in the case of RT and RG pulses are obtained as

$$\Delta W_{RT} = -\frac{ea_2 a_3}{k_p\left(1 - \beta_r\right)}\left\{\sin\eta + \sin\left(\frac{k_p L}{2}\right)\right\} \text{ and } \Delta W_{RG} = \frac{ea_4 a_5}{k_p\left(1 - \beta_r\right)}\left\{\sin\eta + \sin\left(\frac{k_p L}{2}\right)\right\}.$$

Now we examine the effects of pulse duration (plasma density), laser intensity and laser frequency on the electron acceleration for different shapes of the laser pulses and make a comparative study.

4.1.1.1. Effect of pulse duration τ

We have already seen that the wakefield gets enhanced with the increased pulse duration τ for all the shapes of the laser pulses. Therefore, it is obvious that the electron will gain larger energy in the wakefield, which is excited by the pulses of longer durations. The same has been portrayed in Fig.10 for the laser intensity of 3×10^{18} W/m^2 and its frequency as 1.6 PHz. A comparison of the three graphs infers that the energy gains follow the trend $W_{RT} > W_{RG} > W_{GL}$. The increased gain for the longer pulse durations is attributed to the enhanced plasma wavelength λ_p. Since λ_p is independent of the pulse shapes, the electron gets larger energy for the increasing τ for all types of the pulses irrespective of their shapes. This can also be seen from this figure that the change in energy gain is faster when the pulses of longer durations are employed for the wakefield excitation.

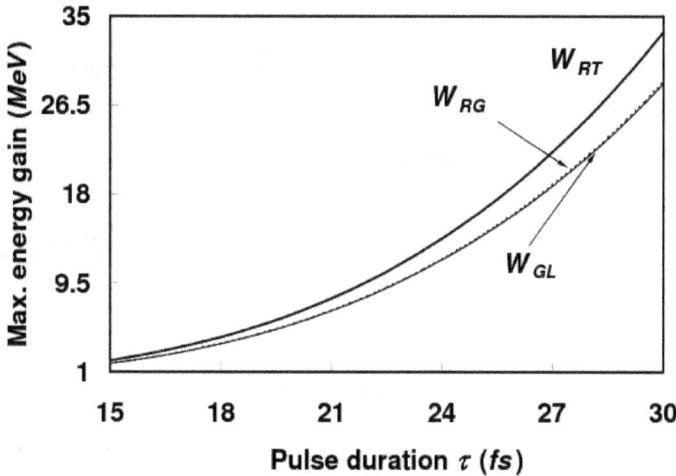

Figure 10. Dependence of maximum energy gain of electron on the laser pulse duration for the same parameters as in Fig.2. W_{RT} is the gain in case of rectangular-triangular pulse, W_{RG} is for rectangular-Gaussian pulse and W_{GL} is for Gaussian-like pulse.

4.1.1.2. Effect of laser frequency f

The effect of laser frequency f on the maximum energy gain attained by an electron is shown in Fig.11, from where it is evident that the gain is larger in case of RT pulse. Moreover, the slopes of the graphs reveal that the effect of laser frequency is more significant in the case of RT pulse in comparison with RG and GL pulses. Since the wakefield and plasma wavelength show weak dependence on the frequency f, it is worth clarifying the main factor that leads to significant increase in the electron energy gain with f. Actually a slight change in v_g due to f causes a greater change in the factor $(1-\beta)$ appearing in the denominator of the energy gain expressions. Since v_g increases for the larger frequencies, the gain gets larger with the increasing laser frequencies.

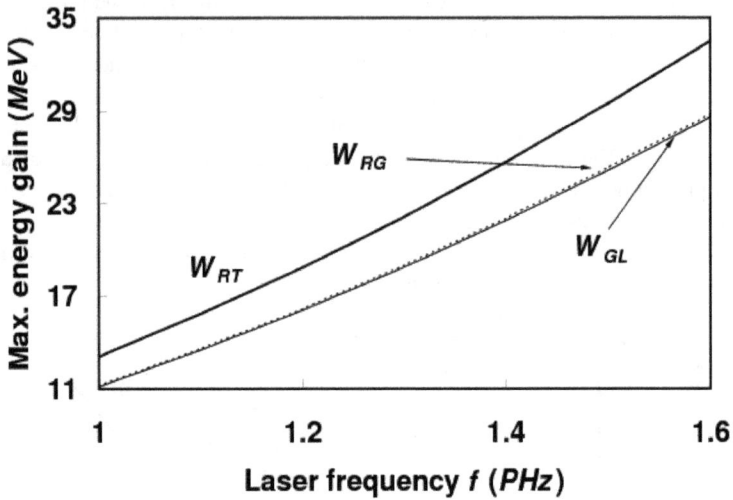

Figure 11. Dependence of maximum energy gain of electron on the laser frequency when the laser intensity is 3×10^{18} W/m^2 and pulse duration is $30\,fs$. W_{RT}, W_{RG} and W_{GL} have the same meaning as in Fig.10.

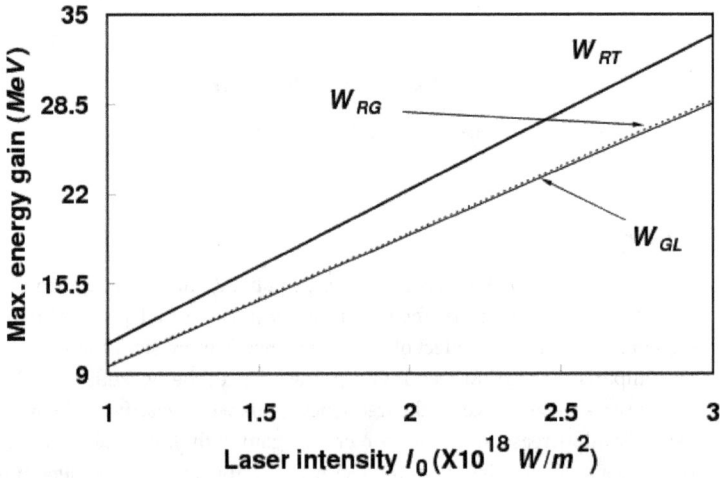

Figure 12. Variation of maximum energy gain of electron with the laser intensity when the pulse duration is $30\,fs$ and laser frequency is $1.6\,PHz$. W_{RT}, W_{RG} and W_{GL} have the same meaning as in Fig.10.

4.1.1.3. Effect of laser intensity I₀

The expressions of wakefield for all types of the pulses show that the wake amplitude is directly proportional to E_0^2, i.e. the intensity of the laser. It means higher intensity pulses will excite larger amplitude wakefield owing to the larger density perturbations in the plasma. However, we cannot indefinitely increase the amplitude of the wakefield because there is a limit on the maximum field that a plasma can support. Figure 12 shows that the maximum energy gain is increased from 9.5 *MeV* to 33.5 *MeV*, when the laser intensity is raised from 1×10^{18} *W/m²* to 3×10^{18} *W/m²* in case of RT pulse. A comparison of all the graphs shows that the RT pulse supersedes and gives the best results. Also the difference in energy gain becomes more and more significant when the intensity of the pulses is increased. The better results, i.e. higher amplitude wakefield and larger energy gain, obtained in case of RT pulse having smooth/fast rising time are consistent with the observations of Bulanov et al. [91] where he observed regular wakefields by a pulse with sharp steepening of its leading front.

4.2. Wakefield by different microwave pulses in waveguides

Here we present some results on wakefield excitation in a waveguide by different shapes of the microwave pulses, i.e. GL, RG and RT pulses.

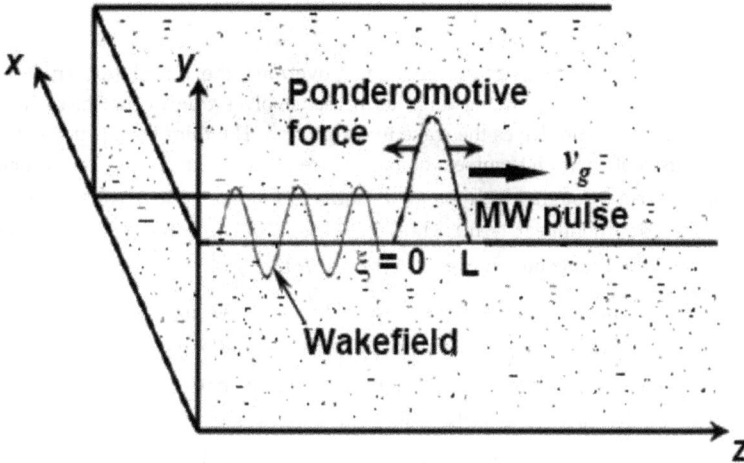

Figure 13. Schematic of wakefield generation in plasma filled rectangular waveguide by microwave pulse. Here $\xi = z - v_g t$ and L is the pulse width.

We consider that a microwave pulse of pulse duration τ at a frequency f propagates in a plasma filled $b \times h$ rectangular waveguide. This pulse resonantly excites the wakefield (corresponding potential ϕ) in the waveguide under the action of ponderomotive force (Fig.13), when the pulse duration τ matches with the inverse of the plasma frequency, i.e.

$\tau = f_{pe}^{-1}$. The electric and magnetic fields associated with the microwave are represented by $\vec{E}$ and $\vec{B}$. We use the Maxwell's equations and obtain the group velocity of the microwave pulse as $v_g = c\left(1 - \omega_p^2 / \omega^2 - \pi^2 c^2 / b^2 \omega^2\right)^{1/2}$ that coincides with the phase velocity of the wakefield.

For the rectangular waveguide, we take the distribution of the microwave field as $\vec{E} = \hat{y}\, E(\xi) \sin\left(\dfrac{\pi x}{b}\right)$ and $\vec{B} = \hat{x}\, B_x(\xi) \sin\left(\dfrac{\pi x}{b}\right) + \hat{z}\, B_z(\xi) \cos\left(\dfrac{\pi x}{b}\right)$. Using these relations in the basic fluid equations, we integrate them under the conditions that all the oscillating quantities tend to zero as $\xi \to \infty$ under the weakly nonlinear theory. With this we get the following equation

$$\frac{\partial^2 \phi}{\partial \xi^2} + \frac{e}{2 m_e v_g^2}\left(\frac{\partial \phi}{\partial \xi}\right)^2 + \frac{\omega_p^2}{v_g^2}\phi - \left(\frac{c^2 e}{m_e v_g^2}\right)\sin\left(\frac{\pi x}{b}\right) F(\xi) = 0, \qquad (42)$$

where

$$F(\xi) = \frac{1}{2}\left[\left(\frac{1}{v_g^2} - \frac{1}{c^2}\right) E^2(\xi) - B_z^2(\xi)\right].$$

This equation can be viewed as the equation governing the forced (driven) harmonic oscillator. Here the last term is the force term that evolves due to the microwave field (ponderomotive force) and drives the wake in the plasma. The third term is proportional to ϕ and hence its coefficient determines the natural frequency of the wake. The second term is the damping term through which the nonlinearity enters the system as it is proportional to square of $\partial \phi / \partial \xi$ (nonlinear term). Thus, the wake with potential ϕ is evolved in the plasma as a combined contribution of each term of Eq.(42).

We can use the information related to the shape of the pulse through the last term of Eq.(42) via the coefficient $F(\xi)$. Using the fourth-order Runge-Kutta method we simulate this equation for the above-mentioned three types of the pulse shapes. Here we take different profiles for the electric field of the pulse keeping in mind its shape and calculate $B_z(\xi)$ with the help of Maxwell's equation. The relation between $B_z(\xi)$ and $E(\xi)$ thereby comes out to be $B_z(\xi) = \dfrac{1}{v_g}\dfrac{\pi}{b}\int E(\xi)\,d\xi$.

4.2.1. Results on wakefield in the waveguide

As mentioned, we solve Eq.(42) numerically and obtain the potential ϕ from which we look for the wakefield amplitude for the mentioned three types of the pulse shapes. Figures 14 –

16 show the profile of wakefield generated by GL pulse, RG pulse and RT pulse, respectively. It can be easily seen that the amplitude of the wakefield is the largest in the case of RT pulse and is the least for the case of GL pulse; in other words, the wakefield amplitude follows the trend $E_{RT} > E_{RG} > E_{GL}$.

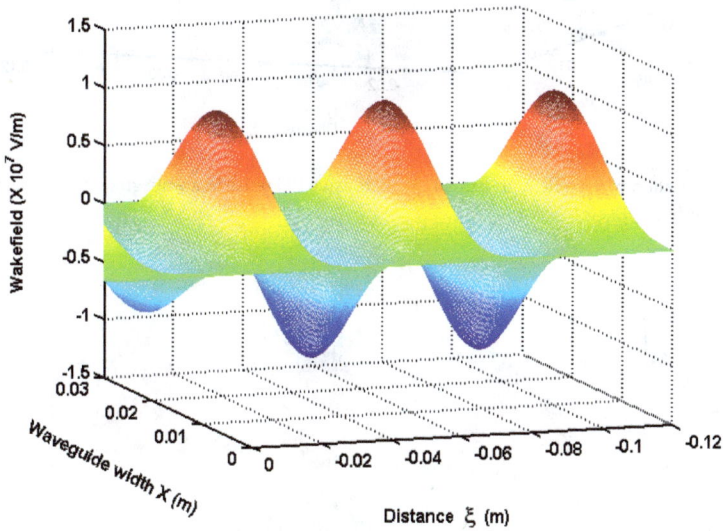

Figure 14. Variation of wakefield generated by microwave GL pulse in a waveguide for microwave intensity $I = 2GW/m^2$, frequency $f = 30GHz$, pulse duration $\tau = 2ns$, plasma density $n_0 = 4.5 \times 10^{17} \, m^{-3}$ and waveguide width $b = 0.03 \, m$.

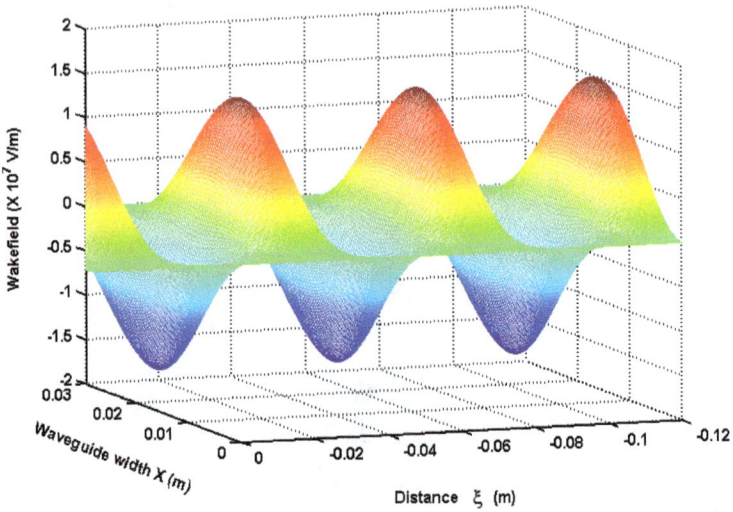

Figure 15. Variation of wakefield generated by microwave RG pulse in a waveguide for the same parameters as in Fig.14.

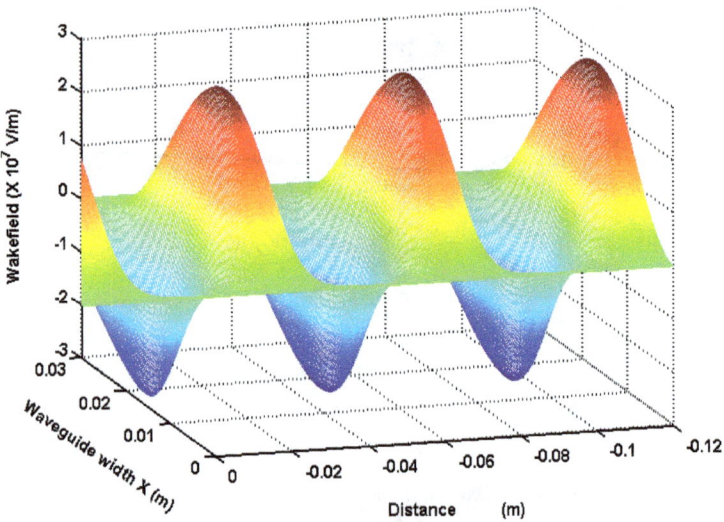

Figure 16. Variation of wakefield generated by microwave RT pulse in a waveguide for the same parameters as in Fig.14.

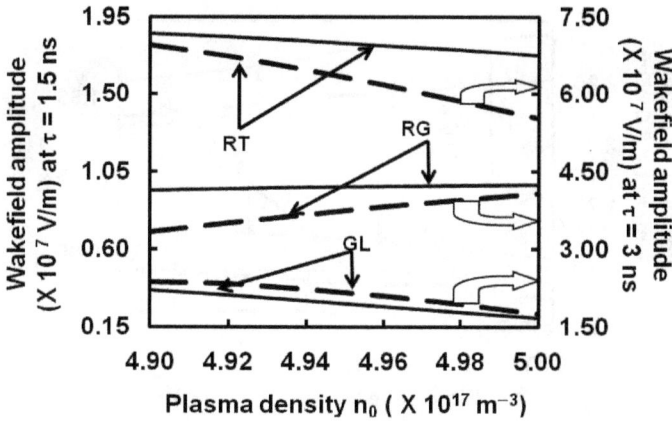

Figure 17. Dependence of wakefield amplitude on plasma density n_0 for two different microwave pulse durations $\tau = 1.5\,ns$ (solid line graphs, left axis) and $\tau = 3.0\,ns$ (dashed line graphs, right axis), when the other parameters are the same as in Fig.14.

Dependence of the wakefield amplitude on the plasma density for two different pulse durations is shown in Fig.17, from where it is found that the amplitude is increased for the higher plasma density in the waveguide for the case of RG pulse. However, the opposite trend is realized for the other types of the pulses (GL and RT pulses); the wakefield amplitude remains the largest in case of the RT pulse. A comparison of slopes of the graphs yields that the RT pulse shows stronger dependence on the plasma density in comparison with the other types of the pulses. With regard to the effect of pulse duration, we notice that the larger wakefield is obtained for the case of longer pulse durations; this is true for all types of the pulses.

In Fig.18, the effects of microwave frequency and its intensity are studied on the wakefield amplitude, where it is seen that an increase in the frequency leads to an enhancement in the wakefield amplitude for the cases of RG pulse and GL pulse; opposite is true for the RT pulse, which also shows a strong dependence (slope 0.097 at $I = 2 \times 10^9\,W/m^2$) on the frequency as compared with the other pulses. With regard to the effect of microwave intensity, we observe that the larger wakefield is obtained for the higher microwave intensity. This is further evident that the amplitude is modified at a faster rate in the case of RT pulse in comparison with the other pulses. Generally, we can conclude that the RT pulse is most sensitive to microwave frequency and intensity.

It is worth noticing from Figs. 14 – 18 that tens of MV/m wakefield is attained with the use of moderate intensity microwave pulses. Therefore, in view of the effect of microwave intensity (Fig.18), it is expected that the wakefield of the order of GV/m can be generated if the microwave pulses of intensity ~ TW/m² are available. Since the wakefield of this order is generally obtained by ultra high intensity lasers in usual wakefield generation schemes, the present mechanism of exciting wakefield in the waveguide by microwave pulses seems to be more effective and feasible as it can reduce the cost of accelerator and also it will provide an additional controlling parameter (the waveguide width).

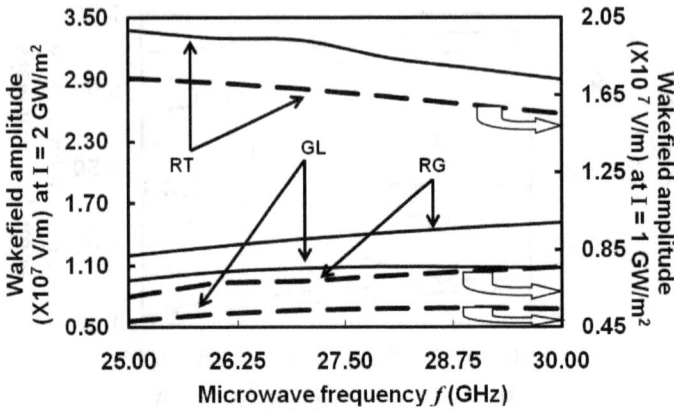

Figure 18. Variation of wakefield amplitude with microwave frequency f for two different microwave intensities $I = 2 \times 10^9 \, W/m^2$ (solid line graphs, left axis) and $I = 1 \times 10^9 \, W/m^2$ (dashed line graphs, right axis), when the other parameters are the same as in Fig. 14.

Variation of the wakefield amplitude with the waveguide width is shown in Fig.19, where it is observed that the amplitude is decreased with the increase of waveguide width; same result was obtained in an analytical calculation [38]. It means the larger wakefield can be obtained for the case of plasma filled narrower waveguide. A comparison of the slopes of the graphs reveals that the wakefield amplitude changes at a faster rate in the case of RT pulse. Therefore, the RT pulse is found to be more sensitive to the waveguide width. Thus, we can conclude that a plasma filled narrower waveguide is best suited for an effective wakefield excitation and the significant particle acceleration if the RT pulse is used.

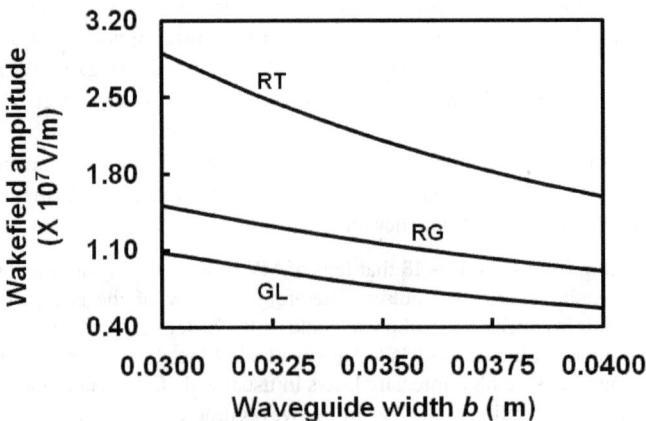

Figure 19. Variation of wakefield amplitude with waveguide width. The values of intensity, frequency, pulse duration and plasma density are given in the caption of Fig. 14.

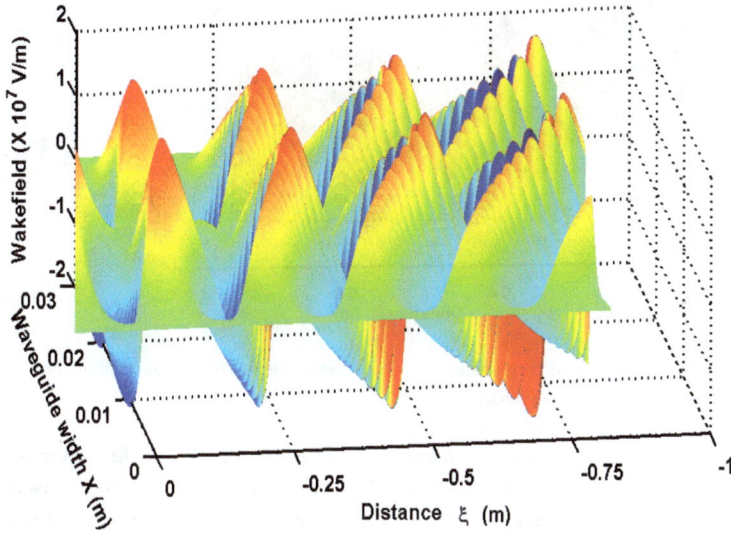

Figure 20. Profile of wakefield E_W in a plasma filled waveguide near cutoff conditions when $f = 6.76GHz$, $\tau = 0.9ns$, $f_c = 5.11GHz$, $I = 1 \times 10^9 \, W/m^2$ and $b = 0.03m$.

4.2.1.1. Wakefield near cutoff conditions

This has already been explored that the wakefield of larger amplitude is obtained for the smaller waveguide width and the longer pulse duration. For fixed microwave frequency, the cutoff frequency f_c gets higher under the effect of decreased width b and plasma density n_0. These effects can be viewed as if the microwave frequency is brought near the cutoff frequency f_c. Therefore, it is of much importance to investigate the wakefield structure near cutoff conditions, i.e. when f is near f_c. These results are presented in Figs.20 and 21 for $f = 6.76GHz$ whereas $f_c = 5.11GHz$. Fig.20 shows that the amplitude of wake wave gets increased under this situation as we move away from the microwave pulse, i.e. for decreasing values of ξ. This is further noticed that various peaks develop along the waveguide width during the growth of wakefield amplitude and it becomes unstable. Thus, it is plausible that some instability develops near the cutoff conditions. In order to further investigate this effect, we show in Fig.21 the maximum distance by which this growth occurs. Here we observe that the amplitude gets terminated around $\xi = 3.7\,L$ in the plasma and the field breaks down.

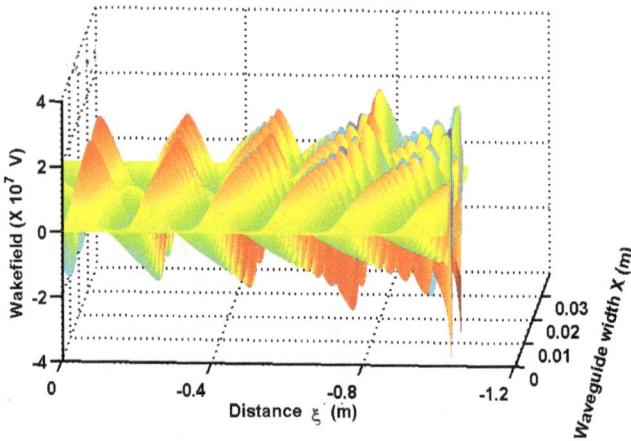

Figure 21. Profile of wakefield corresponding to Fig.20, showing its cutoff around maximum distance of $\xi = 3.7\,L$ from the microwave pulse.

These results suggest that the microwave of higher frequency should be employed for avoiding any instability in the system and the wakefield can be effectively used for the purpose of particle acceleration. The present mechanism of wakefield generation can be realized experimentally if we use wider waveguide filled with higher plasma density. The high density plasma can be produced in the waveguide under the action of Electron Cyclotron Resonance (ECR). However, under such situation short microwave pulses would be more effective in order to resonantly excite the plasma wake wave.

5. Concluding remarks

The electromagnetic waves were classified based on their frequency and a small region (10^{11}Hz to $\sim10^{13}$ Hz) that remained the last unexplored region was introduced as the THz rays. While explaining the propagation of EM waves, it was shown that their propagation velocity depends on the properties of the medium, and unlike the cases of vacuum and dielectrics, their electric field and magnetic field vectors do not remain in phase in the case of conductors. Very fascinating phenomenon of skin depth was discussed in the conductors where the wave vector was found to become a complex quantity and its imaginary part led to the attenuation of the wave. It was mentioned that the EM waves can propagate through the plasma medium if their frequency is larger than the plasma frequency. In the case of plasmas, another interesting phenomenon of wakefield excitation by the laser or microwave pulses was talked about in detail. Two case studies were conducted using the laser pulses and microwave pulses with different envelopes. It was shown that moderate intensity microwave pulses can also generate the wakefield effectively and accelerate the particles to sufficiently large energies. Moreover, the high cost laser systems can be replaced with microwave systems if the microwave pulses can be tailored properly.

6. Applications and future prospects

The electromagnetic wave and plasma interaction has diverse applications in different fields such as nuclear fusion, particle acceleration, heating of ionospheric and laboratory plasmas by radio waves etc. along with controlled fusion applications to ITER (International Thermonuclear Experimental Reactor), frequency upshifting, resonance absorption, laser focusing and defocusing, material processing, generation of X-ray, THz and microwave radiations, higher order harmonic generation, laser filamentation etc. With the inclusion of plasma, the performance of some devices such as backward wave oscillator (BWO), travelling wave tube (TWT) amplifiers, gyrotrons and other microwave tubes have been found to increase.

The use of laser plasma accelerators has been made in radioisotope production through (γ, n) reactions with laser accelerated electron bunches in the range tens of MeV [92, 93]. The short pulse nature and high charge of the accelerated bunches also has applications in the production of coherent THz radiation, which is achieved when femtosecond electron bunches cross the plasma vacuum boundary and emit transition radiation [94 – 96]. The generation of THz radiation has interesting applications in nonlinear THz spectroscopy, material characterization, imaging, topography, remote sensing, chemical and security identification [97, 98] etc. Another application of accelecerated electron beams / bunches is in the generation of femtosecond X-ray pulses produced by the betatron radiation emitted when the electron beam propagates through the plasma. By making an array of nanoholes on an alumina target, X-ray emission from laser produced plasma can be greatly enhanced even in soft X-ray energy regions (< 0.25 keV). The enhancement increases as the ionization level of Al becomes higher and the X-ray wavelength becomes shorter. Over 50 fold enhancement was obtained at a soft X-ray wavelength around 6 nm, which corresponds to the emission from $Al^{8+,9+}$ ions. X-ray pulse duration was 17 ps, which is much shorter than that obtained by using the prepulse technique [99]. Towards the generation of other types of electromagnetic radiation, Tripathi and Liu [82] have proposed a dielectric-lined waveguide for the free-electron laser emission in millimeter wavelength band. Farokhi et al. [83] have presented a linear theory for a free electron laser with a three-dimensional helical wiggler and axial magnetic field in the collective regime in a configuration consisting of an annular electron beam propagating inside a cylindrical waveguide. For the generation of high power (140 MW) subnanosecond (75 ps) microwave pulses in the range of 38 – 150 GHz, Yalandin et al. [84] have done experiments on coherent stimulated radiation from intense, subnanosecond electron bunches moving through a periodic waveguide and interacting with a backward propagating TM_{01} wave. Hayashi et al. [85] have also designed a two-stage ferroelectric electron gun and a peak power of 5.9 MW microwave radiation was observed when a 100A 450 kV electron beam was used.

Research directed towards the development of high power electromagnetic radiation sources accounts for much of the current interest in the plasma filled waveguides. Plasma filled waveguides may also be used for the transportation of electromagnetic energy and charged particles, and in the basic study of plasma phenomena. In spite of such extensive work over the years, the understanding of the physics of *wave plasma interaction* is still an active area of research, which also finds additional applications in plasma based focused ion beams (FIB), plasma sources for negative ion beams for neutral beam injection, rf-based plasma thrusters, etc. With regard to the particle acceleration, we have carried out analytical and numerical studies on the wakefield excitation by different types of the pulses in a rectangular waveguide filled with homogeneous plasma. Our analyses reveal that moderate intensity (~ 10^9 W/m^2) microwave pulses can produce up to 100 MV/m wakefield in the waveguide if the nanosecond pulses are used. Since the amplitude of resonantly excited wakefield changes at a faster rate with the waveguide width, pulse duration and microwave intensity and it is larger for the smaller waveguide width, longer pulse duration and the higher microwave intensity in the case of rectangular triangular pulse, the significant wakefield can be excited in the waveguide and effective particle acceleration can be achieved with the use of RT pulses for which the parameters can be optimized using the present studies. This theoretical work on the contribution of different microwave and laser pulses for the purpose of particle acceleration and the THz generation [100, 101] shall induce experimentalists to develop rectangular, rectangular-Gaussian, rectangular-triangular, sawtooth and triagular pulses of appropriate lengths for accomplishing various experiments on wakefield generation, particle acceleration, and the THz generation. Through such efforts our researchers would be able to benfit the society more via the medical, scientific, and technological applications of the subject *electromagnetic wave and plasma interaction*.

Author details

Hitendra K. Malik
Department of Physics, Indian Institute of Technology Delhi, New Delhi, India

7. References

[1] V. L. Ginzburg, The propagation of electromagnetic waves in plasmas, Pengamon, New York (1970).

[2] N. G. Denisov, Zh. Eksp. Teor. Fiz. 31, 609 (1956); Engi. Transi. Sov. Phys. JETP 4, 544 (1957).

[3] P. Koch and J. Albritton, Phys. Rev. Left. 32, 1420 (1974); J. Aibritton and P. Koch, Phys. Fluids 18, 1136 (1975).

[4] V. B. Gildenburg, Sov. Phys. JETP 18, 1359 (1964).

[5] T. H. Stix, Phys. Fluids 1, 308 (1958).

[6] B. N. Breizman and A. V. Arefiev, Phys. Plasmas 8, 907 (2001);

[7] A. F. Kuckes, Plasma Phys. 10, 367 (1968)

[8] M. N. Rosenbluth and C. S. Liu, Phys. Rev. Lett. 29, 701 (1972).

[9] C. E. Clayton, C. Joshi, C. Darrow and D. Umstadter, Phys. Rev. Lett. 54, 2343 (1985).

[10] W. Yu and Z. -Z. Xu, Phys. Rev. A. 36, 285 (1987).

[11] R. P. H. Chang, M. Porkolab and B. Grek, Phys. Rev. Lett. 28, 206 (1972)

[12] A. B. Kitsenkov, I. Panchenko and K. N. Stepanov, Plasma Phys. 16 1109 (1974).

[13] R. Fedosejevs, X. F. Wang and G. D. Tsakiris, Phys. Rev. E 56, 4615 (1997)

[14] C. E. Max and J. Arons Phys. Rev. Lett. 33, 209 (1974)

[15] P. K. Kaw, G. Schmidt and T. Wilcox, Phys. Fluids 16, 1522 (1973)

[16] D. J. Nicholas and S. G. Sajjadi, J. Phys. D: Appl. Phys. 19, 737 (1986)

[17] D. A. Jones, E. L. Kane, P. Lalousis, P. R. Wiles and H. Hora, Appl. Phys. B 27, 157 (1982)

[18] T. Tajima and J. M. Dawson, Phys. Rev. Lett. 43, 267 (1979).

[19] P. Chen, J. M. Dawson, R. W. Huff and T. Katsouleas, Phys. Rev. Lett. 54, 693 (1985).

[20] J. B. Rosenzweig, D. B. Cline, B. Cole, H. Figueroa, W. Gai, R. Konecny, J. Norem, P. Schoessow and J. Simpson, Phys. Rev. Lett. 61, 98 (1988).

[21] J. B. Rosenzweig, P. Schoessow, B. Cole, W. Gai, R. Konecny, J. Norem and J. Simpson, Phys. Rev. A 39, 1586 (1989).

[22] K. Nakajima, D. Fisher, T. Kawakubo, H. Nakanishi, A. Ogata, Y. Kato, Y. Kitagawa, R. Kodama, K. Mirna, H. Shiraga, K. Suzuki, K. Yamakawa, T. Zhang, Y. Sakawa T. Shoji, Y. Nishida, N. Yugarni, M. Downer and T. Tajima, Phys. Rev. Lett. 74, 4428 (1995).

[23] H. K.Malik, S. Kumar and Y. Nishida Electron. Opt. Comm. 280, 417 (2007).

[24] K. V. Lotov, Laser Part. Beams 19, 219 (2001).

[25] A. J. W. Reitsma and D. A. Jaroszynski, Laser Part. Beams 22, 407 (2004).

[26] A. F. Lifschitz, J. Faure, Y. Glinec, V. Malka and P. Mora, Laser Part. Beams 24, 255 (2006).

[27] K. Flippo, B. M. Hegelich, B. J. Albright, I. Yin, D. C. Gautier and S. Letzring, M. Schollmeier, J. Schreiber, R. Schulze, J. C. Fernandez. Laser Part. Beams 25, 3 (2007).

[28] P. V. Nickles, S. Ter-avetisyan, M. Schnuerer, T. Sokollik, W. Sandner, J. Schreiber, D. Hilscher, U. Jahnke, A. Andreev and V. Tikhonchuk, Laser Part. Beams 25, 347 (2007).

[29] C. T. Zhou, M. Y. Yu and X. T. He, Laser Part. Beams 25, 313 (2007).

[30] Y. Nishida, T. Okazaki, N. Yugami and T. Nagasawa, Phys. Rev. Lett. 66, 2328 (1991).

[31] D. W. Aossey, J. E. Williams, H.-S. Kim, J. Cooney and Y.-C. Hsu and K. E. Lonngren, Phys. Rev. E 47, 2759 (1993).

[32] B. B. Balakirev, V. I. Karas, I. V. Karas and V. D. Levchenko, Laser Part. Beams 19, 597 (2001).

[33] T. B. Zhang, J. L. Hirshfield, T. C. Marshall and B. Hafizi, Phys. Rev. E 56, 4647 (1997) .

[34] S. Y. Parka and J. L. Hirshfield, Phys. Rev. E 62, 1266 (2000).

[35] C. Jing, W. Liu, I. Xiao, W. Gai and P. Schoessow, Phys. Rev. E 68, 016502 (2003).

[36] Y. Nishida, S. Kusaka and N. Yugami, Physica Scripta T52, 65 (1994).

[37] Z. M. Sheng, K. Mima, Y. Sentoku, K. Nishihara and J. Zhang, Phys. Plasmas 9, 3147 (2002).

[38] H.K. Malik, J. Appl. Phys. 104, 053308 (2008).

[39] R. J. Kingham and A. R. Bell, Phys. Rev. Lett. 79, 4810 (1997).

[40] N. E. Andreev, M. V. Chegotov and M. E. Veisman, IEEE Trans. Plasma Sci. 28, 1098 (2000).

[41] B. Cros, C. Courtois, G. Malka, G. Matthieussent, J. R. Marques, F. Dorchies, F. Amiranoff, S. Rebibo, G. Hamoniaux, N. Blanchot and J. L. Miquel, IEEE Trans. Plasma Sci. 28, 1071 (2000).

[42] P. Sprangle, B. Hafizi, J. R. Peñano, R. F. Hubbard, A. Ting, C. I. Moore, D. F. Gordon, A. Zigler, D. Kaganovich and T. M. Antonsen Jr., Phys. Rev. E 63, 056405 (2001).

[43] P. Manz, M. Ramisch and U. Stroth, Phys. Rev. Lett. 103, 165004 (2009).

[44] T. Happel, F. Greiner, N. Mahdizadeh, B. Nold, M. Ramisch and U. Stroth, Phys. Rev. Lett. 102, 255001 (2009).

[45] U. Stroth, Plasma Phys. Control. Fusion 40, 9 (1998).

[46] U. Stroth, F. Greiner, C. Lechte, N. Mahdizadeh, K. Rahbarnia and M. Ramisch, Phys. Plasmas 11, 2558 (2004).

[47] P. Manz, M. Ramisch, U. Stroth, V. Naulina and B. D. Scott, Plasma Phys. Control. Fusion 50, 035008 (2008).

[48] M. Ramisch, E. Häberle, N. Mahdizadeh and U. Stroth, Plasma Sources Sci Technol. 17, 024007 (2008).

[49] F. Aziz and U. Stroth, Phys. Plasmas 16, 032108 (2009).

[50] H. D. Hochheimer, K. Weishaupt and M. Takesada, J. Chem. Phys. 105, 374 (1996).

[51] F. Widulle, J. T. Held, M. Huber, H. D. Hochheimer, R. T. Kotitschke and A. R. Adams, Rev. Sci. Instrum. 68, 3992 (1997).

[52] A. Asenbaum, O. Blaschko and H. D. Hochheimer, Phys. Rev. B 34, 1968 (1986).

[53] A. B. Garg, V. Vijayakumar, B. K. Godwal, A. Choudhury and H. D. Hochheimer, Solid State Comm. 142, 369 (2007).

[54] R. Singh and M. P. Bora, Phys. Plasmas 7, 2335 (2000).

[55] N. Singh and R. Singh, Phys. Plasmas 11, 5475 (2004).

[56] R. Singh, V. Tangri, P. Kaw and P. N. Guzdar, Phys. Plasmas 12, 092307 (2005).

[57] N. Chakrabarti, R. Singh, P. K. Kaw and P. N. Guzdar, Phys. Plasmas 14, 052308 (2007).

[58] M. Starodubtsev and C. Krafft, Phys. Rev. Lett. 83, 1335 (1999)

[59] M. Starodubtsev and C. Krafft, Phys. Plasmas 6, 2598 (1999)

[60] A. K. Attri, U. Kumar and V. K. Jain, Nature 411, 1015 (2001).

[61] S. H. Kim, E. Agrimson, M. J. Miller, N. D'Angelo, R. L. Merlino and G. I. Ganguli, Phys. Plasmas 11, 4501 (2004).

[62] S. C. Sharma and M. P. Srivastava, Phys. Plasmas 8, 679 (2001).

[63] Y. Sakawa, C. Joshi, P. K. Kaw, F. F. Chen and V. K. Jain, Phys. Fluids B 5, 1681 (1993).

[64] A. Neogi and R. K. Thareja, Phys. Plasmas 6, 365 (1999).

[65] T. A. Peyser, C. K. Manka, B. H. Ripin and G. Ganguli, Phys. Fluids B 4, 2448 (1992).

[66] C. J. McKinstrie and E. A. Startsev, Phys. Rev. E 54, R1070 (1996).

[67] A. Pukhov, Phys.Rev. Lett. 86, 3562 (2001).

[68] M. Y. Yu, W. Yu, Z. Y. Chen, J. Zhang, Y. Yin, L. H. Cao, P. X. Lu and Z. Z. Xu, Phys. Plasmas, 10, 2468 (2003).

[69] H. Suk, J. Appl. Phys. 91, 487 (2002).

[70] Y. Sentoku, T. E. Cowan, A. Kemp and H. Ruhl, Phys. Plasmas 10, 2009 (2003).

[71] K. P. Singh and V. K. Tripathi, Phys. Plasmas 11, 743 (2004).

[72] M. Kado, H. Daido, A. Fukumi, Z. Li, S. Orimo, Y. Hayashi, M. Nishiuchi, A. Sagisaka, K. Ogura, M. Mori, S. Nakamura, A. Noda, Y. Iwashita, T. Shirai, H. Tongu, T. Takeuchi, A. Yamazaki, H. Itoh, H. Souda, K. Nemoto, Y. Oishi, T. Nayuki, H. Kiriyama, S. Kanazawa, M. Aoyama, Y. Akahane, N. Inoue, K. Tsuji, Y. Nakai, Y.

Yamamoto, H. Kotaki, S. Kondo, S. Bulanov, T. Esirkepov, T. Utsumi, A. Nagashima, T. Kimura and K. Yamakawa, Laser Part. Beams 24, 117 (2006).

[73] A. Karmakar and A. Pukhov, Laser Part. Beams 25, 371 (2007).

[74] J. J. Xu, Q. Kong, Z. Chen, P. X. Wang, W. Wang, D. Lin and Y. K. Ho, Laser Part. Beams 25, 253 (2007).

[75] Y. Nishida and T. Shinozaki, Phys. Rev. Lett. 65, 2386 (1990).

[76] Y. Nishida and N. Sato, Phys. Rev. Lett. 59, 653 (1987).

[77] Y. Nishida, M. Yoshizumi and R. Sugihara, Phys. Fluids 28, 1574 (1985).

[78] C.G. Durfee III, A. R. Rundquist, S. Backus, C. Herne, M. M. Murnane and H. C. Kapteyn, Phys. Rev. Lett. 83, 2187 (1999).

[79] X. Letartre, C. Seassal, C. Grillet, P. Rojo-Romeo, P. Viktorovitch, M. Le V. d'Yerville, D. Cassagne and C. Jouanin, Appl. Phys. Lett. 79, 2312 (2001).

[80] J. L. Hirshfield, M. A. LaPointe, A. K. Ganguly, R. B. Yoder and C. Wang Phys. Plasmas 3, 2163 (1996).

[81] E. Carlsten, Phys. Plasmas 9, 5088 (2002).

[82] V. K. Tripathi and C. S. Liu, IEEE Trans. Plasma Sci. 17, 583 (1989).

[83] B. Farokhi, Z. Family and B. Maraghechi, Phys. Plasmas 10, 2566 (2003).

[84] M. I. Yalandin, V. G. Shpak, S. A. Shunailov, M. R. Oulmaskoulov, N. S. Ginzburg, I. V. Zotova, Y. V. Novozhilova, A. S. Sergeev, A. D. R. Phelps, A. W. Cross, S. M. Wiggins and K. Ronald, IEEE Trans. Plasma Sci. 28, 1615 (2000).

[85] Y. Hayashi, X. Song, J. D. Ivers, D. D. Flechtner, J. A. Nation and L. Schächter, IEEE Trans. Plasma Sci. 29, 599 (2001).

[86] R. B. Yoder, T. C. Marshall and J. L. Hirshfield, Phys. Rev. Lett. 86, 1765 (2001).

[87] S. Kumar and H. K. Malik, J. Plasma Phys. 72, 983 (2006).

[88] H. K. Malik, Opt. Comm. 278, 387 (2007).

[89] A.K. Aria and H. K. Malik, The Open Plasma Phys. J. 1, 1 (2008).

[90] A.K. Aria, H. K. Malik and K.P. Singh, Laser Part. Beams 27, 41 (2009).

[91] S.V. Bulanov, T.J. Esirkepov, N.M. Naumova, F. Pegoraro, I.V. Pogorelsky, A.M. Pukhov, IEEE Trans. Plasma Sci. 24, 393 (1996).

[92] W. P. Leemans, D. Rodgers, P. E. Catravas, C. G. R. Geddes, G. Fubiani, E. Esarey, B. A. Shadwick, R. Donahue and A. Smith, Phys. Plasmas 8, 2510 (2001).

[93] M. I. K. Santala, M. Zepf, F. N. Beg, E. L. Clark, A. E. Dangor, K. Krushelnick, M. Tatarakis, I. Watts, K. W. D. Ledingham, T. McCanny, I. Spencer, A. C. Machacek, R. Allott, R. J. Clarke and P. A. Norreys, Appl. Phys. Lett. 78, 19 (2001).

[94] W. P. Leemans, C. G. R. Geddes, J. Faure, C. Tóth, J. V. Tilborg, C. B. Schroeder, E. Esarey, G. Fubiani, D. Auerbach, B. Marcelis, M. A. Carnahan, R. A. Kaindl, J. Byrd and M. Martin, Phys. Rev. Lett. 91, 074802 (2003).

[95] C. B. Schroeder, E. Esarey, J. V. Tilborg and W. P. Leemans, Phys. Rev. E 69, 016501 (2004).

[96] W. P. Leemans, J. V. Tilborg, J. Faure, C. G. R. Geddes, C. To'th, C. B. Schroeder, E. Esarey, G. Fubiani and G. Dugan, Phys. Plasmas 11, 2899 (2004).

[97] Y. C. Shen, T. W. P. F. Today, B. E. Cole, W. R. Tribe, and M. C. Kemp, Appl. Phys. Lett. 86, 241116 (2005).

[98] H. Zhong, A. Redo-Sanchez, and X.-C. Zhang, Opt. Express 14, 9130 (2006).

[99] T. Nishikawa, H. Nakano, N. Uesugi, M. Nakao and H. Masuda, Appl. Phys. Lett. 75, 4079 (1999).

[100] A.K. Malik, H.K. Malik and S. Kawata, J. Appl. Phys. 107, 113105 (2010).

[101] A.K. Malik, H.K. Malik and Y. Nishida, Phys. Lett. A 375, 1191 (2011).

Chapter 7

Optical Wave Propagation in Kerr Media

Michal Čada, Montasir Qasymeh and Jaromír Pištora

Additional information is available at the end of the chapter

1. Introduction

Optical wave propagation and interaction are important effects usable in designing and implementing various photonic devices ranging from passive splitters to active switches to light amplifiers. The material aspects are crucial as strong effects are desirable for efficient and robust devices. Electronics has its silicon that is an amazing rather universal material that makes it possible to implement microelectronics chips of unthinkable performance and functionalities. Photonics does not have such a common material, and therefore one has to choose suitable material system for a given application. However, with silicon being the best technologically mastered material, attempts have been made to employ it also in the implementation of photonic functions. Examples include electro-optic modulators and, of course, high speed photodetectors.

Recently we have investigated physical effects in silicon that are usable for photonic functionalities[1]. Since silicon is a cubic material, it does not possess the classical electro-optic effect exploited in other material systems (e.g. GaAs, InP) for high-speed switching and modulation. On the other hand, as basically all materials, silicon does possess the third-order nonlinear effect, originally known as the Kerr effect, discovered by J. Kerr in 1875[2]. This is one of the most interesting phenomena for potential exploitation. The two facts, i.e. universality of silicon and the existence of a nonlinear effect in it, led to our thorough exploration of the possibilities that Kerr effect[3] in silicon can offer in terms of potential future photonic devices[4]. As expected, the theoretical, numerical and design studies have been dominated by the optical wave propagation issues[5]. The results are general enough to apply to a wide range of materials that do not possess the classical linear electro-optic effect.

This chapter describes the original results obtained from those studies. Optical wave propagation in a nonlinear medium with a Kerr-type nonlinearity (third-order susceptibility) is analyzed theoretically. New features are found where not only waves and their polarization interactions are present as a result of the nonlinearity of the medium, but also an interplay between the optical and electrical Kerr effects contributes to the resulting functionality. Several novel wave propagation effects are discovered and discussed. They include cross-polarized wave conversion, optical multistability, nonlinear tunability of periodic structures, ultra-fast electro-optic switching, and a new photorefractive effect. Possible applications of these functionalities are addressed as well. The fine physical and mathematical details of our unified treatment are well reviewed in[6].

2. Electro-optic effects

Electro-optic effects are reviewed[7] pertaining to cubic (e.g. silicon) and isotropic (e.g. glass) materials; therefore, the well-studied and widely exploited (e.g. lithium niobate, gallium arsenide or indium phosphide) linear electro-optic phenomenon is not discussed here.

2.1. Electro-absorption

The Franz-Keldysh effect in semiconductors alters the absorption spectrum of a material. The effect is due to field-induced tunneling between valence and conduction band states. The electric field affects the overlap of electron and hole wavefunctions, which leads to increased absorption at energies lower than the bandgap. This electro-optic effect is thus normally referred to as electro-absorption. The associated electro-refraction effect is coupled via the Kramer-Kronig relation. Both effects depend on the applied electric field, the wavelength, and the carrier density.

Electro-absorption has been used in switches and modulators in various materials including III-V semiconductors. Electro-refraction is quite weak; for example in an undoped silicon at the telecommunication wavelength of $1.55\ \mu m$, where it is caused mostly by indirect gap electro-absorption, the value of the refractive index change is[8] $\Delta n = 1.5\ x\ 10^{-6}$ at $V_{app} = 10$ $V/\mu m$. The effect is polarization dependent and it is a factor of two stronger when the optical field is parallel to the applied field. It is a pure electric-field effect and as such, its speed is high (sub-picosecond range) and determined by the tunneling speed between the conduction and valence bands.

2.2. Quantum-confined Stark effect

The quantum-confined Stark effect is the similar phenomenon occurring in semiconductor quantum-well structures. In quantum wells when close to the exciton resonances, absorption changes and Kramer-Kronig-related refractive index changes behave in a Kerr-like fashion while the medium response is enhanced due to the electron-hole confinement. The quantum well confinement increases the overlap of electron and hole wave-functions while the applied electric field reduces this overlap. This results in a corresponding reduction in optical absorption. The direct change in the light intensity resulting from this electro-absorption effect has been used in efficient bulk as well waveguide modulators. Waveguide modulators achieve better performance overall due to the confinement of light. Device details are beyond the scope of this chapter and can be found in literature[9].

2.3. Free-carrier plasma dispersion effect

Injection of charge carriers into an undoped material or removal of free carriers from a doped material, changes the refractive index (generally optical properties, e.g. absorption). Generally, three carrier effects are involved: free-carrier absorption, Burstein-Moss band-filling (shifting the absorption spectrum to shorter wavelengths), and Coulombic interaction of carriers with impurities (shifting the absorption spectrum to longer wavelengths).

The refractive index increases when carriers are depleted from a doped material and it decreases when they are injected into an undoped material. This is the largest effect compared to the electro-refraction and the Kerr effects (see below). It is polarization independent, but generally the slowest of all effects. In the injection case, the switch-off time is limited by minority carrier lifetime (tens of picoseconds at best due to recombination). In the depletion mode, the response time is determined by carrier sweep-out (picoseconds at best due to carrier drift over a finite distance of a sample or a device).

A change in refractive index is always accompanied by a change in absorption, therefore a trade-off is required when utilizing this effect for applications. The residual linear loss is usually negligible at the telecommunications wavelength; the two-photon absorption is normally a concern[10]. Successful devices have been implemented using this effect in combination with a Mach-Zehnder waveguide configuration, one example being a reverse-biased pn junction (silicon)[11], the other being a forward-biased pn junction (indium phosphide)[12], and the third one being a MOS capacitor (fully compatible with standard CMOS)[13]. The designs exploited the free-carrier plasma dispersion effect in efficient ways.

2.4. Kerr effect

This effect is a pure electric field phenomenon and it is of interest in this work. It is a quadratic electro-optic effect caused by displacement of bound electrons under the influence of an external electric field. It is basically a nonlinear polarization generated in a medium, which results in changes of its refractive index. It exists in crystals, glasses, gasses, basically all materials including the isotropic ones, i.e. also in the cubic silicon and silicon nanocrystals. It is one of the several different phenomena (e.g. self-focusing, soliton generation, four-wave mixing, phase conjugation, etc.) associated with the third-order nonlinearity in a given material, usually described by the third-order susceptibility, $\chi^{(3)}$.

The susceptibility $\chi^{(3)}$ is, generally, dispersive. Depending on the frequency region, it describes the nonlinear response in a phenomenological way, which means that it includes combination of all physical effects that contribute to the response in that particular frequency range and usually on different time scales. Normally, the resonant effects are the strongest and thus may dominate the behavior and the values of $\chi^{(3)}$. On the other hand, in a lossless medium and/or far away from any resonant frequencies (absorption lines) of a material, the dispersion of $\chi^{(3)}$ is insignificant and the response is instantaneous; thus $\chi^{(3)}$ can be considered dispersion-less. This was shown using the classical anharmonic oscillator model[14], whereby, basically, the electro-optic Kerr effect (DC Kerr effect) is a quasi-static limit of the optical one (AC Kerr effect).

The AC Kerr effect is responsible for what is generally known as all-optical effects (e.g. self-phase and cross-phase modulation, four-wave mixing). The corresponding refractive index change, Δn, is a linear function of light intensity, I:

$$\Delta n_{opt} = n_{NL} I = 3 \chi^{(3)} Z_0 / 4(n_L)^2, \tag{1}$$

where n_{NL} is the nonlinear refractive index coefficient, n_L is the linear refractive index, and Z_0 is the free space impedance. The external-electric-field-controlled refractive index change is a result of the DC Kerr effect and can be described by:

$$\Delta n_{ext} = 3 \chi^{(3)} / 8 \, n_L = n_L \, n_{NL} / 2 \, Z_0. \tag{2}$$

The pure Kerr effect is very fast, well in the sub-picosecond range. It is polarization dependent; there is a factor of one-third involved whereby parallel optical and electrical fields display a stronger interaction. The Kerr effect makes an isotropic material behave as a uniaxial crystal once the voltage is applied, with an optical axis being in the direction of the external field. The Kerr effect depends on the bandgap energy, thus it is much stronger in, for example, semiconductors than in silica glass. At wavelengths far enough away from the band-edge the effect may be considered as a pure Kerr effect in a moreless lossless medium, although multi-photon absorption might have to be considered in some cases.

In order to avoid large absorption losses required to obtain enhanced resonant nonlinearities, it is preferable to propagate waves at around a half-gap wavelength, as successfully exploited in the past in the III-V semiconductor technology[15]. At that wavelength range in semiconductors, the nonlinear refractive index, arising from the real part of the third-order susceptibility, is still relatively large to be usable. At the same time the two-photon absorption that contributes to the imaginary part of the third-order susceptibility, is relatively low to obtain reasonable propagation lengths. The real and imaginary parts are related by causality expressed by the well-known Kramers-Kronig relation.

Since the Kerr effect is bound-electron related, it is very fast and usually dominant. The refractive index change is positive. However, for higher intensities, the nonlinear two-photon absorption will start generate more free carriers as intensities increase. The free-carrier refraction is negative and slow; therefore, it is desirable to avoid such ranges of optical intensities that may lead to large nonlinear losses of propagating waves. A strong and low-loss interaction of propagating waves is the key to designing and developing efficient and robust optoelectronic devices such as switches or modulators.

The fundamental problem with the third-order nonlinearity is that the effect is very weak in most materials. The promise of the development of new materials that fall under the umbrella of nanotechnology is quite attractive. Materials are being developed on the nanometer scale, thus promising a potential to open a new world of scalability and integration. Reducing the size of the optical material structures to a nanoscale leads to significant (orders of magnitudes) enhancements of the third-order optical susceptibility due to the confinement effect. Combined with optical waveguide enhancing effects (e.g.

photonic crystals or high-contrast slot waveguides), a much stronger nonlinear interaction of propagating waves and modes is achieved. It is this promise that led us to the studies of nonlinear wave propagation in the Kerr media, with integrated optoelectronics and nanophotonics being the main area for potential applications.

3. Optical wave propagation in nonlinear media

The interaction of a light wave with a propagating medium and additionally with an applied external electric field is described by the nonlinear wave equation. When considering the general vector nonlinear wave equation, the polarization components are mutually coupled nonlinearly thus yielding coupled differential equations. Finding solutions to such a nonlinear system, which possess physical meaning, is a challenging problem even with today's available powerful computing technology. In order to obtain at least an approximate analytical solution that would offer an insight into the complexity of the problem, simplifying assumptions have to be made[3].

In a third-order nonlinear medium, the relationship between the nonlinear polarization and the electric field vector of an optical wave is governed by a fourth-ranked susceptibility tensor, $\chi^{(3)}$. For materials of interest here (cubic, isotropic), the tensor is much simpler having most of its components zero. The wave propagation can then be simplified to the point that after neglecting second-order coupling between the polarization components, nonlinear wave equations can be solved as individual scalar equations (Helmholz equations). Such a scalar equation can be solved approximately for some situations, the most known being the spatial soliton in optical fibers[16]. For this approximate case specifically, the equation is in literature incorrectly called the nonlinear Schrödinger equation due to its similar form. However, Schrödinger himself did not derive any nonlinear equation. In 1925 he formulated a linear motion equation for a free particle[17].

In the Cartesian coordinates (x, y, z) with z being the propagation direction, the optical wave field components solution can be written in a form:

$$E_x = e_x(0)\, e^{j\omega t}\, e^{-jk_0(n_L + n_{NL} I + n_{EXT} E_{ext}^2)z}$$

$$E_{y,z} = e_{y,z}(0)\, e^{j\omega t}\, e^{-jk_0(n_L + n_{NL} I + \frac{1}{3}n_{EXT} E_{ext}^2)z} \tag{3}$$

where $e_x(0)$ and $e_{y,z}(0)$ are the field's initial amplitudes, $k_0 = \omega/c$ is the free-space wavevector, c is the speed of light in vacuum, n_L is the linear refractive index, n_{NL} is the nonlinear index coefficient that describes the Kerr-like all-optical effects, n_{EXT} is what we call the nonlinear electrical index coefficient that describes the Kerr electrical effect. It is related to the original Kerr constant, K, by $n_{EXT} = 2\pi\, K/k_0$. With I being the intensity of the propagating optical wave, the second term in the exponents of Eq. (1) represents known all-optical effects of self-phase and cross-phase modulation. The third term in the exponents is a new addition to the mutual wave interaction control via an applied external electric field. Obviously, both effects now being incorporated into the overall interaction of the propagating waves, interplay between the all-optical and electro-optic phenomena significantly affects the propagation properties.

The second-order correction[16] to the solutions above yields more complex field expressions that, however, provide interesting insight into the waves/components interaction:

$$e_x = e^{-j\frac{k_0}{n_L}\chi\frac{3}{8}\{|e_x|^2+|e_y|^2+E_{ext}^2-\frac{1}{3}|e_y|^2[1-sinc(2\phi z)]\}z}$$
$$\times e^{\frac{k_0}{n_L}\chi\frac{1}{8}|e_y|^2 \cdot \frac{1-cos(2\phi z)}{2\phi}}$$

(4)

In Eq. (2), only the transversal x-component is shown; the transversal y-component is symmetrically identical. The parameter $\Phi = k_0\chi^{(3)}(E_{ext})^2/4n_L$ and is a determining factor in the overall propagation interaction; when $\Phi = 0$, only all-optical effects of self-phase and cross-phase modulation remain with the waves amplitude being constant. Eq. (2) clearly indicates periodic exchange of power between both components along the propagation path, as the second multiplicative term is an amplitude rather than a phase term.

The power densities of both components, p_x and p_y, with the total power being p_T, can then be found as:

$$p_x = \frac{1}{f} ln[1 + 2 sinh(f p_T / 2) p_x^N e^{f p_T/2}]$$
$$p_y = \frac{1}{f} ln[1 - 2 sinh(f p_T / 2) p_y^N e^{-f p_T/2}]$$

(5)

The function $f = Z_0 [1-cos(2\Phi z)]/n_L(E_{ext})^2$ ($Z_0 = 377$ Ω) is the key variable controlling the interaction. As can be seen form Eq. (3), the power exchange is controlled by both, the optical power in the interacting waves (all-optical effect) and the applied external electric field (electrical Kerr effect). It should be pointed out that the nonlinear susceptibility of various materials is not a unique quantity as several physical effects contribute to a material response on different time scales[18]. Also, the third-order susceptibility possesses normally real and imaginary parts that correspond to the nonlinear phase and loss, respectively. A summary of nonlinear parameters of a number of materials relevant to the optical wave propagation issues is given in[19].

4. Re-configurable all-optical switching

The power exchange described by Eq. (3) represents basically cross-polarized wave conversion controlled optically as well as electrically. The optical control is obtained via the optical Kerr effect whereby the intensity of the wave changes the refractive index of the material. In silicon nanocrystal, for example, $\Delta n = 2 \times 10^{-6}$ at an intensity of 10^6 W/cm^2. The price paid for this all-optical control is absorption that leads to the total power loss. Linear absorption is usually negligible since it is desirable that the waves propagate in the transparent region of a given material. The nonlinear absorption however can and does play a negative role due to two-photon or even three-photon absorption. The more common two-photon absorption coefficient causes the nonlinear absorption increase with a square of the intensity thus becoming detrimental at higher wave power densities such as those in optical waveguides and optical fibers.

The electrical control is achieved via the electrical Kerr effect whereby bound electrons in the material are displaced by an electric field, which leads to changes in refractive index with the square of the voltage. In silicon nanocrystal, for example, $\Delta n = 4.2 \times 10^{-5}$ at an electric field of $10 \ V/\mu m$. The attractiveness in exploiting this electro-optic control is in its extremely high speed in the subpicosecond range. An example of electrically controlled periodic power exchange in silicon nanocrystal[20] is shown in figure 1. The applied electric field values are $E_{ext(1)} = 0.8 \ V/\mu m$, $E_{ext(2)} = 2 \ V/\mu m$, $E_{ext(1)} = 3 \ V/\mu m$, respectively. The total optical power is $0.11 \ W/cm^2$.

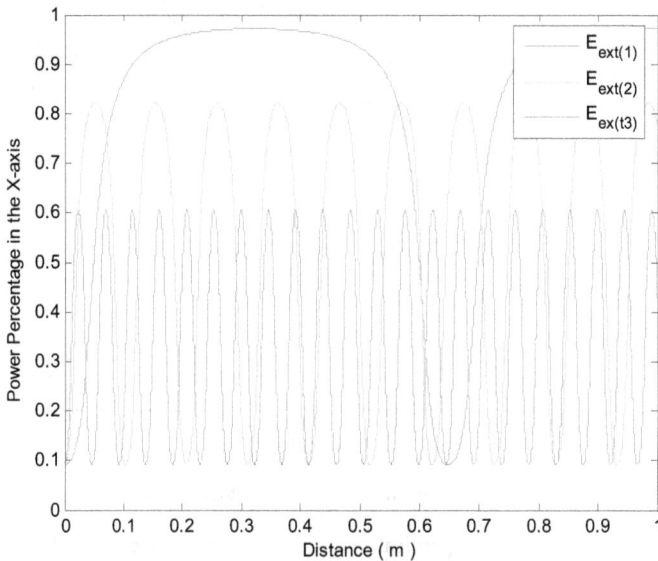

Figure 1. Electrically controlled cross-polarized wave-conversion power exchange

5. Electrically controlled optical multistability

Waves propagating in a resonator filled with an optical Kerr medium that is also subject to an external electric field establish nonlinear behaviour characterized by hysteresis. The shape and the size of the hysteresis transfer function of such a nonlinear optical resonator[21] are controlled by the electric field as well as by the power inside the resonator. Interplay between self-modulation optical and quadratic electro-optic effects is conveniently described using the concept of an effective nonlinear refractive index[22], $n_{eff} = n_L + 3 \chi^{(3)} (E_{ext})^2 +3 \chi^{(3)} (E_{opt})^2/4$. The transfer function of the resonator is written as:

$$F(\gamma) = \frac{1}{1 + \dfrac{4R}{T^2} \sin^2(\gamma)}. \tag{6}$$

It is controlled by the phase parameter, γ, which is a function of the intensity, the electric field, and the wavelength. Figure 2 is an example of the nonlinear behaviour of the input-output characteristic as it depends on the external electric field. A material with $\chi^{(3)} = -9 \times 10^{-16}$ cm^2/V^2 (silicon) was used and $E_{ext\,1,2,3,4,5} = 0.5, 0.8, 7, 8, 9$ $V/\mu m$, respectively.

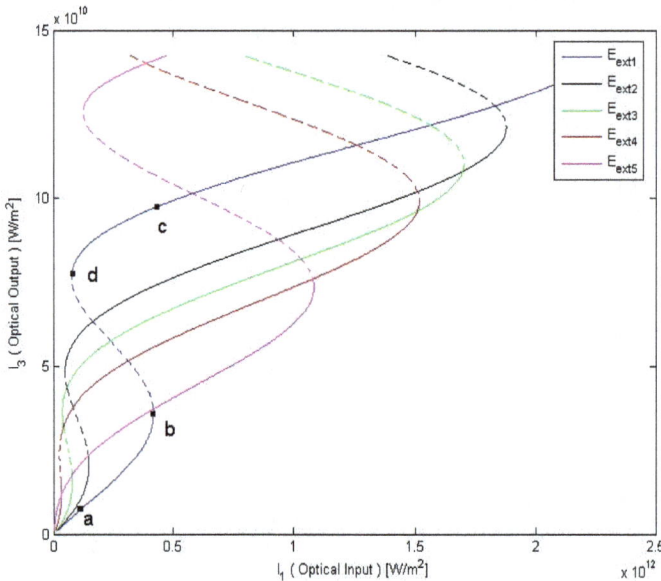

Figure 2. Input-output characteristic of optical nonlinear resonator for different external electric fields

The dashed lines are the unstable optical outputs for given dc electric fields that cause the multistabilities. The electric field controls the required optical input for bistable switching as well as the corresponding output values. Some outputs, which are unstable, become stable, and vice versa, for different biases. This is a tuning property that combined with the previously discussed switching/power-exchange behavior suggests electrically controlled tunable reconfigurability. One can construct a phase diagram of all possible stable optical outputs for a given input as a function of the external electric field [23].

The analysis concludes that the dependence of the system output evolution on the external electric field exhibits a hysteresis-like character as much as with respect to the optical input/output intensity levels. For example, the value of the optical output, at a certain electric field, depends on the history of that field. Such an optically stored electric hysteresis control is a novel feature that can be potentially utilized in the future applications. For example, this hysteresis effect can be used to store an electrical signal (information) optically since the optical system remembers and stores the action of the past electrical signal behavior.

6. Electrically tunable Bragg grating

The rich dispersion properties of Bragg gratings offer many interesting wave propagation features when a nonlinear Kerr medium is incorporated into a grating structure. The key feature is the well-known dispersion property of Bragg gratings whereby their dispersive response is very strong when the operating wavelength is sufficiently close to the Bragg resonance[24] and even if the refractive index changes are very small. An electronically tunable Bragg grating can be constructed based on the third-order nonlinearity discussed here. The nonlinear wave propagation characteristics become interesting and attractive for potential applications.

A homogenous electrical field is known to control, via the Kerr quadratic electro-optic effect, the average refractive index and the birefringence[25]. An inhomogeneous (spatially profiled) electric field bias has been proposed to mediate a linear electro-optic waveguide, by which an effective electro-optic grating is induced[26]. We proposed a novel scheme in which a spatially modulated electric field is applied to a Kerr-nonlinear periodic structure[27]. It was found that several phenomena, including the modulation instability gain, the amplitude (and the width) of the gap soliton, and the band gap, can be efficiently electrically controlled, as long as a proper spatial profile of the electric field is formed.

This can be explained by noting that the electrical spatial profile needs to be designed to act as an extension of the linear perturbation of the periodic structure. An illustrative example of an electrical field bias that has a quasi square-wave shape, which is further modulated by a slow profile, was studied. It was found that, besides the functionality of the periodic part in inducing/controlling all above mentioned phenomena, the slow part of the spatial electrical bias was able to manipulate the linear and nonlinear switching parameters of the band gap. The action of this inhomogeneous electric field is facilitated via the Kerr quadratic electro-optic effect such that the structure's coupling coefficient as well as the average refractive index are controlled.

The effective average refractive index, $\bar{n}_{eff}$, and the effective coupling coefficient, κ_{eff}, of the nonlinear grating were derived[27] showing as they depend on the character of the applied electric field, including its period and shape. The bandwidth of the main reflectivity peak of the grating, $\Delta\omega_{gap}$, was found to be approximately:

$$\Delta\omega_{gap} = \frac{2c}{\bar{n}_{eff}}\left|\kappa_{eff}\right|. \tag{7}$$

It can be seen from Eq. (5) that the band gap width is electrically controlled through the effective coupling coefficient κ_{eff} and the effective average refractive index $\bar{n}_{eff}$. The controllability through the coupling coefficient is more significant. The reason is that the perturbation in the refractive index, which causes the coupling dynamics, is much smaller than the linear refractive index and thus it is more sensitive to a small change in the refractive index that is induced by the electric field. For example, for a fiber Bragg grating at $\lambda = 1550\ nm$, with $n_{NL} = 2.6\ x\ 10^{-16}\ cm^2/W$ and grating linear refractive index variations of $n_1 = 10^{-4}$, the periodically shaped applied electric field will initiate the band gap width change of $287.5\ MHz$ for a field value of $1\ V/\mu m$. As a comparison, if the applied field is spatially constant, the bandwidth of the band gap will change by only $15\ kHz$ for the same $1\ V/\mu m$. The average refractive index is not as sensitive to the external electric field as the grating coupling coefficient is.

We note that intensive optical excitations can detune themselves (totally or partially) out from a band gap of a periodic medium[28]. As the external electric field also controls the periodic structure dispersion properties, one possesses a Bragg grating band gap that has a dually controllable reflection/transmission characteristic. Using a silicon nanocrystal material[29], figure 3 shows a full simulation of a Bragg grating in a waveguide subjected to an electric field. Both the optical and electrical controls are demonstrated.

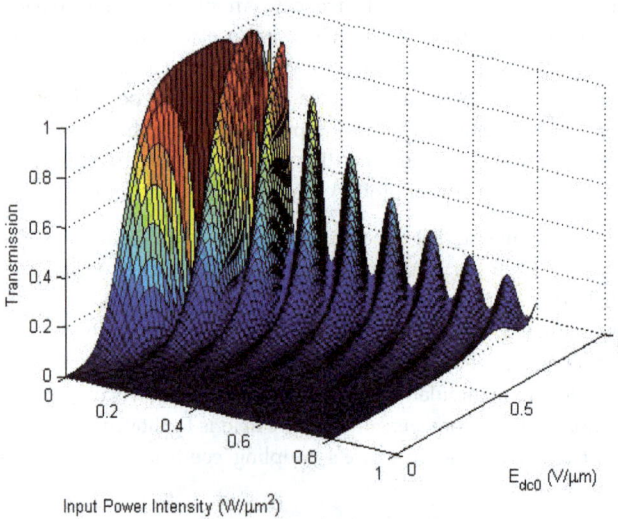

Figure 3. The transmitivity of 5-mm nonlinear waveguide with grating, as function of external electric field and input optical power density; waveguide filled with silicon nanocrystal

7. Kerr switch

The Kerr nonlinearity is usually weak in most materials, although some new materials being developed with nanotechnologies offer promise of significantly enhanced third-order nonlinear coefficients. Silicon nanocrystals can serve as a good example of such a promising material with a stronger nonlinearity of at least one order of magnitude larger than in the bulk counterpart[19]. In order to obtain strong interaction of propagating waves, it is not only the material's nonlinearity that is important. The interaction can be drastically enhanced when optical waves are confined to within a small interaction volume, possibly on the order of less than the wavelength of the waves. Photonic crystal structures, nanoresonators or slot optical waveguides offer such enhanced light confinement.

A combination of all nonlinear interaction enhancing approaches is demonstrated in a Kerr switch design[7] where a ring nanoresonator structure coupled with a slot waveguide filled with silicon nanocrystal is used. For a nominal design wavelength of $\lambda = 1.55\ \mu m$ and the

resonator's length of $38\ \mu m$, the free spectral range of $15.5\ nm$ was obtained with the linewidth of $0.043\ nm$. The transmission characteristics are shown in figure 4 for zero refractive index change and for a $10\ dB$ extinction ratio, respectively.

This shift in figure 4 requires an effective index change of $\Delta n_{eff} = 1.9\ x\ 10^{-5}$, which in turn calls for the material index change of $\Delta n_{si\text{-}nc} = 3.8\ x\ 10^{-5}$. Taking an experimental value for silicon nanocrystal as $\chi^{(3)} = 2\ x\ 10^{-14}\ cm^2/V^2$ (29% of Si in SiO$_2$), the refractive index change for a voltage of $1V/100\ nm$ is $\Delta n_{si\text{-}nc} = 4.2\ x\ 10^{-5}$. The nonlinear loss coefficient is $\beta_2 = 70\ cm/GW$. Taking the power inside the waveguide as $4\ mW$ and with the cross-section being $4\ x\ 10^{-10}\ cm^2$, the nonlinear absorption per one round trip is less than $0.03\ dB$. This is less than in[30], where the loss is due to free carriers. The Kerr effect is as fast as sub-picoseconds. The practical speed of the switch is limited by the capacitance of the electrical contacts onto the resonator/waveguide configuration. For realistic parameters chosen[7], the capacitance is approximately $0.01pF$, which with $50\ \Omega$ yields a theoretical bandwidth of $300\ GHz$.

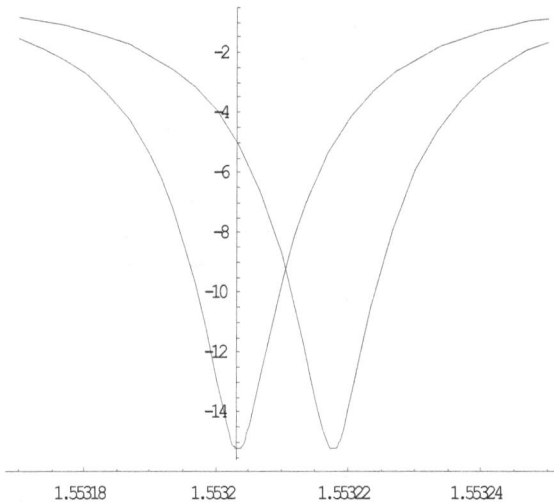

Figure 4. The spectra of Kerr switch.

8. Electrically induced birefringence

Polarization dynamics of ultra-short pulses propagating in an electrically biased silicon waveguide[31] showed interesting features related to wave propagation in nonlinear media. An external electric filed applied to a cubic material induces birefringence via the quadratic

electro-optic effect (DC Kerr effect). Transmitted optical pulse polarization can thus be controlled by adjusting the magnitude of the external electric field. When studying propagation of waves in semiconductor materials, the free-carrier induced susceptibility needs to be accounted for by including the free-carrier index changes and the free-carrier absorption into the analysis[31, 32]. The birefringence coefficient as it depends on the applied external electric field can be written as:

$$\kappa_{eff} = [\Delta\beta + \varepsilon_0 k_0 c \, n_L \, n_{NL} \, (E_{ext})^2]/2, \tag{8}$$

where $\Delta\beta$ is the material linear birefringence. Figure 5 illustrates the electrical-control effect by showing the polarization component transmission coefficient of a 6-mm long waveguide as it changes with the input optical power of a 70-fs long Gaussian pulse.

It should be pointed out that the quantities E_{ext1} and E_{ext2} are amplitudes of the electric field that has, generally, a certain profile along the propagation direction (along the waveguide). Shaping the electric field offers an additional control parameter. The temporal profile of the ultra-short pulse can be governed by a field profile properly designed[31]; an example of a profile is an exponential dependence such that:

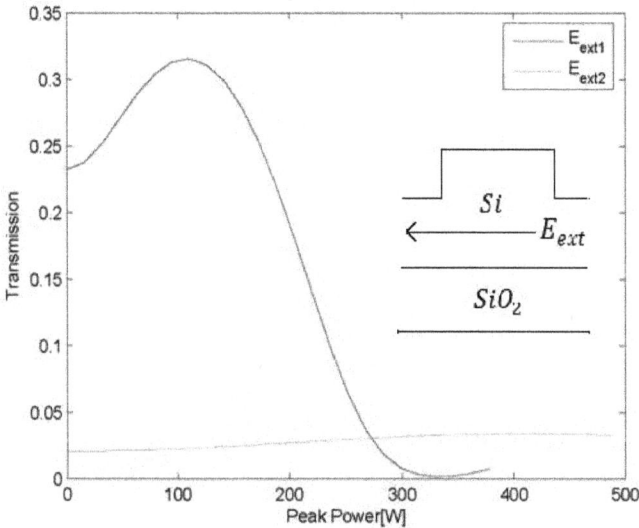

Figure 5. Fig. 5: Transmission coefficient of an electrically induced birefringence waveguide; $E_{ext1}=12.5$ $V/\mu m$, $E_{ext2}=0$ $V/\mu m$, $\Delta\beta=2x10^{-5}k_0$.

$$E_{ext}(z) = \left(\frac{\Delta\beta}{\varepsilon_0 k_0 n_L n_{NL}}\left(e^{\alpha(L-z)} - 1\right)\right)^{1/2}, \qquad (9)$$

where L is the length of the electric field profiling along the propagation direction, and α is a design parameter determining the rate of decay of the exponential profile. Figure 6 illustrates the effect of a properly designed spatial profile of the control electric field for a Gaussian 70-fs pulse traveling along a 2-cm long waveguide. The azimuths of the pulse are chosen within the polarization instability regime. The field shape design parameter $\alpha = 57.5 \ m^{-1}$.

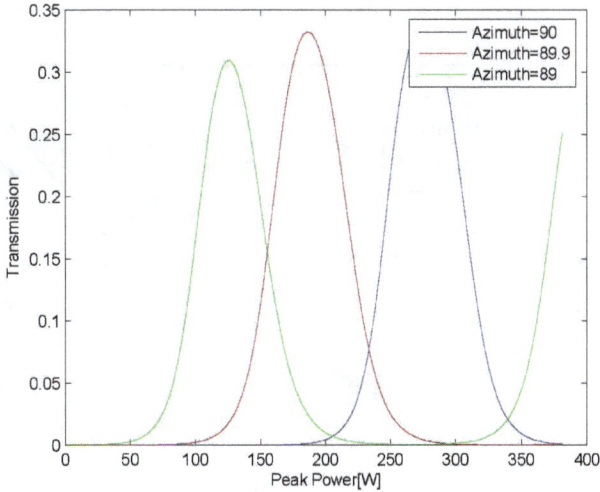

Figure 6. Transmission of Gaussian pulse through electric-field profiled nonlinear waveguide

9. Photorefractive effect in silicon

The crystal symmetry in cubic or isotropic materials, e.g. silicon or glass, etc., can be broken via the third-order nonlinearity by applying an external electric field. This causes such materials behave as if they possessed the linear electro-optic effect. The photorefractive effect has been known since the early 1960's[33]; it was observed in many electro-optic crystals, including $LiNbO_3$, $GaAs$, InP, or $CdTe$[34]. The photorefractive effect is an automatically phase-matched nonlinear phenomenon, whereby interfering light modes can generate a spatially phase-shifted electric field in a host material. This spatially phase-shifted electric field, in turn, couples the interfering light modes in a phase-matched fashion.

It is therefore interesting to investigate the existence and properties of the photorefractive effect in materials without the natural linear electro-optic effect, and exploit the third-order nonlinearity to establish it[6]. The detailed and complex spatio-temporal nonlinear analysis of two contra-propagating waves, including the free carriers in a semiconductor, shows that the photorefractive effect can lead to gain or loss of one of the waves (signal or probe). The power exchange with the other wave (pump) is controlled by the polarity of the external electric field. This is electronically controlled unidirectional power transfer. It may also be considered as a parametric process.

Figure 7 shows an example of such power transfer in a silicon waveguide at $\lambda = 1.55$ μm with an effective cross-section of $0.3 \ \mu m^2$, linear loss of $0.57 \ cm^{-1}$, and an n-doping of 10^{19} cm^{-3}.

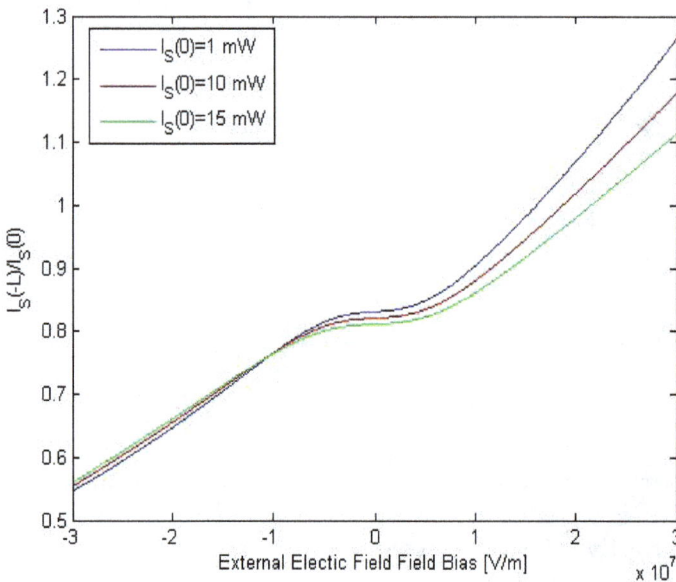

Figure 7. Net signal gain in electrically induced photorefractive silicon waveguide

As can be seen in the figure, the net signal gain can be achieved despite material losses. Also, as is expected in any photorefractive medium, the net signal gain increases with decreasing input signal power for a constant pump power.

If the frequency detuning, Ω, is different from zero, gain enhancement will be experienced at a certain frequency detuning, i.e. at which the amplitude of the photo-induced space charge electric field is maximized[6]. The net signal gain versus frequency detuning is exhibited in figure 8.

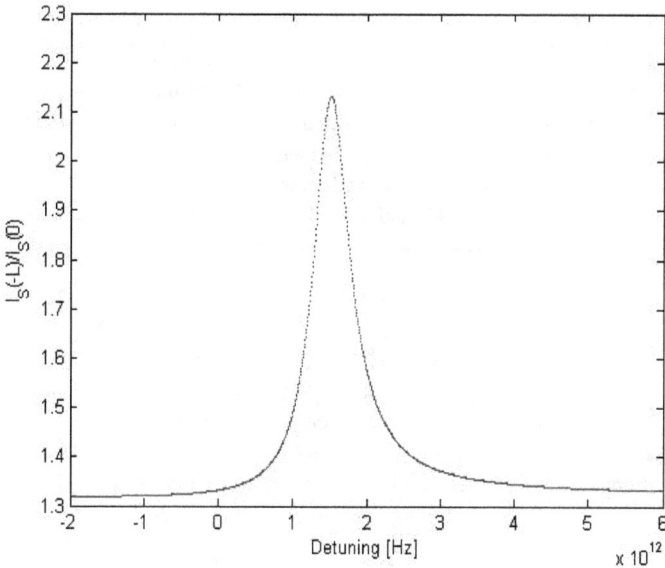

Figure 8. Net signal gain enhancement versus frequency detuning; $E_{ext} = 25\ V/\mu m$

10. Discussion

This work focused on investigating optical wave propagation properties and associated potential devices functionalities in Kerr-type media with applied external electric field. The assisting external fields induce a quadratic electro-optic effect in a centro-symmetric (cubic, isotropic) third-order nonlinear materials (e.g. glass, silicon, silicon nanocrystal). If the optical field of propagating waves is sufficiently intense, the all-optical effects start to appear as well. The interplay between these two effects (the Kerr electro-optic and all-optical effects) was the main focus with the goal to demonstrate phenomena potentially useful in the design of novel photonic devices. Although the few presented numerical examples are for silicon, silicon nanocrystal or silicon nanowires, chosen due to the silicon's attractiveness in its integreability with standard microelectronics technologies, the obtained results are applicable to a variety of other optical materials, including silica glass, GaAs bulk, CdTe bulk, GaAs and InP quantum wells, CdTe nanocrystal, CdS nanocrystal, poly (β-pinene), fullerene-containing polyurethane films, natural rubber, and many others.

Optical waves propagating in an electro-optic-Kerr-effect-induced birefringence medium were studied showing electrical and optical control of power exchange between their components. The birefringence is proportional to the square of an applied external electric field. The concept of an effective refractive index containing nonlinear optical as well as electrical dependencies was introduced to model Kerr nonlinear wave propagation behaviour when the material is subjected to external electric fields The wave properties suggest that one can exploit them in designing novel photonic devices, such as, for example, an optically controlled electro-optical switch or an electronically controlled all-optical modulator.

An optical Fabry–Pérot resonator filled with a Kerr nonlinear material and subjected to an external electric field was investigated. As expected, the optical input-output transfer function displays hysteresis that is controlled by the applied electric field. The stability analysis revealed electronically-tunable optical multi-stability of the resonator. This means that the state of the optical intensity implies a desired information for a given external electric field. Such a feature offers a new functionality whereby electrical information can be stored optically.

When a Bragg grating is made in a Kerr material, the electrically induced control of the spatial inhomogeneity is established. As a result, the modulation instability gain was found to be electrically controlled. This suggests a possible realization of an electrically controlled pulse generator. Also, the amplitude and the width of the gap soliton can be, in such a case, electrically adjustable, and thus a tunable soliton channel is possible. The external electric field having a spatial profile modulated by a slow varying profile makes it possible to control the nonlinear grating band gap electrically very efficiently compared to other known means, the reason being that by profiling the electric field one gains a direct control over the grating coupling coefficients.

Ultra-short optical pulses propagating in a nonlinear Kerr medium while an external electric field was applied, was studied. A silicon waveguide was considered. Several realistic effects were taken into account, including large linear loss, nonlinear anisotropy, two-photon absorption, and associated free carriers. It was shown that the silicon waveguide can be used as a practical platform for all-optical applications, including polarization switching and pulse shaping. A properly designed of the external electric spatial profile was shown to help achieve polarization instability regime, which is important for realizing sensitive polarization discriminating devices.

The new photorefractive effect in cubic materials (e.g. silicon nanocrystal) was established and investigated. As cubic materials do not possess the linear electro-optic effect, the photorefractive effect is not readily obtained. We demonstrated that a proper external electric field can assist in realizing the effect in such materials and structures, for example in silicon waveguides. Despite the linear and nonlinear losses, it was shown that a weak signal counter-propagating with a strong pump can experience a net gain. One may suggest that integrated photorefractive devices that are optically and electronically controlled can be designed based on this new phenomenon.

The four-wave mixing (FWM) phenomenon was not examined in this work. However, since it is a nonlinear process that takes place in a third-order nonlinear medium, resulting in the generation of a new optical wave with a new frequency (parametric process), it is realizable in the Kerr media considered here. As it is known, a readily efficient FWM process cannot be achieved because of the phase-mismatching dilemma. One way to achieve efficient FWM is to utilize a nonlinear periodic structure[35]. Based on the work presented here, a properly profiled external electric field could be employed to achieve a tunable quasi-phase matching operation. It is thus interesting to investigate the FWM phenomenon in the presence of an external electric field, as this is very important in realizing tunable devices for all-optical communications and processing.

As a conclusion, owing to its potential for integration with micro-electronics, the silicon-based technology is considered one of the most important means for photonic applications. The dual electro-optical and all-optical functionality studied here in materials of the same symmetry as silicon and its derivatives, resulting from the Kerr effect, offers a promise and a potential for realizing technologically compatible and implementation friendly electrically-controlled all-optical devices and/or optically controlled electro-optical devices that can be readily integrated onto a common material platform.

11. Conclusion

Optical wave propagation in a Kerr-type nonlinear medium has been analyzed theoretically and studied numerically. New features were found where not only waves and their polarization interactions are present as a result of the nonlinearity of the medium, but also an interplay between the optical and electrical Kerr effects contributes to the resulting functionality. Several novel wave propagation effects were discovered. They include cross-polarized wave conversion, optical multistability, nonlinear tunability of periodic structures, ultra-fast electro-optic switching, and a new photorefractive effect. Possible applications of such novel functionalities were discussed. Examples utilizing silicon as the common semiconductor material were given.

Author details

Michal Čada
Dalhousie University, Halifax,
Canada

Montasir Qasymeh
Abu Dhabi University, Abu Dhabi,
United Arab Emirates

Jaromír Pištora
VŠB - Technical University of Ostrava, Ostrava,
Czech Republic

Acknowledgement

Authors acknowledge generous support for this work from the National Science and Engineering Council (NSERC), The Mathematics of Information Technology and Complex Systems (MITACS) Network of Excellence, and OZ Optics, Ltd. of Kanata, Ontario, all of Canada; and from the European Union project NANOBASE, # CZ.1.07/2.3.00/20.0074, of the Czech Republic.

12. References

[1] M. Cada, "Electro-optic devices", report, NTC, UPV, Spain, 2008

[2] J. Kerr, Phil. Mag. J.Sci., ser. Fourth, vol.50, 1875

[3] M. Cada, M. Qasymeh, J. Pistora, Opt. Express, 16, 5, 2008

[4] M. Qasymeh, M. Cada, S. A. Ponomarenko, IEEE JQE-44, 8, 2008

[5] G. Torrese, J. Taylor, H. P. Schriemer, M. Cada, J. Opt. A, 8, 2006

[6] M. Qasymeh, Ph.D. thesis, Dalhousie University, Halifax, Canada, 2010

[7] M. Cada, J. Pistora, ISMOT 2011, June, Prague, Czech Republic

[8] R. A. Soref, B. R. Bennett, IEEE JQE-23, 1, 1987

[9] J.-H. Ryou et all, IEEE JSTQE-15, 4, 2009

[10] L. Liao et all, Electron. Lett. 43, 22, 2007

[11] A. Liu et all, Opt. Express, 15, 2, 2007

[12] M. Cada et all, Electron. Lett. 28, 23, 1992

[13] A. Liu et all, Nature 427, 2004

[14] T. S. Moss et all, "Semiconductor opto-electronics", Butterworths, London, 1973

[15] C. Rolland et all, Electron. Lett. 29, 5, 1993

[16] E. Infeld, G. Rowlands, "Nonlinear waves, solitons and chaos", Cambridge University Press, London, 2000

[17] S. Brandt, H. D. Dahmen, "The picture book of quantum mechanics", Springer Verlag, New York, 1995

[18] M. J. Weber, "Handbook of optical materials", CRC Press, Washington D.C., 2003

[19] M. Cada, "Nonlinear Kerr materials", report, NTC, UPV, Spain, 2008

[20] M. Cada, Ostrava Nano-2011, March, Ostrava, Czech Republic

[21] S. H. Shehadeh et all, IEEE Sensors J., 11, 9, 2011

[22] M. Qasymeh et all, IEEE JQE, 44, 8, 2008

[23] M. Qasymeh et all, IJMOT, 3, 3, 2008

[24] A. Yariv, P. Yeh, "Optical waves in crystals", John Wiley & Sons, 1984

[25] M. Li et all, Opt. Express, 16, 9, 2008.

[26] M. Kulishov et all, JOSA B, 18, 2, 2001

[27] M. Cada, M. Qasymeh, Adv. Sci. Eng. Med., 3, 1/2, 2011

[28] A. Melloni et all, IEEE Photon. Tech. Lett., 12, 1, 2000

[29] R. Spano et all, "Nonlinear optical properties of silicon nanocrystal for applications in photonic logic gates devices" Winter Topical Meeting Series, IEEE/LEOS, January, 2008

[30] Xu Q. et all, Nature 435, May, 2005

[31] M. Qasymeh et all, Opt. Express, 17, 10, 2009

[32] Q. Lin et all, Opt. Express, 15, 6, 2007

[33] A. Ashkin et all, Appl. Phys. Lett., 9, 2, 1966

[34] P. Yeh," Introduction to Photorefractive Nonlinear Optics", Wiley, New York, 1993

[35] G. Bartal et all, Phys. Rev. Lett., 97, 10, 2006

Analyzing Wave Propagation in Helical Waveguides Using Laplace, Fourier, and Their Inverse Transforms, and Applications

Z. Menachem and S. Tapuchi

Additional information is available at the end of the chapter

1. Introduction

Various methods for the analysis of wave propagation in the curved waveguides have been studied in the literature. Two interesting methods of investigation of propagation along the curved waveguides are based on the ray model and the mode model. A review of the hollow waveguide technology [1-2] and a review of IR transmitting, hollow waveguides, fibers and integrated optics [3] were published. The first theoretical analysis of the problem of hollow cylindrical bent waveguides was published by Marcatili and Schmeltzer [4], where the theory considers the bending as a small disturbance and uses cylindrical coordinates to solve Maxwell equations. They derive the mode equations of the disturbed waveguide using the ratio of the inner radius r to the curvature radius R as a small parameter ($r/R \ll 1$). Their theory predicts that the bending has little influence on the attenuation of a hollow metallic waveguide. Marhic [5] proposed a mode-coupling analysis of the bending losses of circular metallic waveguide in the IR range for large bending radii. In the circular guide it is found that the preferred TE_{01} mode can couple very effectively to the lossier TM_{11} mode when the guide undergoes a circular bend. For circular waveguides, the microwave approximation has been used for the index of refraction and the straight guide losses, and the results indicate very poor bending properties due to the near degeneracy of the TE_{01} and TM_{11} modes, thereby offering an explanation for the high losses observed in practice.

Miyagi et al. [6] suggested an improved solution, which provided agreement with the experimental results, but only for $r/R \ll 1$. A different approach [5,7] treats the bending as a perturbation that couples the modes of a straight waveguide. That theory explains the large difference between the metallic and metallic-dielectric bent waveguide attenuation. The reason for this difference is that in metallic waveguides the coupling between the TE and TM modes caused by the bending mixes modes with very low attenuation and modes with very high attenuation, whereas in metallic-dielectric waveguides, both the TE and TM modes have low attenuation. Hollow waveguides with both metallic and dielectric internal layers were proposed to reduce the transmission losses. Hollow-core waveguides have two possibilities. The inner core materials have relative refractive indices greater than one (namely, leaky waveguides) or the inner wall material has a relative refractive index of less than one. A hollow waveguide can be made, in principle, from any flexible or rigid tube (plastic, glass, metal, etc.) if its inner hollow surface (the core) is covered by a metallic layer and a dielectric overlayer. This layer structure enables us to transmit both the TE and TM polarization with low attenuation [5,7].

A method for the electromagnetic (EM) analysis of bent waveguides [8] is based on the expansion of the bend mode in modes of the straight waveguides, including the modes under the cutoff. A different approach to calculate the bending losses in curved dielectric waveguides [9] is based on the well-known conformal transformation of the index profile and on vectorial eigenmode expansion combined with perfectly matched layer boundary conditions to accurately model radiation losses. An improved ray model for simulating the transmission of laser radiation through a metallic or metallic dielectric multibent hollow cylindrical waveguide was proposed [10-11]. It was shown theoretically and proved experimentally that the transmission of CO_2 laser radiation is possible even through bent waveguide.

The propagation of EM waves in a loss-free inhomogeneous hollow conducting waveguide with a circular cross section and uniform plane curvature of the longitudinal axis was considered [12]. For small curvature the field equations can be solved by means of an analytical approximation method. In this approximation the curvature of the axis of the waveguide was considered as a disturbance of the straight circular cylinder, and the perturbed torus field was expanded in eigenfunctions of the unperturbed problem. An extensive survey of the related literature can be found especially in the book on EM waves and curved structures [13]. The radiation from curved open structures is mainly considered by using a perturbation approach, that is by treating the curvature as a small perturbation of the straight configuration. The perturbative approach is not entirely suitable for the analysis of relatively sharp bends, such as those required in integrated optics and especially short millimeter waves. The models based on the perturbation theory consider the bending as a perturbation ($r/R \ll 1$), and solve problems only for a large radius of curvature.

Several methods of investigation of propagation were developed for study of empty curved waveguide and bends [14-17]. The results of precise numerical computations and extensive analytical investigation of the angular propagation constants were presented for various electromagnetic modes which may exit in waveguide bends of rectangular cross section [14]. A new equivalent circuit for circular E-plane bends, suitable for any curvature radius and rectangular waveguide type was presented in Ref. [15]. An accurate and efficient method of moments solution together with a mode-matching technique for the analysis of curved bends in a general parallel-plate waveguide was described in the case of a rectangular waveguides [16]. A rigorous differential method describing the propagation of an electromagnetic wave in a bent waveguide was presented in Ref. [17].

Several methods of propagation along the toroidal and helical waveguides were developed, based on Maxwell's equations. The method for the analysis of EM wave propagation along the toroidal waveguide [18] has been derived with arbitrary profiles, and with rectangular metal tubes. An improved approach has been derived for the propagation of EM field along a toroidal dielectric waveguide with a circular cross-section [19]. The meaning of the improved approach is that the method employs helical coordinates (and not cylindrical coordinate, such as in the methods that considered the bending as a perturbation). Thus the Laplacian of the wave equations is based on the metric coefficients in the case of the helical waveguide with a circular cross section. The method for the propagation of EM field along a helical dielectric waveguide with a circular cross section [20] has been proposed. The method for the propagation of EM field along a helical dielectric waveguide with a rectangular cross section has been proposed [21]. It is very interesting to compare between the mode model methods for wave propagation in the curved waveguide with a rectangular cross section and with a circular cross section. The methods [18-19] have been derived for one bending of the toroidal waveguide (approximately a plane curve) in the case of small values of step angle of the helix. The methods [20-21] have been derived for one bending of the helical waveguide (a space curved waveguide) for an arbitrary value of the step's angle of the helix. These methods were generalized from a toroidal dielectric waveguide (approximately a plane curve) with one bending to a helical waveguide (a space curved waveguide for an arbitrary value of the step's angle of the helix) with one bending. The two above methods employ toroidal or helical coordinates (and not cylindrical coordinates, such as in the methods that considered the bending as a perturbation $(r/R \ll 1)$), and the calculations are based on using Laplace and Fourier transforms, and the output fields are computed by the inverse Laplace and Fourier transforms. Laplace transform on the differential wave equations is needed to obtain the wave equations (and thus also the output fields) that are expressed directly as functions of the transmitted fields at the entrance of the waveguide at $\zeta = 0^+$. Thus, the Laplace transform is necessary to obtain the comfortable and simple *input-output* connections of the fields.

This chapter presents two improved methods for the propagation of EM fields along a helical dielectric waveguide with a circular cross section and a rectangular cross section. The two different methods employ helical coordinates (and not cylindrical coordinates, such as in the methods that considered the bending as a perturbation). The calculations are based on using Laplace and Fourier transforms, and the output fields are computed by the inverse Laplace and Fourier transforms. Laplace transform on the differential wave equations is needed to obtain the wave equations and the output fields that are expressed directly as functions of the transmitted fields at the entrance of the waveguide. Thus, the Laplace transform is necessary to obtain the comfortable and simple *input-output* connections of the fields. The output power transmission and the output power density are improved by increasing the step's angle or the radius of the cylinder of the helix, especially in the cases of space curved waveguides. These methods can be a useful tool to improve the output results in all the cases of the hollow helical waveguides in medical and industrial regimes (by the first method) and in the microwave and millimeter-wave regimes, for the diffused optical waveguides in integrated optics (by the second method).

2. The derivation of the two different methods

This chapter presents two improved methods for the propagation of EM fields along a helical dielectric waveguide with a circular cross section (by the first method) and a rectangular cross section (by the second method). A general scheme of the helical coordinate system (r, θ, ζ) is shown in Fig. 1(a) and the circular helical waveguide is shown in Fig. 1(b), where $0 \leq r \leq a + \delta_m$, and 2a is the internal diameter of the cross-section. A general scheme of the helical coordinate system (x, y, ζ) is shown in Fig. 1(c) and the rectangular helical waveguide is shown in Fig. 1(d), where $0 \leq x \leq a, 0 \leq y \leq b$, and a and b are the dimensions in the cross section. In these figures, R is the radius of the cylinder, and ζ is the coordinate along the axis of the helical waveguide.

It is very interesting to compare between the mode model methods for wave propagation in the helical waveguide with a circular cross section and in the helical waveguide with a rectangular cross section. These the two kinds of the different methods enable us to solve practical problems with different boundary conditions. The two methods employ helical coordinates (and not cylindrical coordinates, such as in the methods that considered the bending as a perturbation (r/R)). The calculations are based on using Laplace and Fourier transforms, and the output fields are computed by the inverse Laplace and Fourier transforms. Laplace transform on the differential wave equations is needed to obtain the wave equations (and thus also the output fields) that are expressed directly as functions of the transmitted fields at the entrance of the waveguide at $\zeta = 0^+$. Thus, the Laplace transform is necessary to obtain the comfortable and simple *input-output* connections of the fields. The derivation for a helical waveguide with a circular cross section is given in detail in [20]. The derivation for a helical waveguide with a rectangular cross section is given in detail in [21]. Let us repeat these difference methods, in brief.

2.1 Formulation of the problem for the helical coordinate system (r, θ, ζ) and for the helical coordinate system (x, y, ζ).

We start by finding the metric coefficients from the helical transformation of the coordinates. The helical transformation of the coordinates is achieved by two rotations and one translation, and is given in the form:

$$\begin{pmatrix} X \\ Y \\ Z \end{pmatrix} = \begin{pmatrix} \cos(\phi_c) & -\sin(\phi_c) & 0 \\ \sin(\phi_c) & \cos(\phi_c) & 0 \\ 0 & 0 & 1 \end{pmatrix} \begin{pmatrix} 1 & 0 & 0 \\ 0 & \cos(\delta_p) & -\sin(\delta_p) \\ 0 & \sin(\delta_p) & \cos(\delta_p) \end{pmatrix} \begin{pmatrix} r\sin\theta \\ 0 \\ r\cos\theta \end{pmatrix} + \begin{pmatrix} R\cos(\phi_c) \\ R\sin(\phi_c) \\ \zeta\sin(\delta_p) \end{pmatrix}, \quad (1)$$

where ζ is the coordinate along the helix axis, R is the radius of the cylinder, δ_p is the step's angle of the helix (see Figs. (2(a))-(2(b))), and $\phi_c = (\zeta \cos(\delta_p))/R$. Likewise, $0 \leq r \leq a + \delta_m$, where 2a is the internal diameter of the cross-section of the helical waveguide, and δ_m is the thickness of the metallic layer, as shown in Fig. 3(a).

Figure 2(a) shows the rotations and translation of the orthogonal system $(\overline{X}, \overline{\zeta}, \overline{Z})$ from point A to the orthogonal system (X,Y,Z) at point K. Figure 2(b) shows the deployment of the helix depicted in Fig. 2(a).

According to Equation (1), the helical transformation of the coordinates with a circular cross section becomes

$$X = (R + r\sin\theta)\cos(\phi_c) + r\sin(\delta_p)\cos\theta\sin(\phi_c), \qquad (2a)$$

$$Y = (R + r\sin\theta)\sin(\phi_c) - r\sin(\delta_p)\cos\theta\cos(\phi_c), \qquad (2b)$$

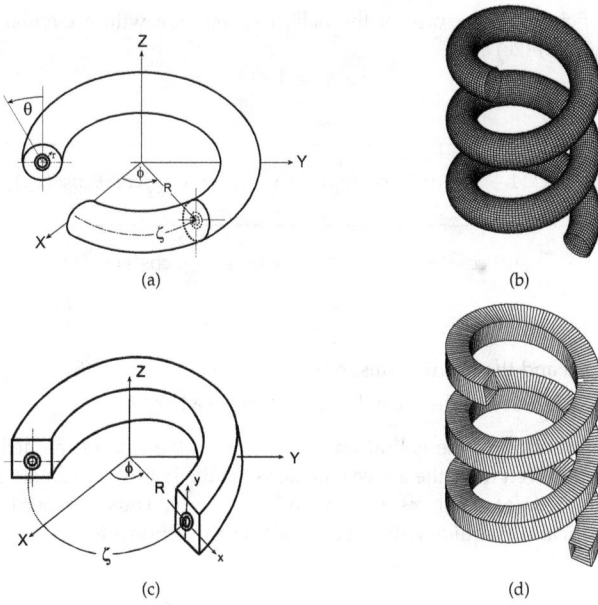

Figure 1. (a) A general scheme of the helical coordinate system (r, θ, ζ). (b) The circular helical waveguide. (c) A general scheme of the helical coordinate system (x, y, ζ). (d) The rectangular helical waveguide.

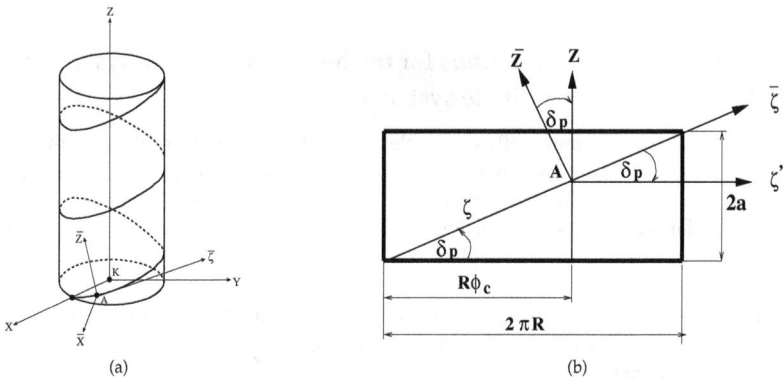

Figure 2. (a) Rotations and translation of the orthogonal system ($\overline{X}$, $\overline{\zeta}$, $\overline{Z}$) from point A to the orthogonal system (X,Y,Z) at point K. (b) Deployment of the helix.

$$Z = r\cos\theta\cos(\delta_p) + \zeta\sin(\delta_p), \qquad (2c)$$

where $\phi_c = (\zeta/R)\cos(\delta_p)$, R is the radius of the cylinder, and (r, θ) are the parameters of the cross-section. Note that $\zeta\sin(\delta_p) = R\phi_c\tan(\delta_p)$.

The metric coefficients in the case of the helical waveguide with a circular cross section, according to Eqs. (2a)-(2c) are:

$$h_r = 1, \tag{3a}$$

$$h_\theta = r, \tag{3b}$$

$$h_\zeta = \sqrt{\left(1 + \frac{r}{R}\sin\theta\right)^2 \cos^2(\delta_p) + \sin^2(\delta_p)\left(1 + \frac{r^2}{R^2}\cos^2\theta\cos^2(\delta_p)\right)}$$

$$= \sqrt{1 + \frac{2r}{R}\sin\theta\cos^2(\delta_p) + \frac{r^2}{R^2}\sin^2\theta\cos^2(\delta_p) + \frac{r^2}{R^2}\cos^2\theta\cos^2(\delta_p)\sin^2(\delta_p)}$$

$$\simeq 1 + \frac{r}{R}\sin\theta\cos^2(\delta_p). \tag{3c}$$

Note that the third and the fourth terms in the root of the metric coefficient h_ζ are negligible in comparison to the first and the second terms when $(r/R)^2 \ll 1$.

The metric coefficients, and the helical transformation in the case of the helical coordinate system (x, y, ζ) are given from the above equations for the helical coordinate system (r, θ, ζ) and according to Fig. 1(a), where $r\sin\theta = x$, and $r\cos\theta = y$. Thus, the metric coefficients in the case of the helical waveguide with a rectangular cross section are:

$$h_x = 1, \tag{4a}$$

$$h_y = 1, \tag{4b}$$

$$h_\zeta \simeq 1 + \frac{x}{R}\cos^2(\delta_p). \tag{4c}$$

3. Solution of the wave equations for the helical coordinate system (r, θ, ζ) and for the helical coordinate system (x, y, ζ).

The two kinds of the different methods enable us to solve practical problems with different boundary conditions. The two methods employ helical coordinates (and not cylindrical coordinates, such as in the methods that considered the bending as a perturbation (r/R $\ll$ 1)). The calculations are based on using Laplace and Fourier transforms, and the output fields are computed by the inverse Laplace and Fourier transforms. Laplace transform on the differential wave equations is needed to obtain the wave equations (and thus also the output fields) that are expressed directly as functions of the transmitted fields at the entrance of the waveguide at $\zeta = 0^+$. Thus, the Laplace transform is necessary to obtain the comfortable and simple *input-output* connections of the fields.

3.1 Solution of the wave equations for the helical coordinate system (r, θ, ζ).

The derivation is based on an arbitrary value of the step's angle of the helix (δ_p). The derivation is based on Maxwell's equations for the computation of the EM field and the radiation power density at each point during propagation along a helical waveguide, with a radial dielectric profile. The longitudinal components of the fields are developed into the

Analyzing Wave Propagation in Helical Wave
guides Using Laplace, Fourier, and Their Inverse Transforms, and Applications

199

Fourier-Bessel series. The transverse components of the fields are expressed as a function of the longitudinal components in the Laplace transform domain. Finally, the transverse components of the fields are obtained by using the inverse Laplace transform by the residue method, for an arbitrary value of the step's angle of the helix (δ_p).

The derivation is given for the lossless case to simplify the mathematical expressions. In a linear lossy medium, the solution is obtained by replacing the permitivity ϵ by $\epsilon_c = \epsilon - j(\sigma/\omega)$ in the solutions for the lossless case, where ϵ_c is the complex dielectric constant, and σ is the conductivity of the medium. The boundary conditions for a lossy medium are given after the derivation. For most materials, the permeability μ is equal to that of free space ($\mu = \mu_0$). The wave equations for the electric and magnetic field components in the inhomogeneous dielectric medium $\epsilon(r)$ are given by

$$\nabla^2 E + \omega^2 \mu \epsilon E + \nabla \left(E \cdot \frac{\nabla \epsilon}{\epsilon} \right) = 0, \tag{5a}$$

and

$$\nabla^2 H + \omega^2 \mu \epsilon H + \frac{\nabla \epsilon}{\epsilon} \times (\nabla \times H) = 0, \tag{5b}$$

respectively. The transverse dielectric profile ($\epsilon(r)$) is defined as $\epsilon_0(1 + g(r))$, where ϵ_0 represents the vacuum dielectric constant, and $g(r)$ is its profile function in the waveguide. The normalized transverse derivative of the dielectric profile (g_r) is defined as $(1/\epsilon(r))(\partial \epsilon(r)/\partial r)$.

From the transformation of Eqs. (3a)-(3c) we can derive the Laplacian of the vector E (i.e., $\nabla^2 E$), and obtain the wave equations for the electric and magnetic fields in the inhomogeneous dielectric medium. It is necessary to find the values of $\nabla \cdot E$, $\nabla(\nabla \cdot E)$, $\nabla \times E$, and $\nabla \times (\nabla \times E)$ in order to obtain the value of $\nabla^2 E$, where $\nabla^2 E = \nabla(\nabla \cdot E) - \nabla \times (\nabla \times E)$. All these values are dependent on the metric coefficients (3a,b,c).

The ζ component of $\nabla^2 E$ is given by

$$(\nabla^2 E)_\zeta = \nabla^2 E_\zeta + \frac{2}{Rh_\zeta^2} \left[\sin \theta \frac{\partial}{\partial \zeta} E_r + \cos \theta \frac{\partial}{\partial \zeta} E_\theta \right] - \frac{1}{R^2 h_\zeta^2} E_\zeta, \tag{6}$$

where

$$\nabla^2 E_\zeta = \frac{\partial^2}{\partial r^2} E_\zeta + \frac{1}{r^2} \frac{\partial^2}{\partial \theta^2} E_\zeta + \frac{1}{r} \frac{\partial}{\partial r} E_\zeta + \frac{1}{h_\zeta} \left[\frac{\sin \theta}{R} \frac{\partial}{\partial r} E_\zeta + \frac{\cos \theta}{rR} \frac{\partial}{\partial \theta} E_\zeta + \frac{1}{h_\zeta} \frac{\partial^2}{\partial \zeta^2} E_\zeta \right], \tag{7}$$

and in the case of $h_\zeta = 1 + (r/R) \sin \theta \cos^2(\delta_p)$.

The longitudinal components of the wave equations (5a) and (5b) are obtained by deriving the following terms

$$\left[\nabla(E \cdot \frac{\nabla \epsilon}{\epsilon}) \right]_\zeta = \frac{1}{h_\zeta} \frac{\partial}{\partial \zeta} \left[E_r g_r \right], \tag{8}$$

and

$$\left[\frac{\nabla\epsilon}{\epsilon}\times(\nabla\times H)\right]_\zeta = jw\epsilon\left[\frac{\nabla\epsilon}{\epsilon}\times E\right]_\zeta = jw\epsilon g_r E_\theta. \tag{9}$$

The longitudinal components of the wave equations (5a) and (5b) are then written in the form

$$\left(\nabla^2 E\right)_\zeta + k^2 E_\zeta + \frac{1}{h_\zeta}\frac{\partial}{\partial\zeta}\left(E_r g_r\right) = 0, \tag{10}$$

$$\left(\nabla^2 H\right)_\zeta + k^2 H_\zeta + jw\epsilon g_r E_\theta = 0, \tag{11}$$

where $(\nabla^2 E)_\zeta$, for instance, is given in Eq. (6). The *local* wave number parameter is $k = w\sqrt{\mu\epsilon(r)} = k_0\sqrt{1+g(r)}$, where the free-space wave number is $k_0 = w\sqrt{\mu_0\epsilon_0}$.

The transverse Laplacian operator is defined as

$$\nabla_\perp^2 \equiv \nabla^2 - \frac{1}{h_\zeta^2}\frac{\partial^2}{\partial\zeta^2}. \tag{12}$$

The Laplace transform

$$\tilde{a}(s) = \mathcal{L}\{a(\zeta)\} = \int_{\zeta=0}^{\infty} a(\zeta)e^{-s\zeta}d\zeta \tag{13}$$

is applied on the ζ-dimension, where $a(\zeta)$ represents any ζ-dependent variables, where $\zeta = (R\phi_c)/\cos(\delta_p)$.

The next four steps are given in detail in Ref. [19], as a part of our derivation. Let us repeat these four steps, in brief.

1). By substituting Eq.(6) into Eq.(10) and by using the Laplace transform (13), the longitudinal components of the wave equations (Eqs. (10)-(11)) are described in the Laplace transform domain, as *coupled* wave equations.

2). The transverse fields are obtained directly from Maxwell's equations, and by using the Laplace transform (13), and are given by

$$\tilde{E}_r(s) = \frac{1}{s^2 + k^2 h_\zeta^2}\left\{-\frac{jw\mu_0}{r}\left[\frac{r}{R}\cos\theta\cos^2(\delta_p)\tilde{H}_\zeta + h_\zeta\frac{\partial}{\partial\theta}\tilde{H}_\zeta\right]h_\zeta + s\left[\frac{\sin\theta}{R}\cos^2(\delta_p)\tilde{E}_\zeta + h_\zeta\frac{\partial}{\partial r}\tilde{E}_\zeta\right]\right.$$

$$\left. +sE_{r_0} - jw\mu_0 H_{\theta_0}h_\zeta\right\}, \tag{14a}$$

$$\tilde{E}_\theta(s) = \frac{1}{s^2 + k^2 h_\zeta^2}\left\{\frac{s}{r}\left[\frac{r}{R}\cos\theta\cos^2(\delta_p)\tilde{E}_\zeta + h_\zeta\frac{\partial}{\partial\theta}\tilde{E}_\zeta\right] + jw\mu_0 h_\zeta\left[\frac{\sin\theta}{R}\cos^2(\delta_p)\tilde{H}_\zeta + h_\zeta\frac{\partial}{\partial r}\tilde{H}_\zeta\right]\right.$$

$$\left. +sE_{\theta_0} + jw\mu_0 H_{r_0}h_\zeta\right\}, \tag{14b}$$

Analyzing Wave Propagation in Helical Wave
guides Using Laplace, Fourier, and Their Inverse Transforms, and Applications

201

$$\tilde{H}_r(s) = \frac{1}{s^2 + k^2 h_\zeta^2} \left\{ \frac{jw\epsilon}{r} \left[\frac{r}{R} \cos\theta \cos^2(\delta_p) \tilde{E}_\zeta + h_\zeta \frac{\partial}{\partial\theta} \tilde{E}_\zeta \right] h_\zeta + s \left[\frac{\sin\theta}{R} \cos^2(\delta_p) \tilde{H}_\zeta + h_\zeta \frac{\partial}{\partial r} \tilde{H}_\zeta \right] \right.$$

$$\left. + sH_{r_0} + jw\epsilon E_{\theta_0} h_\zeta \right\}, \tag{14c}$$

$$\tilde{H}_\theta(s) = \frac{1}{s^2 + k^2 h_\zeta^2} \left\{ \frac{s}{r} \left[\frac{r}{R} \cos\theta \cos^2(\delta_p) \tilde{H}_\zeta + h_\zeta \frac{\partial}{\partial\theta} \tilde{H}_\zeta \right] - jw\epsilon h_\zeta \left[\frac{\sin\theta}{R} \cos^2(\delta_p) \tilde{E}_\zeta + h_\zeta \frac{\partial}{\partial r} \tilde{E}_\zeta \right] \right.$$

$$\left. + sH_{\theta_0} - jw\epsilon E_{r_0} h_\zeta \right\}. \tag{14d}$$

Note that the transverse fields are dependent only on the longitudinal components of the fields and as function of the step's angle (δ_p) of the helix.

3). The transverse fields are substituted into the *coupled* wave equations.

4). The longitudinal components of the fields are developed into Fourier-Bessel series, in order to satisfy the metallic boundary conditions of the circular cross-section. The condition is that we have only ideal boundary conditions for r=a. Thus, the electric and magnetic fields will be zero in the metal.

5).   Two sets of equations are obtained by substitution the longitudinal components of the fields into the wave equations.   The first set of the equations is multiplied by $\cos(n\theta)J_n(P_{nm}r/a)$, and after that by $\sin(n\theta)J_n(P_{nm}r/a)$, for $n \neq 0$. Similarly, the second set of the equations is multiplied by $\cos(n\theta)J_n(P'_{nm}r/a)$, and after that by $\sin(n\theta)J_n(P'_{nm}r/a)$, for $n \neq 0$.

6).   In order to find an algebraic system of four equations with four unknowns, it is necessary to integrate over the area (r, θ), where $r = [0, a]$, and $\theta = [0, 2\pi]$, by using the orthogonal-relations of the trigonometric functions.

7). The propagation constants β_{nm} and β'_{nm} of the TM and TE modes of the hollow waveguide [22] are given, respectively, by $\beta_{nm} = \sqrt{k_0^2 - (P_{nm}/a)^2}$ and $\beta'_{nm} = \sqrt{k_0^2 - (P'_{nm}/a)^2}$, where the transverse Laplacian operator ($\nabla_\perp^2$) is given by $-(P_{nm}/a)^2$ and $-(P'_{nm}/a)^2$ for the TM and TE modes of the hollow waveguide, respectively.

The separation of variables is obtained by using the preceding orthogonal-relations. Thus the algebraic equations ($n \neq 0$) are given by

$$\alpha_n{}^{(1)} A_n + \beta_n{}^{(1)} D_n = \frac{1}{\pi} \widehat{(BC1)}_n, \tag{15a}$$

$$\alpha_n{}^{(2)} B_n + \beta_n{}^{(2)} C_n = \frac{1}{\pi} \widehat{(BC2)}_n, \tag{15b}$$

$$\beta_n{}^{(3)} B_n + \alpha_n{}^{(3)} C_n = \frac{1}{\pi} \widehat{(BC3)}_n, \tag{15c}$$

$$\beta_n{}^{(4)} A_n + \alpha_n{}^{(4)} D_n = \frac{1}{\pi} \widehat{(BC4)}_n. \tag{15d}$$

Further we assume $n' = n = 1$. The elements $(\alpha_n^{(1)}, \beta_n^{(1)}, \text{etc})$, on the left side of (15a) for $n=1$ are given for an arbitrary value of the step's angle (δ_p) by:

$$\alpha_1^{(1)mm'} = \pi \left(s^2 + \beta_{1m'}^2 \right) \left[\left(s^2 + k_0^2 \right) G_{00}^{(1)mm'} + k_0^2 G_{01}^{(1)mm'} \right]$$

$$+ \pi \frac{1}{R^4} k_0^2 s^2 \left(\frac{1}{4} \cos^4(\delta_p) G_{02}^{(1)mm'} + \frac{1}{2} \cos^4(\delta_p) G_{03}^{(1)mm'} \right)$$

$$+ \pi k_0^2 \left\{ s^2 G_{01}^{(1)mm'} + G_{05}^{(1)mm'} + \frac{1}{R^2} \left(G_{00}^{(1)mm'} + G_{01}^{(1)mm'} \right) + \frac{3}{2R^2} \beta_{1m'}^2 \cos^4(\delta_p) \left(G_{02}^{(1)mm'} + G_{03}^{(1)mm'} \right) \right.$$

$$\left. + \frac{1}{4R^4} \cos^4(\delta_p) \left(G_{02}^{(1)mm'} + G_{03}^{(1)mm'} \right) + \frac{1}{8R^4} \cos^8(\delta_p) \left(G_{06}^{(1)mm'} + G_{07}^{(1)mm'} \right) \right\}$$

$$+ \pi s^2 \left[G_{08}^{(1)mm'} + \frac{1}{2R^2} \cos^2(\delta_p) G_{00}^{(1)mm'} + \frac{1}{4R^2} \left(\cos^4(\delta_p) \beta_{1m'}^2 G_{02}^{(1)mm'} + \cos^2(\delta_p) G_{09}^{(1)mm'} \right) \right.$$

$$\left. + \frac{1}{2R^2} \frac{P_{1m'}}{a} \cos^2(\delta_p) \left(G_{10}^{(1)mm'} + \frac{1}{2} \cos^2(\delta_p) G_{11}^{(1)mm'} \right) \right]$$

$$+ \pi k_0^4 \cos^4(\delta_p) \left[\frac{3}{2R^2} \left(G_{03}^{(1)mm'} + G_{04}^{(1)mm'} \right) + \frac{1}{8R^4} \cos^8(\delta_p) \left(G_{07}^{(1)mm'} + G_{12}^{(1)mm'} \right) \right], \quad (16a)$$

$$\beta_1^{(1)mm'} = -j\omega\mu_0 \pi s \left\{ G_{13}^{(1)mm'} + \left(\frac{1}{2} \cos^2(\delta_p) + \frac{3}{4} \cos^4(\delta_p) \right) \frac{1}{R^2} G_{14}^{(1)mm'} \right.$$

$$\left. + \left(\frac{1}{2} + \cos^2(\delta_p) \right) \frac{1}{R^2} G_{15}^{(1)mm'} - \frac{1}{2R^2} G_{00}^{(1)mm'} - \cos^2(\delta_p) \frac{1}{R^2} \frac{P'_{1m'}}{a} G_{16}^{(1)mm'} \right\}, \quad (16b)$$

where the elements of the matrices $(G_{00}^{(1)mm'}, \text{etc.})$ are given in [20]. Similarly, the rest of the elements on the left side in Eqs. (15a)-(15d) are obtained. We establish an algebraic system of four equations with four unknowns. All the elements of the matrices in the Laplace transform domain are dependent on the step's angle of the helix (δ_p), the Bessel functions; the dielectric profile $g(r)$; the transverse derivative $g_r(r)$; and (r, θ).

The elements of the boundary conditions (e.g., $\widehat{(BC2)}_1$) at $\zeta = 0^+$ on the right side in (15b) are dependent on the step's angle δ_p as follows :

$$\widehat{(BC2)}_1 = \int_0^{2\pi} \int_0^a (BC2) \sin\theta J_1(P_{1m}r/a) r dr d\theta,$$

where

$$(BC2) = \left[\left(s^2 + k^2 h_\zeta^2 \right) \left(s E_{\zeta_0} + E'_{\zeta_0} \right) \right] + j\omega\mu_0 H_{\theta_0} s g_r h_\zeta^2$$

$$+\frac{2}{R}h_\zeta\sin\theta\left(j\omega\mu_0 H_{\theta_0}s+k^2 E_{r_0}h_\zeta\right)+\frac{2}{R}h_\zeta\cos\theta\left(-j\omega\mu_0 H_{r_0}s+k^2 E_{\theta_0}h_\zeta\right)+k^2 h_\zeta^3 E_{r_0}g_r,$$

and for $h_\zeta = 1 + (r/R)\sin\theta\cos^2(\delta_p)$.

The boundary conditions at $\zeta = 0^+$ for TEM_{00} mode in excitation become to:

$$\widehat{(BC2)}_1 = 2\pi\left\{\int_0^a Q(r)(k(r)+js)J_{1m}(P_{1m}r/a)rdr\right\}\delta_{1n}$$

$$+\frac{4js\pi}{R^2}\cos^2(\delta_p)\left\{\int_0^a Q(r)k(r)J_{1m}(P_{1m}r/a)r^2 dr\right\}\delta_{1n}$$

$$+\frac{9\pi}{2R^2}\cos^4(\delta_p)\left\{\int_0^a Q(r)k^2(r)J_{1m}(P_{1m}r/a)r^3 dr\right\}\delta_{1n}$$

$$+\frac{3js\pi}{2R^2}\cos^4(\delta_p)\left\{\int_0^a Q(r)k(r)J_{1m}(P_{1m}r/a)r^3 dr\right\}\delta_{1n}$$

$$+\frac{8\pi}{R^2}\cos^2(\delta_p)\left\{\int_0^a Q(r)k^2(r)J_{1m}(P_{1m}r/a)r^2 dr\right\}\delta_{1n} \qquad (17)$$

where :

$$Q(r) = \frac{E_0}{n_c(r)+1}g_r\exp\left(-(r/w_o)^2\right).$$

Similarly, the remaining elements of the boundary conditions at $\zeta = 0^+$ are obtained. The matrix system of Eqs.(15a)-(15d) is solved to obtain the coefficients (A_1, B_1, etc).

According to the Gaussian beams [23] the parameter w_0 is the minimum spot-size at the plane z=0, and the electric field at the plane z=0 is given by $E = E_0\exp[-(r/w_0)^2]$. The modes excited at $\zeta = 0$ in the waveguide by the conventional CO_2 laser IR radiation (λ=10.6 μm) are closer to the TEM polarization of the laser radiation. The TEM_{00} mode is the fundamental and most important mode. This means that a cross-section of the beam has a Gaussian intensity distribution. The relation between the electric and magnetic fields [23] is given by $E/H = \sqrt{\mu_0/\epsilon_0} \equiv \eta_0$, where η_0 is the intrinsic wave impedance. Suppose that the electric field is parallel to the y-axis. Thus the components of E_y and H_x are written by the fields $E_y = E_0\exp[-(r/w_0)^2]$ and $H_x = -(E_0/\eta_0)\exp[-(r/w_0)^2]$.

After a Gaussian beam passes through a lens and before it enters to the waveguide, the waist cross-sectional diameter ($2w_0$) can then be approximately calculated for a parallel incident beam by means of $w_0=\lambda/(\pi\,\theta)\simeq(f\,\lambda)/(\pi\,w)$. This approximation is justified if the parameter w_0 is much larger than the wavelength λ. The parameter of the waist cross-sectional diameter ($2w_0$) is taken into account in our method, instead of the focal length of the lens (f). The initial fields at $\zeta = 0^+$ are formulated by using the Fresnel coefficients of the transmitted fields [24] as follows

$$E_{r_0}^+(r) = T_E(r)(E_0 e^{-(r/w_o)^2}\sin\theta), \qquad (18a)$$

$$E_{\theta_0}^+ = T_E(r)(E_0 e^{-(r/w_0)^2} \cos\theta), \tag{18b}$$

$$H_{r_0}^+ = -T_H(r)((E_0/\eta_0)e^{-(r/w_0)^2} \cos\theta), \tag{18c}$$

$$H_{\theta_0}^+ = T_H(r)((E_0/\eta_0)e^{-(r/w_0)^2} \sin\theta), \tag{18d}$$

where $E_{\zeta_0}^+ = H_{\zeta_0}^+ = 0$, $T_E(r) = 2/[(n(r)+1]$, $T_H(r) = 2n(r)/[(n(r)+1]$, and $n(r) = [\epsilon_r(r)]^{1/2}$. The index of refraction is denoted by $n(r)$.

The transverse components of the fields are finally expressed in a form of *transfer matrix functions* for an arbitrary value of δ_p as follows:

$$E_r(r,\theta,\zeta) = E_{r0}^+(r)e^{-jkh_\zeta \zeta} - \frac{j\omega\mu_0}{R}h_\zeta \cos^2\theta\cos^2(\delta_p)\sum_{m'} C_{S1}^{m'}(\zeta)J_1(\psi)$$

$$-\frac{j\omega\mu_0}{R}h_\zeta \sin\theta\cos\theta\cos^2(\delta_p)\sum_{m'} D_{S1}^{m'}(\zeta)J_1(\psi) + \frac{j\omega\mu_0}{r}h_\zeta^2 \sin\theta\sum_{m'} C_{S1}^{m'}(\zeta)J_1(\psi)$$

$$-\frac{j\omega\mu_0}{r}h_\zeta^2 \cos\theta\sum_{m'} D_{S1}^{m'}(\zeta)J_1(\psi) + \frac{1}{R}\sin\theta\cos\theta\cos^2(\delta_p)\sum_{m'} A_{S2}^{m'}(\zeta)J_1(\xi)$$

$$+\frac{1}{R}\sin^2\theta\cos^2(\delta_p)\sum_{m'} B_{S2}^{m'}(\zeta)J_1(\xi) + h_\zeta \cos\theta\sum_{m'} A_{S2}^{m'}(\zeta)\frac{dJ_1}{dr}(\xi)$$

$$+h_\zeta \sin\theta\sum_{m'} B_{S2}^{m'}(\zeta)\frac{dJ_1}{dr}(\xi), \tag{19}$$

where $h_\zeta = 1 + (r/R)\sin\theta\cos^2(\delta_p)$, R is the radius of the cylinder, δ_p is the the step's angle, $\psi = [P'_{1m'}(r/a)]$ and $\xi = [P_{1m'}(r/a)]$. The coefficients are given in the above equation, for instance

$$A_{S1}^{m'}(\zeta) = \mathcal{L}^{-1}\left\{\frac{A_{1m'}(s)}{s^2 + k^2(r)h_\zeta^2}\right\}, \tag{20a}$$

$$A_{S2}^{m'}(\zeta) = \mathcal{L}^{-1}\left\{\frac{sA_{1m'}(s)}{s^2 + k^2(r)h_\zeta^2}\right\}, \tag{20b}$$

where

$$m' = 1,...N, \qquad 3 \le N \le 50. \tag{20c}$$

Similarly, the other transverse components of the output fields are obtained. The first fifty roots (zeros) of the equations $J_1(x) = 0$ and $dJ_1(x)/dx = 0$ may be found in tables [25-26].

The inverse Laplace transform is performed in this study by a direct numerical integration in the Laplace transform domain by the residue method, as follows

$$f(\zeta) = \mathcal{L}^{-1}[\tilde{f}(s)] = \frac{1}{2\pi j}\int_{\sigma-j\infty}^{\sigma+j\infty} \tilde{f}(s)e^{s\zeta}ds = \sum_n Res[e^{s\zeta}\tilde{f}(s); S_n]. \tag{21}$$

By using the inverse Laplace transform (21) we can compute the output transverse components in the real plane and the output power density at each point at $\zeta=(R\,\phi_c)/cos(\delta_p)$.

Analyzing Wave Propagation in Helical Wave
guides Using Laplace, Fourier, and Their Inverse Transforms, and Applications

205

The integration path in the right side of the Laplace transform domain includes all the singularities according to Eq.(21). All the points S_n are the poles of $\tilde{f}(s)$ and $Res[e^{s\zeta}\tilde{f}(s); S_n]$ represent the residue of the function in a specific pole. According to the residue method, two dominant poles for the helical waveguide are given by

$$s = \pm j\, k(r) h_\zeta = \pm j\, k(r) \left(1 + \frac{r}{R} \sin\theta \cos^2(\delta_p) \right).$$

Finally, knowing all the transverse components, the ζ component of the average-power density Poynting vector is given by

$$S_{av} = \frac{1}{2} Re\left\{ E_r H_\theta{}^* - E_\theta H_r{}^* \right\} \quad , \tag{22}$$

where the asterisk indicates the complex conjugate.

The total average-power transmitted along the guide in the ζ direction can now be obtained by the integral of Eq.(22) over the waveguide cross section. Thus, the output power transmission is given by

$$T = \frac{1}{2} \int_0^{2\pi} \int_0^a Re\left\{ E_r H_\theta{}^* - E_\theta H_r{}^* \right\} r\, dr\, d\theta \quad . \tag{23}$$

3.2 Solution of the wave equations for the helical coordinate system (x, y, ζ).

The method is based on Fourier coefficients of the transverse dielectric profile and those of the input wave profile. Laplace transform is necessary to obtain the comfortable and simple input-output connections of the fields. This model is useful for the analysis of dielectric waveguides in the microwave and the millimeter-wave regimes, for diffused optical waveguides in integrated optics. The output power transmission and the output power density are improved by increasing the step's angle or the radius of the cylinder of the helical waveguide, especially in the cases of space curved waveguides.

We assume that for most materials, the permeability μ is equal to that of free space ($\mu = \mu_0$). The wave equations for the electric and magnetic field components in the inhomogeneous dielectric medium $\epsilon(x,y)$ are given by the wave equations (5a) and (5b), respectively. The transverse dielectric profile $(\epsilon(x,y))$ is defined as $\epsilon_0(1 + \chi_0 g(x,y))$, where ϵ_0 represents the vacuum dielectric constant, $g(x,y)$ is its profile function in the waveguide, and χ_0 is the susceptibility of the dielectric material. The normalized transverse derivatives of the dielectric profile $g(x,y)$ are defined as $(1/\epsilon(x,y))[(\partial/\partial x)\epsilon(x,y)]$ and $(1/\epsilon(x,y))[(\partial/\partial y)\epsilon(x,y)]$, respectively. From the helical transformation of Eqs. 1 we can derive the Laplacian of the vector E (i.e., $\nabla^2 E$), and obtain the wave equations for the electric and magnetic fields in the inhomogeneous dielectric medium. It is necessary to find the values of $\nabla \cdot E$, $\nabla(\nabla \cdot E)$, $\nabla \times E$, and $\nabla \times (\nabla \times E)$ in order to obtain the value of $\nabla^2 E$, where $\nabla^2 E = \nabla(\nabla \cdot E) - \nabla \times (\nabla \times E)$. All these values are dependent on the metric coefficients (4a,b,c).

The components of $\nabla^2 E$ are given by

$$(\nabla^2 E)_x = \nabla^2 E_x - \frac{1}{R^2 h_\zeta^2}\cos^2(\delta_p)E_x - 2\frac{1}{R h_\zeta^2}\cos^2(\delta_p)\frac{\partial}{\partial\zeta}E_\zeta, \tag{24a}$$

$$(\nabla^2 E)_y = \nabla^2 E_y, \tag{24b}$$

$$(\nabla^2 E)_\zeta = \nabla^2 E_\zeta - \frac{1}{R^2 h_\zeta^2}\cos^2(\delta_p)E_\zeta + 2\frac{1}{R h_\zeta^2}\cos^2(\delta_p)\frac{\partial}{\partial\zeta}E_x, \tag{24c}$$

where

$$\nabla^2 = \frac{\partial^2}{\partial x^2} + \frac{\partial^2}{\partial y^2} + \frac{1}{h_\zeta^2}\frac{\partial^2}{\partial\zeta^2} + \frac{1}{R h_\zeta}\cos^2(\delta_p)\frac{\partial}{\partial x}, \tag{24d}$$

and for $h_\zeta = 1 + (x/R)\cos^2(\delta_p)$.

The wave equations (5a) and (5b) are written in the form

$$(\nabla^2 E)_i + k^2 E_i + \partial_i(E_x g_x + E_y g_y) = 0, \tag{25a}$$

$$(\nabla^2 H)_i + k^2 H_i + \partial_i(H_x g_x + H_y g_y) = 0, \tag{25b}$$

where i=x, y, ζ. The *local* wavenumber parameter is given by $k = \omega\sqrt{\mu\epsilon(x,y)} = k_0\sqrt{1 + \chi_0 g(x,y)}$, and the free-space wavenumber is given by $k_0 = \omega\sqrt{\mu_0\epsilon_0}$. The expression $(\nabla^2 E)_x$, for instance, is given according to Eq. (24a).

The transverse Laplacian operator is defined according to Eq. (12), where

$$h_\zeta^2 = 1 + \frac{2x}{R}\cos^2(\delta_p) + \left(\frac{x}{R}\right)^2\cos^4(\delta_p).$$

The metric coefficient h_ζ is a function of x, thus we defined

$$h_\zeta = 1 + p_\zeta(x) \ , \ p_\zeta(x) = \cos^2(\delta_p)(x/R), \tag{26a}$$

$$h_\zeta^2 = 1 + q_\zeta(x) \ , \ q_\zeta(x) = \cos^2(\delta_p)(2/R)x. \tag{26b}$$

The Laplace transform (Eq. (13)) is applied on the ζ-dimension, where $a(\zeta)$ represents any ζ-dependent variables and $\zeta = (R\phi_c)/\cos(\delta_p)$. Laplace transform on the differential wave equations is needed to obtain the wave equations (and thus also the output fields) that are expressed directly as functions of the transmitted fields at the entrance of the waveguide at $\zeta = 0^+$. Thus, the Laplace transform is necessary to obtain the comfortable and simple *input-output* connections of the fields.

By substitution of Eqs. (24a)-(24c) into Eqs. (25a), by using the Laplace transform (13), and multiply by h_ζ^2, Eqs. (5a) are described in the Laplace transform domain in the form

$$h_\zeta^2\left(\nabla_\perp^2 + \frac{s^2}{h_\zeta^2} + k^2\right)\tilde{E}_x + h_\zeta^2\partial_x\left(\tilde{E}_x g_x + \tilde{E}_y g_y\right) + h_\zeta\frac{1}{R}\cos^2(\delta_p)\partial_x\left(\tilde{E}_x\right)$$

Analyzing Wave Propagation in Helical Wave
guides Using Laplace, Fourier, and Their Inverse Transforms, and Applications

207

$$-\frac{2}{R}\cos^2(\delta_p)s\tilde{E}_\zeta = \left(sE_{x_0} + E'_{x_0}\right) - \frac{2}{R}\cos^2(\delta_p)E_{\zeta_0}, \tag{27a}$$

$$h_\zeta^2\left(\nabla_\perp^2 + \frac{s^2}{h_\zeta^2} + k^2\right)\tilde{E}_y + h_\zeta^2\partial_y\left(\tilde{E}_x g_x + \tilde{E}_y g_y\right) + h_\zeta\frac{1}{R}\cos^2(\delta_p)\partial_x\left(\tilde{E}_y\right) = \left(sE_{y_0} + E'_{y_0}\right), \tag{27b}$$

$$h_\zeta^2\left(\nabla_\perp^2 + \frac{s^2}{h_\zeta^2} + k^2\right)\tilde{E}_\zeta + sh_\zeta^2\left(\tilde{E}_x g_x + \tilde{E}_y g_y\right) + h_\zeta\frac{1}{R}\cos^2(\delta_p)\partial_x\left(\tilde{E}_\zeta\right) + \frac{2}{R}\cos^2(\delta_p)s\tilde{E}_x =$$

$$\left(sE_{\zeta_0} + E'_{\zeta_0}\right) + \frac{2}{R}\cos^2(\delta_p)E_{x_0} + h_\zeta^2\left(E_{x_0}g_x + E_{y_0}g_y\right), \tag{27c}$$

where the transverse Laplacian operator is defined according to (12), E_{x_0}, E_{y_0}, E_{ζ_0} are the initial values of the corresponding fields at $\zeta = 0$, i.e. $E_{x_0} = E_x(x,y,\zeta = 0)$ and $E'_{x_0} = \frac{\partial}{\partial\zeta}E_x(x,y,\zeta)|_{\zeta=0}$.

A Fourier transform is applied on the transverse dimension

$$\bar{g}(k_x,k_y) = \mathcal{F}\{g(x,y)\} = \int_x\int_y g(x,y)e^{-jk_x x - jk_y y}dxdy, \tag{28}$$

and the differential equation (27(a)) is transformed to an algebraic form in the (ω, s, k_x, k_y) space, as follows

$$k_\zeta^2\tilde{\bar{E}}_x + s^2\tilde{\bar{E}}_x + k_0^2\chi_0\bar{g}*\tilde{\bar{E}}_x + jk_x\left(\bar{g}_x*\tilde{\bar{E}}_x + \bar{g}_y*\tilde{\bar{E}}_y\right) - \frac{2}{R}\cos^2(\delta_p)s\tilde{\bar{E}}_\zeta + \bar{q}_\zeta*\left(k_\zeta^2\right)\tilde{\bar{E}}_x +$$

$$k_0^2\chi_0\bar{q}_\zeta*\left(\bar{g}*\tilde{\bar{E}}_x\right) + j\bar{q}_\zeta*\left[k_x\left(\bar{g}_x*\tilde{\bar{E}}_x + \bar{g}_y*\tilde{\bar{E}}_y\right)\right]$$

$$+\frac{1}{R}\cos^2(\delta_p)\left(jk_x\right)\tilde{\bar{E}}_x + \frac{1}{R}\cos^2(\delta_p)j\bar{p}_\zeta*\left(k_x\tilde{\bar{E}}_x\right)$$

$$= \left(s\tilde{\bar{E}}_{x_0} + \tilde{\bar{E}}'_{x_0}\right) - \frac{1}{sR}\cos^2(\delta_p)\left(s\tilde{\bar{E}}_{\zeta_0} + \tilde{\bar{E}}'_{\zeta_0}\right), \tag{29}$$

where $k_\zeta = \sqrt{k_0^2 - k_x^2 - k_y^2}$. Similarly, the other differential equations are obtained. The asterisk symbol denotes the convolution operation $\bar{g}*\tilde{\bar{E}} = \mathcal{F}\{g(x,y)E(x,y)\}$. The method of images is applied to satisfy the conditions $\hat{n}\times E = 0$ and $\hat{n}\cdot(\nabla\times E) = 0$ on the surface of the ideal metallic waveguide walls, where $\hat{n}$ is a unit vector perpendicular to the surface. The metric coefficient h_ζ is a function of x (Eqs. (26a) and (26b)). Thus the elements of the matrices $\mathbf{P}^{(0)}$ and $\mathbf{Q}^{(0)}$ are defined as:

$$\bar{p}_{\zeta(n,m)}^{(o)} = \frac{1}{4ab}\int_{-a}^a\int_{-b}^b p_\zeta(x)\ e^{-j(n\frac{\pi}{a}x + m\frac{\pi}{b}y)}\ dxdy, \tag{30a}$$

$$\bar{q}_{\zeta(n,m)}^{(o)} = \frac{1}{4ab}\int_{-a}^a\int_{-b}^b q_\zeta(x)\ e^{-j(n\frac{\pi}{a}x + m\frac{\pi}{b}y)}\ dxdy, \tag{30b}$$

and the matrices $\mathbf{P}^{(1)}$ and $\mathbf{Q}^{(1)}$ are defined as:

$$\mathbf{P}^{(1)} = \left(\mathbf{I} + \mathbf{P}^{(0)}\right), \quad \mathbf{Q}^{(1)} = \left(\mathbf{I} + \mathbf{Q}^{(0)}\right), \tag{30c, d}$$

where $\mathbf{I}$ is the unity matrix.

Equation (29) and similarly, the two other equations are rewritten in a matrix form as follows

$$\mathbf{K}^{(0)} E_x + \frac{k_o^2 \chi_0}{2s} \mathbf{Q}^{(1)} \mathbf{G} E_x + \frac{jk_{ox}}{2s} \mathbf{Q}^{(1)} \mathbf{N} \left(\mathbf{G_x} E_x + \mathbf{G_y} E_y \right)$$

$$-\frac{1}{R}\cos^2(\delta_p) E_\zeta + \mathbf{Q}^{(0)} \mathbf{K1}^{(0)} E_x + \frac{1}{2sR}\cos^2(\delta_p) jk_{ox} \mathbf{P}^{(1)} \mathbf{N} E_x = \hat{E}_{x_0} - \frac{1}{sR}\cos^2(\delta_p) \bar{E}_{\zeta_0}, \tag{31a}$$

$$\mathbf{K}^{(0)} E_y + \frac{k_o^2 \chi_0}{2s} \mathbf{Q}^{(1)} \mathbf{G} E_y + \frac{jk_{oy}}{2s} \mathbf{Q}^{(1)} \mathbf{M} \left(\mathbf{G_x} E_x + \mathbf{G_y} E_y \right)$$

$$+\mathbf{Q}^{(0)} \mathbf{K1}^{(0)} E_y + \frac{1}{2sR}\cos^2(\delta_p) jk_{ox} \mathbf{P}^{(1)} \mathbf{N} E_y = \hat{E}_{y_0}, \tag{31b}$$

$$\mathbf{K}^{(0)} E_\zeta + \frac{k_o^2 \chi_0}{2s} \mathbf{Q}^{(1)} \mathbf{G} E_\zeta + \frac{1}{2}\mathbf{Q}^{(1)} \left(\mathbf{G_x} E_x + \mathbf{G_y} E_y \right) + \frac{1}{R}\cos^2(\delta_p) E_x$$

$$+\mathbf{Q}^{(0)} \mathbf{K1}^{(0)} E_\zeta + \frac{1}{2sR}\cos^2(\delta_p) jk_{ox} \mathbf{P}^{(1)} \mathbf{N} E_\zeta = \hat{E}_{\zeta_0} + \frac{1}{sR}\cos^2(\delta_p) E_{x_0} + \frac{1}{2s}\mathbf{Q}^{(1)} \left(\mathbf{G_x} E_{x_0} + \mathbf{G_y} E_{y_0} \right),$$

$$\tag{31c}$$

where the initial-value vectors, $\hat{E}_{x_0}$, $\hat{E}_{y_0}$, and $\hat{E}_{\zeta_0}$ are defined from the terms $(s\bar{E}_{x_0} + \bar{E}'_{x_0})/2s$, $(s\bar{E}_{y_0} + \bar{E}'_{y_0})/2s$, and $(s\bar{E}_{\zeta_0} + \bar{E}'_{\zeta_0})/2s$, respectively.

The elements of the diagonal matrices $\mathbf{K}^{(0)}$, $\mathbf{M}$, $\mathbf{N}$ and $\mathbf{K}^{(1)}$ are defined as

$$K^{(0)}_{(n,m)(n',m')} = \left\{ \left[k_o^2 - (n\pi/a)^2 - (m\pi/b)^2 + s^2 \right] /2s \right\} \delta_{nn'} \delta_{mm'}, \tag{32a}$$

$$M_{(n,m)(n',m')} = m\delta_{nn'} \delta_{mm'}, \tag{32b}$$

$$N_{(n,m)(n',m')} = n\delta_{nn'} \delta_{mm'}, \tag{32c}$$

$$K^{(1)}_{(n,m)(n',m')} = \left\{ \left[k_o^2 - (n\pi/a)^2 - (m\pi/b)^2 \right] /2s \right\} \delta_{nn'} \delta_{mm'}, \tag{32d}$$

where $\delta_{nn'}$ and $\delta_{mm'}$ are the Kronecker delta functions.

The modified wave-number matrices are defined as

$$\mathbf{D_x} \equiv \mathbf{K}^{(0)} + \mathbf{Q}^{(0)} \mathbf{K1}^{(0)} + \frac{k_o^2 \chi_0}{2s} \mathbf{Q}^{(1)} \mathbf{G} + \frac{jk_{ox}}{2s} \mathbf{Q}^{(1)} \mathbf{NG_x} + \frac{1}{2sR}\cos^2(\delta_p) jk_{ox} \mathbf{P}^{(1)} \mathbf{N}, \tag{33a}$$

$$\mathbf{D_y} \equiv \mathbf{K}^{(0)} + \mathbf{Q}^{(0)} \mathbf{K1}^{(0)} + \frac{k_o^2 \chi_0}{2s} \mathbf{Q}^{(1)} \mathbf{G} + \frac{1}{2sR}\cos^2(\delta_p) jk_{ox} \mathbf{P}^{(1)} \mathbf{N} + \frac{jk_{oy}}{2s} \mathbf{Q}^{(1)} \mathbf{MG_y}, \tag{33b}$$

Analyzing Wave Propagation in Helical Wave
guides Using Laplace, Fourier, and Their Inverse Transforms, and Applications

209

$$\mathbf{D}_\zeta \equiv \mathbf{K}^{(0)} + \mathbf{Q}^{(0)}\mathbf{K1}^{(0)} + \frac{k_0^2\chi_0}{2s}\mathbf{Q}^{(1)}\mathbf{G} + \frac{1}{2sR}\cos^2(\delta_p)jk_{ox}\mathbf{P}^{(1)}\mathbf{N}. \qquad (33c)$$

Thus, Eqs. (31a)-(31c) result in

$$\mathbf{D_x}E_x = \hat{E}_{x_0} - \frac{jk_{ox}}{2s}\mathbf{Q}^{(1)}\mathbf{NG_y}E_y - \frac{1}{sR}\cos^2(\delta_p)E_{\zeta_0} + \frac{1}{R}\cos^2(\delta_p)E_\zeta, \qquad (34a)$$

$$\mathbf{D_y}E_y = \hat{E}_{y_0} - \frac{jk_{oy}}{2s}\mathbf{Q}^{(1)}\mathbf{MG_x}E_x, \qquad (34b)$$

$$\mathbf{D}_\zeta E_\zeta = \hat{E}_{\zeta_0} + \frac{1}{2s}\mathbf{Q}^{(1)}\left(\mathbf{G_x}E_{x_0} + \mathbf{G_y}E_{y_0}\right) - \frac{1}{2}\mathbf{Q}^{(1)}\left(\mathbf{G_x}E_x + \mathbf{G_y}E_y\right) + \frac{1}{sR}\cos^2(\delta_p)E_{x_0} - \frac{1}{R}\cos^2(\delta_p)E_x.$$
$$(34c)$$

After some algebraic steps, the components of the electric field are formulated as follows:

$$E_x = \left\{\mathbf{D_x} + \alpha_1\mathbf{Q}^{(1)}\mathbf{M_1}\mathbf{Q}^{(1)}\mathbf{M_2} + \frac{1}{R}\cos^2(\delta_p)\mathbf{D}_\zeta^{-1}\right.$$

$$\left. \cdot\left(-\frac{1}{2}\mathbf{Q}^{(1)}\mathbf{G_x} + \frac{1}{2}\alpha_2\mathbf{Q}^{(1)}\mathbf{M_3}\mathbf{Q}^{(1)}\mathbf{M_2} - \frac{1}{R}\cos^2(\delta_p)\mathbf{I}\right)\right\}^{-1}$$

$$\left(\hat{E}_{x_0} - \frac{1}{sR}\cos^2(\delta_p)E_{\zeta_0} - \alpha_3\mathbf{Q}^{(1)}\mathbf{M_1}\hat{E}_{y_0} + \frac{1}{R}\cos^2(\delta_p)\mathbf{D}_\zeta^{-1}\left(\hat{E}_{\zeta_0} + \frac{1}{sR}\cos^2(\delta_p)E_{x_0} + \right.\right.$$

$$\left.\left.\frac{1}{2s}\mathbf{Q}^{(1)}(\mathbf{G_x}E_{x_0} + \mathbf{G_y}E_{y_0}) - \frac{1}{2}\mathbf{Q}^{(1)}\mathbf{M_3}\hat{E}_{y_0}\right)\right), \qquad (35a)$$

$$E_y = \mathbf{D_y}^{-1}\left(\hat{E}_{y_0} - \frac{jk_{oy}}{2s}\mathbf{Q}^{(1)}\mathbf{MG_x}E_x\right), \qquad (35b)$$

$$E_\zeta = \mathbf{D}_\zeta^{-1}\left\{\hat{E}_{\zeta_0} + \frac{1}{2s}\mathbf{Q}^{(1)}\left(\mathbf{G_x}E_{x_0} + \mathbf{G_y}E_{y_0}\right) - \frac{1}{2}\mathbf{Q}^{(1)}\left(\mathbf{G_x}E_x + \mathbf{G_y}E_y\right)\right.$$

$$\left. - \frac{1}{R}\cos^2(\delta_p)E_x + \frac{1}{sR}\cos^2(\delta_p)E_{x_0}\right\}, \qquad (35c)$$

where:

$$\alpha_1 = \frac{k_{ox}k_{oy}}{4s^2}, \quad \alpha_2 = \frac{jk_{oy}}{2s}, \quad \alpha_3 = \frac{jk_{ox}}{2s}, \quad \mathbf{M_1} = \mathbf{NG_y}\mathbf{D_y}^{-1}, \quad \mathbf{M_2} = \mathbf{MG_x}, \quad \mathbf{M_3} = \mathbf{G_y}\mathbf{D_y}^{-1}.$$

These equations describe the transfer relations between the spatial spectrum components of the output and input waves in the dielectric waveguide. Similarly, the other components of the magnetic field are obtained. The transverse field profiles are computed by the inverse

Laplace and Fourier transforms, as follows

$$E_y(x,y,\zeta) = \sum_n \sum_m \int_{\sigma-j\infty}^{\sigma+j\infty} E_y(n,m,s)e^{jnk_{ox}x+jmk_{oy}y+s\zeta}ds. \tag{36}$$

The inverse Laplace transform is performed in this study by a direct numerical integration on the s-plane by the method of Gaussian Quadrature. The integration path in the right side of the s-plane includes all the singularities, as proposed by Salzer [27-28]

$$\int_{\sigma-j\infty}^{\sigma+j\infty} e^{s\zeta}E_y(s)ds = \frac{1}{\zeta}\int_{\sigma-j\infty}^{\sigma+j\infty} e^p E_y(p/\zeta)dp = \frac{1}{\zeta}\sum_{i=1}^{15} w_i E_y(s=p_i/\zeta), \tag{37}$$

where w_i and p_i are the weights and zeros, respectively, of the orthogonal polynomials of order 15. The Laplace variable s is normalized by p_i/ζ in the integration points, where $\text{Re}(p_i) > 0$ and all the poles should be localized in their left side on the Laplace transform domain. This approach of a direct integral transform does not require as in other methods, to deal with each singularity separately.

The ζ component of the average-power density of the complex Poynting vector is given by

$$S_{av} = \frac{1}{2}Re\left\{ E_x H_y{}^* - E_y H_x{}^* \right\}, \tag{38}$$

where the asterisk indicates the complex conjugate. The active power is equal to the real part of the complex Poynting vector. The total average-power transmitted along the guide in the ζ direction is given by a double integral of Eq. (38). A Fortran code is developed using NAG subroutines [29]. Several examples computed on a Unix system are presented in the next section.

4. Numerical results

An example of the circular cross section of the helical waveguide is shown in Fig. 3(a). An example of the rectangular dielectric slab of the helical waveguide is shown in Fig. 3(b), and an example of the rectangular cross section with a circular dielectric profile of the helical waveguide is shown in Fig. 3(c). The results of the output transverse components of the fields and the output power density ($|S_{av}|$) (e.g., Fig. 4(a)) show the behavior of the solutions for the TEM_{00} mode in excitation, for the straight waveguide ($R \rightarrow \infty$). The result of the output power density (Fig. 4(a)) is compared also to the result of published experimental data [30] as shown also in Fig. 4(b). This comparison shows good agreement (a Gaussian shape) as expected, except for the secondary small propagation mode. The experimental result (Fig. 4(b)) is affected by the additional parameters (e.g., the roughness of the internal wall of the waveguide) which are not taken theoretically into account. In this example with the circular cross section (Fig. 3(a)), the length of the straight waveguide is 1 m, the diameter (2a) of the waveguide is 2 mm, the thickness of the dielectric layer [$d_{(AgI)}$] is 0.75 μm, and the minimum spot-size (w_0) is 0.3 mm. The refractive indices of the air, the dielectric layer (AgI) and the metallic layer (Ag) are $n_{(0)} = 1$, $n_{(AgI)} = 2.2$, and $n_{(Ag)} = 13.5 - j75.3$, respectively.

The value of the refractive index of the material at a wavelength of λ=10.6 μm is taken from the table compiled by Miyagi, et al. [6]. The toroidal dielectric waveguide is demonstrated in Fig. 4(c). The experimental result is demonstrated in Fig. 4(d). This experimental result was obtained from the measurements of the transmitted CO_2 laser radiation (λ=10.6 μm) propagation through a hollow tube covered on the bore wall with silver and silver-iodide layers (Fig. 3(a)), where the initial diameter (ID) is 1 mm (namely, small bore size).

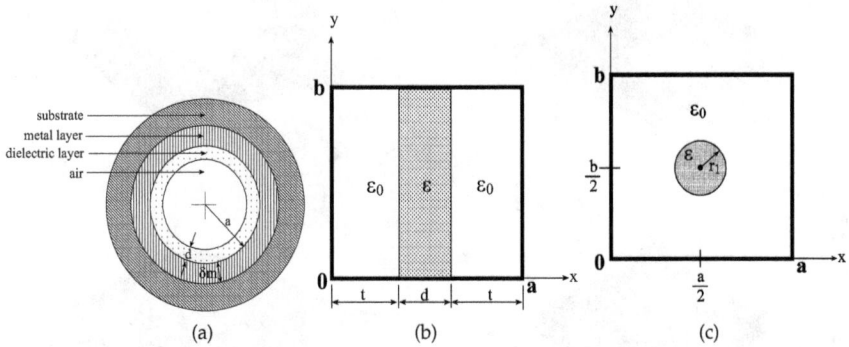

Figure 3. (a) An example of the circular cross section of the helical waveguide. (b) An example of the rectangular dielectric slab of the helical waveguide. (c) An example of the rectangular cross section with a circular dielectric profile of the helical waveguide.

The output modal profile is greatly affected by the bending, and the theoretical and experimental results (Figs. 4(c)-4(d)) show that in addition to the main propagation mode, several other secondary modes and asymmetric output shape appear. The amplitude of the output power density ($|S_{av}|$) is small as the bending radius (R) is small, and the shape is far from a Gaussian shape. This result agrees with the experimental results, but not for all the propagation modes. The experimental result (Fig. 4(d)) is affected by the bending and additional parameters (e.g., the roughness of the internal wall of the waveguide) which are not taken theoretically into account. In this example, a=0.5 mm, R=0.7 m, $\phi=\pi/2$, and $\zeta=$ 1 m. The thickness of the dielectric layer [$d_{(AgI)}$] is 0.75 μm (Fig. 3(a)), and the minimum spot size (w_0) is 0.2 mm. The values of the refractive indices of the air, the dielectric layer (AgI) and the metallic layer (Ag) are $n_{(0)} = 1$, $n_{(AgI)} = 2.2$, and $n_{(Ag)} = 13.5 - j75.3$, respectively. In both theoretical and experimental results (Figs. 4(c)-4(d)) the shapes of the output power density for the curved waveguide are not symmetric. The output modal profile is greatly affected by the bending, and the theoretical and experimental results (Figs. 4(c)-4(d)) show that in addition to the main propagation mode, several other secondary modes and asymmetric output shape appear. The amplitude of the output power density ($|S_{av}|$) is small as the bending radius (R) is small, and the shape is far from a Gaussian shape. This result agrees with the experimental results, but not for all the propagation modes. The experimental result (Fig. 4(d)) is affected by the bending and additional parameters (e.g., the roughness of the internal wall of the waveguide) which are not taken theoretically into account. In this example, a=0.5 mm, R=0.7 m, $\phi=\pi/2$, and $\zeta=$ 1 m. The thickness of the dielectric layer [$d_{(AgI)}$] is 0.75 μm (Fig. 3(a)), and the minimum spot size (w_0) is 0.2 mm. The values of the refractive indices of the air, the dielectric layer (AgI) and the metallic layer (Ag) are $n_{(0)} = 1$, $n_{(AgI)} = 2.2$, and $n_{(Ag)} = 13.5 - j75.3$, respectively. In both theoretical and experimental results (Figs. 4(c)-4(d)) the shapes of the output power density for the curved waveguide are not symmetric.

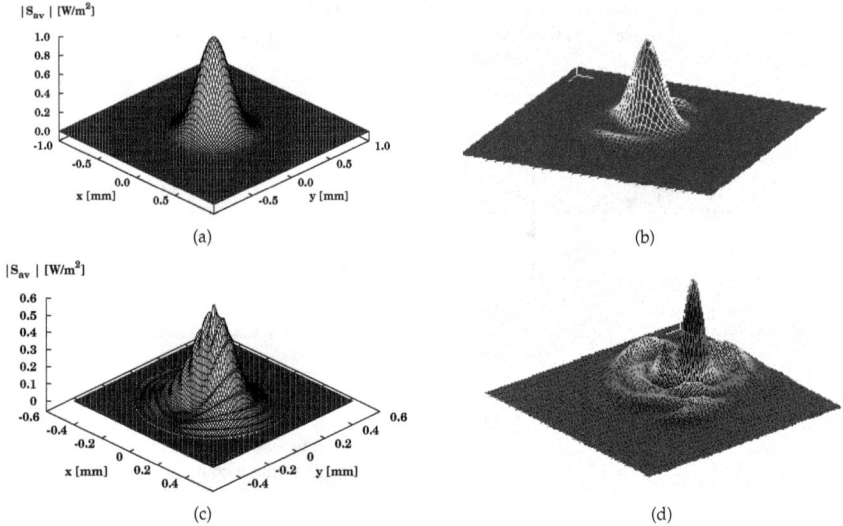

(a)

(b)

(c)

(d)

Figure 4. The output power density for R → ∞, where a=1 mm, w_0 = 0.3 mm, and the length of the straight waveguide is 1 m. (a) theoretical result; (b) experimental result. The output power density for the toroidal dielectric waveguide, where a=0.5 mm, w_0 = 0.2 mm, R = 0.7 m, $\phi = \pi/2$, and ζ =1 m; (c) theoretical result; (d) experimental result. The other parameters are: $d_{(AgI)}$= 0.75 μm, λ = 10.6 μm, $n_{(0)}$ = 1, $n_{(AgI)}$ = 2.2, and $n_{(Ag)}$ = 13.5 - j 75.

Figures 5(a)-(b) show the results of the output power density as functions of the step's angle (e.g., δ_p =0.4, 0.8) and the radius of the cylinder (e.g., R=0.7 m). For these results ζ= 1 m, where a=1 mm, w_0 = 0.3 mm, n_d =2.2, and $n_{(Ag)}$ = 13.5 - j 75.3 (Fig. 3(a)). Fig. 5(a) shows that in addition to the main propagation mode, several other secondary modes appear, where δ_p = 0.4 and R=0.7 m. By increasing only the step's angle from δ_p = 0.4 to δ_p = 0.8 where R = 0.7 m, the amplitude of the output power density is greater (e.g., ($|S_{av}|$ = 0.7 W/m^2) and also the output shape is changed (Fig. 5(b)).

Let us compare the second theoretical model (e.g., Eq. 35(b)) with the known analytical theory [22] for the rectangular dielectric slab (Fig. 3(b)). For the given dimensions a and d, we find the values Λ and Ω according to the next transcendental equation for a dielectric slab (Fig. 3(b)).

According to our theoretical model we can calculate $E_{y_0}(n,m)$ and $g(n,m)$ as follows:

$$E_{y_0}(n,m) = \frac{1}{4ab} \int_{-a}^{a} \int_{-b}^{b} E_y(x,y,z=0) e^{-j(n\frac{\pi}{a}x + m\frac{\pi}{b}y)} dxdy,$$

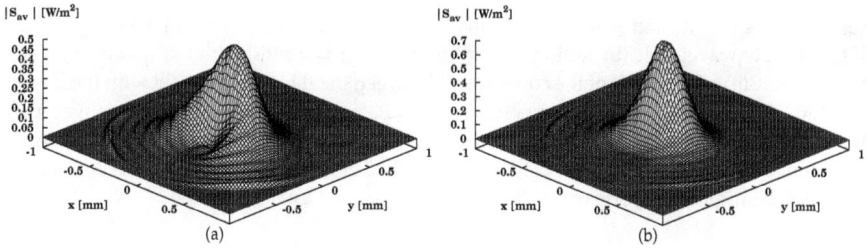

Figure 5. The results of the output power density as functions of the step's angle (δ_p) and the radius of the cylinder (R), where $\zeta = 1$ m, a=1 mm, $w_0 = 0.3$ mm, $n_d = 2.2$, and $n_{(Ag)} = 13.5 - j\ 75.3$: (a). $\delta_p = 0.4$, and R = 0.7 m; (b). $\delta_p = 0.8$, and R = 0.7 m.

and

$$g(n,m) = \frac{1}{4ab} \int_{-a}^{a} \int_{-b}^{b} g(x,y) e^{-j(n\frac{\pi}{a}x + m\frac{\pi}{b}y)} dx\, dy.$$

The known solution for the dielectric slab modes based on transcendental equation [22] is given as follows

$$\begin{cases} E_{y1} = j\frac{k_z}{\epsilon_0} \sin(vx) & 0 < x < t \\ E_{y2} = j\frac{k_z}{\epsilon_0} \frac{\sin(vt)}{\cos(\mu(t-a/2))} \cos[\mu(x-a/2)] & t < x < t+d \ , \\ E_{y3} = j\frac{k_z}{\epsilon_0} \sin[v(a-x)] & t+d < x < a \end{cases} \tag{39}$$

where $v \equiv \sqrt{k_0^2 - k_z^2}$ and $\mu \equiv \sqrt{\epsilon_r k_0^2 - k_z^2}$ result from the transcendental equation

$$\left(\frac{a}{d} - 1\right) \frac{d\mu}{2} \tan\left(\frac{d\mu}{2}\right) - (tv)\cot(tv) = 0.$$

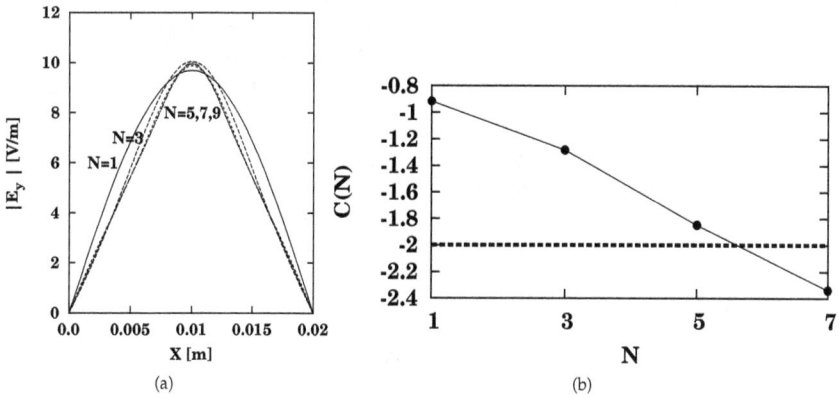

Figure 6. (a). A comparison between amplitude results of the theoretical model and the transcendental equation (a=2b=2 cm, d=3.3 mm, $\epsilon_r = 9$, and $\lambda = 6.9$ cm; (b). The convergence of our theoretical results.

Eqs. (39) were substituted as the initial fields into the Eq. (35(b)) at $z = 0^+$ in the practical case of the straight waveguide (by letting R $\rightarrow \infty$ or by taking $\delta_p = \pi/2$) with the symmetrical slab profile (Fig. 3(b)). The result of the comparison between the theoretical model with the known solution [22] is shown in Fig. 6(a), where $\epsilon_r = 9$, d=3.3 mm, and $\lambda = 6.9$ cm. The convergence of the numerical results as a function of the matrix order is shown in Fig. 6(b). The comparison is demonstrated for every order (N=1, 3, 5, 7, and 9). The order N determines the accuracy of the solution, and the convergence of the solution is verified by the criterion

$$C(N) \equiv log\left\{ \frac{max(|S_{av}^{N+2} - S_{av}^N|)}{|max(S_{av}^{N+2}) - min(S_{av}^N)|} \right\}, \qquad N \geq 1, \qquad (40)$$

for the E_y component of the fields (instead of S_{av}), where the number of the modes is equal to $(2N + 1)^2$. The method of this model is based on Fourier coefficients, thus the accuracy of the method is dependent on the number of the modes in the system. Further we assume N = M. If the value of the criterion is less then -2, then the numerical solution is well converged. When N increases, then $E_y(N)$ approaches E_y. The value of the criterion between N=7 and N=9 is equal to -2.38 $\simeq$ -2, namely a hundredth part. The comparison between the theoretical mode-model and the known model [22] has shown good agreement. Note that we have two ways to compare between the results of our mode model with the other methods. The first way is to compare between the results of the output fields for every order (N=1, 3, 5, 7, and 9) with the final solution of the known method. The second way is to compare between the results of the output fields (according to our model) for every two orders (N=1,3, N=3,5, N=5,7, and N=7,9), until our numerical solution is well converged. This way is efficient in the cases that we have complicated problems that we cannot compare with the final solution of the known method.

The geometrical shape of a circular dielectric profile loaded rectangular waveguide is demonstrated in Fig. 3(c) for an inhomogeneous dielectric profile in the cross section. The radius of the circle is denoted as r_1 and the dimensions of the waveguide in the cross-section are denoted as a and b. The refractive index of the core (dielectric profile) is greater than that of the cladding (air). The results of the solution in this case will demonstrate for $r_1 = 0.5$ mm and for a=b=2 cm. Let us assume that the center of the circle located at the point (a/2, b/2), as shown in Fig. 3(c).

Figures 7(a)-(b) show the results of the output power density (S_{av}) as functions of the step's angle $(\delta_p=1)$ and the radius of the cylinder (R=0.5 m). The other parameters are: $\zeta = 15$ cm, a=b= 2 cm, $r_1 = 0.5$ mm, and $\lambda = 3.75$ cm. The output fields are dependent on the input wave profile $(TE_{10}$ mode) and the circular dielectric profile of the rectangular cross section (Fig. 3(c)). Fig. 7(c) shows the output amplitude and the Gaussian shape of the central peak in the same cross section of Figs. 7(a-b), where y=b/2 = 1 cm, and for five values of ϵ_r =2, 5, 6, 8, and 10, respectively. By changing the value of the parameter ϵ_r of the core in the cross section (Fig. 3(c)) with regard to the cladding (air) from 2 to 10, the output transverse profile of the power density (S_{av}) is changed. For small values of ϵ_r, the half-sine (TE_{10}) shape of the output power density appears, with a little influence of the Gaussian shape in the center of the output profile.

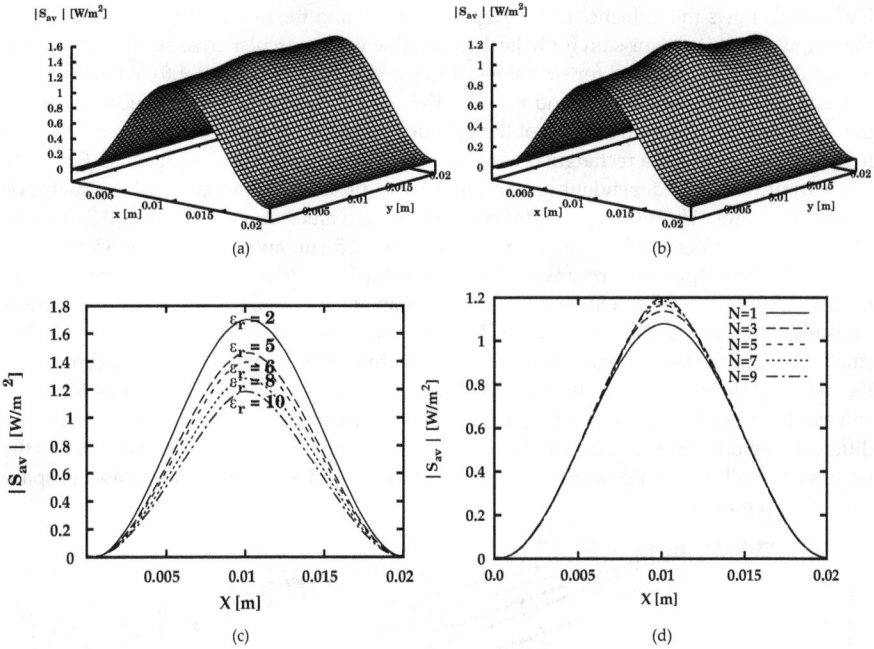

Figure 7. The results of the output power density as functions of the step's angle ($\delta_p= 1$) and the radius of the cylinder (R=0.5 m), where ζ= 15 cm, a=b= 2 cm, r_1 = 0.5 mm, λ = 3.75 cm: (a). ϵ_r = 6; (b). ϵ_r = 10. (c). The output amplitude and the Gaussian shape of the central peak in the same cross section where y=b/2 = 1 cm, and for different values of ϵ_r. (d). The output profile for N=1, 3, 5, 7 and 9, where ϵ_r = 10.

On the other hand, for large values of ϵ_r (e.g., ϵ_r=10) the Gaussian shape of the output power density appears in the center of the output profile (Fig. 7(b)), with a little influence of the half-sine (TE_{10}) shape in the center of the output profile. By increasing only the parameter ϵ_r from 2 to 10, the result of the output power density shows a Gaussian shape and the amplitude of the output power density is changed from 1.6 W/m^2 to 1 W/m^2, as shown in Fig. 7(c). In this case, the output Gaussian profile increases with increasing the value of ϵ_r. These examples demonstrate the influence of the dielectric profile for an inhomogeneous cross section, for arbitrary step's angle and the radius of the cylinder of the helical waveguide. Figure 7(d) shows an example for the output profiles with ϵ_r = 10, and for the same other parameters of Figs. 7(a)-(c). The output results are demonstrated for every order (N=1, 3, 5, 7, and 9). By increasing only the parameter of the order from N=1 to N=9, then the output profile approaches to the final output profile.

The other main contributions of the proposed methods are demonstrated in Fig. 8(a) and in Fig. 8(b), in order to understand the influence of the step's angle (δ_p) and the radius of the cylinder (R) on the output power transmission, for helical waveguide with a circular cross section (Fig. 3(a)) and with a rectangular cross section (Fig. (3(c)), respectively.

Figure 8(a) shows the influence of the step's angle (δ_p) and the radius of the cylinder (R) on the output power transmission for helical waveguide with a circular cross section (Fig. 3(a)). Six results are demonstrated for six values of δ_p (δ_p =0.0, 0.4, 0.7, 0.8, 0.9, 1.0), where ζ = 4 m, a=1 mm, w_0 = 0.06 mm, n_d =2.2 and $n_{(Ag)} = 13.5 - j75.3$. Figure 8(b) shows the influence of the step's angle (δ_p) and the radius of the cylinder (R) on the output power transmission for helical waveguide with a rectangular cross section with a circular dielectric profile (Fig. 3(c)). The output fields are dependent on the input wave profile (TE_{10} mode) and the dielectric profile (Fig. 3(c). Six results are demonstrated for six values of δ_p (δ_p =0, 0.4, 0.7, 0.8, 0.9, 1.0), where ζ = 15 cm, a=b=2 cm, r_1 = 0.5 mm, λ = 3.75 cm, and ϵ_r = 10. For an arbitrary value of R, the output power transmission is large for large values of δ_p and decreases with decreasing the value of δ_p. On the other hand, for an arbitrary value of δ_p, the output power transmission is large for large values of R and decreases with decreasing the value of R. For small values of the step's angle, the radius of curvature of the helix can be approximated by the radius of the cylinder (R). In this case, the output power transmission is large for small values of the bending (1/R), and decreases with increasing the bending. Thus, these two different methods can be a useful tool to find the parameters (δ_p and R) which will give us the improved results (output power transmission) of the curved waveguide in the cases of space curved waveguides.

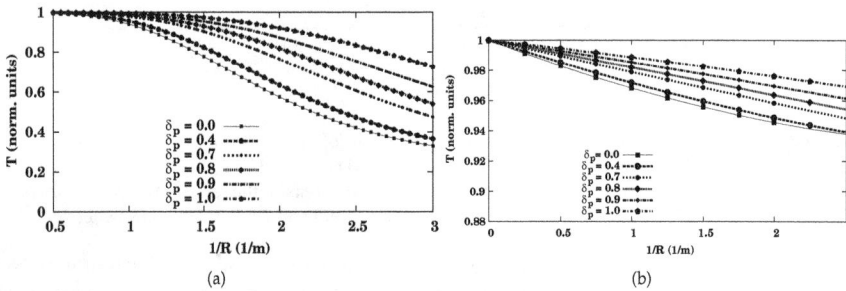

Figure 8. The results of the output power transmission of the helical waveguide as a function of 1/R, where R is the radius of the cylinder. Six results are demonstrated for six values of δ_p (δ_p =0.0, 0.4, 0.7, 0.8, 0.9, 1.0) (a) for circular cross section (Fig. 3(a)), where ζ = 4 m, a=1 mm, w_0 = 0.06 mm, n_d =2.2, and $n_{(Ag)} = 13.5 - j75.3$. (b) for a rectangular cross section with a circular dielectric profile (Fig. 3(c)), where a=20 mm, b=20 mm, r_1 = 0.5 mm, λ = 3.75 cm, and ϵ_r = 10.

5. Conclusions

Two improved methods have been presented for the propagation of EM fields along a helical dielectric waveguide with a circular cross section and a rectangular cross section. The two different methods employ helical coordinates (and not cylindrical coordinates, such as in the methods that considered the bending as a perturbation), and the calculations are based on using Laplace and Fourier transforms. The output fields are computed by the inverse Laplace and Fourier transforms. An example of the circular cross section of the helical waveguide is shown in Fig. 3(a). An example of the rectangular dielectric slab of the helical waveguide is

Analyzing Wave Propagation in Helical Wave
guides Using Laplace, Fourier, and Their Inverse Transforms, and Applications

217

shown in Fig. 3(b), and an example of the rectangular cross section with a circular dielectric profile of the helical waveguide is shown in Fig. 3(c).

The results of the output transverse components of the fields and the output power density ($|S_{av}|$) (e.g., Fig. 4(a)) for the circular cross section (Fig. 3(a)) show the behavior of the solutions for the TEM_{00} mode in excitation, for the straight waveguide. The result of the output power density (Fig. 4(a)) is compared also to the result of published experimental data [30] as shown also in Fig. 4(b). This comparison shows good agreement (a Gaussian shape) as expected, except for the secondary small propagation mode. The experimental result (Fig. 4(b)) is affected by the additional parameters (e.g., the roughness of the internal wall of the waveguide) which are not taken theoretically into account.

The toroidal dielectric waveguide is demonstrated in Fig. 4(c), and the experimental result is demonstrated in Fig. 4(d). This experimental result was obtained from the measurements of the transmitted CO_2 laser radiation ($\lambda=10.6$ μm) propagation through a hollow tube covered on the bore wall with silver and silver-iodide layers (Fig. 3(a)), where the initial diameter (ID) is 1 mm (namely, small bore size). The output modal profile is greatly affected by the bending, and the theoretical and experimental results (Figs. 4(c)-4(d)) show that in addition to the main propagation mode, several other secondary modes and asymmetric output shape appear. The amplitude of the output power density ($|S_{av}|$) is small as the bending radius (R) is small, and the shape is far from a Gaussian shape. This result agrees with the experimental results, but not for all the propagation modes. The experimental result (Fig. 4(d)) is affected by the bending and additional parameters (e.g., the roughness of the internal wall of the waveguide) which are not taken theoretically into account. In both theoretical and experimental results (Figs. 4(c)-4(d)) the shapes of the output power density for the curved waveguide are not symmetric. Fig. 5(a) shows that in addition to the main propagation mode, several other secondary modes appear, where $\delta_p = 0.4$ and R=0.7 m. By increasing only the step's angle, the amplitude of the output power density is greater and also the output shape is changed (Fig. 5(b)).

Figure 6(a) shows the comparison between the theoretical model with the known solution[22] for the rectangular dielectric slab (Fig. 3(b)), for every order (N=1, 3, 5, 7, and 9), where the order N determines the accuracy of the solution. The comparison has shown good agreement. Note that we have two ways to compare between the results of our mode model with the other methods. The first way is to compare between the results of the output fields for every order (N=1, 3, 5, 7, and 9) with the final solution of the known method. The second way is to compare between the results of the output fields (according to our model) for every two orders (N=1,3, N=3,5, N=5,7, and N=7,9), until our numerical solution is well converged. This way is efficient in the cases that we have complicated problems that we cannot compare with the final solution of the known method.

Figures 7(a)-(b) show the results of the output power density (S_{av}) as functions of the step's angle (δ_p=1) and the radius of the cylinder (R=0.5 m). The output fields are dependent on the input wave profile (TE_{10} mode) and the circular dielectric profile of the rectangular cross section (Fig. 3(c)). For small values of ϵ_r, the half-sine (TE_{10}) shape of the output power density appears, with a little influence of the Gaussian shape in the center of the output profile. On the other hand, for large values of ϵ_r (e.g., ϵ_r=10) the Gaussian shape of the output power density appears in the center of the output profile (Fig. 7(b)), with a little influence of the half-sine (TE_{10}) shape in the center of the output profile. By increasing only the parameter ϵ_r from 2 to 10, the result of the output power density shows a Gaussian shape and the amplitude of the output power density is changed, as shown in Fig. 7(c). In this case, the output Gaussian profile increases with increasing the value of ϵ_r. These examples demonstrate the influence of the dielectric profile for an inhomogeneous cross section, for arbitrary step's angle and the radius of the cylinder of the helical waveguide. Figure 7(d) shows an example for the output profiles with $\epsilon_r = 10$. The output results are demonstrated for every order (N=1, 3, 5, 7, and 9). By increasing only the parameter of the order from N=1 to N=9, then the output profile approaches to the final output profile.

The other main contributions of the proposed methods are demonstrated in Fig. 8(a) and in Fig. 8(b), in order to understand the influence of the step's angle (δ_p) and the radius of the cylinder (R) on the output power transmission, for helical waveguide with a circular cross section (Fig. 3(a)) and with a rectangular cross section (Fig. 3(c)), respectively. Six results are demonstrated for six values of δ_p (δ_p =0.0, 0.4, 0.7, 0.8, 0.9, 1.0), in all case. For an arbitrary value of R, the output power transmission is large for large values of δ_p and decreases with decreasing the the value of δ_p. On the other hand, for an arbitrary value of δ_p, the output power transmission is large for large values of R and decreases with decreasing the value of R. For small values of the step's angle, the radius of curvature of the helix can be approximated by the radius of the cylinder (R). In this case, the output power transmission is large for small values of the bending (1/R), and decreases with increasing the bending. Thus, these two different methods can be a useful tool to find the parameters (δ_p and R) which will give us the improved results (output power transmission) of the curved waveguide in the cases of space curved waveguides.

The output power transmission and the output power density are improved according to the two proposed methods by increasing the step's angle or the radius of the cylinder of the helix, especially in the cases of space curved waveguides. These methods can be a useful tool to improve the output results in all the cases of the hollow helical waveguides in medical and industrial regimes (by the first method) and in the microwave and millimeter-wave regimes (by the second method), for the diffused optical waveguides in integrated optics.

Author details

Z. Menachem and S. Tapuchi
Department of Electrical Engineering, SCE-Shamoon College of Engineering, Israel

6. References

[1] Harrington J.A, Matsuura Y (1995) Review of hollow waveguide technology, *SPIE*, 2396.

[2] Harrington J.A, Harris D.M, Katzir A (1995) *Biomedical Optoelectronic Instrumentation*, pp. 4-14.

[3] Harrington J.A (2000) A review of IR transmitting, hollow waveguides, *Fiber and Integrated Optics*, 19, pp. 211-228.

[4] Marcatili E.A.J, Schmeltzer R.A (1964) Hollow metallic and dielectric waveguides for long distance optical transmission and lasers, *Bell Syst. Tech. J.*, 43, pp. 1783-1809.

[5] Marhic M.E (1981) Mode-coupling analysis of bending losses in IR metallic waveguides, *Appl. Opt.*, 20, pp. 3436-3441.

[6] Miyagi M, Harada K, Kawakami S (1984) Wave propagation and attenuation in the general class of circular hollow waveguides with uniform curvature, *IEEE Trans. Microwave Theory Tech.*, 32, pp. 513-521.

[7] Croitoru N, Goldenberg E, Mendlovic D, Ruschin S, Shamir N (1986) Infrared chalcogenide tube waveguides, *SPIE*, 618, pp. 140-145.

[8] Melloni A, Carniel F, Costa R, Martinelli M (2001) Determination of bend mode characteristics in dielectric waveguides, *J. Lightwave Technol.*, 19, pp. 571-577.

[9] Bienstman P, Roelens M, Vanwolleghem M, Baets R (2002) Calculation of bending losses in dielectric waveguides using eigenmode expansion and perfectly matched layers, *IEEE Photon. Technol. Lett.*, 14, pp. 164-166.

[10] Mendlovic D, Goldenberg E, Ruschin S, Dror J, Croitoru N (1989) Ray model for transmission of metallic-dielectric hollow bent cylindrical waveguides, *Appl. Opt.*, 28, pp. 708-712.

[11] Morhaim O, Mendlovic D, Gannot I, Dror J, Croitoru N (1991) Ray model for transmission of infrared radiation through multibent cylindrical waveguides, *Opt. Eng.*, 30, pp. 1886-1891.

[12] Kark K.W (1991) Perturbation analysis of electromagnetic eigenmodes in toroidal waveguides, *IEEE Trans. Microwave Theory Tech.*, 39, pp. 631-637.

[13] Lewin L, Chang D.C, Kuester E.F (1977) *Electromagnetic Waves and Curved Structures*, Peter Peregrinus Ltd., London, Chap. 6, pp. 58-68.

[14] Cochran J.A, Pecina R.G (1966) Mode propagation in continuously curved waveguides, *Radio Science*, vol. 1 (new series), No. 6, pp. 679-696.

[15] Carle P.L (1987) New accurate and simple equivalent circuit for circular E-plane bends in rectangular waveguide, *Electronics Letters*, vol. 23, No. 10, pp. 531-532.

[16] Weisshaar A, Goodnick S.M, Tripathi V.K (1992) A rigorous and efficient method of moments solution for curved waveguide bends, *IEEE Trans. Microwave Theory Tech.*, vol. MTT-40, No. 12, pp. 2200-2206.

[17] Cornet P, Duss'eaux R, Chandezon J (1999) Wave propagation in curved waveguides of rectangular cross section, *IEEE Trans. Microwave Theory Tech.*, vol. MTT-47, pp. 965-972.

[18] Menachem Z (2003) Wave propagation in a curved waveguide with arbitrary dielectric transverse profiles, *Journal of Electromagnetic Waves and Applications*, 17, No. 10, pp. 1423-1424, and *Progress In Electromagnetics Research, PIER*, 42, pp. 173-192.

[19] Menachem Z, Croitoru N, Aboudi J (2002) Improved mode model for infrared wave propagation in a toroidal dielectric waveguide and applications, *Opt. Eng.*, 41, pp. 2169-2180.

[20] Menachem Z, and Mond M (2006) Infrared wave propagation in a helical waveguide with inhomogeneous cross section and applications, *Progress In Electromagnetics Research, PIER* 61, pp. 159-192.

[21] Menachem Z, Haridim M (2009) Propagation in a helical waveguide with inhomogeneous dielectric profiles in rectangular cross section, *Progress In Electromagnetics Research B,* 16, pp. 155-188.

[22] Collin R.E (1996) Foundation for microwave engineering, New York: McGraw-Hill.

[23] Yariv A (1985) *Optical Electronics*, 3rt edition, Holt-Saunders Int. Editions.

[24] Baden Fuller A.J (1969) *Microwaves*, Pergamon Press, A. Wheaton and Co. Ltd, Oxford, Chap. 5, pp. 118-120.

[25] Olver F.W.J (1960) *Royal Society Mathematical Tables, Zeros and Associated Values*, University Press Cambridge, pp. 2-30.

[26] Jahnke E, Emde F (1945) *Tables of Functions with Formulae and Curves*, Dover Publications, New York, Chap. 8, 166.

[27] Salzer H.E (1955) Orthogonal polynomials arising in the numerical evaluation of inverse Laplace transforms, *Math. Tables and Other Aids to Comut.*, Vol. 9, pp. 164-177.

[28] Salzer H.E (1961) Additional formulas and tables for orthogonal polynomials originating from inversion integrals, *J. Math. Phys.*, Vol. 39, pp. 72-86.

[29] The Numerical Algorithms Group (NAG) Ltd., Wilkinson House, Oxford, U.K..

[30] Croitoru N, Inberg A, Oksman M, Ben-David M (1997) Hollow silica, metal and plastic waveguides for hard tissue medical applications, *SPIE*, 2977, pp. 30-35.

Chapter 9

Electromagnetic Wave Propagation Modeling for Finding Antenna Specifications and Positions in Tunnels of Arbitrary Cross-Section

Jorge Avella Castiblanco, Divitha Seetharamdoo, Marion Berbineau,Michel Ney and François Gallée

Additional information is available at the end of the chapter

1. Introduction

Electromagnetic wave propagation in confined environments, such as mine, road or railway tunnels, has inadmissible quality due to multi-path and high scattering effects of receivable signals. The ever-increasing work difficulties and the large demand for systems operating in these scenarios, requires more and more innovative wireless systems to guarantee a reliable communication. From the radiating element point of view, some investigations have been given over to the development of novel technologies to provide radio communication in these environments. In 1956, a paper [46] disclosed what is now generally known as the leaky-feeder principle for propagation of VHF or UHF signals through a tunnel. Leaky coaxial cables (LCXs) have been employed since then [18, 38]. Nevertheless, high cost, maintenance and installation difficulties are the main disadvantages of leaky feeders. These problems are easily solved in short tunnels, but in the case of longer tunnels the leaky feeder solution becomes too expensive. For these reasons, solutions that are based on the use of distributed antennas are becoming more interesting [6].

Antenna design for these scenarios is a challenge : Current tunnel cross-sections can have arbitrary shapes and antennas have to operate close to tunnel walls. These antennas have to be discreet and small and propose higher operating frequencies. Moreover, possible interference among systems may be suffered due to the growing number of wireless systems working simultaneously. Contrary to free-space applications, the specifications and parameters involved for antennas working in these environments have not yet been thoroughly established. The understanding and modeling of radio-wave propagation are essential steps toward deriving optimum radio communication systems and deployment in these scenarios. Various studies have been done considering several models to find or analyze radiation characteristics and positioning of antennas in tunnels [21, 27, 28] or simply to model the wave-propagation in these environments [13, 15, 19, 34, 41, 57]. The models used to study the wave-propagation in these environments are limited due to the intrinsic principles, applicability and assumptions on which they are based. Asymptotic methods are mostly used [55]. However, asymptotic methods cannot predict well the fields for antennas associated with systems which are strongly affected by the surrounding environment. Near-field considerations have to be accounted for and, thus full-wave methods should be used.

The problem of antenna operation in tunnel environments has been addressed very briefly. From the positioning point of view, the U.K. guidance (GK/GN06602) on train rooftop antenna positioning as well as some manufacturers provide some obstruction, separation, installation and maintenance specifications for train antennas. Alternative solutions offer improvements in the channel capacity by means of the transmit, receive or spatial diversity, *i.e.*, by using multiple antennas in the transmission/reception located at different positions [44, 45, 56]. A modal interpretation of the principles of these solutions are studied in [37, 43]. The determination of field specifications to excite correct modes to avoid causing interference between systems, can be also analyzed by using the modal interpretation. Due to multi-modal propagation in these environments, different positions and types of excitation are involved. A modal approach is considered to provide a straightforward description of the wave propagation and the degrees of freedom in these environments [1, 54]. These modes may be combined in some way to produce a desired effect, *e.g.*, the maximization of the transfered power by the modes. Thus, modal-based analysis and optimization techniques for mode-weight adjustment are used to determine correct co-excitation of these modes, and fulfill the desired specifications. This approach is often used in adaptive array theory [26, 47]. However, to the best of the authors' knowledge, no such treatment has been reported to deal with multi-modal propagation in tunnel environments.

This chapter is organized as follows : Section II introduces the modal approach for guiding structures. It is based on a full-wave method, namely the Transmission Line Matrix (TLM) method. These methods has been hampered by their large computational time when compared to asymptotic methods when large size environments are considered. Thus, a suitable 2.5 D TLM implementation to reduce the computational time and to include lossy dielectric walls of tunnels is briefly presented [2]. The computation cost is reduced compared to typical solutions by using the concept of Surface Impedance Boundary Condition (SIBC). Section III is devoted to the description of a methodology for the determination of antenna field specifications and positioning in operational scenarios at high frequencies. Section IV presents the validation of this methodology for a simple canonical case. Lastly, section V describes the analysis and results for a real scenario representative of tunnel environments. Finally, discussions and conclusions are developed.

2. Modeling approach for guiding structures

The high complexity of today's relevant EM problems has popularized the use of numerical approaches to approximate exact solutions [30]. The main limitation of these methods is the high computation time when electrical-large structures are considered. With the increasing performance of computers, research in electromagnetic modeling through numerical methods has been continuously advancing. New techniques have merged and the capability of the existing methods has been improved. A modal approach based in the *Transmission Line Matrix*

(TLM) method, is considered to analyze the wave propagation in tunnels. A reduced TLM formulation, is employed to find the mode field distribution of the uniform guiding structure. Adequate boundary conditions are additionally used to to limit the computation volume. The modeling approach can be applied for a tunnel with arbitrary cross section, including lossy walls.

2.1. Brief overview

The modal approach is a rigorous technique, allowing us to have an appropriate physical insight of the propagation in tunnel environments. It allows us to express the fields in terms of modes and generalize the applicability of our algorithm for tunnels with arbitrary cross section. This theory is based on functional analysis. The total fields in a guiding structure, traveling in the $\pm z$ direction, can be expanded by the sum of a set of linearly-independent functions Ψ. Their mutual orthogonality is not necessarily required. The space spanned by these functions is given by:

$$
\begin{aligned}
\vec{E}_{guided}\left(\vec{r}\right) &= \sum_{n=1}^{N} w_n\left(\vec{r}\right) \Psi_{\vec{E}}\left(\vec{r}\right) = \sum_{n=1}^{N} w_n\left(\vec{r}\right) \hat{E}_n\left(x,y\right) e^{-(\alpha_n \pm j\beta_n)z} \\
\vec{H}_{guided}\left(\vec{r}\right) &= \sum_{n=1}^{N} w_n\left(\vec{r}\right) \Psi_{\vec{H}}\left(\vec{r}\right) = \sum_{n=1}^{N} w_n\left(\vec{r}\right) \hat{H}_n\left(x,y\right) e^{-(\alpha_n \pm j\beta_n)z},
\end{aligned}
\tag{1}
$$

where $\vec{E}_{guided}$ and $\vec{H}_{guided}$ are the total guided- electric and magnetic fields, respectively, N is the total number of modes. $\hat{E}_n\left(\vec{r}\right)$ and $\hat{H}_n\left(x,y\right)$ are the normalized mode profiles, $w_n\left(\vec{r}\right)$ are the mode weight coefficients. Finally, α and β are the attenuation and phase constants, respectively. Considering a particular guiding structure, these modes are solutions for the fields which satisfy both Maxwell's equations and the boundary conditions of the waveguide.

To fully characterize the modes, field profiles and the attenuation and phase constants have to be determined. A modal approach based on the TLM method is employed to this end. The TLM is a time-domain explicit algorithm based on local wave decomposition. The computational domain is simulated by a network of interconnected transmission-lines. All fields components are computed at the center of the nodes (connexions) by linear combination of impulse voltages traveling in the transmission-lines at discrete time Δt. The explicit scheme means that voltages at a later time from the system are calculated at the current time. The method is carried out basically through two processes: scattering and connection. In the scattering process, voltage pulses $_k V^i$ are incident upon the node from each of the link-lines (halfway between two nodes) at each time step k. These pulses are then scattered to produce a set of scattered voltages, $_k V^r$, which become incident on adjacent nodes at the next time step $k+1$. In the connection process, pulses are simply exchanged among immediate neighbors. A simplified approach based on this method and by using the procedure described in [39, 58], is briefly introduced in the next subsection.

2.2. EM Modeling of arbitrary cross-section over-sized waveguides: The 2.5-D TLM approach

The simplification of guiding problems in TLM was first introduced in [29] to obtain the dispersion properties of guiding structures. They proposed the reduction of the calculation region by introducing the field dependence along the propagation direction z, which can be expressed by the terms $e^{-j\beta z}$, where β is the phase constant. Two points along the longitudinal distance of the guide can be characterized by the phase difference $\beta(z_2 - z_1)$. Then, it is possible to find a relationship between the reflected voltages of the node at the time $k\Delta t$ and the incident ones at the time $(k+1)\Delta t$. The *Symmetrical Condensed Node* (SCN), shown in Fig. 1, constitutes the most widely used TLM node for 3D structures. The expressions relating to these voltages for this node are given by equations (2, 5), where $_kV_4^r$, $_kV_2^r$, $_kV_9^r$, $_kV_8^r$ are the reflected voltages at the time $k\Delta t$ and $_{k+1}V_4^i$, $_{k+1}V_2^i$, $_{k+1}V_9^i$, $_{k+1}V_8^i$ are the incident voltages at the time $(k+1)\Delta t$ respectively, on lines 4, 2 9 and 8.

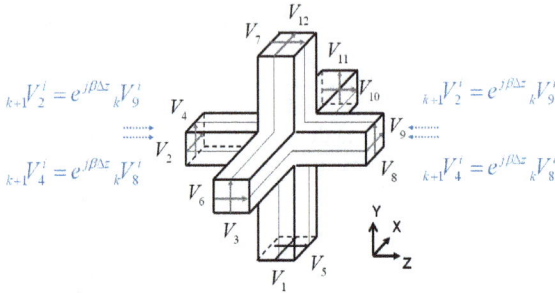

Figure 1. The TLM node for guiding structures.

$$_{k+1}V_8^r = e_k^{-j\beta\Delta z} V_4^r \tag{2}$$

$$_{k+1}V_9^r = e_k^{-j\beta\Delta z} V_2^r \tag{3}$$

$$_{k+1}V_2^r = e_k^{-j\beta\Delta z} V_9^r \tag{4}$$

$$_{k+1}V_4^r = e_k^{-j\beta\Delta z} V_8^r \tag{5}$$

Note that the characteristic parameter β must be imposed at the beginning of each calculation. Then, the resulting fields correspond to the solution of the fields for this value. Two traveling waves are injected in opposite directions in some of the nodes forming a standing wave. A time-domain signal, such as a Dirac or a Gaussian pulse is usually employed to analyze the structure in a wide-frequency range. After a considerable number of time step iterations, the response is taken at some nodes and correspond to the superposition of the modal fields. By performing a *Fast Fourier Transform* (FFT), the response in frequency domain can be obtained. The peaks in the spectrum correspond to the modes. Several β values have to be imposed if the response of the modes for different phase constants is desired. Based on this idea, a 2.5 Dimensional SCN for guiding structures formulation was employed to determine the fields [39, 58]. The term 2.5D is used as the 3D computational domain is reduced to a 2D mesh in the guide cross-section. However, cells are 3D ones and account for all 6 electromagnetic field components.

In addition to the reduced node for guiding structures, the concept of Surface Impedance Boundary Condition (SIBC) is also considered to efficiently model the tunnel walls, while reducing the computational domain, as seen in Fig. 2 and explained below.

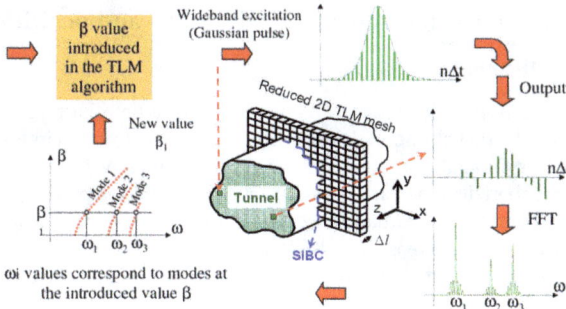

Figure 2. 2.5 D TLM approach for guiding structures.

2.3. SIBC boundary conditions to model the tunnel walls

Tunnels are made with dielectric walls and may contain other materials (trains, cars, objects etc.). For volumic methods, such as TLM, Finite Difference Time Domain (FDTD) or Finite Element Method (FEM), the limitation of the computational domain within dielectric materials is usually achieved by using Perfect Matched Layer (PML) as boundary conditions. However, this technique is too expensive in terms of complexity and consequently computation time, when dealing with very large electrical objects. Since we are interested only in the EM fields within the hollow region of the tunnel, an appropriate boundary conditions, namely, the Surface Impedance Boundary Conditions (SIBC), constitutes the mathematical artifice to limit the computation volume. For detailed description of its formulation, validation and results obtained with the TLM method, the reader is referred to [2, 3, 58]. Here, an introduction of this concept is briefly presented.

Boundary conditions in TLM are simulated by introducing a relationship between incident and reflected voltages at the boundary. Its expression depends on the nature of the materials on both sides. An interface of air and a lossy dielectric, as in the case of tunnel walls, can be simulated by introducing a reflection coefficient Γ, given by [10]:

$$_{k+1}V_{armj}^i = \Gamma_k V_{armj}^r = \frac{Zs - Z_0}{Zs - Z_0}V_{armj}^r, \tag{6}$$

Γ is in general complex and would alter the shape of the excitation pulses, which normally cannot be accounted for in the TLM method [10]. V_{armj}^i and V_{armj}^r represent the incident and reflected voltages of the j-arm of the node, respectively, and Zs correspond to the Surface Impedance of a lossy dielectric [50]. Thus, the SIBC concept allows the efficient modeling of lossy dielectrics, replacing one medium by a local reflection coefficient at the interface. This avoids meshing and calculating fields beyond the interface. For tunnel wall modeling applications, these field values are not required for mode parameter computation. However, the presence of the the dielectric beyond the interface is accounted for by the SIBC. This approach is applied in the next sections to study the radio wave propagation for waveguide with arbitrary cross-section. However, it is limited to walls of permittivity $e_r \gg 1$ but without any restrictions on the conductivity σ. However, the problem of finding antenna specifications is examined and thoroughly discussed first.

3. Methodology to find antenna field specifications and positioning

3.1. Justification of the choice

One of the most important properties of an antenna is that it radiates or receives radio waves in a way related to field distribution. In the simplest scenario of a wireless propagation channel, radio waves propagate in free space, expand spherically from the TX to the RX antenna. Antennas are then designed according to a desired radiation pattern and considering far fields. A more complicated situation occurs if the radio waves propagate through more complex environments, such as tunnels. Waves find several paths through a complex environment with a variety of several scattering obstacles. Antenna specifications for these environments are needed.

Modal theory is usually employed to study the correct excitation in guiding structures [11, 40, 51]. In the particular case of radio-wave propagation in tunnel environments, the modal theory has been employed in the past. Some of these approaches have considered canonical geometries and excitation sources [15–17, 43]. Thus, in the next subsections, a methodology to find the field specifications and positioning in these scenarios, based on some recent research, is discussed.

3.2. Basic principles

In this subsection, essential principles used with this methodology to obtain the field specifications and antenna positioning in confined environments, are described. These principles make use of field modal expansion, the mode orthogonality. They will allow the optimum antenna positioning for mode excitation.

3.2.1. Modal expansion of the fields

The modal expansion (1) is only valid for closed source-free regions. Inside the source region, besides the discrete mode series expansion, an additional term is needed to account for the fields of the sources [4]. Additionally, for open waveguides, there is a continuous part of the spectrum related to radiation modes [4, 59]. Thus, the total fields for wave-guiding structures can be generally expressed by (7):

$$\vec{E}(\vec{r}) = \vec{E}_{guided}(\vec{r}) + \vec{E}_{sources}(z) + \vec{E}_{radiated}(\vec{r}), \tag{7}$$

where $\vec{E}_{sources}(z)$ and $\vec{E}_{radiated}(\vec{r})$ are orthogonal complements of the functions $\Psi_{\vec{E}}$ and $\Psi_{\vec{H}}$. An analogous expression can be written for the magnetic fields. Inside the source region, the weight coefficients $w_n(\vec{r})$ in (1) represent the mode amplitude and take into account the z-dependence of the modes and the exciting sources, if any. They depend on the coordinate z as a result of the source actions, modeling the transverse field components of the sources. However, for source-free regions and assuming that modes are orthogonal, they are constants $w_n(\vec{r}) = w_n$. The longitudinal components of the source are modeled by the term $\vec{E}_{sources}(\vec{r})$ [4].

From the point of view of the fields, modes can be excited and, hence, treated separately. However, from the point of view of the total power carried by the modes, they may be coupled. This point will be discussed in the next subsection by means of the quasi-orthogonality relationship in hollow waveguides.

3.2.2. Quasi-orthogonality relation for hollow-lossy waveguides

Mode orthogonality is directly related to the properties of functions defined on Hilbert spaces H. The elements of these spaces are defined with an inner product. Consider a set of functions $\Psi_1, \Psi_2, \Psi_3 ... \Psi_N$ in a Hilbert space H. The inner product of $\Psi_k(r)$ and $\Psi_l(r)$ for k not necessarily equal to l is defined by:

$$< \Psi_k(r), \Psi_l(r) > = \int_S \Psi_k(r) \cdot \Psi_l^*(r) dr, \tag{8}$$

where $*$ indicates the conjugate. The functions $\Psi_k(r)$ and $\Psi_l(r)$ are said to be orthogonal if their inner product for $k \neq l$ is zero. Otherwise they are said to be non-orthogonal.

The existence of the orthogonality property allows us to simplify the analysis of a given waveguide. This property plays a key role in solving problems of excitation and scattering in waveguides [40]. The orthogonality between modes means that each mode carries its own power independently of other modes. In lossless waveguides with perfectly electric or magnetic walls, the orthogonality is an inherent property. In waveguides with lossy dielectric walls, such as tunnels, orthogonality no longer holds and so-called quasi-orthogonality can prevail. A complete treatment of this problem can be found in [4]. To better understand this consequence, consider the k-th and l-th modes propagating in a source-free region. The curl Maxwell equations applied to these modes are given by:

$$\begin{aligned}
\nabla \times \vec{E}_{k,l} &= -\sigma_m \vec{H}_{k,l} = -j\omega\mu\vec{H}_{k,l} \\
\nabla \times \vec{H}_{k,l} &= \sigma_e \vec{E}_{k,l} = (\sigma_c + j\omega e)\,\vec{E}_{k,l},
\end{aligned} \tag{9}$$

where σ_m and σ_e are the electric and magnetic conductivities, respectively. Using Maxwell's equations, one finds:

$$\nabla . \left(\vec{E}_k^* \times \vec{H}_l + \vec{E}_l \times \vec{H}_k^* \right) = -2 \left(\vec{E}_k^* \sigma_e \vec{E}_l + \vec{H}_k^* . \sigma_m \vec{H}_l \right) \tag{10}$$

Application of the two-dimensional divergence theorem:

$$\int_S \nabla . F dS = \frac{\partial}{\partial z} \int_S \hat{z} . F dS + \oint_L \hat{n} . F dl \tag{11}$$

into (10) yields:

$$\frac{\partial}{\partial z} \int_S \left(\vec{E}_k^* \times \vec{H}_l + \vec{E}_l \times \vec{H}_k^* \right) . \hat{z} dS = -2 \int_S \left(\vec{E}_k^* \sigma_e \vec{E}_l + \vec{H}_k^* . \sigma_m \vec{H}_l \right) dS - 2 \oint_L Z_s \tilde{H}_k^* . \tilde{H}_l dl, \tag{12}$$

where S is a closed surface with interface contour L, Z_s denotes the surface impedance, $\tilde{H}$ denotes the tangential component of $\vec{H}$ at the boundaries and:

$$\begin{aligned}
\vec{E}_\xi &= \hat{E}_\xi (x,y)\, e^{-\gamma_\xi z} \\
\vec{H}_\xi &= \hat{H}_\xi (x,y)\, e^{-\gamma_\xi z}
\end{aligned} \tag{13}$$

where γ is the propagation constant and ξ is equal to either k or l. Substitution of (13) into equation (12) yields to:

$$(\gamma_k^* + \gamma_l) \int_S (\hat{E}_k^* \times \hat{H}_l + \hat{E}_l \times \hat{H}_k^*) . \hat{z} dS = 2 \int_S (\hat{E}_k^* \sigma_e \hat{E}_l + \hat{H}_k^* . \sigma_m \hat{H}_l)\, dS + 2 \oint_L Z_s \hat{H}_k^* . \hat{H}_l dl \tag{14}$$

The real (time-averaged) total power transmitted $P(z)$ and dissipated $Q(z)$ can be calculated by equations (15).

$$P(z) = \frac{1}{2} Re \left\{ \int_S \left(\vec{E}^* \times \vec{H} \right) . \hat{z} d\vec{S} \right\} = \frac{1}{4} \int_S \left(\vec{E}^* \times \vec{H} + \vec{E} \times \vec{H}^* \right) . \hat{z} d\vec{S}$$
$$Q(z) = \frac{1}{2} \int_S \left(\vec{E}^* \sigma_e \vec{E} + \vec{H}_k^* \sigma_m \vec{H}_l \right) dS + \oint_L Z_s \vec{H}_k^* . \vec{H}_l dl \tag{15}$$

Substitution of the modal expansion (1) yields to the expression (16) and (17).

$$P(z) = \frac{1}{2} \sum_{k=1}^{N} \sum_{l=1}^{N} P_{kl} = \frac{1}{2} \sum_{k=1}^{N} \sum_{l=1}^{N} Re \left\{ w_k^* w_l \int_S \left(\hat{E}_k \times \hat{H}_l^* \right) . \hat{z} d\vec{S} \right\} \tag{16}$$

$$Q(z) = \frac{1}{2} \sum_{k=1}^{N} \sum_{l=1}^{N} Q_{kl} = \frac{1}{2} \sum_{k=1}^{N} \sum_{l=1}^{N} w_k^* w_l \left[\int_S \left(\hat{E}_k^* \sigma_e \hat{E}_l + \hat{H}_k^* . \sigma_m \hat{H}_l \right) dS + \oint_L Z_s \hat{H}_l^* . \hat{H}_l dl \right] \tag{17}$$

Finally, by comparing expressions (16), (17) and (14), the *quasi-orthogonality relation* (18), can be obtained.

$$\left(\gamma_k^* + \gamma_l \right) P_{kl} = Q_{kl}. \tag{18}$$

Every single mode carries a real self-power flow $P_k = P_{kk}$ and self-power loss $Q_k = Q_{kk}$. If self powers are coupled to each other, *i.e.* $Q_{kl} \neq 0$, it leads to the non-orthogonality between modes. Then, equation (18) allows us to analyze the cases where the non-orthogonality between modes occurs. It is worth noting that, for a hollow-dielectric waveguide, such a tunnel, the surface integral in $Q(z)$ vanishes, and the expression reduces to:

$$Q(z) = \frac{1}{2} \sum_{k=1}^{N} \sum_{l=1}^{N} w_k^* w_l \oint_L Z_s \hat{H}_k^* . \hat{H}_l dl \tag{19}$$

where L represent a closed contour around the surface of the hollow-waveguide. In this case, modes are non-orthogonal due to the fact that fields do not vanish at the boundaries leading to $Q_{kl} \neq 0$. Another interesting case occurs when a waveguide with perfectly electric walls and loaded with a lossy dielectric, is considered. In this case, the expression (18) reduces to:

$$Q(z) = \frac{1}{2} \sum_{k=1}^{N} \sum_{l=1}^{N} w_k^* w_l \left[\int_S \left(\hat{E}_k^* \sigma_e \hat{E}_l + \hat{H}_k^* . \sigma_m \hat{H}_l \right) dS \right], \tag{20}$$

The field components are obtained by multiplying the expressions for the fields for an unloaded waveguide with $e^{\alpha z}$, where α is the attenuation constant [7]. Evaluating expression (20) for this case leads to zero. Thus, the modes remains orthogonal even when losses are present in the waveguide. However, the integration surface S can be chosen such that the integral in (20) does not vanish, obtaining the non-orthogonality between modes. This simple case will be considered later to evaluate our methodology for waveguides with non-orthogonal modes.

The direct consequence of the non-orthogonality of modes is that the total power is not additive, *i.e.*, it is incorrect to state that the total power loss is the sum of the power loss of the modes propagating independently [4, 59]. Thus, the powers of the modes in lossy-hollow-dielectric waveguides, such as tunnels, are in general, mutually coupled and this fact has to be considered.

3.2.3. *Optimum positioning for mode excitation*

In the most general case, a wave-guiding structure can be excited by a current source or by an external field. Several modes can be excited and the main objective consists in deriving the weight coefficients due to a given source. For lossless waveguides, the orthogonality property implies that there is no power exchange between modes. The formalism for determining these coefficients is well-known and is usually done by considering current sheets [40, 51]. In practice, such currents are approximated by probes or loops. The orthogonality in this case simplifies the analysis and modes can be selectively excited or excluded.

Nevertheless, in lossy-dielectric waveguides, energy exchange between modes can take place. The theory that describes the interaction between them is known as the *Coupled Mode Theory*. This theory allows us to determine the mode amplitudes. The problem of the weight calculation is rather complicated and lies outside the scope of this chapter. The reader is referred on this point to the works described in [4, 35, 59]. The mutual coupling between modes causes the mode amplitudes to be coupled and the resulting coefficients are no longer constant but become dependent on the propagation direction. The desired set of the equations of mode excitation for $k = 1, 2, ...,$ can be written in the form (21).

$$\Sigma_l N_{kl} \frac{dw_l(z)}{dz} e^{\gamma_l z} = R_k(z),\qquad(21)$$

where:

$$N_{kl} = \int_S \left(\hat{E}_k^* \times \hat{H}_l + \hat{E}_l \times \hat{H}_k^*\right) . \hat{z} dS\qquad(22)$$

$$R_k(z) = -\int_{S_s} \left(J_a^e . \hat{E}_k^* + J_a^m . \hat{H}_k^*\right) dS - \int_{C_s} \left(J_b^e . \hat{E}_k^* + J_b^m . \hat{H}_k^*\right) dl,\qquad(23)$$

S_s and C_s correspond to the surface and contour for the source, respectively. J_a^e, J_b^e, J_a^m, J_b^m are electric and magnetic currents to model the source. The subscripts a and b denotes the external bulk and surface sources. The ideal excitation position can be studied by considering two modes, the resulting expression for the mode amplitude of one of the modes is given by:

$$w_1(z) = \int e^{-\gamma_1 z} \left(\frac{N_{22} R_1 - N_{12} R_2}{N_{11} N_{22} - N_{12} N_{21}}\right) dz,\qquad(24)$$

where N_{12} and N_{21} are the coupling factors between the modes, γ_1 is the propagation constant of the first mode, R_1 and R_2 correspond to the excitation terms (23) for the first two modes. In [43], it was demonstrated that these values can be relatively small. By comparing (24) and (23), it can be noted that the ideal excitation for a given mode is predominantly dominated by R_1. In fact, the dot product in (23) indicates that the excitation should coincide with the mode profile. As a result, to maximize the excitation of a given eigenfunction, the source should be located in regions where it is predominant and avoid those where it vanishes. In a more general case, the eigenfunction of the electric and magnetic field should be maximized. Thus, a proper excitation should be placed in regions where the power is maximum. This analysis may be extended for the expression of the weight coefficient for all modes, which is given by:

$$\mathbf{w}(z) = \int e^{-\gamma z} \mathbf{N}^{-1} \mathbf{R} dz,\qquad(25)$$

where $\mathbf{w}(z)$ is a vector containing the mode amplitudes, $e^{-\gamma z}$ is a vector containing the exponential terms of the mode propagation constant, $\mathbf{N}$ is a matrix of the coupling factors given by (22) and $\mathbf{R}$ is a matrix with the source terms given by (23). Appropriate mode excitation should be located in regions where the sum of the mode field profiles is maximum and avoid those where it vanishes. More generally speaking, locations where the total power carried by the modes maximizes, correspond to those where the electric and magnetic fields are simultaneously better excited.

3.2.4. Optimum field excitation for mode excitation

Field specifications for correct mode excitation has mainly been studied in the literature for applications in optical communications. The selective mode excitation in a hollow waveguide is studied in [42, 48, 49]. A linearly polarized input light is converted to a TE_{01}-like profile to excite this mode in a circular waveguide. Several papers on the design of mode converters can be consulted by the interested reader in [24, 36]. A Gaussian beam launched axially into a circular fiber is used in [20], for exciting low-order modes. Finally, only canonical cases have been reported for the integration of antennas in guiding structures [9, 12, 14, 22, 23, 51].

An optimum source excitation in dielectric waveguides, such as tunnels, is still to be determined and its characteristics are determined by the nature of the weight coefficients given by (1). As mentioned before, multi-modal propagation in these environments can cause detrimental effects to the received power. A similar problem arises in adaptive array theory where the array output power has to be minimized by choosing adequate weight coefficients of an N-element array. Optimization techniques are employed to find a proper set of coefficients. A complete treatment of this problem can be found in [26, 47]. Optimum mode coefficients for correct mode excitation in tunnels can be determined by following a similar optimization procedure as in adaptive array theory. The total power carried out by the modes, given by (16), must be optimized. The power must be restricted to a given range, and only finite values of the weight coefficients should be considered. The problem can be stated into different ways. First, by maximizing the transmitted power through M receivable modes in Rx, such that its value is bounded. Its formulation can be expressed mathematically as the maximization of:

$$P\left(x, y, z\right) = \mathbf{w}^{H}\mathbf{P}_{\Delta Tx}\mathbf{w} \tag{26}$$

subject to the constrains:

$$\mathbf{w}_{Tx}^{H}\mathbf{P}_{\Delta Tx}\mathbf{w}_{Tx} \leq \delta_{max_{Tx}}$$

$$\mathbf{w}\left(x, y, z\right)^{H}\mathbf{w}\left(x, y, z\right) = c. \tag{27}$$

The second choice consist of maximizing the received power through M transmittable modes in Tx, such that its value is higher than a given threshold. The problem can be reduced to the maximization of:

$$P\left(x, y, z\right) = \mathbf{w}^{H}\mathbf{P}_{\Delta Rx}\mathbf{w} \tag{28}$$

subject to the constrains:

$$\mathbf{w}_{Rx}^{H}\mathbf{P}_{\Delta Rx}\mathbf{w}_{Rx} \geq \delta_{min_{Rx}}$$

$$\mathbf{w}\left(x, y, z\right)^{H}\mathbf{w}\left(x, y, z\right) = c, \tag{29}$$

Equation (26) and (28) are the matrix form of (16) in Tx and Rx. The vector $\mathbf{w}$ contains the weight coefficients and $\mathbf{w}^{H}$ denotes the conjugate transpose of $\mathbf{w}$. $\mathbf{P}$ is a hermitian $N \times N$

Electromagnetic Wave Propagation Modeling for
Finding Antenna Specifications and Positions in Tunnels of Arbitrary Cross-Section

231

matrix (*i.e.* $\mathbf{P} = \mathbf{P}^H$) formed from the cross-Poynting powers of the N modes in the structure, $\mathbf{P}$ is a non-diagonal matrix, the (l, k) element of this matrix is given by:

$$\mathbf{P}_{\Delta\Omega}(x, y, z)_{(l,k)} \approx \frac{1}{2} \Re \frac{\left\{ \left(\vec{E}_l(x, y, z) \times \vec{H}_k^*(x, y, z) \right) \cdot \Delta\hat{\Omega} \right\}}{\Delta\Omega} \tag{30}$$

Hence, $\mathbf{P} = \mathbf{P}_{\Delta Tx}$ or $\mathbf{P} = \mathbf{P}_{\Delta Rx}$ correspond to the power densities in Wm^{-2} for an element of area $\Delta\Omega$ and represent the evaluation of (30) in Tx or Rx. The terms $\delta_{max_{Tx}} = P_{max_{Tx}}/Ae_{Tx}$ and $\delta_{min_{Rx}} = P_{min_{Rx}}/Ae_{Rx}$ are the maximum and minimum powers densities in Tx and Rx, respectively. $P_{max_{Tx}}$ and $P_{min_{Rx}}$ correspond to the maximum and minimum powers that can be transmitted and received in an effective area Ae_{Tx} in Tx and Ae_{Rx} in Rx. These power densities will be clear later on. The first constraint in (27) ensures that the power density (or simply the power) is bounded, whereas in (29), it ensures that the power is higher than a given threshold. In general, only one of these problem has to be solved, however the treatment of both will be considered here for better understanding. The third constraint avoids the solution $\mathbf{w} = 0$ insuring that the weight coefficient vector has a nonzero magnitude and that c is necessarily a real constant. Finding $\mathbf{w}$ optimum to satisfy (27) can be accomplished by the method of Lagrange Multipliers. The partial derivative of the unconstrained problem (31) with respect to each variable yields a system of equations to find the values of the weight coefficients. However, $\mathbf{w}$ is a complex vector and it is not clear how the derivative operates on the real and imaginary parts. A complex gradient operator defined in [5] is rather employed to this end. This operator or its complex conjugate are suitable for determining stationary points of a real function, such as $\mathbf{P}$. The necessary Kuhn–Tucker conditions [53], to maximize the following cost function:

$$\Lambda_{\Delta\Omega} = \mathbf{w}^H \mathbf{P}_{\Delta\Omega} \mathbf{w} + \lambda \left(c - \mathbf{w}^H \mathbf{w} \right) + \rho \left(\delta - \mathbf{w}_\Omega^H \mathbf{P}_{\Delta\Omega} \mathbf{w}_\Omega \right) \tag{31}$$

are given by:

$$\begin{aligned}
\nabla_{w*} \Lambda_{\Delta\Omega} = \mathbf{P}_{\Delta\Omega} \mathbf{w} - \lambda \mathbf{w} \mp \rho \mathbf{P}_{\Delta\Omega} \mathbf{w}_\Omega &= 0 \\
\nabla_\lambda \Lambda_{\Delta\Omega} = c - \mathbf{w}^H \mathbf{w} &= 0 \\
\rho \nabla_\rho \Lambda_{\Delta\Omega} = \rho \left(\delta - \mathbf{w}_\Omega^H \mathbf{P}_{\Delta\Omega} \mathbf{w}_\Omega \right) &= 0 \\
\left(\delta \mp \mathbf{w}_\Omega^H \mathbf{P}_{\Delta\Omega} \mathbf{w}_\Omega \right) &\geq 0 \\
\rho &\geq 0,
\end{aligned} \tag{32}$$

where ∇_v denotes the complex gradient operator with respect to v, $\Omega = Tx$ or Rx and $\delta = \delta_{max_{Tx}}$ or $\delta = -\delta_{min_{Rx}}$.

The first, third and fifth equation in (32) are satisfied for $\mathbf{P}_{\Delta\Omega} \mathbf{w} = \lambda \mathbf{w}$, $\rho = 0$. Thus, the optimum vector is an eigenvector of the Poynting matrix $\mathbf{P}_{\Delta\Omega}$. The eigenvectors are usually scaled so that the norm of each is 1, satisfying the second equation in (32), *i.e.* $\mathbf{w}^H \mathbf{w} = c = 1$. Lastly, the fourth equation in (32) are satisfied for $\delta_{max_{Tx}} - \mathbf{w}(x, y, z_{Tx})^H \mathbf{P}_{\Delta Tx} \mathbf{w}(x, y, z_{Tx}) = \delta_{max_{Tx}} - \lambda \geq 0$ or $\mathbf{w}(x, y, z_{Rx})^H \mathbf{P}_{\Delta Rx} \mathbf{w}(x, y, z_{Rx}) - \delta_{min_{Rx}} = \lambda - \delta_{min_{Rx}} \geq 0$. Thus, at each position on the propagation direction z, the optimum solution corresponds to the eigenvector(s) associated to $\delta_{min_{Rx}} \leq \lambda$ or $\lambda \leq \delta_{max_{Tx}}$:

$$\mathbf{P}_{\Delta\Omega} \mathbf{w} = \lambda \mathbf{w} \tag{33}$$

The sufficient condition for (26) to have a constrained relative maximum, is that the determinant of the Hessian matrix formed by the second partial derivatives of (31), is positive definite. The application of this condition in this case leads to zero. Thus, the critical point is called a saddle point. The characteristic of a saddle point is that it corresponds to a relative minimum or maximum. By considering the solution of the problem (26) or (28) subject only to the second constraint in (27) or (29), it is possible to discriminate further. In this case, the optimum solution vector can be found using (33) and the determinant of the Hessian matrix is given by (34). Thus, the eigenvectors of $\mathbf{P}_{\Delta\Omega}$ correspond to the maxima of (26).

$$H = \begin{vmatrix} 0 & -\mathbf{w} \\ -\mathbf{w} & 0 \end{vmatrix} = \mathbf{w}^2 > 0 \tag{34}$$

The additional constraint can be included in the solution by examining the norm of the eigenvectors of $\mathbf{P}_{\Delta\Omega}$. It is noteworthy that the equality in the first constraint in (27) or (29) can be reached by evaluating the weight vectors (35) or (36) at z_{Tx} or z_{Rx}, respectively.

$$\mathbf{w}_{opt \to \delta_{max_{Tx}}}(z) = \sqrt{\frac{\delta_{max_{Tx}}}{\lambda_{max_{Tx}}}} \mathbf{w}_{max}(z), \tag{35}$$

$$\mathbf{w}_{opt \to \delta_{min_{Rx}}}(z) = \sqrt{\frac{\delta_{min_{Rx}}}{\lambda_{max_{Rx}}}} \mathbf{w}_{max}(z), \tag{36}$$

where $\lambda_{max}(z)$ and $\mathbf{w}_{max}(z)$ correspond to the z-dependent maximum eigenvalues and eigenvectors of (33). The first term on the right-hand side of (35) and (36) normalizes the eigenvalues to the maximum or minimum desired powers. These magnitudes are modified so that $\mathbf{w}_{Tx}^H \mathbf{w}_{Tx} = \delta_{max_{Tx}}/\lambda_{max_{Tx}}$ and $\mathbf{w}_{Rx}^H \mathbf{w}_{Rx} = \delta_{min_{Rx}}/\lambda_{max_{Rx}}$, which still satisfies the second equation in (32). In other words, optimal weight vectors in (35) and (36) ensure that the maximum and minimum powers in (33) coincide with the desired output power densities $\delta_{max_{Tx}}$ or $\delta_{min_{Rx}}$ at Tx and Rx, by modifying their norm. Thus, these vectors correspond to the maxima of (26). It should be noted that all the eigenvectors of (33) are solutions and here, only the maximum one was considered. Finally, the optimum fields at z_{Tx} and z_{Rx} can be found by using (1). They can be written in the matrix form:

$$\begin{aligned} \vec{E}_{opt}(z) &= \sum_{n=1}^N w_n^{opt}(z) \, \widehat{E}_n(x,y) \, e^{-\gamma z} = \mathbf{w}_{opt} \mathbf{\Psi}_{\vec{E}} \\ \vec{H}_{opt}(z) &= \sum_{n=1}^N w_n^{opt}(z) \, \widehat{H}_n(x,y) \, e^{-\gamma z} = \mathbf{w}_{opt} \mathbf{\Psi}_{\vec{H}} \end{aligned} \tag{37}$$

3.3. Assumptions

In this subsection, the assumptions considered in this methodology are examined. First, to apply the modal theory, fields in these structures are assumed to vary as $e^{-(j\omega t \pm j\beta)z}$. Electromagnetic waves propagate along a uniform infinitely long structure and field solutions must satisfy the boundary conditions. Tunnel walls are supposed infinitely thick and hence fields are considered to be at negligible levels when they reach the end of the wall. As a result, the term in (7) corresponding to the radiated fields, is assumed to be zero. The geometry of a hollow dielectric waveguide with arbitrary cross-section is shown in Fig. 3. Tunnels are assumed to be straight. Corners and objects can be simulated as long as the tunnel cross section remains constant.

To calculate the mode parameters, a few points over the cross-section should be excited by considering the 2.5-Dimensional TLM approach for guiding structures. Then, the source term in (7) should be considered. These sources are employed to obtain all the possible modes

Electromagnetic Wave Propagation Modeling for
Finding Antenna Specifications and Positions in Tunnels of Arbitrary Cross-Section

233

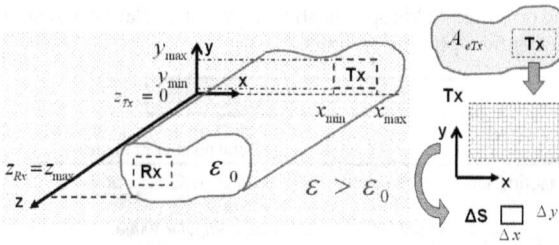

Figure 3. Geometry of an uniform section infinitely long hollow waveguide.

that can be excited, but their description is not needed. Since this term depends on the z-field component of the source, it can be neglected by exciting the structure only by the transversal components. This means that the guide is excited on a slice of the tunnel and remains applicable for Transversal-Electric (TE), Transversal-Magnetic (TM) and hybrid mode excitation. It is due to the fact that at least the value of one transversal component is non-zero.

The modal expansion (1), allow us to obtain all the modes that propagate in the structure. However, assuming that the objective is to design the transmitter antenna, in a more realistic scenario, only the modes excited by this antenna (source) and captured by the receiver should be considered. This can be accomplished by exciting the structure in the transmitter region, as shown in Fig. 3. Ae_{Tx} is the area where the power would be radiated if the source is placed somewhere in Tx. For an unloaded tunnel it corresponds to the whole tunnel cross-section. The excitation is simulated by an electric field probe concentrated in a differential of area ΔS in Tx. The power density of the i-th element has to be calculated over ΔS to apply the restrictions given by (27).

Finally, the orthogonality property for the modal functions in (1) is not mandatory. Its existence guarantees that the total power is the addition of individual modal power. However, as explained before, in general for lossy hollow-dielectric waveguides, such as tunnels, modes are non-orthogonal and the weight coefficients depend on z. Thus, these cross-power interactions have to be considered.

3.4. Algorithm flow chart

The antenna synthesis problem can be stated as the inverse of the analysis problem, *i.e.*, given a set of design specifications, such as required fields, type of excitation and positioning, determine an optimum antenna. Optimization techniques can be employed to solve a constrained problem and fulfill the design requirements. The flow chart in Fig. 4 illustrates the process for finding the optimum field specifications and positioning in tunnel environments. It is divided into five steps. The first one, concerns the definition of the inputs to simulate the tunnel. The characteristic parameters of the modes α, β, $\widehat{\mathbf{E}}(x,y)$ and $\widehat{\mathbf{H}}(x,y)$, are calculated at this point. In the second step, the modes are discriminated by their power carried through the tunnel, so that only modes among those with the highest power, are considered. Then, in the third step, an optimization procedure on the weight coefficients is carried out to obtain an optimal vector that maximizes the transmitted power by the modes. In the fourth step, the optimum weight coefficients to maximize the mean power along the propagation direction are obtained. Finally, the optimum fields specifications, type of excitation and source positioning are determined by using the optimum weights.

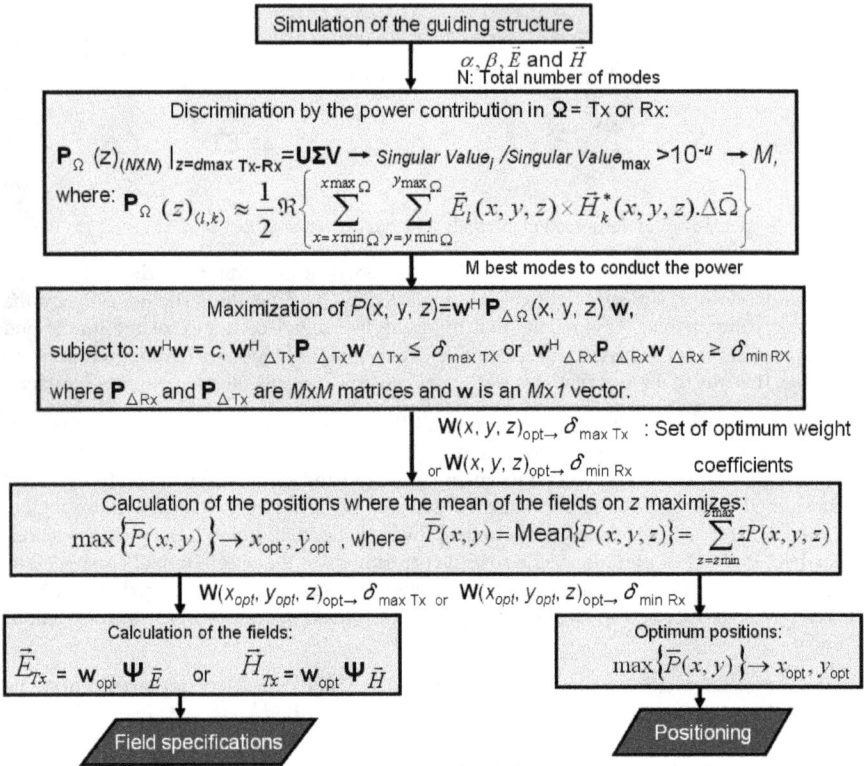

Figure 4. Flowchart of the process for determining the field specifications and positioning in tunnel environments.

3.4.1. Description

Finding antenna design specifications is a synthesis problem. It is of the utmost importance for correct antenna integration in over-sized guiding structures considered here. The fundamental problem facing antenna engineers is to improve the power reception in a certain desired region by using all possible parameters, so that the communication system meets the required specifications. An optimization algorithm is proposed and described below. It is divided into five steps.

The starting point is to solve this electromagnetic problem with the 2.5-Dimensional TLM for guiding structures. It consists in defining the geometry of the structure and constitutive parameters of the involved materials. The solution region is divided into a number of

Electromagnetic Wave Propagation Modeling for
Finding Antenna Specifications and Positions in Tunnels of Arbitrary Cross-Section

235

elementary nodes and their maximum size is defined by the minimum signal wavelength (or maximum frequency). Its value must be much smaller than the minimum wavelength to avoid the so-called numerical dispersion errors; a rule of thumb is to consider this value as one-tenth of the wavelength at the maximum frequency. The analysis is restricted to regions where the transmitter (Tx) and receiver (Rx) are usually located, and only the fields in these regions should be sampled. The sampling step is defined by the operation frequency and it should be chosen small enough to guarantee that fields are almost constant, for instance by considering a value of one-tenth of the wavelength at the operation frequency. Then, by using the previous entries in the 2.5-D TLM for guiding structures, the characteristic parameters for the modes can be calculated.

It is worth to remember that this study is meant to define the main factors affecting radio communication in tunnel environments, and to establish some criteria to diminish the undesired effects. Multi-modal propagation is one of the most detrimental factors degrading communication in tunnel environments. The problem of mode excitation has already been studied by means of modal theory [15]. High-order modes are responsible for rapid fluctuations in power, and their effects should be mitigated as much as possible. Moreover, a high number of modes leads to a considerable calculation time. Thus, in the second step of the proposed methodology, the number of considered modes is reduced. The strategy is straightforward: Assuming that the transmitter Tx is placed at $z = 0$, modes are classified by their power contribution in Rx located at z along the longitudinal distance. Since the calculation domain has been discretized, the (l, k) element of the matrix of crossed powers with dimensions $N \times N$ and given by (30), is approximated by using:

$$\mathbf{P}_\Omega(z)_{(l,k)} \approx \frac{1}{2} \Re \left\{ \sum_{x=x_{min_\Omega}}^{x_{max_\Omega}} \sum_{y=y_{min_\Omega}}^{y_{max_\Omega}} \left(\vec{E}(x,y)_l \times \vec{H}(x,y)_k^* \right) \cdot \Delta\vec{\Omega} \right\}, \tag{38}$$

where $\Delta\Omega$ is the approximation of the differential element of area in Ω and (x, y) correspond to the coordinates in the region for the receiver where the power has to be optimized. A singular-value decomposition (SVD) of (38), $\mathbf{P}_\Omega = \mathbf{U}\Sigma\mathbf{V}$, is done at each point along the z-direction. $\mathbf{U}$ and $\mathbf{V}$ are unitary matrices composed of eigenvectors of $\mathbf{P}_\Omega^H\mathbf{P}_\Omega$, and Σ is a diagonal matrix containing the singular values of $\mathbf{P}_\Omega$. The choice of the number of selected modes M is made at this stage. We look at the ratio of the various singular values with respect to the largest one. Consider the ratio for the i-th singular value ς_i, given by:

$$\varsigma_i = \frac{i\text{-th Singular Value}}{\text{Maximum Singular Value}} \approx 10^{-u} \tag{39}$$

A threshold value for the parameter u, indicating the number of significant digits that pertains to the i-th mode, is established. The first M modes with singular values along z above this threshold, where $M \leq N$, are considered. Hence, those have some significant contribution to the response at Ω, while others should be neglected. This technique is usually employed to best estimate the order of significant poles in a function that can be expressed as a sum of complex exponentials, as in (1). Note that the procedure accounts for the interaction among all modes. A similar parameter is used in MIMO systems for confined environments as a figure of merit to evaluate its performances [56].

In the third step, the set of optimum weight coefficients to excite the modes is defined by using the optimization procedure presented in subsection 3.2.4. In practice, the maximum power density that can be transmitted by Tx and/or the minimum power density that can be detected at Rx, denoted as $\delta_{max_{Tx}}$ and $\delta_{min_{Rx}}$, respectively, are defined by the industrial specifications. The optimum weight coefficients for all the distances $0 \leq z \leq Z_{max}$ are calculated by using (33). The approximation is done in a differential element of area in Tx or Rx where the excitation is concentrated and (x, y) correspond to the coordinates where it is located. Equation (33) is evaluated for all the points in Ω. A set of the optimum weight coefficients for each point in the (x, y)-plane, are obtained. The upper limit for the weight amplitudes can be found by scaling the power density using (35) or (36), so that the extreme values in Tx or Rx are satisfied. For $\Omega = Tx$, the weight amplitudes in (35) have to be scaled with respect to the maximum eigenvalue in Tx, and for $\Omega = Rx$, the weights in (36) have to be scaled with respect to the maximum eigenvalue in Rx. It should be noticed that, according to the previous step, the number of contributing modes is M and only the most contributing eigenvalues and eigenvectors are considered, so that the amplitudes of the remaining eigenvectors can be adjusted such that $Pmin_{Rx} \leq \mathbf{w}_{Rx}^H \mathbf{P}_{\Delta Rx} \mathbf{w}_{Rx}$ in (27) or $\mathbf{w}_{Tx}^H \mathbf{P}_{\Delta Tx} \mathbf{w}_{Tx} \leq Pmax_{Tx}$ for (29).

Hitherto, the optimum weight coefficients were calculated at each point in Ω, obtaining a set of values that depends on the (x, y, z) position. The z-dependence of these coefficients is due to the non-orthogonality between modes. The (x, y)-dependence indicates which coefficients should be used if the structure is excited in the (x, y)-plane. Thus, the fourth step consists in finding the optimum position in the (x, y)-plane to calculate these coefficients. In doing so, the mean of the total power is calculated at each point (x, y) in Ω for the distances z along the axis of the tunnel. This procedure is repeated for all the points in Ω, obtaining a matrix containing the means for all the points (40). Next, the criterion is similar to that already explained. It consists in finding the positions where the mean of the power in Ω versus distance is maximum.

$$\bar{P}(x, y) = \sum_{z=z_{min}}^{z_{max}} z \bar{P}(x, y, z) \tag{40}$$

Lastly, in the fifth step, the field specifications to excite the tunnel and the best transmitter location are determined. The total fields are calculated by using the optimum weight coefficients obtained in the third step at the position where the mean (40) is maximum and by using the mode characteristic parameters obtained in the first step. Expression (37) gives the optimum fields at any cross section $z \geq 0$. This expression can also be employed to define the privileged polarization by evaluating the total field components separately, and observing the predominant one. Finally, points where the matrix (40) is maximized belong to the best excitation points for the M considered modes according to subsection 3.2.3 and ,thus, the best source location.

3.5. Validation

The first three stages are essentially the core of the methodology presented in the previous subsection: The modal approach, the non-orthogonality between modes and an optimization technique were employed to determine the optimum weight coefficients. To validate these steps, the study of a simple theoretical *"reference solution"* for a canonical geometry was considered. A metallic-rectangular waveguide was studied in this subsection to help understand the treatment of a realistic scenario.

A dielectric-loaded WR-90 rectangular waveguide was considered as an exact field solution exists. The loading lossy material has a conductivity of $\sigma = 0.01 Sm^{-1}$ and a relative dielectric constant equals to the unity. The waveguide has a width $w = 2.286$ cm and height $h = 4$ cm and the operation frequency was chosen at $f = 14$ GHz. The objective consists in maximizing the power conducted by the first two modes in a region on the left side over the cross-section of the guide. Figure 5 illustrates the geometry of the waveguide and the cutoff frequencies of the first three mode. For the sake of simplicity, the gray region represent the possible locations for the transmitter and receiver, and consequently, where the power has to be maximized. As it was explained at the end of section 3.2.2, it should be pointed out that because only a part of the cross-section of the waveguide is considered, the orthogonality property is destroyed, so that this example serves as validation case of a real tunnel environment.

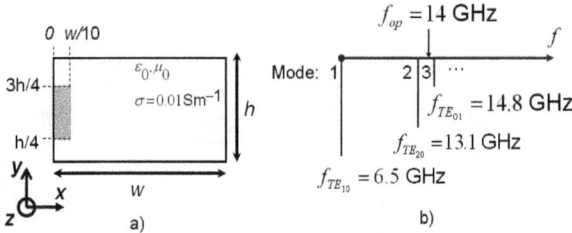

Figure 5. a) Geometry of the rectangular waveguide and region where the power has to be maximized and b) Cutoff frequencies of the first three modes.

Regarding the first step of the proposed methodology, the exact expression for the fields can be found in [51] for this case. So, the 2.5-D TLM is not required to determine the mode characteristic parameters. For the second step, it can be demonstrated that the matrix of crossed powers is given by:

$$
\mathbf{P} = \begin{bmatrix} P_{1,1} & P_{1,2} & P_{1,3} \\ P_{2,1} & P_{2,2} & P_{2,3} \\ P_{3,1} & P_{3,2} & P_{3,3} \end{bmatrix} = \Re \begin{bmatrix} 0.162\varsigma e^{-2\alpha_{10}z}\beta_{10} & 0.157\varsigma e^{-(\gamma_{10}+\gamma_{20}^*)z}\beta_{20} & 0 \\ 0.157\varsigma e^{(\gamma_{10}^*+\gamma_{20})z}\beta_{10} & \varsigma e^{-2\alpha_{20}z}\beta_{20} & 0 \\ 0 & 0 & \frac{-3\varsigma h^2}{2w^2}e^{-2(\alpha_{01})z}\beta_{01} \end{bmatrix},
$$

(41)

where:

$$
\varsigma = \frac{\mu \omega w^3 h}{32\pi^3},
$$

(42)

ω is the angular frequency, $\mu = \mu_0$ is the permeability inside the waveguide, w and h are the waveguide dimensions and $\gamma_{10} = \alpha_{10} + j\beta_{10}$ and $\gamma_{20} = \alpha_{20} + j\beta_{20}$ are the propagation constant of the first and the second modes, respectively; α_{10}, α_{20} and β_{10}, β_{20} are their corresponding attenuation and phase constants. Lastly, the $P_{l,k}$-element in (41) can be calculated by using (30). For reasons that will become clear later, one can consider the first two singular values of $\mathbf{P}$. Additionally, it is worth noting that the term $P_{3,3}$ is zero for frequencies below 14.8 GHz. In that case, the matrix (41) has only two non-zero singular values. Thus, the number of significant modes is set to $M = 2$ and only the terms $P_{1,1}$, $P_{1,2}$, $P_{1,1}$ and $P_{2,2}$ in (41), are considered.

The third step is to obtain the solution for the constrained problem (26). The maximum power density that can be transmitted by Tx or the minimum detectable power density in Rx location, may be specified. Suppose that the maximum and minimum desired power densities are $\delta_{max_{Tx}} = 1\,\mathrm{Wm}^{-2}$ and $\delta_{min_{Rx}} = 10\,\mu\mathrm{Wm}^{-2}$, respectively. Equations (35) and (36) are employed to calculate the optimum weight coefficients and the results are shown in Fig. 6. These coefficients fulfill the power constraints at the maximum and minimum distances. Any choice for the weight amplitudes between the solutions for the maximum and minimum powers could be considered. By observing these figures, one can intuitively suppose that they can be approximated by a sum of complex exponentials due to their dependence with the modes. The Matrix Pencil method is a very popular technique to estimate the parameters of a sum of complex exponentials [25]. By using this method, the weight coefficients can be determined and are listed in (43) for the case of $\delta_{max_{Tx}}$. This result demonstrates that some coupling between non-orthogonal modes exists. Similar results can be obtained for $\delta_{min_{Rx}}$.

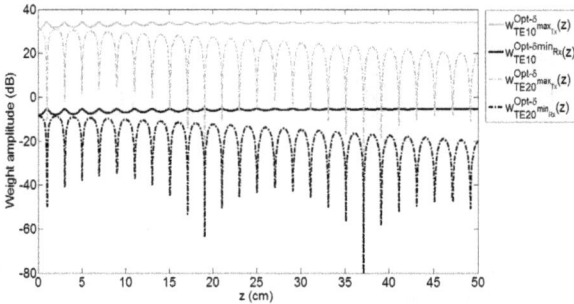

Figure 6. Optimum weight coefficients satisfying the maximum and the minimum required power densities.

$$\mathbf{w}_{TE10}^{Opt-\delta_{max}} \approx -50.4e^{-0.03z} + 10.1e^{-5.09z} +$$

$$(5.9 - j2.2)e^{-2(\alpha_{20}+j(\beta_{10}-\beta_{20}))z} + (5.9 + j2.2)e^{-2(\alpha_{20}+j(\beta_{10}-\beta_{20}))z} +$$

$$(-9 - j2.3)e^{-(\alpha_{10}+\alpha_{20})-j(\beta_{10}-\beta_{20})z} + (-9 + j2.3)e^{-(\alpha_{10}+\alpha_{20})+j(\beta_{10}-\beta_{20})z} \tag{43}$$

$$\mathbf{w}_{TE20}^{Opt-\delta_{max}} \approx (20.7 - j1.6)e^{-(\alpha_{10}-j(\beta_{10}-\beta_{20}))z} + (20.7 + j1.6)e^{-(\alpha_{10}+j(\beta_{10}-\beta_{20}))z}$$

Figure 7 illustrates the amplitudes of the inner terms of the matrix $\mathbf{P}$ as a function of the longitudinal distance. The fluctuations observed in the crossed terms coincide with those of the weight amplitudes, confirming the mutual dependence between modes, as shown in (43). Finally, the black curves in Fig. 7 illustrate the constrained solutions for the maximum power at $z = 0$ and minimum one at $z = z_{zmax} = 50$ cm. It is interesting to observe that the influence of the crossed terms P_{12} and P_{12} is higher at closer distances to the source, *i.e.* at $z = 0$. This is due to the strong coupling close to Tx, as can be seen in Fig. 6.

To verify this result, the intersection of the curves for the total power (26) and the constraints (27) has to be found. Different weight amplitudes $\mathbf{w}$ were considered in the power function at the maximum and minimum distances and the points for which the constraints were achieved were plotted. It is important to note that, thanks to the fact that only two

Electromagnetic Wave Propagation Modeling for
Finding Antenna Specifications and Positions in Tunnels of Arbitrary Cross-Section

239

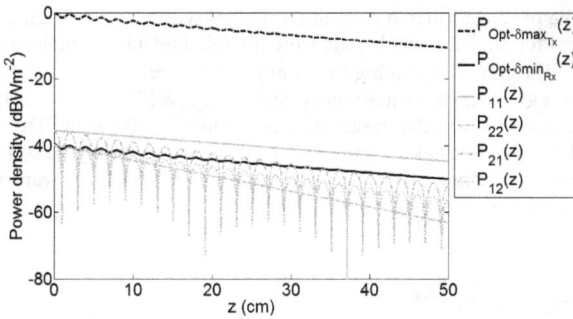

Figure 7. Optimum power density conducted by the modes and terms of the matrix **P** versus distance.

modes were considered, a 3D plot of the power versus the amplitudes of both modes can be obtained, as shown in Fig. 8. This result confirms that the solution given by equations (35) and (36) maximize (27) the power at Tx and Rx and their values correspond to $Pmax_{Tx}$ and $Pmin_{Rx}$.

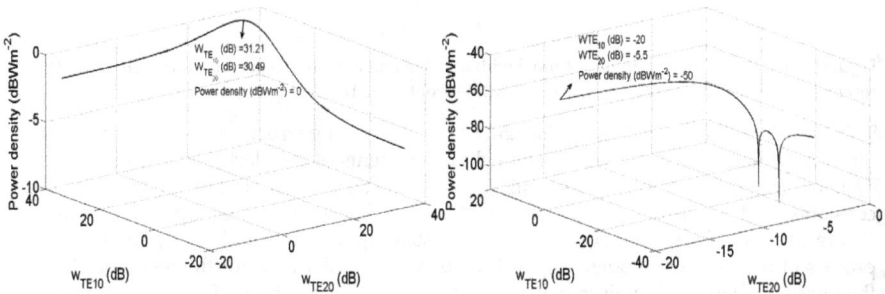

Figure 8. Optimum weight coefficients satisfying the maximum and the minimum required power densities.

4. Results for a rectangular tunnel

The determination of optimum-field specifications and antenna positioning in tunnel environments following the basic principles and procedure outlined in the previous subsections is now straightforward. Multi-modal propagation is experimented in these scenarios and the correct excitation of these modes determines the efficiency of a given source. Tunnels are made with dielectric walls and contain other materials (trains, cars, objects etc.). The use of a full-wave approach for mode parameters calculation now becomes evident due to the necessity of modeling the tunnels walls, near-fields and arbitrary cross sections. The calculation time is the main disadvantage of full-wave methods compared with commonly used ray tracing techniques. The reduced 2.5-D TLM node and the SIBC concept avoid large region meshing and hence, reduce the required computer cost. In this section, the source optimization procedure is applied in the case of a real tunnel environment.

First of all, the wave propagation in a rectangular tunnel with transverse dimension $w = 7$m $\times h = 5$m was considered. The tunnel walls were modeled by a lossy dielectric with $\epsilon_r = 12$ and $\sigma = 0.02$Sm^{-1}. The central operating frequency and the maximum frequency to simulate for the 2.5 D TLM for guiding structures were established at 2.4 GHz and 3 GHz, respectively. Thus, for 2.5-D TLM simulation the mesh size was set to $\Delta l = 0.009$ m. The regions for the Tx and Rx were discretized by taking samples at each $\lambda/10$ at the operation frequency, i.e., $\Delta S = c_0/f_{op}$. The tunnel transverse section and the zones in which the transmitter (Tx) and receiver (Rx) can be located, are shown in (Fig. 9).

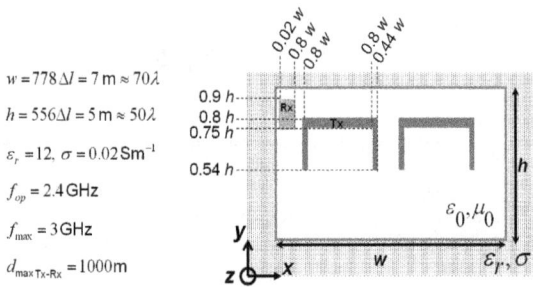

Figure 9. Schema of the rectangular tunnel with $w = 7$m and $b = 5$m. Rx denotes the position for the receiver (fixed) and Tx position of the receiver (to be optimized).

This tunnel was simulated using the 2.5-D TLM for guiding structures and the time-domain SIBC implementation. The transversal field components of the electric field were excited in Tx in order to calculate the field specifications for the transmitter. The attenuation and phase constants, and the field profiles of the modes were determined at the operation frequency. Figure 10 shows the attenuation and phase constant up to 3 GHz and Fig. 11 shows the field profiles of the first two modes. Slight discrepancies are observed for the mode profiles and the phase constant β. The discrepancies in the attenuation constant α for low frequencies are explained by two facts: First, the comparison was made with a commonly used theory, namely Marcatilli's theory, which neglects the fields on the corners. Secondly, because the calculation of the attenuation constant α is derived from β. One can show that errors on β can produce large calculation error on α. However at the operation frequency, this error can be considered to be negligible.

In the next step, the matrix of crossed powers $\mathbf{P}_\Omega$ of dimensions $N \times N$ was calculated for the first $N = 32$ modes and a singular value decomposition of $\mathbf{P}_\Omega$ was carried out. So, the region of the receiver was sampled at one-tenth of the wavelength of the operation frequency and the (l, k)-element of (30) was approximated by the Riemmann sum (38). The choice of the M best modes to carry the power was made at this stage. The normalized amplitudes of the singular values were calculated and only those accurate up to 4 significant digits were considered. The amplitudes were obtained for the first 1,000 m in the tunnel, which constitute the maximum distance between Tx and Rx. The singular values for $\mathbf{P}_{\Delta Rx}$ are shown in Fig. 12, the value M was set to 11. The remaining singular values for which the ratio in equation (39) is below $10^{-u} = 10^{-4}$ are neglected and are not be used in the calculation of the reduced matrix $\mathbf{P}$ of dimension $M \times M$. One can note that the propagation in the tunnel is mainly dominated by the first two modes.

Electromagnetic Wave Propagation Modeling for
Finding Antenna Specifications and Positions in Tunnels of Arbitrary Cross-Section

241

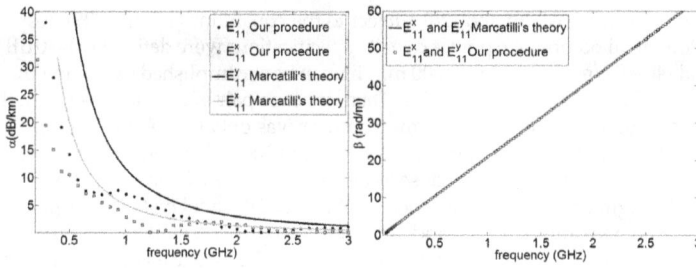

Figure 10. a) Attenuation constant in dB/km, and b) Dispersion curves for the E_{11}^y and E_{11}^x modes, calculated in the frequency range from 0.4 GHz to 3 GHz with Marcatilli's theory [31, 33] and the procedure in this paper [3].

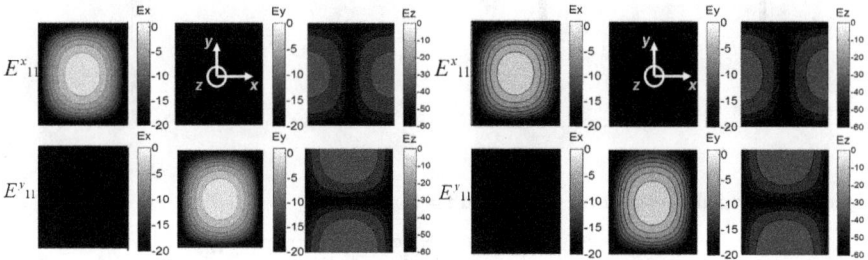

Figure 11. Field configuration in dBV/m of the E_{11}^y and E_{11}^x modes: a) Calculated with Marcatilli's theory and b) Calculated with the procedure in this paper [3].

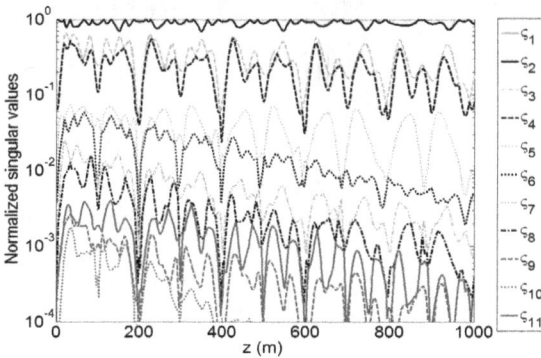

Figure 12. Normalized amplitudes of the first eleven singular values versus distance.

Next, the matrix $\mathbf{P}_{\Delta\Omega}$ of dimensions $M \times M$ was calculated at each point on the cross section of the tunnel and all distances along the z-direction. The optimum weight coefficients that maximize the cost functions (26) and (28) subject to the constrains (27) and (29), respectively, were calculated for all points. The power density specifications were defined to be $0 \, \text{dB} \, \text{Wm}^{-2}$ at $z_{Tx} = 0$ and $-40 \, \text{dBWm}^{-2}$ at $z_{Rx} = 1,000$ m. This can be accomplished by scaling the weight coefficients, as explained in subsection 3.2.4. Then, for each pair of coordinates (x, y), the field evolution was calculated along z and the mean power was obtained by using (40). Figure 13 shows the mean power for different positions in Tx and Rx. It is interesting to note that the mean power is maximum for positions close to the center for both Tx and Rx. The evolution of the normalized weight coefficients as a function of the longitudinal distance for points where the mean power is maximized in Tx and Rx, were computed (see Fig. 14). To satisfy the condition at $z_{Tx} = 0$ and $z_{Rx} = 1,000$ m, the levels should be increased or decreased by 26.78 dB or -35.53 dB, respectively.

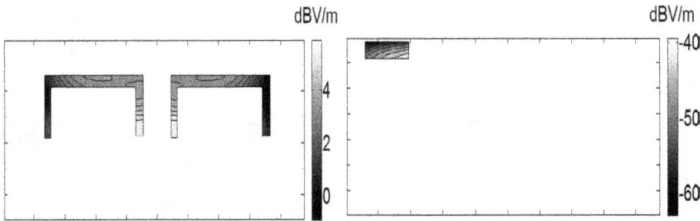

Figure 13. Mean of the power at $z = 0$ for Tx and $z = z_{max}$ for Rx.

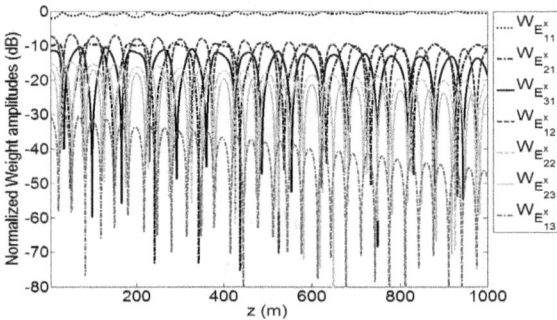

Figure 14. Normalized optimum weight coefficients for the non-zero valued modes in dB versus distance to satisfy the required specifications in Tx and Rx.

Using these coefficients, the optimum power in Tx and Rx were calculated. The best and worst case for the mean power were also computed and compared. As expected, the required specifications were satisfied, as illustrated in Fig. 15. Moreover, from this figure, one can observe that, by correctly locating the source, improvements in the received power of the order of the order of 10 dB, can be obtained.

Finally, the optimum electric fields were obtained by using the optimum weight coefficients in (37). By way of illustration, the three field components of the electric field that satisfy the power density specifications in Tx were calculated. They were computed for the overall tunnel cross section at $z = 0$, as shown in Fig. 16. Intuitively, one could expect that optimum

Electromagnetic Wave Propagation Modeling for
Finding Antenna Specifications and Positions in Tunnels of Arbitrary Cross-Section

243

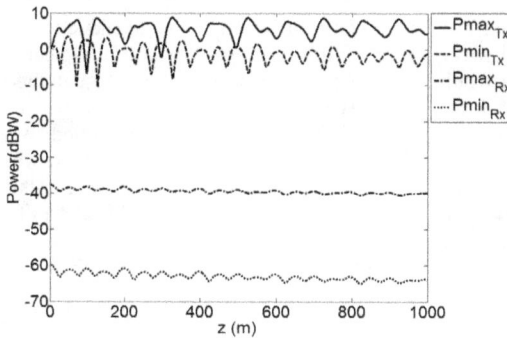

Figure 15. Power versus distance for the best and worst points in Tx and Rx.

fields are most influenced by the lowest attenuated modes. This result can be confirmed by observing that the points where the x-component of the electric field in Fig. 16 is minimum at corners, and the z-component is minimum at half of the width as it occurs for the E_{11}^x mode. This observation can also be confirmed as the x-component is maximum at one-third, as well as one-quarter of the width as for both E_{21}^x and E_{31}^x modes.

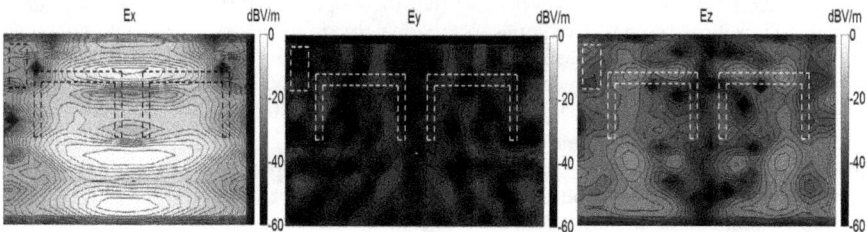

Figure 16. Optimum electric fields at $z = 0$ for optimum excitation of the rectangular tunnel.

It is important to clarify the fact that the weight coefficients vary along the longitudinal distance is due to the non-orthogonality between modes. Note that this variation does not imply that the tunnel must be excited with different weights at each point. Therefore, coefficients should be given only at $z = 0$. Finally, the optimum positions to locate the transmitter are those where the mean power is maximum, *i.e.*, for the points close to the white regions in Fig. 13.

5. Analysis and discussion

A procedure for determining the optimum antenna field specification and location in guiding structures was developed and presented in details. The understanding and modeling of radio-wave propagation through the 2.5 D TLM for guiding structures, were essential steps toward deriving this approach. Radio wave propagation inside tunnels was traditionally modeled using ray tracing and ray launching techniques based on geometrical optics (GO) [13, 15, 19, 34, 41, 57]. However, the analysis can be rather complicated at long ranges due to the fast growing number of contributing rays and breaks down completely at and near caustics [32]. Moreover, the main disadvantage consist in that near-field effects cannot be modeled, and as a result, antenna performances cannot be guaranteed by using these methods. This explains why full-wave techniques are preferred. In the past, full-wave techniques, such us Finite Element Method (FEM), were employed to analyze the wave propagation in these structures [8]. However, these solutions were very limited due to the complexity of the resulting Electromagnetic problem and computation burden. The fundamental strength of the proposed approach is the simplification of the calculation to deal with this electromagnetic problem. Nowadays other recent attempts to deal with these structures are emerging and they are still under development [52, 55].

The modal representation of the fields and the non-orthogonality between modes were the main assumptions involved in developing this methodology. A metallic-rectangular waveguide was considered to validate this procedure. The validation was done by comparing the power density obtained with the optimum weight coefficients obtained with our procedure and comparing with those obtained for different combinations of the weight amplitudes. The results confirm that this approach allows us to obtain optimal weight coefficients to excite the modes. Lastly, the study of a rectangular tunnel was considered. The characteristic parameters of the modes were computed and the most contributing modes in Ω were determined. The resulting modes are coupled to each other and an isolated excitation of them is not possible, as observed in Fig. 14. This fact can also be seen in Fig. 6 for the validation case. In Fig. 14, one can observe that the amplitude of the first mode is almost constant. Thus it can be considered quasi-orthogonal due to its independence with the other modes. Large fluctuations are observed for higher order modes, which is in agreement with the theory. Analysis of the elements of the weight vector $\mathbf{w}$ with the corresponding mode functions $\Psi_{\vec{E}}(\vec{r})$ and $\Psi_{\vec{H}}(\vec{r})$ in (1) reveals that, for a correct excitation of the structure, modes E_{11}^y, E_{12}^y and E_{13}^y should vanish and only the modes E_{11}^x, E_{21}^x, E_{31}^x, E_{12}^x, E_{22}^x, E_{23}^x and E_{13}^x must be considered. Regarding this figure, it can be noted that the wave-propagation is predominantly dominated by the first three modes. This result is in agreement with the works in [43] for a rectangular tunnel. The weighted sum of these coupled modes constitute the optimum fields to excite the structure. On the one hand, from these results of the optimum fields, one can conclude that, for this case, an excitation by the x-component of the electric field is preferred. In [15], it was also verified that, for rectangular tunnels, the vertically polarized modes are much more affected than the horizontally polarized ones. In practice, these fields are difficult to generate and they are usually approximated with dipoles or loops. Consequently, a deeper study to link these solutions with typical antennas are currently under investigation. On the other hand, from the result for the optimum positions, one can conclude that there is a significant influence of the first mode and the transmitting source should be located as close as possible to the center.

Generalization of these concepts to analyze arbitrary-shaped tunnels may be possible by using the procedure outlined in this chapter. Full characterization of propagation in different railway scenarios (several tunnel cross-section and dimensions, loaded and unloaded tunnels) may be analyzed with the 2.5-Dimensional TLM for guiding structures. Useful information, such as field distribution, mode propagation characteristics (attenuation and phase constants) and depolarization effects for various scenarios, can be obtained. The treatment of a more realistic scenario where different cross sections are present, can be studied by establishing a common region to calculate the source specifications. This region should correspond to the biggest possible common area for all the cases considered. A well-adapted antenna for these environments should match as well as possible all results obtained for each case. Loaded or unloaded tunnels with different cross sections may be then analyzed with the proposed procedure.

6. Conclusions

A methodology for finding optimum antenna field specifications and positioning in tunnel environments has been presented and discussed. A TLM approach suitable for the analysis of radio-wave propagation in tunnels was employed and briefly presented. By this approach, the phase constant, attenuation constant and field profiles can be determined. The concept of Surface Impedance Boundary Condition (SIBC) was additionally employed to reduce the TLM computational domain. Combined techniques can be applied to electrically large uniform guiding structures such as lossy tunnels with arbitrary cross-section.

The optimality criterion of the proposed methodology is the maximization of the power density in a given region, as required in industrial applications. The optimum field specifications and positioning in a guiding structure are obtained by means of the modal fields and the optimum weight coefficients satisfying the power density requirements. A metallic waveguide was used to validate this approach and results for a rectangular tunnel with lossy dielectric walls were obtained. Results in terms of field distribution and positioning allow us to analyze the relevant modes and the coupling between them, as well as the privileged polarizations and enhancements in the performances by using a proper excitation.

Concerning future work, results of this approach were presented for a Single Input and Single Output (SISO) system. However, the procedure can be extended for Multiple Input Multiple Output (MIMO) systems to further improve communication system in tunnels.

Acknowledgment

This work was funded by the French National Agency within the ANR-VTT 2009 program in the framework of the METAPHORT project. The authors also acknowledge support from the I-TRANS railway cluster of the Nord Region of France, as well as the Images et Reseaux cluster of the Brittany region.

Authors details

Jorge Avella Castiblanco, Divitha Seetharamdoo and Marion Berbineau
French institute of science and technology for transport, France

Michel Ney and François Gallée
Institute Telecom Bretagne, France

7. References

[1] Avella Castiblanco, J., Seetharamdoo, D., Berbineau, M., Ney, M. & Gallee, F. [2011a]. Determination of antenna specification and positioning for efficient railway communication in tunnels of arbitrary cross section, *Intelligent Transport Systems Telecommunications* pp. 678 –683.

[2] Avella Castiblanco, J., Seetharamdoo, D., Berbineau, M., Ney, M. & Gallee, F. [2011b]. Surface boundary conditions for lossy dielectrics to model electromagnetic wave propagation in tunnels, *European Conference on Antennas and Propagation (EUCAP)* .

[3] Avella Castiblanco, J., Seetharamdoo, D., Berbineau, M., Ney, M. & Gallee, F. [2012]. Surface impedance boundary conditions in time domain for guided structures of arbitrary cross section with lossy dielectric walls (submitted for publication), *Antennas and Propagation, IEEE Transactions on* .

[4] Barybin, A. [1998]. Modal expansions and orthogonal complements in the theory of complex media waveguide excitation by external sources for isotropic, anisotropic, and bianisotropic media, *Journal of Electromagnetic Waves and Applications* 12(8): 1085 – 1086.

[5] Brandwood, D. [1983]. Complex gradient operator and its application in adaptive array theory, *IEE Proceedings, Part F: Communications, Radar and Signal Processing* 130(1): 11 – 16.

[6] C. Briso-Rodriguez, J. M. C. & Alonso, J. I. [2007]. Measurements and modeling of distributed antenna systems in railway tunnels, *IEEE Transactions on Vehicular Technology* vol. 56, no. 5: 2870 –Ú 2879.

[7] Chen, D. K. [1983]. *Field and wave electromagnetics*, Addison-Wesley.

[8] Chiba, J., S. K. [1982]. Effects of trains on cuttoff frequency and field in rectangular tunnel as waveguide, *IEEE Transactions on microwave theory and techniques* 30(5): 757–759.

[9] Collin, R. E. [1960]. *Field theory of guided waves*, McGraw-Hill.

[10] Cristopoulos, C. [1995]. *The Transmission-Line Modelling Method.*, IEEE PRESS, New York, USA.

[11] Das, B., Prasad, K. & Rao, K. [1986]. Excitation of waveguide by stripline- and microstrip-line-fed slots, *Microwave Theory and Techniques, IEEE Transactions on* 34(3): 321 – 327.

[12] Demidov, V. E., Kostylev, M. P., Rott, K., Krzysteczko, P., Reiss, G. & Demokritov, S. O. [2009]. Excitation of microwaveguide modes by a stripe antenna, *Applied Physics Letters* 95(11): 112509 –112509-3.

[13] Didascalou, D. [2000]. *Ray-Optical Wave Propagation Modelling in Arbitrarily Shaped Tunnels*, PhD thesis, Institut fur Hochstfrequenztechnik und Elektronik der Universitat Karlsruhe (TH).

[14] Dowling, J. P., Scully, M. O. & DeMartini, F. [1991]. Radiation pattern of a classical dipole in a cavity, *Optics Communications* 82(5-6): 415 – 419. Atomic Decay;Atomic Dipole Radiation Patterns;.
URL: *http://dx.doi.org/10.1016/0030-4018(91)90351-D*

[15] Dudley, D. G., Lienard, M., Mahmoud, S. F. & Degauque, P. [2007]. Wireless propagation in tunnels, 49(2): 11–26.

[16] Dudley, D. G. & Mahmoud, S. F. [2006]. Linear source in a circular tunnel, 54(7): 2034–2047.

[17] Emslie, A., Lagace, R. & Strong, P. [1975]. Theory of the propagation of uhf radio waves in coal mine tunnels, 23(2): 192–205.

[18] Engelbrecht, J., Collmann, R., Birkel, U. & Weber, M. [2010]. Methodical leaky feeder design for indoor positioning considering multipath environments, *Radio and Wireless Symposium (RWS), 2010 IEEE*, pp. 164 –167.

[19] Fujimori, K. & Arai, H. [1996]. Ray tracing analysis of propagation characteristics in tunnels including transmitting antenna, *Proc. AP-S Antennas and Propagation Society International Symposium Digest*, Vol. 2, pp. 1222–1225.

[20] Gambling, W., Payne, D. & Matsumura, H. [1973]. Mode excitation in a multimode optical-fibre waveguide, *Electronics Letters* 9(18): 412 –414.

[21] Gao, M. [2009]. Analysis for the radiation characteristics of symmetrical dipole antenna in mine tunnel, *Information Engineering, 2009. ICIE '09. WASE International Conference on*, Vol. 1, pp. 652 –655.

[22] Gorobets, N., Ovsyannikova, E. & Shishkova, A. [2007]. Near-field plane distribution of rectangular waveguide excited by dominant and higher-order modes, *Physics and Engineering of Microwaves, Millimeter and Submillimeter Waves and Workshop on Terahertz Technologies, 2007. MSMW '07. The Sixth International Kharkov Symposium on*, Vol. 2, pp. 687 –689.

[23] Han-Qing, M., Tao, F. & Yuan, F. [2010]. Antenna with omnidirectional e-plane radiation pattern based on te01 mode circular waveguide for ka band applications, *Microwave and Millimeter Wave Technology (ICMMT), 2010 International Conference on*, pp. 944 –946.

[24] Haq, T., Webb, K. & Gallagher, N. [1994]. Scattering optimization synthesis of compact mode converters for waveguides, *Antennas and Propagation Society International Symposium, 1994. AP-S. Digest*, Vol. 3, pp. 1668 –1671 vol.3.

[25] Hua, Y. & Sarkar, T. [1988]. Matrix pencil method and its performance., pp. 2476 –2479 vol.4.

[26] Hudson, J. E. [1981]. *Adaptive array principles*, Peter Peregrinus Ltd. on behalf of the Institution of Electrical Engineers.

[27] Huo, Y., Xu, Z., dang Zheng, H. & Zhou, X. [2009]. Effect of antenna on propagation characteristics of electromagnetic waves in tunnel environments, *Microelectronics Electronics, 2009. PrimeAsia 2009. Asia Pacific Conference on Postgraduate Research in*, pp. 268 –271.

[28] Ji-ping, S., Jisheng, L. & Shuying, L. [2007]. Analysis of radiation characteristics of the electric dipole in the rectangular tunnel, *Wireless Communications, Networking and Mobile Computing, 2007. WiCom 2007. International Conference on*, pp. 1124 –1126.

[29] Jin, H. & Vahldieck, R. [1993]. Full-wave analysis of guiding structures using a 2-d array of 3-d tlm nodes, *Microwave Theory and Techniques, IEEE Transactions on* 41(3): 472 –477.

[30] Johns, D. [2008]. Ensuring antennas perform correctly in their environment [application notes], *Microwave Magazine, IEEE* 9(6): 171 –175.

[31] Krammer, H. [1976]. Field configurations and propagation constants of modes in hollow rectangular dielectric waveguides, *Quantum Electronics, IEEE Journal of* 12(8): 505 – 507.

[32] Kravtsov, Y. A. & Orlov, Y. I. [Berlin, 1990]. *Geometrical Optics of Inhomogeneous Media*, Springer-Verlag.

[33] Laakmann, K. & W., S. [May, 1976]. Waveguides: characteristic modes of hollow rectangular dielectric waveguides, *Applied optics* 15, no 5.

[34] Lamminmaki, J. & Lempiainen, J. [1998]. Radio propagation characteristics in curved tunnels, *Microwaves, Antennas and Propagation, IEE Proceedings* 145(4): 327–331.

[35] Lifante, G. [2003]. *Integrated photonics: Fundamentals*, John Wiley & Sons Ltd.

[36] Low, A., Yong, Y. S., You, A. H., Chien, S. F. & Teo, C. F. [2004]. A five-order mode converter for multimode waveguide, *Photonics Technology Letters, IEEE* 16(7): 1673 –1675.

[37] Loyka, S. [2005]. Multiantenna capacities of waveguide and cavity channels, 54(3): 863–872.

[38] M. M. F. Sayadi, M. I. & Jumari, K. [2009]. radio coverage inside tunnel utilizing leaky coaxial cable base station, *Journal of Applied Sciences* 9: 2887 Ű– 2895.

[39] Maguer, S. L. [France 1998]. *Developpement de nouvelles procedures numeriques pour la modelisation TLM*, PhD thesis, L'Universite de Bretagne Occidentale.

[40] Mahmoud, S. F. [1991]. *Electromagnetic Waveguides: Theory and applications*, Peter Peregrinus Ltd. on behalf of the Institution of Electrical Engineers.

[41] Masson, E. [France 2010]. *Etude de la propagation des ondes electromagnetiques dans les tunnels courbes de section non droite pour des applications metros et ferroviaires*, PhD thesis, University of Poitiers.

[42] Mohammed, W., Mehta, A., Pitchumani, M. & Johnson, E. [2005]. Selective excitation of the te01 mode in hollow-glass waveguide using a subwavelength grating, *Photonics Technology Letters, IEEE* 17(7): 1441 –1443.

[43] Molina-Garcia-Pardo, J., Lienard, M., Degauque, P., Dudley, D. G. & Juan-Llacer, L. [2008]. Interpretation of mimo channel characteristics in rectangular tunnels from modal theory, 57(3): 1974–1979.

[44] Molina-Garcia-Pardo, J.-M., Lienard, M., Degauque, P., Simon, E. & Juan-Llacer, L. [2009]. On mimo channel capacity in tunnels, *Antennas and Propagation, IEEE Transactions on* 57(11): 3697 –3701.

[45] Molina-Garcia-Pardo, J.-M., Lienard, M., Stefanut, P. & Degauque, P. [2009]. Modeling and understanding mimo propagation in tunnels, *Journal of Communications* 4(4): 241 – 247.

[46] Monk, N. & Winbigler, H. [1956]. Communication with moving trains in tunnels, *Vehicular Communications, IRE Transactions on* 7: 21 –Ŭ 28.

[47] Mozingo, R. [1980]. *Introduction to adaptive arrays*, Jhon Wiley-Interscience, John Wiley and Sons, Inc.

[48] Niv, A., Yirmiyahu, Y., Biener, G., Kleiner, V. & Hasman, E. [2008a]. Inhomogeneous anisotropic subwavelength structures for the excitation of single hollow waveguide modes, pp. 1 –2.

[49] Niv, A., Yirmiyahu, Y., Biener, G., Kleiner, V. & Hasman, E. [2008b]. Inhomogeneous anisotropic subwavelength structures for the excitation of single hollow waveguide modes, *Lasers and Electro-Optics, 2008 and 2008 Conference on Quantum Electronics and Laser Science. CLEO/QELS 2008. Conference on*, pp. 1 –2.

[50] Pao, H.-Y., Zhu, Z. & Dvorak, S. [2004]. Exact, closed-form representations for the time-domain surface impedances of a homogeneous, lossy half-space, *Antennas and Propagation, IEEE Transactions on* 52(10): 2659 – 2665.

[51] Pozar, D. [1998]. *Microwave Engineering*, Wiley-Interscience, John Wiley and Sons, Inc., New York.

[52] Rana, M. & Mohan, A. [2012]. Segmented-locally-one-dimensional-fdtd method for em propagation inside large complex tunnel environments, *Magnetics, IEEE Transactions on* 48(2): 223 –226.

[53] Rao, S. S. [2009]. *Engineering Optimization*, John Wiley & Sons, Inc.

[54] Seetharamdoo, D., Avella Castiblanco, J. & Berbineau, M. [2011]. Metamaterial for trainborne antenna integration and reduction of emi between onboard systems in the railway environment, *World Congress on railway research* .

[55] Valcarce, A., de la Roche, G., Nagy, L., Wagen, J.-F. & Gorce, J.-M. [2011]. A new trend in propagation prediction, *Vehicular Technology Magazine, IEEE* 6(2): 73 –81.

[56] Valdesueiro, J. A., Izquierdo, B. & Romeu, J. [2010]. Multiple element antenna placement and structure studies in subway environments, *Antennas and Propagation (EuCAP), 2010 Proceedings of the Fourth European Conference on*, pp. 1 –5.

[57] Wang, T. S. & Yang, C. F. [2006]. Simulations and measurements of wave propagations in curved road tunnels for signals from gsm base stations, 54(9): 2577–2584.

[58] *Wave Propagation (To be published)* [2012]. Academy Publish.

[59] Yeh, C. & Shimabukuro, F. I. [2008]. *The Essence of Dielectric Waveguides*, Springer., New York, USA.

Chapter 10

Efficient CAD Tool for Noise Modeling of RF/Microwave Field Effect Transistors

Shahrooz Asadi

Additional information is available at the end of the chapter

1. Introduction

Efficient models are the key of successful designs. Widely used in modern wireless communication systems, active devices such as field-effect transistors (FETs) require up-to-date models to achieve reliable circuit/system design especially in terms of noise performance since most of communication systems operate in noisy environments. [1]-[2]. Among existing FET modeling techniques, the full-wave modeling approach can be considered as the most reliable but is computationally expensive in terms of CPU time and memory [3]-[5]. On the other side, circuit equivalent models are fast but cannot accurately integrate EM effects. Therefore, a hybrid transistor model, called the semi-distributed model (Sliced model) has been proposed [6]. With the assumption of a quasi transverse electromagnetic (TEM) approximation, this model can be seen as a finite number of cascaded cells, each of them representing a unit transistor equivalent circuit. However, this model presents some limitations. In fact, in mm-wave frequencies, it cannot precisely take into account some EM effects that can significantly degrade the overall device behavior, like the wave propagation and the phase cancellation phenomena. To efficiently include such effects more general distributed models need to be developed. In this chapter, a new distributed FET model is proposed. In this model[7]- [8], each infinitely unit segment of the device electrodes was divided into two parts namely, active and passive. The passive part describes the behavior of the transistor as a set of three coupled transmission lines while the active part that can be modeled by an electrical equivalent distributed circuit whose elements are all per-unit length.

To demonstrate the efficiency of our model in terms of noise, we applied the Laplace transformation to the device as an active multi-conductor transmission line structure and successfully compared its simulated response to measurements. Furthermore, by easily including the effects of scaling, the proposed algorithm is suitable for integration in computer-aided-design (CAD) packages for MMIC design.

2. Signal modeling of high-frequency FET

A typical millimeter-wave field effect transistor (FET) is shown in Fig.1. It consists on three coupled electrodes (i.e., three active transmission lines).

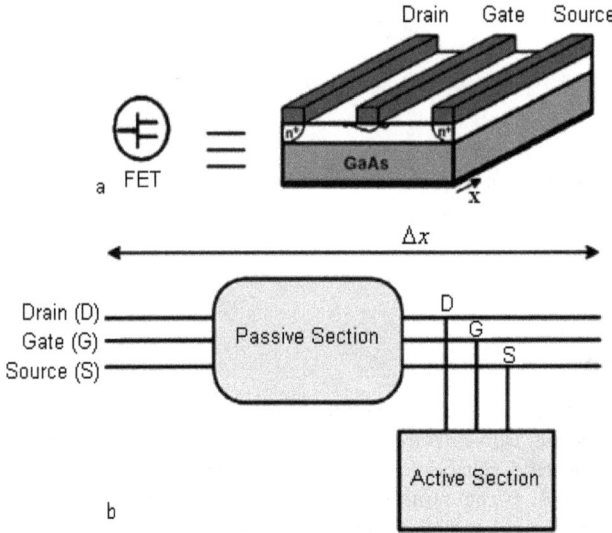

Figure 1. (a) 3D structure of FET used in millimeter-wave frequency. (b) a segment of distributed model along the wave propagation direction.

In the lower part of the microwave spectrum, the longitudinal electromagnetic (EM) field is very small in magnitude as compared to the transverse field [9]-[10]. Therefore, a quasi-TEM mode can be considered to obtain the generalized active multi-conductor transmission line equation. An equivalent circuit of a section of the transistor is shown in Fig. 2. Each segment is represented by a 6-port equivalent circuit which combines a conventional FET small-signal equivalent circuit model with a distributed circuit to account for the coupled transmission line effect of the electrode structure where the all parameters are per unit length. By applying Kirchhoff's current laws to the left loop of the circuit in Fig. 2 with the condition $\Delta x \rightarrow 0$, we obtain the following system of equations [11]-[12]:

$$\frac{\partial I_d(x,t)}{\partial x} + C_{11}\frac{\partial V_d(x,t)}{\partial t} - C_{12}\frac{\partial V_g(x,t)}{\partial t} - C_{13}\frac{\partial V_s(x,t)}{\partial t} + G_m V_g'(x,t) + G_{ds}(V_d(x,t) - V_s(x,t)) = 0 \quad (1)$$

$$\frac{\partial I_g(x,t)}{\partial x} + C_{22}\frac{\partial V_g(x,t)}{\partial t} - C_{12}\frac{\partial V_d(x,t)}{\partial t} + C_{gs}\frac{\partial V_g'(x,t)}{\partial t} = 0 \quad (2)$$

$$\frac{\partial I_s(x,t)}{\partial x} + C_{33}\frac{\partial V_s(x,t)}{\partial t} - C_{13}\frac{\partial V_d(x,t)}{\partial t} - C_{gs}\frac{\partial V_g'(x,t)}{\partial t} - G_m V_g'(x,t) + G_{ds}(V_s(x,t) - V_d(x,t)) = 0 \quad (3)$$

$$\frac{\partial V_d(x,t)}{\partial x} + R_d I_d(x,t) + L_{dd}\frac{\partial I_d(x,t)}{\partial t} + M_{gd}\frac{\partial I_g(x,t)}{\partial t} + M_{ds}\frac{\partial I_s(x,t)}{\partial t} = 0 \tag{4}$$

$$\frac{\partial V_g(x,t)}{\partial z} + R_g I_g(x,t) + L_{gg}\frac{\partial I_g(x,t)}{\partial t} + M_{gd}\frac{\partial I_d(x,t)}{\partial t} + M_{gs}\frac{\partial I_s(x,t)}{\partial t} = 0 \tag{5}$$

$$\frac{\partial V_s(x,t)}{\partial z} + R_s I_s(x,t) + L_{ss}\frac{\partial I_s(x,t)}{\partial t} + M_{ds}\frac{\partial I_d(x,t)}{\partial t} + M_{gs}\frac{\partial I_g(x,t)}{\partial t} = 0 \tag{6}$$

with

$$C_{11} = C_{dp} + C_{ds} + C_{dg} \qquad C_{22} = C_{gp} + C_{dg} \quad C_{33} = C_{sp} + C_{ds} \quad C_{12} = C_{dg} \quad C_{13} = C_{ds}$$

where V_d, V_g, and V_s, are the drain, gate and source voltages, respectively, V'_g is the voltage across gate-source capacitor, while I_d, I_g, and I_s are the drain, gate and source currents, respectively. These variables are time-dependant and function of the position x along the device width. Also, M_{ds}, M_{gd}, and M_{gs} represent the mutual inductances between drain-source, gate-drain and gate-source, respectively; In the above system, we have an extra unknown parameter, i.e., the gate-source capacitance voltage V_g'. Therefore, the following equation should be included to complete the system of equations

$$V'_g(x,t) + V_s(x,t) + R_i C_{gs}\frac{\partial V'_g(x,t)}{\partial t} - V_g(x,t) = 0 \tag{7}$$

which can be then reformatted into two matrix equations

$$\frac{\partial}{\partial x}\begin{pmatrix} I_d(x,t) \\ I_g(x,t) \\ I_s(x,t) \\ 0 \end{pmatrix} + \frac{\partial}{\partial t}\begin{pmatrix} C_{11} & -C_{12} & -C_{13} & 0 \\ -C_{12} & C_{22} & 0 & C_{gs} \\ -C_{13} & 0 & C_{33} & -C_{gs} \\ 0 & 0 & 0 & R_i C_{gs} \end{pmatrix}\begin{pmatrix} V_d(x,t) \\ V_g(x,t) \\ V_s(x,t) \\ V'_g(x,t) \end{pmatrix} + \begin{pmatrix} G_{ds} & 0 & -G_{ds} & G_m \\ 0 & 0 & 0 & 0 \\ -G_{ds} & 0 & G_{ds} & -G_m \\ 0 & -1 & 1 & 1 \end{pmatrix}\begin{pmatrix} V_d(x,t) \\ V_g(x,t) \\ V_s(x,t) \\ V'_g(x,t) \end{pmatrix} = 0 \tag{8}$$

$$\frac{\partial}{\partial x}\begin{pmatrix} V_d(x,t) \\ V_g(x,t) \\ V_s(x,t) \end{pmatrix} + \frac{\partial}{\partial t}\begin{pmatrix} L_{dd} & M_{gd} & M_{ds} \\ M_{gd} & L_{gg} & M_{gs} \\ M_{ds} & M_{gs} & L_{ss} \end{pmatrix}\begin{pmatrix} I_d(x,t) \\ I_g(x,t) \\ I_s(x,t) \end{pmatrix} + \begin{pmatrix} R_d & 0 & 0 \\ 0 & R_g & 0 \\ 0 & 0 & R_s \end{pmatrix}\begin{pmatrix} I_d(x,t) \\ I_g(x,t) \\ I_s(x,t) \end{pmatrix} = 0 \tag{9}$$

3. Noise modeling of high-frequency FETs

The transmission line structure, exciting by noise equivalent sources distributed on the conductors as a new noise model of the high-frequency FET is shown in Fig. 3.

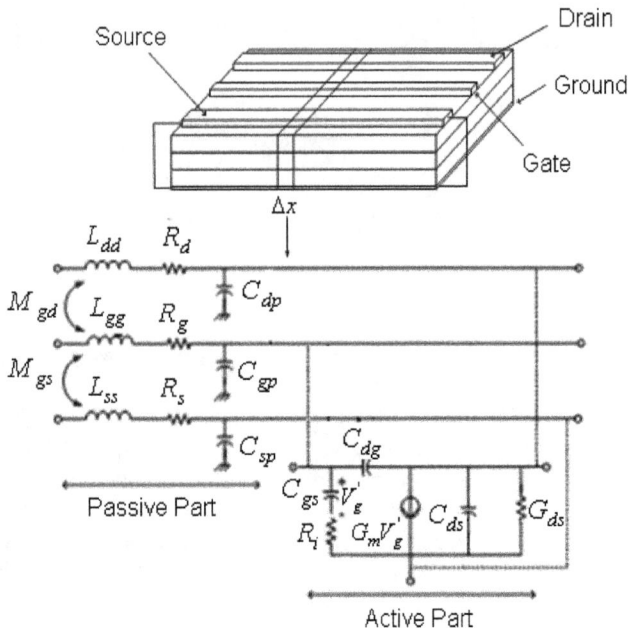

Figure 2. The different parts of a segment in the distributed model.

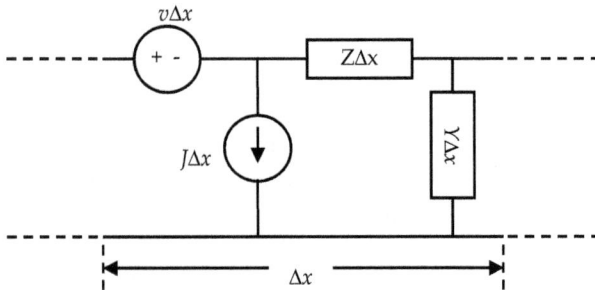

Figure 3. Differential subsection of an excited transmission line

Applying Kirchhoff's laws in time domain leads to

$$\frac{\partial}{\partial x}\mathbf{I'} + \mathbf{C}\frac{\partial}{\partial t}\mathbf{V'} + \mathbf{G}\mathbf{V'} + \mathbf{j_n} = 0 \qquad \text{(a)}$$

$$\frac{\partial}{\partial x}\mathbf{V} + \mathbf{L}\frac{\partial}{\partial t}\mathbf{I} + \mathbf{R}\mathbf{I} + \mathbf{v_n} = 0 \qquad \text{(b)}$$

(10)

where

$$\mathbf{I}'(x,t) = \begin{pmatrix} I_d(x,t) \\ I_g(x,t) \\ I_s(x,t) \\ 0 \end{pmatrix} \quad \mathbf{V}'(x,t) = \begin{pmatrix} V_d(x,t) \\ V_g(x,t) \\ V_s(x,t) \\ V'_g(x,t) \end{pmatrix} \quad \mathbf{I}(x,t) = \begin{pmatrix} I_d(x,t) \\ I_g(x,t) \\ I_s(x,t) \end{pmatrix} \quad \mathbf{V}(x,t) = \begin{pmatrix} V_d(x,t) \\ V_g(x,t) \\ V_s(x,t) \end{pmatrix}$$

$$\mathbf{L} = \begin{pmatrix} L_{dd} & M_{gd} & M_{ds} \\ M_{gd} & L_{gg} & M_{gs} \\ M_{ds} & M_{gs} & L_{ss} \end{pmatrix} \quad \mathbf{R} = \begin{pmatrix} R_d & 0 & 0 \\ 0 & R_g & 0 \\ 0 & 0 & R_s \end{pmatrix}$$

$$\mathbf{C} = \begin{pmatrix} C_{11} & -C_{12} & -C_{13} & 0 \\ -C_{12} & C_{22} & 0 & C_{gs} \\ -C_{13} & 0 & C_{33} & -C_{gs} \\ 0 & 0 & 0 & R_i C_{gs} \end{pmatrix} \quad \mathbf{G} = \begin{pmatrix} G_{ds} & 0 & -G_{ds} & G_m \\ 0 & 0 & 0 & 0 \\ -G_{ds} & 0 & G_{ds} & -G_m \\ 0 & -1 & 1 & 1 \end{pmatrix}$$

Note that vectors $\mathbf{v_n}$ and $\mathbf{j_n}$ are the linear density of exciting voltage and current noise sources, respectively. To evaluate the noise sources, we considered a noisy FET subsection with gate width Δx, as shown in Fig. 4. Thus, the unit-per-length noise correlation matrix for chain representation of the transistor ($\mathbf{CA_{UPL}}$) can be deduced as

$$\mathbf{CA_{UPL}} = \left\langle \begin{pmatrix} \mathbf{v_n} \\ \mathbf{j_n} \end{pmatrix} \begin{pmatrix} \mathbf{v_n} \\ \mathbf{j_n} \end{pmatrix}^+ \right\rangle = \begin{pmatrix} C_{11} & C_{12} \\ C_{21} & C_{22} \end{pmatrix} \tag{11}$$

Where $\langle \rangle$ denotes the ensemble average and + the transposed complex conjugate. According to the correlation matrix definition, we can calculate $\mathbf{v_n}$ and $\mathbf{j_n}$ knowing ($\mathbf{CA_{UPL}}$), to completely describe the proposed FET noise model. Indeed, by solving (11), the noise parameters of the transistor can be obtained.

4. The FDTD formulation

The FDTD technique was used to solve the above equations. Applications of the FDTD method to the full-wave solution of Maxwell's equations have shown that accuracy and stability of the solution can be achieved if the electric and magnetic field solution points are chosen to alternate in space and be separated by one-half the position discretization, e.g., $\Delta x/2$, and to also be interlaced in time and separated by $\Delta t/2$ [13]. To incorporate these constraints into the FDTD solution of the transmission-line equations, we divided each line into Nx sections of length Δx, as shown in Fig. 5. Similarly, we divided the total solution time into segments of length Δt. In order to insure the stability of the discretization process and to insure second-order accuracy, we interlaced the $Nx + 1$ voltage points, $V_1, V_2 \ldots V_{Nx+1}$ and the Nx current points, $I_1, I_2 \ldots I_{Nx}$. Each voltage and adjacent current solution points were separated by $\Delta x/2$. In addition, the time points are also interlaced, and each voltage time point and adjacent current time point were separated by $\Delta t/2$. Then, (10) can lead to

$$\frac{{}_dI_k^{n+1/2} - {}_dI_{k-1}^{n+1/2}}{\Delta x} + C_{11}\frac{{}_dV_k^{n+1} - {}_dV_k^n}{\Delta t} - C_{12}\frac{{}_gV_k^{n+1} - {}_gV_k^n}{\Delta t} - C_{13}\frac{{}_sV_k^{n+1} - {}_sV_k^n}{\Delta t} + G_m\frac{{}_gV_k^{n+1} + {}_gV_k^n}{2}$$

$$+G_{ds}\frac{({}_dV_k^{n+1} + {}_dV_k^n - {}_sV_k^{n+1} - {}_sV_k^n)}{2} + \sum_{m=1}^{Nx+1}\frac{{}_{n1}v_m^{n+3/2} + {}_{n1}v_m^{n+1/2}}{2} = 0 \tag{12}$$

$$\frac{{}_gI_k^{n+1/2} - {}_gI_{k-1}^{n+1/2}}{\Delta x} + C_{22}\frac{{}_gV_k^{n+1} - {}_gV_k^n}{\Delta t} - C_{12}\frac{{}_dV_k^{n+1} - {}_dV_k^n}{\Delta t} + C_{gs}\frac{{}_gV_k^{n+1} - {}_gV_k^n}{\Delta t}$$

$$+ \sum_{m=1}^{Nx+1}\frac{{}_{n2}v_m^{n+3/2} + {}_{n2}v_m^{n+1/2}}{2} = 0 \tag{13}$$

$$\frac{{}_sI_k^{n+1/2} - {}_sI_{k-1}^{n+1/2}}{\Delta x} + C_{33}\frac{{}_sV_k^{n+1} - {}_sV_k^n}{\Delta t} - C_{13}\frac{{}_dV_k^{n+1} - {}_dV_k^n}{\Delta t} - G_m\frac{{}_gV_k^{n+1} + {}_gV_k^n}{2}$$

$$-G_{ds}\frac{({}_dV_k^{n+1} + {}_dV_k^n - {}_sV_k^{n+1} - {}_sV_k^n)}{2} + \sum_{m=1}^{Nx+1}\frac{{}_{n3}v_m^{n+3/2} + {}_{n3}v_m^{n+1/2}}{2} = 0 \tag{14}$$

$$\frac{{}_dV_{k-1}^{n+1} - {}_dV_k^{n+1}}{\Delta x} + R_d\frac{{}_dI_k^{n+3/2} + {}_dI_k^{n+1/2}}{2} + L_{dd}\frac{{}_dI_k^{n+3/2} - {}_dI_k^{n+1/2}}{\Delta t} + M_{gd}\frac{{}_gI_k^{n+3/2} - {}_gI_k^{n+1/2}}{\Delta t}$$

$$+M_{gs}\frac{{}_sI_k^{n+3/2} - {}_sI_k^{n+1/2}}{\Delta t} + \sum_{m=1}^{Nx+1}\frac{{}_{n1}j_m^{n+1} + {}_{n1}j_m^n}{2} = 0 \tag{15}$$

$$\frac{{}_gV_{k-1}^{n+1} - {}_gV_k^{n+1}}{\Delta x} + R_g\frac{{}_gI_k^{n+3/2} + {}_gI_k^{n+1/2}}{2} + L_{gg}\frac{{}_gI_k^{n+3/2} - {}_gI_k^{n+1/2}}{\Delta t} + M_{gd}\frac{{}_dI_k^{n+3/2} - {}_dI_k^{n+1/2}}{\Delta t}$$

$$+M_{gs}\frac{{}_sI_k^{n+3/2} - {}_sI_k^{n+1/2}}{\Delta t} + \sum_{m=1}^{Nx+1}\frac{{}_{n2}j_m^{n+1} + {}_{n2}j_m^n}{2} = 0 \tag{16}$$

$$\frac{{}_sV_{k-1}^{n+1} - {}_sV_k^{n+1}}{\Delta x} + R_s\frac{{}_sI_k^{n+3/2} + {}_sI_k^{n+1/2}}{2} + L_{ss}\frac{{}_sI_k^{n+3/2} - {}_sI_k^{n+1/2}}{\Delta t} + M_{ds}\frac{{}_dI_k^{n+3/2} - {}_dI_k^{n+1/2}}{\Delta t}$$

$$+M_{gs}\frac{{}_gI_k^{n+3/2} - {}_gI_k^{n+1/2}}{\Delta t} + \sum_{m=1}^{Nx+1}\frac{{}_{n3}j_m^{n+1} + {}_{n3}j_m^n}{2} = 0 \tag{17}$$

Applying the finite difference approximation to (7) gives

$$R_iC_{gs}\frac{({}_gV_k^{n+1})' - ({}_gV_k^n)'}{\Delta t} + \frac{({}_gV_k^{n+1})' - ({}_gV_k^n)'}{2} + \frac{{}_sV_k^{n+1} + {}_sV_k^n}{2} = \frac{{}_gV_k^{n+1} + {}_gV_k^n}{2} \tag{18}$$

with

$$_dV_i^j = {_dV}((i-1)\Delta x, j\Delta t) \quad \text{and} \quad _dI_i^j = {_dI}((i-1/2)\Delta x, j\Delta t) \text{ for the drain electrode} \qquad \text{(a)}$$

$$_gV_i^j = {_gV}((i-1)\Delta x, j\Delta t) \quad \text{and} \quad _gI_i^j = {_gI}((i-1/2)\Delta x, j\Delta t) \text{ for the gate electrode} \qquad \text{(b) (19)}$$

$$_sV_i^j = {_sV}((i-1)\Delta x, j\Delta t) \quad \text{and} \quad _sI_i^j = {_sI}((i-1/2)\Delta x, j\Delta t) \text{ for the source electrode} \qquad \text{(c)}$$

and where k, m and n are integers. Solving these equations give the required recursion relations

$$V_k^{m+1} = \left(\frac{C}{\Delta t} + \frac{G}{2}\right)^{-1} \left\{ \left(\frac{C}{\Delta t} - \frac{G}{2}\right) V_k^m - \frac{I_k^{m+1/2} - I_{k-1}^{m+1/2}}{\Delta x} + \frac{\Delta x}{2} \sum_{m=1}^{Nx+1} \left(j_m^{n+1} + j_m^n\right) \right\} \qquad (20)$$

$$I_k^{n+3/2} = \left(\frac{L}{\Delta t} + \frac{R}{2}\right)^{-1} \left\{ \left(\frac{L}{\Delta t} - \frac{R}{2}\right) I_k^{n+1/2} - \frac{V_{k+1}^{n+1} - V_k^{n+1}}{\Delta x} + \frac{\Delta x}{2} \sum_m^{Nx+1} \left(v_m^{n+3/2} + v_m^{n+1/2}\right) \right\} \qquad (21)$$

Superposing all the distributed noise sources is equivalent to a summation in (20) and (21) over the gate width for $m = 1\ldots Nx+1$. Because of its simplicity, the leap-frog method was used to solve the above equations. First the voltages along the line were solved for a fixed time using (20) then the currents were determined using (21). The solution starts with an initially relaxed line having zero voltage and current [13].

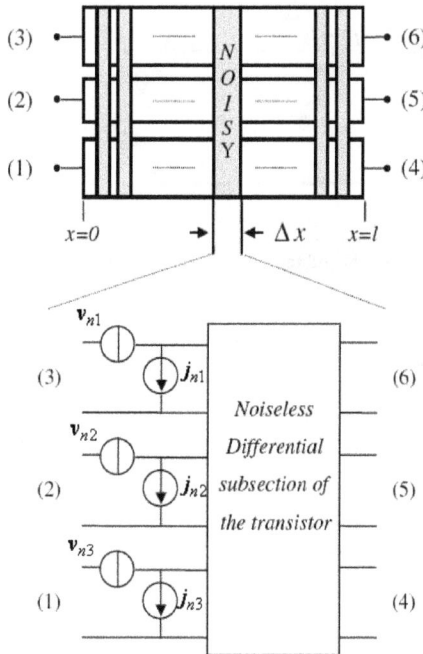

Figure 4. Noise-equivalent voltage and current sources

5. Noise correlation matrix of transistor

To find the noise correlation matrix for admittance representation of the transistor as a noisy six-port active network (as in Fig. 2), the values of port currents should be determined when they are all assumed short-circuited simultaneously. Equation (20) for $k = 0$ and $k = Nx + 1$ becomes

$$V_1^{m+1} = \left(\frac{C}{\Delta t} + \frac{G}{2}\right)^{-1} \left\{\left(\frac{C}{\Delta t} - \frac{G}{2}\right)V_1^m - \frac{I_1^{m+1/2} - I_0^{m+1/2}}{\Delta x / 2} + \frac{\Delta x}{2}\sum_{m=1}^{Nx+1}\left(j_m^{n+1} + j_m^n\right)\right\} \tag{22}$$

$$V_{Nx+1}^{m+1} = \left(\frac{C}{\Delta t} + \frac{G}{2}\right)^{-1} \left\{\left(\frac{C}{\Delta t} - \frac{G}{2}\right)V_{Nx+1}^m - \frac{I_{Nx+1}^{m+1/2} - I_{Nx}^{m+1/2}}{\Delta x / 2} + \frac{\Delta x}{2}\sum_{m=1}^{Nx+1}\left(j_m^{n+1} + j_m^n\right)\right\} \tag{23}$$

By considering Fig. 3, this equation requires that we replace Δx with $\Delta x/2$ only for $k = 1$ and $k = Nx+1$.

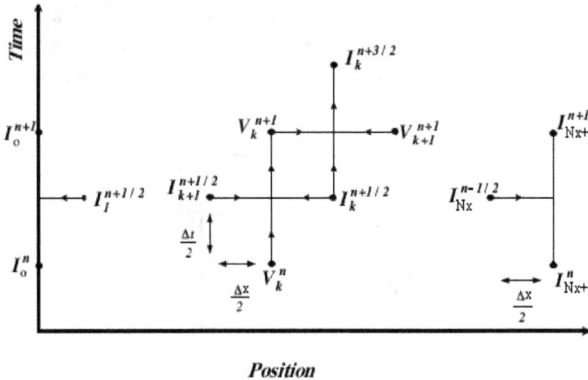

Figure 5. Relation between the spatial and temporal discretization to achieve second-order accuracy in the discretization of the derivatives.

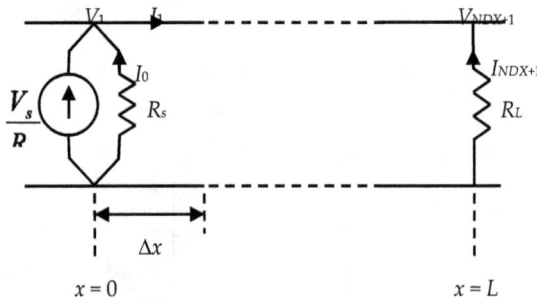

Figure 6. Voltage and current solution points. Spatial discretization of the line showing location of the interlaced points

In order to determine the transistor noise parameters, we set the input voltage source as zero ($V_s = 0$) [8]- [9]. Referring Fig. 6 we denoted the currents at the source point ($x = 0$) as I_0 and at the load point ($x = L$) as I_{Nx+1}. By substituting this notation into (22) we obtain

$$
\begin{pmatrix} I_{0d} \\ I_{0g} \\ I_{0s} \end{pmatrix} = \begin{pmatrix} \dfrac{V_{1d}^{n} - V_{1d}^{n+1}}{2R_{sd}} & 0 & 0 \\ 0 & \dfrac{V_{1g}^{n} - V_{1g}^{n+1}}{2R_{sg}} & 0 \\ 0 & 0 & \dfrac{V_{1s}^{n} - V_{1s}^{n+1}}{2R_{ss}} \end{pmatrix} \tag{24}
$$

Similarly, we imposed the terminal constraint at $x = L$ by substituting I_{Nx+1} into (23) as follow:

$$
\begin{pmatrix} I_{Nx+1,d} \\ I_{Nx+1,g} \\ I_{Nx+1,s} \end{pmatrix} = \begin{pmatrix} \dfrac{V_{Nx+1,d}^{n} - V_{Nx+1,d}^{n+1}}{2R_{Ld}} & 0 & 0 \\ 0 & \dfrac{V_{Nx+1,g}^{n} - V_{Nx+1,g}^{n+1}}{2R_{Lg}} & 0 \\ 0 & 0 & \dfrac{V_{Nx+1,s}^{n} - V_{Nx+1,s}^{n+1}}{2R_{Ls}} \end{pmatrix} \tag{25}
$$

To determine the currents I_1 and I_{Nx} at short-circuited ports ($x=0$ and $x=L$), we set $V_1 = V_{Nx+1} = 0$. The finite difference approximation of (21) for $k = 1$ and $k = Nx$ can be then written as (26) and (27), respectively.

$$
I_1^{n+3/2} = \left(\frac{L}{\Delta t} + \frac{R}{2} \right)^{-1} \left\{ \left(\frac{L}{\Delta t} - \frac{R}{2} \right) I_1^{n+1/2} - \frac{V_1^{n+1}}{\Delta x} + \frac{\Delta x}{2} \sum_{m=1}^{Nx+1} \left(v_m^{n+3/2} + v_m^{n+1/2} \right) \right\} \tag{26}
$$

$$
I_{Nx}^{n+3/2} = \left(\frac{L}{\Delta t} + \frac{R}{2} \right)^{-1} \left\{ \left(\frac{L}{\Delta t} - \frac{R}{2} \right) I_{Nx}^{n+1/2} - \frac{V_{Nx}^{n+1}}{\Delta x} + \frac{\Delta x}{2} \sum_{m=1}^{Nx+1} \left(v_m^{n+3/2} + v_m^{n+1/2} \right) \right\} \tag{27}
$$

Replacing $I_1^{n+1/2}$ and $I_{Nx}^{n+1/2}$ into (26) and (27), respectively, leads to short-circuit currents at input and output terminals.

$$
I_1^{n+3/2} = \left(\frac{L}{\Delta t} + \frac{R}{2} \right)^{-1} \left\{ \left(\frac{L}{\Delta t} - \frac{R}{2} \right) \left(\frac{\Delta x}{2} \right)^2 \sum_{m=1}^{Nx+1} \left(j_m^{n+1} + j_m^{n} \right) + \frac{\Delta x}{2} \sum_{m=1}^{Nx+1} \left(v_m^{n+3/2} + v_m^{n+1/2} \right) - \frac{V_2^{n+1}}{\Delta x} \right\} \tag{28}
$$

$$
I_{Nx}^{n+3/2} = \left(\frac{L}{\Delta t} + \frac{R}{2} \right)^{-1} \left\{ \left(\frac{L}{\Delta t} - \frac{R}{2} \right) \left(\frac{\Delta x}{2} \right)^2 \sum_{m=1}^{Nx+1} \left(j_m^{n+1} + j_m^{n} \right) + \frac{\Delta x}{2} \sum_{m=1}^{Nx+1} \left(v_m^{n+3/2} + v_m^{n+1/2} \right) - \frac{V_{Nx}^{n+1}}{\Delta x} \right\} \tag{29}
$$

Finally, the currents of the short-circuited ports can be determined as

$$
\begin{bmatrix} I_1^{n+1/2} \\ I_{Nx}^{n+3/2} \end{bmatrix} \cong \begin{bmatrix} A & B \\ A & B \end{bmatrix} \begin{bmatrix} \sum_{m=1}^{Nx+1} (j_m^{n+1} + j_m^n) \\ \sum_{m=1}^{Nx+1} (v_m^{n+3/2} + v_m^{n+1/2}) \end{bmatrix} = K \begin{bmatrix} \sum_{m=1}^{Nx+1} (j_m^{n+1} + j_m^n) \\ \sum_{m=1}^{Nx+1} (v_m^{n+3/2} + v_m^{n+1/2}) \end{bmatrix} \tag{30}
$$

with

$$
A = \left(\frac{L}{\Delta t} + \frac{R}{2} \right)^{-1} \left\{ \left(\frac{L}{\Delta t} - \frac{R}{2} \right) \left(\frac{\Delta x}{2} \right)^2 \right\} \qquad\qquad B = \left(\frac{L}{\Delta t} + \frac{R}{2} \right)^{-1} \left(\frac{\Delta x}{2} \right)
$$

The admittance noise correlation matrix of the six-port FET noise model is then equal to

$$
CY_{tr} = \left\langle \begin{bmatrix} I_1^{n+1/2} \\ I_1^{n+3/2} \end{bmatrix} \begin{bmatrix} I_1^{n+1/2} \\ I_1^{n+3/2} \end{bmatrix}^+ \right\rangle = \left\langle \left(K \begin{bmatrix} \sum j_n \\ \sum v_n \end{bmatrix} \right) \left(K \begin{bmatrix} \sum j_n \\ \sum v_n \end{bmatrix} \right)^+ \right\rangle = K \times CA_{UPL} \times K^+ \tag{31}
$$

6. CAD algorithms for noise analysis of mm-wave FETs

6.1. Multi-port network connection

In Fig. 7, a noisy multiport sub-network S of scattering matrix [S] is embedded in a noisy sub-network T of scattering matrix [T], with respective noise wave correlation matrices noted [C_s] and [C_t]. Let [S_{net}] and [C_{net}] be the scattering and noise wave correlation matrices of the total network called N. The scattering matrix [T] of the embedding network T can be partitioned into sub-matrices that satisfy

$$
\begin{bmatrix} b_e \\ b_i \end{bmatrix} = \begin{bmatrix} [T_{ee}] & [T_{ei}] \\ [T_{ie}] & [T_{ii}] \end{bmatrix} \begin{bmatrix} a_e \\ a_i \end{bmatrix} + \begin{bmatrix} c_e \\ c_i \end{bmatrix} \tag{32}
$$

where subscript i designates the internal waves at the connections between the two-networks S and T while subscript e designates the external waves at the S_{net} terminals. The noise wave correlation matrix of network T is similarly partitioned such that

$$
[C_t] = \begin{bmatrix} \overline{c_e c_e^*} & \overline{c_e c_i^*} \\ \overline{c_i c_e^*} & \overline{c_i c_i^*} \end{bmatrix} \tag{33}
$$

The resulting noise wave correlation matrix is then given by [12]:

$$
[C_{net}] = \left[[I] \,|\, T_{ei}([\Gamma] - T_{ii})^{-1} \right] [C_s] \left[[I] \,|\, T_{ei}([\Gamma] - T_{ii})^{-1} \right]^+ \tag{34}
$$

where [I] is the identity matrix and [Γ] the connection matrix expressed as

$$[b_i] = [\Gamma][a_i] \tag{35}$$

The scattering matrix of the total network N is then given by the well known expression [11]

$$[S_{net}] = [T_{ee}] + [T_{ei}]([\Gamma] - [T_{ii}])^{-1}[T_{ie}] \tag{36}$$

Note that this result gives a complete noise characterization of the network. A direct calculation of the new scattering matrix is now possible using (36). Note that the order of the matrix to be inverted was reduced by an amount equals to the number of the external ports.

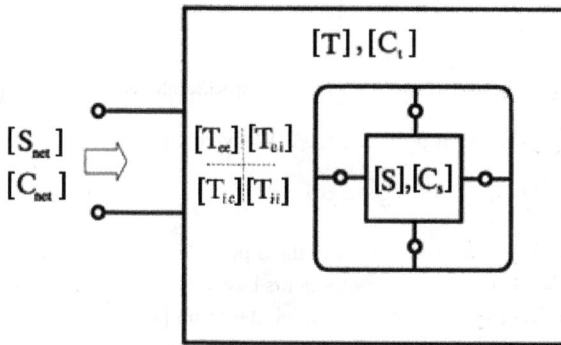

Figure 7. A multiport sub-network S is embedded into a sub-network T. The resulted network N is characterized by the scattering and correlation matrices [S_{net}] and [C_{net}], respectively.

6.2. Scattering and correlation noise matrices

According to the algorithm described above, let us consider the network shown in Fig. 8. In this figure, the ports of the transistor model are numbered from 1 to 24. Ports 23 and 24 are external ports while the rest are internal ports. Since most of the FETs are symmetrical, we can split their geometry into two identical parts. Figure 5 can be then decomposed into two equal parts of $w/2$ each (where w is the gate width) of respective scattering matrix [$S^{(1)}$] and [$S^{(2)}$]. Ports 13, 14 and 15 (the drain, the gate and the source) are terminated by the respective impedances Z_d, Z_g, and Z_s, whose reflection coefficients can be expressed as

$$S^{(3)} = \frac{Z_d - 1}{Z_d + 1} \tag{37}$$

$$S^{(4)} = \frac{Z_g - 1}{Z_g + 1} \tag{38}$$

$$S^{(5)} = \frac{Z_s - 1}{Z_s + 1} = -1 \tag{39}$$

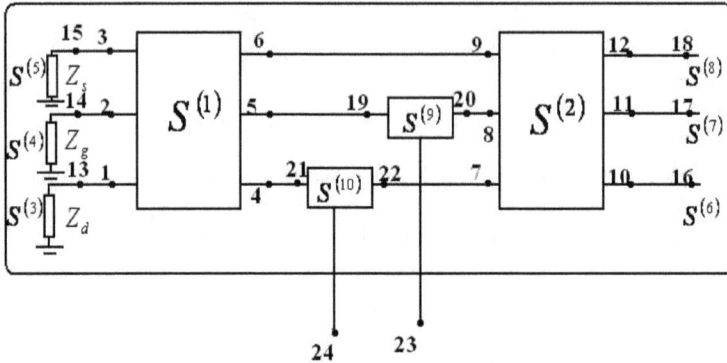

Figure 8. Circuit model of the half structure of a FET with specific internal and external ports

Let us now consider open circuit ports at $x = w/2$. We have then,

$$S^{(6)} = S^{(7)} = S^{(8)} = 1 \tag{40}$$

The only remaining components in Fig. 8 are the 3-port elements $S^{(9)}$ and $S^{(10)}$. Referring to that figure, we can observe that these components basically form the gate line and the drain line, respectively, in the transmission line model. Based on [12], their scattering matrix can be written as

$$[S^{(9)}] = [S^{(10)}] = [S_{con}] = \frac{1}{3}\begin{bmatrix} -1 & 2 & 2 \\ 2 & -1 & 2 \\ 2 & 2 & -1 \end{bmatrix} \tag{41}$$

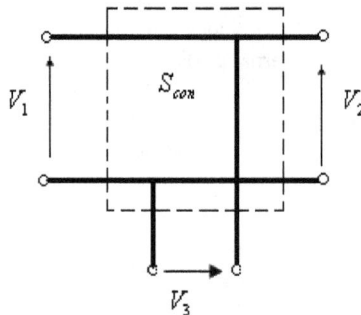

Figure 9. Connection of the series network

In order to define $[C_s]$, we need to know the noise correlation matrices in the form of scattering matrices for all circuit elements. The correlation noise matrix for the 6-port network representing half of the transistor gate width, i.e., $w/2$, can be computed using the techniques described in [9] and [10]. As a result, we can use the proposed CAD algorithm to obtain the scattering and noise correlation matrices of the half-circuit structure.

The scattering matrix of a device is usually computed by partitioning its ports into two groups namely, external and internal ports. Thus, by separating the incoming and outgoing waves in (34), the computation of the connection matrix leads to the resulting scattering matrix

$$[S] = \begin{bmatrix}
[s_{33}^{(9)}] & 0 & 0 & 0 & 0 & 0 & 0 & 0 & 0 & 0 & [s_{31}^{(9)}] & [s_{32}^{(9)}] & 0 & 0 \\
[0_{6\times6}] & [s_{33}^{(10)}] & 0 & 0 & 0 & 0 & 0 & 0 & 0 & 0 & 0 & 0 & [s_{31}^{(10)}] & [s_{32}^{(10)}] \\
[0_{6\times6}] & [0_{6\times6}] & [S^{(1)}] & [0_{6\times6}] & [0_{6\times6}] & [0_{6\times6}] & [0_{6\times6}] & [0_{6\times6}] & [0_{6\times6}] & [0_{6\times6}] & [0_{6\times6}] & [0_{6\times6}] & [0_{6\times6}] & [0_{6\times6}] \\
0 & [0_{6\times6}] & [0_{6\times6}] & [S^{(2)}] & [0_{6\times6}] & [0_{6\times6}] & [0_{6\times6}] & [0_{6\times6}] & [0_{6\times6}] & [0_{6\times6}] & [0_{6\times6}] & [0_{6\times6}] & [0_{6\times6}] & [0_{6\times6}] \\
0 & 0 & 0 & 0 & [S^{(3)}] & 0 & 0 & 0 & 0 & 0 & 0 & 0 & 0 & 0 \\
0 & 0 & 0 & 0 & 0 & [S^{(4)}] & 0 & 0 & 0 & 0 & 0 & 0 & 0 & 0 \\
0 & 0 & 0 & 0 & 0 & 0 & [S^{(5)}] & 0 & 0 & 0 & 0 & 0 & 0 & 0 \\
0 & 0 & 0 & 0 & 0 & 0 & 0 & [S^{(6)}] & 0 & 0 & 0 & 0 & 0 & 0 \\
0 & 0 & 0 & 0 & 0 & 0 & 0 & 0 & [S^{(7)}] & 0 & 0 & 0 & 0 & 0 \\
0 & 0 & 0 & 0 & 0 & 0 & 0 & 0 & 0 & [S^{(8)}] & 0 & 0 & 0 & 0 \\
[s_{13}^{(9)}] & 0 & 0 & 0 & 0 & 0 & 0 & 0 & 0 & 0 & [s_{11}^{(9)}] & [s_{12}^{(9)}] & 0 & 0 \\
[s_{23}^{(9)}] & 0 & 0 & 0 & 0 & 0 & 0 & 0 & 0 & 0 & [s_{21}^{(9)}] & [s_{22}^{(9)}] & 0 & 0 \\
0 & [s_{13}^{(10)}] & 0 & 0 & 0 & 0 & 0 & 0 & 0 & 0 & 0 & 0 & [s_{11}^{(10)}] & [s_{12}^{(10)}] \\
0 & [s_{13}^{(10)}] & 0 & 0 & 0 & 0 & 0 & 0 & 0 & 0 & 0 & 0 & [s_{12}^{(10)}] & [s_{22}^{(10)}]
\end{bmatrix} \quad (42)$$

Then, $[C_s]$ can be written as

$$[C_s] = \begin{bmatrix}
[0_{2\times2}] & [0_{2\times6}] & [0_{2\times6}] & [0_{2\times10}] \\
[0_{6\times2}] & [C_s^{(1)}] & [0_{6\times6}] & [0_{6\times10}] \\
[0_{6\times2}] & [0_{6\times6}] & [C_s^{(2)}] & [0_6^{(10)}] \\
[0_{10\times2}] & [0_{10\times6}] & [0_{10\times6}] & [0_{10\times10}]
\end{bmatrix} \quad (43)$$

Note that based on the proposed algorithm, a designer can easily obtain the scattering matrices of any microwave transistor, highlighting the ease of implementation of the proposed model into existing commercial simulators.

7. Numerical results

The proposed approach was used to model a sub micrometer-gate GaAs transistor (NE710) [14]. The device has a 0.3 μm × 280 μm gate. The first step consisted to characterize the transistor. In this work, we used a bench from Focus microwave that consists on a probing station, the HP 8340B synthesized signal generator, the Agilent 8565EC spectrum analyzer, the CMMT1808 tuners, the Anritsu ML2438A power meter, and the Agilent ML2438A power supplies (Fig. 10).

The intrinsic equivalent circuit model (Fig. 11) was obtained using well-known hot and cold modeling techniques [13]. After removing the extrinsic components via de-embedding methods, a hot modeling technique was utilized to obtain the intrinsic elements. Then, an optimization was performed by varying the values of the intrinsic FET elements in the vicinity of 10% of their mean value until the error between measured and modeled S-parameters was found acceptable (i.e., less than 2%). The obtained values of the extrinsic and intrinsic elements are summarized in Table 1.

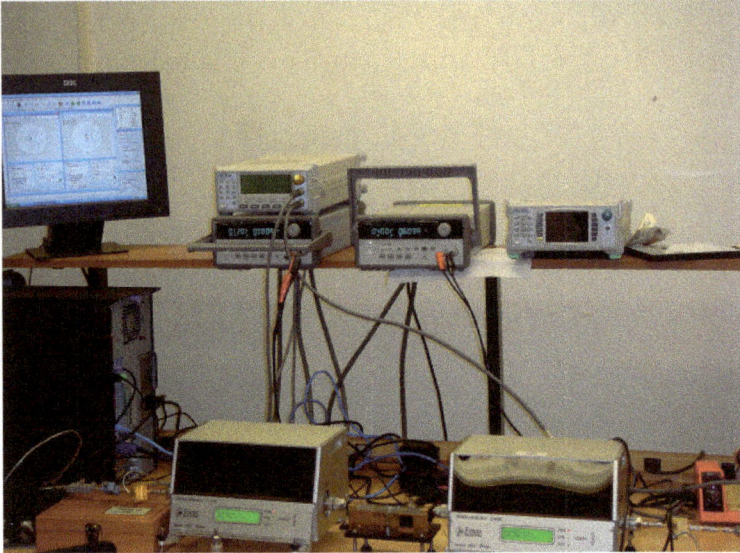

Figure 10. Load-pull bench used to characterize the device

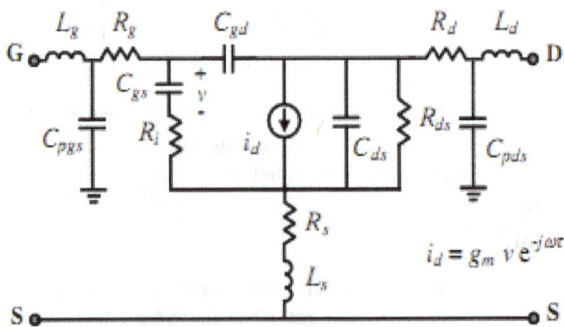

Figure 11. Small-signal equivalent circuit of a FET

Figure 12 shows a good fitting between measured and modeled data for various dc and pulsed voltages while Fig.13 shows the experimental load-pull characteristics of the transistor. When matched, it has an output power of 16 dBm with a 10% PAE at 10GHz. In Fig.10, the output RF power is shown as a function of the complex output impedance matching conditions of the device. The transistor S-parameters over a frequency range of 1-26GHz are plotted in Fig.14. As expected, compared to measurements, our proposed model is more accurate than the slice model [7], especially at the upper part of the frequency spectrum, when the device physical dimensions are comparable to the wavelength. This is due to the fact that our model is based on the full-wave equation while the slice model is based on an electrical equivalent circuit model. Figure 15 shows the noise figure obtained for three different frequencies. Thus, the proposed wave analysis can be applied for accurate noise analysis of FET circuits. To further prove the accuracy of the proposed wave approach in noise analysis, our results were successfully compared to measurements (Fig. 16).

For larger widths, the thermal noise of the gate increases due to the higher gate resistance while for smaller gate widths, the minimum noise figure increases as the capacitances do not scale proportionally with the gate width due to an offset in capacitance at gate width zero [2]. Therefore, we highlighted these effects of gate width on a transistor noise performance by simulating the minimum noise figure and the normalized equivalent noise admittance for three values of the gate width, e.g., 140, 280 and 560 μm (Fig. 17). These values were selected based on the device we modeled. In fact, the NE710 has a gate width of 280 μm, so we took half of that value as well as its double to bound the device behavior and highlight the effect of gate width on a FET performance.

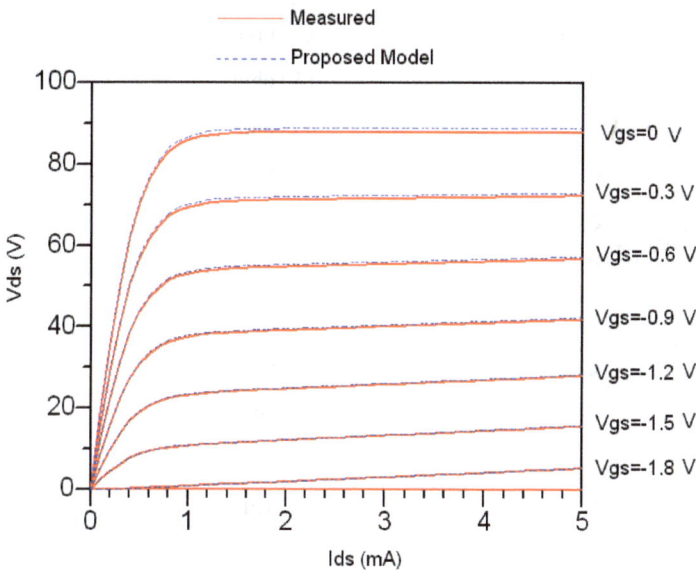

Figure 12. I-V curves for the NE710

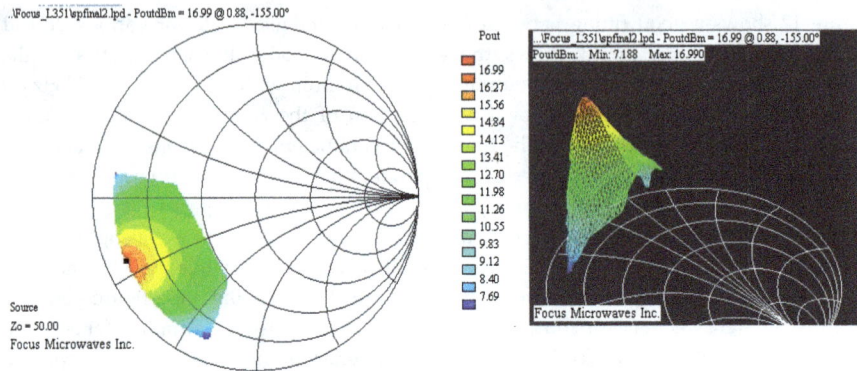

Figure 13. Output power as function of load impedance for an optimized structure at 10 GHz

Lumped Model Values	Numerical Values
L_g	0.383 nH
L_d	0.434 nH
L_s	0.094 nH
R_d	1.77 ohm
R_s	1.74 ohm
R_g	3.29 ohm
C_{pgs}	0.078 pF
C_{pds}	0.092 pF
C_{ds}	0.005 pF
C_{gd}	0.033 pF
g_m	41 mS
R_i	7.3 ohm
R_{ds}	231 ohm
C_{gs}	0.216 pF

Table 1. Values of the lumped elements (The transistor was biased at $V_{ds} = 3$ V and $I_{ds} = 10$ mA)

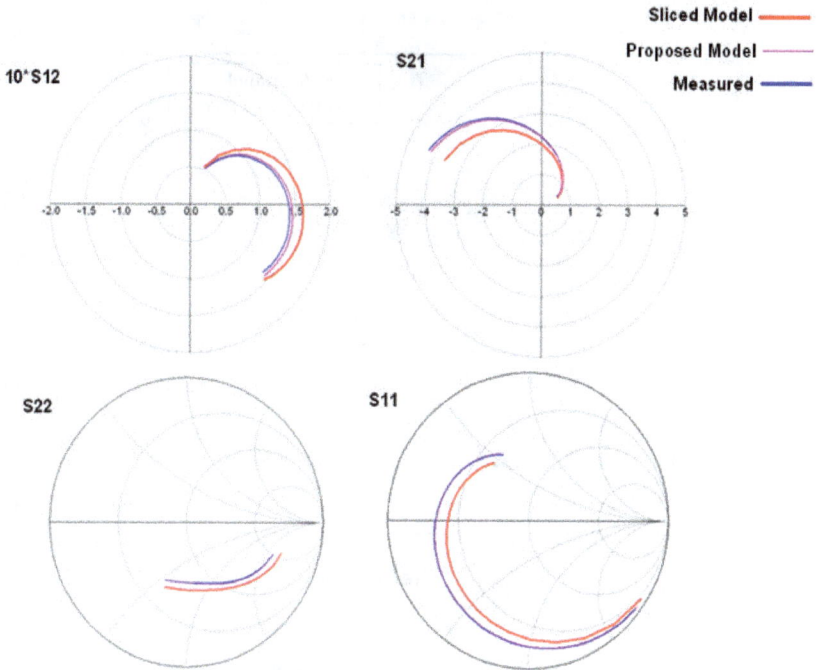

Figure 14. NE710: Comparison between the measured S-parameters and those generated by the sliced and the proposed model.

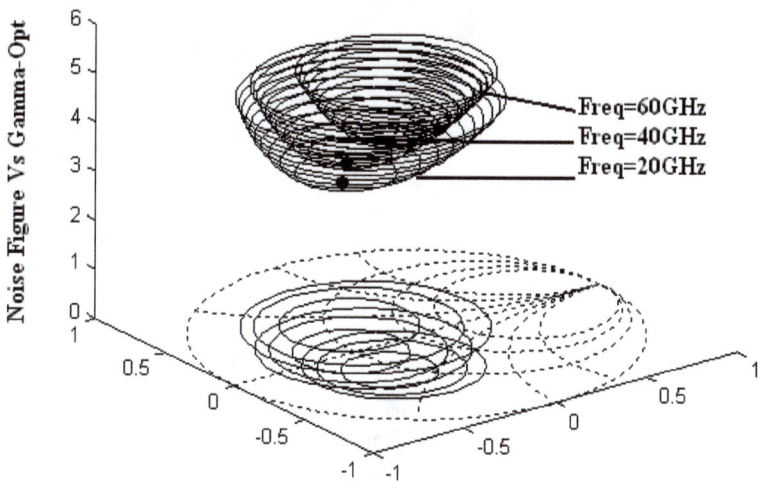

Figure 15. Noise figure circles for three different frequencies versus the source admittance

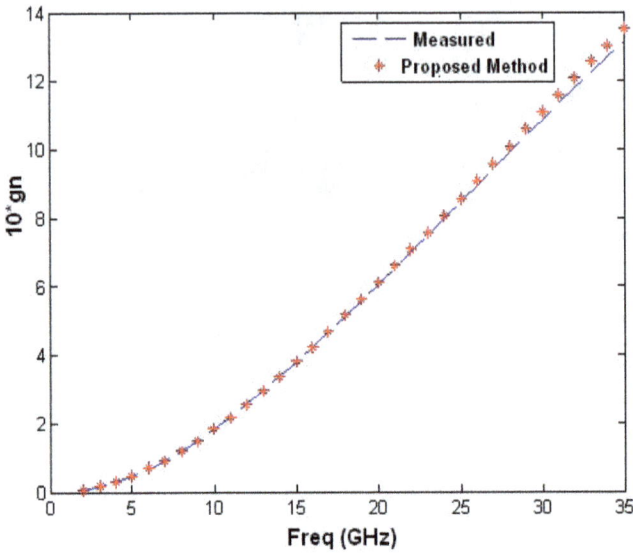

a

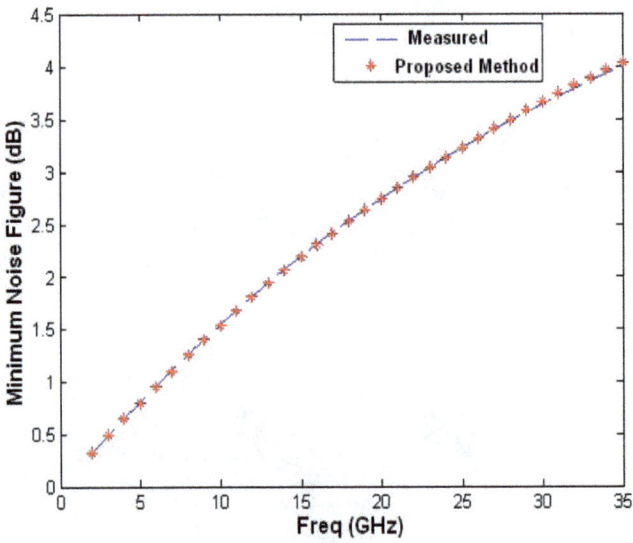

b

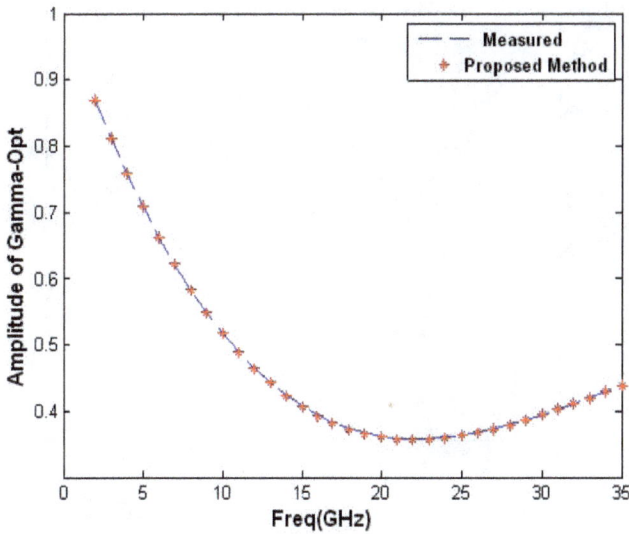

c

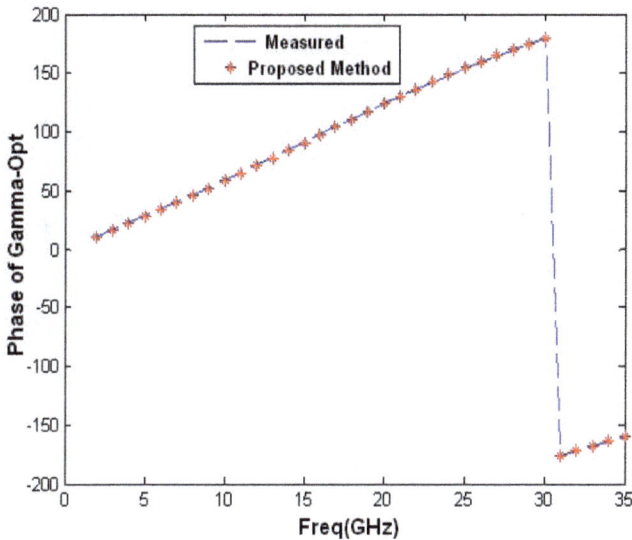

d

Figure 16. a. Normalized equivalent noise admittance and noise figure: Comparison between the proposed method and measurements; b. Amplitude and phase of the optimum reflection coefficient: Comparison between the proposed method and measurements

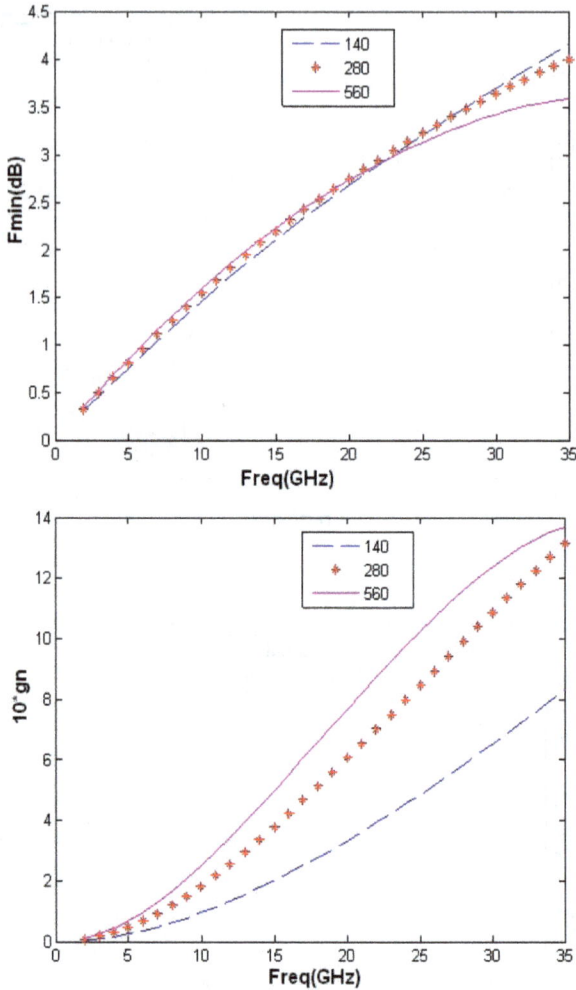

Figure 17. Minimum noise figure and normalized equivalent noise admittance of the transistor for three different values of gate width (μm)

8. Discussions

The transistor modeling approach presented in this chapter is mainly developed for computer-aided design implementation, making it suitable for any FET circuit topology up to the millimeter-wave range and thus, can be easily implemented and used in commercial software. As illustrated in Fig.18, the proposed model was implemented in ADS [15] and the results obtained from the code we developed have been successfully compared with those obtained by the same model after being implemented in the ADS library and used as an internal device. This step shows that the proposed model can be used in any microwave integrated circuit design performed by a commercial simulator. It has also to be noted that even if the proposed model is suitable for any FET structure, large-gate width devices have been targeted in the present work. In fact, this specific type of transistors can handle high output power levels, making them suitable for power amplifier design.

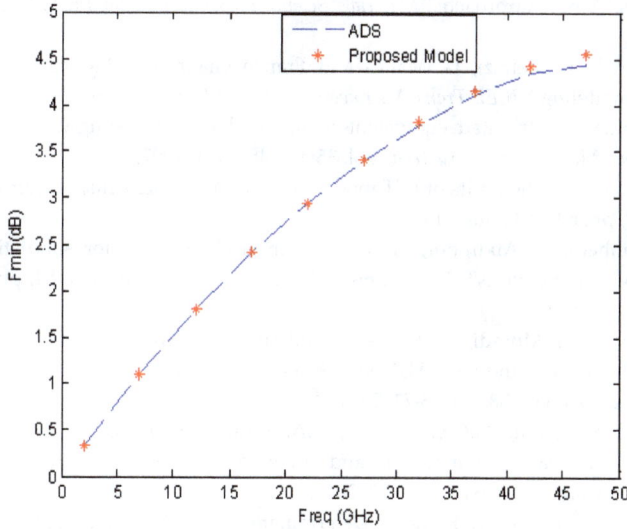

Figure 18. Comparison between simulated minimum noise figure obtained from our developed code and from ADS using our model

9. Conclusion

Using a new CAD algorithm, the noise modeling and analysis of microwave FET have efficiently been studied. In fact, since only half of a FET length is used, instead of the whole structure, the computation time will be significantly affected. Besides, the implementation of this CAD technique in modern microwave and mm-wave simulators is straightforward and will give more reliable results for circuit performance like low-noise amplifiers. Also, as for practical applications, large gate periphery devices are used to generate sufficient output power levels. With the increase of the device gate periphery, the self-heating effect and the defect trapping effect will both be more profound.

Author details

Shahrooz Asadi
University of Ottawa, Canada

10. References

[1] Y.-F. Wu, M. Moore, T. Wisleder, P. Chavarkar, P. Parikh, "Noise Characteristics of Field-Plated HEMTs," *International Journal of High Speed Electronics and Systems*, vol. 14, pp. 192–194, 2004.

[2] B. Boudjelida , A. Sobih, A. Bouloukou, S. Boulay, S. Arshad, J. Sly, M. Missous, "Modelling and simulation of low-frequency broadband LNA using InGaAs/InAlAs structures: A new approach," *Materials Science in Semiconductor Proc.*, Vol.11, pp.398-401, 2008.

[3] A. Cidronali, G. Leuzzi, G. Manes and Franco Giannini, "Physical/Electromagnetic pHEMT modeling," *IEEE Trans. Microwave Theory Tech.*, vol. 51, pp. 830-838, 2003.

[4] W. Heinrich, "Distributed equivalent-circuit model for traveling-wave FET design," *IEEE Trans. Microwave Theory Tech*, Vol. 35, pp. 487–491, 1997.

[5] W. Heinrich, "On the limits of FET modeling by lumped elements," Electron. Lett., Vol. 22, no. 12, pp. 630–632, June 1986.

[6] M. Movahhedi, A. Abdipour, "Efficient numerical methods for simulation of high-frequency active devices", *IEEE Trans. Microwave. Theory Tech.*, Vol. 54, pp. 2636–2645, 2006.

[7] A. Abdipour, G. Moradi, "A CAD-oriented simultaneous signal and noise modeling and analysis of mm-wave FET structures," *AEU- Int. Journal of Electronics and Communications*, Vol. 58, pp. 65-71, 2004.

[8] S. Gaoua, S. Asadi, M.C.E. Yagoub, F.A. Mohammadi, "CAD tools for efficient RF/microwave transistor modeling and circuit design," *Analog Integrated Circuits and Signal Processing J.*, Vol. 63, pp. 59-70, 2010.

[9] M.C. Maya, A. Lazaro, L. Pradell, "Determination of FET noise parameters from 50 Ω noise-figure measurements using a distributed noise model", *Gallium Arsenide applications symposium. GAAS 2002*, September 2002, Milan.

[10] R. Khosravi, A. Abdipour, "A new wave approach for signal and noise modeling of microwave/mm wave FET based on Green's function concept," *Int. J. Electronics*, Vol. 90, pp 303-312, 2003.

[11] J.A. Dobrowolski, *Computer-aided Analysis, Modeling, and Design of Microwave Networks (Wave Approach,)* (Boston: Artech House,1996).

[12] J.A. Dobrowolski, *Introduction to Computer Methods for Microwave Circuit Analysis and Design* (Boston: Artech House, 1991).

[13] W. Pospieszalski, "Modeling of noise parameters of MESFET's and MODFET's and their frequency and temperature dependence," *IEEE Trans. Microwave Theory Tech.*, Vol. 37, pp. 385–388, 1989.

[14] http://www.nec.com.

[15] ADS. (2008). Agilent Technologies, Palo Alto, CA.

Chapter 11

Transient Responses on Traveling-Wave Loop Directional Filters

Kazuhito Murakami

Additional information is available at the end of the chapter

1. Introduction

In design consideration of microwave and millimeter-wave planar circuits, efficient simulation tools have been required for visualizing the operation characteristics of passive circuit components. Also, to clarify the signal propagation in the circuit system is very important in terms of engineering and educational effects. Microwave simulators are widely used as the supporting tool of microwave circuit design and development. For analyzing the circuit components on planar circuits, several numerical techniques are described about the 3-D electromagnetic field analysis modeling [1]. Nowadays, the method of moments (MoM) [2] and the finite-difference time-domain method (FDTD) [3],[4] are mostly used among microwave engineers, and the SNAP simulator [10] is SPICE like one. On the other hand, there are various microwave circuit components using parallel coupled lines and loop resonators [5]. A square loop line with parallel coupled lines is used as ring resonators and traveling-wave loop directional filters. Here, in order to solve the transient problem of the microwave circuit components composed of transmission lines, we provide a numerical analysis method introduced systematic mixed even and odd modes modeling for coupled lines. This method based on a modified central difference method [8] can be applied to the time domain analysis of the microwave circuit components constructed with the parallel tightly coupled lines. For ring resonators and traveling-wave loop directional filters composed of such components [6],[7], to clarify the circuit operation, the transient behavior of the voltage and current waves of transmission line networks will be represented with dynamic expressions. In order to elucidate the mechanism of the circuit operation, the behavior of propagating signals on the transmission line network has been represented in [9]. In this article, we describe the utilization of the modified central difference method incorporating internal boundary treatments. Using this simulation technique, the time and frequency domain properties of the ring resonator and traveling-wave multi-loop directional coupler filters are analyzed and demonstrated.

For obtaining accurate operation characteristics of circuit components, simulation techniques with processing multiple reflections are required. The transient behavior of the voltage and current waves on the ring resonator and traveling-wave loop directional filters is demonstrated to obtain the power division and isolation properties of the directional couplers by using a numerical analysis model. The represented transient phenomena include the effects of multiple reflection waves caused by line discontinuities and parallel coupling. The operating mechanism of the circuit components can be easily confirmed by the visualization of the computed voltage and current solutions. The transient responses along the transmission lines are represented with the variations of the instantaneous voltage and current distributions including all the multiple reflections. Additionally, using the input and output responses extracted from the voltage and current solutions for a Gaussian pulse excitation, the frequency responses are obtained by using the fast Fourier transform.

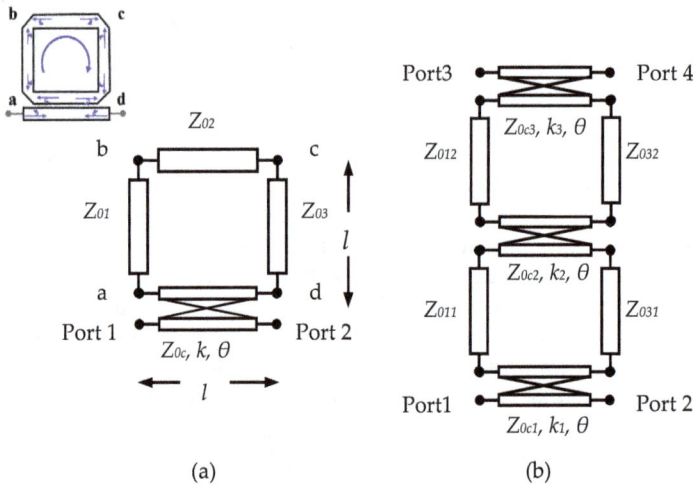

Figure 1. (a) A ring resonator and (b) traveling-wave loop directional filter.

2. Modeling

2.1. Modeling of a single transmission line

The configuration of a ring resonator and traveling-wave loop directional filter is shown in Figure 1. By applying numerical even- and odd-mode analysis model to parts of parallel coupled lines, the solution of the whole system can be solved by single transmission line analysis, with the assumption of TEM mode propagation. The 1-D modeling of the single transmission line has reported in [8]. Telegrapher's equations can be generally written by

$$\frac{\partial}{\partial x}\begin{bmatrix} V(x,t) \\ I(x,t) \end{bmatrix} + \begin{bmatrix} 0 & L \\ C & 0 \end{bmatrix}\frac{\partial}{\partial t}\begin{bmatrix} V(x,t) \\ I(x,t) \end{bmatrix} = -\begin{bmatrix} 0 & R \\ G & 0 \end{bmatrix}\begin{bmatrix} V(x,t) \\ I(x,t) \end{bmatrix} \tag{1}$$

where $V(x,t)$, $I(x,t)$ are the line voltage and current at any time t and at distance x, respectively. L, R, C and G are the inductance, resistance, capacitance and conductance per unit length of the line, respectively. According to the above assumption, the characteristic impedance and phase velocity of the secondary constant parameters are described by $Z_0 = \sqrt{L/C}$ and $v_p = 1/\sqrt{LC}$, respectively. To solve the transmission line equations, the modified central difference approximation is applied to (1). The difference equations can be described as follows:

$$\frac{1}{2\Delta x}\begin{bmatrix} V_{i+1,j} - V_{i-1,j} \\ I_{i+1,j} - I_{i-1,j} \end{bmatrix} + \begin{bmatrix} 0 & L \\ C & 0 \end{bmatrix}\frac{1}{2\Delta t}\begin{bmatrix} 2V_{i,j+1} - (V_{i+1,j} + V_{i-1,j}) \\ 2I_{i,j+1} - (I_{i+1,j} + I_{i-1,j}) \end{bmatrix} = -\begin{bmatrix} 0 & R \\ G & 0 \end{bmatrix}\frac{1}{2}\begin{bmatrix} V_{i+1,j} + V_{i-1,j} \\ I_{i+1,j} + I_{i-1,j} \end{bmatrix} \quad (2)$$

The derived update equations in the case of lossless line are as follows:

$$\begin{bmatrix} V_{i,j+1} \\ I_{i,j+1} \end{bmatrix} = \frac{1}{2}\begin{bmatrix} 1 & Z_0 \\ 1/Z_0 & 1 \end{bmatrix}\begin{bmatrix} V_{i-1,j} \\ I_{i-1,j} \end{bmatrix} + \frac{1}{2}\begin{bmatrix} 1 & -Z_0 \\ -1/Z_0 & 1 \end{bmatrix}\begin{bmatrix} V_{i+1,j} \\ I_{i+1,j} \end{bmatrix} \quad (3)$$

where the suffices $i,j,i+1,i-1$, etc. denote the position at $i=i\Delta x$ $(0 < i \le l)$,$j=j\Delta t$($\Delta t=\Delta x/v_p$) on the $(x-t)$ plane. l is the line length.

2.2. Modeling of coupled transmission lines

Here, as shown in Figure 2, we carry out the even- and odd-mode numerical analysis for the parts of the coupled lines in Figure 1. In the considered parallel coupled lines model, Z_{0e} and Z_{0o}, respectively, are the even- and odd-mode equivalent impedances decided as follows:

$$Z_{0e} = Z_{0c}\sqrt{(1+k)/(1-k)} \quad (4)$$

$$Z_{0o} = Z_{0c}\sqrt{(1-k)/(1+k)} \quad (5)$$

where Z_{0c} and k are the characteristic impedance and coupling coefficient, respectively. The line voltages and currents for each equivalent impedance line are obtained by using (3).

2.3. Boundary treatments

In this modeling, the boundary treatments of both the sides of the coupled line need to calculate at each of time step. For the voltages and currents at the boundaries of the coupled line part as shown in Figure 3, the following scattering matrices expressed by the reflection coefficient are used to compute the reflected and transmitted quantities.

$$\begin{bmatrix} V_{p1}^- \\ V_{p3}^- \\ V_{ea}^- \\ V_{oa}^- \end{bmatrix} = \begin{bmatrix} 0 & \Gamma_a & (1-\Gamma_a)/2 & (1+\Gamma_a)/2 \\ \Gamma_a & 0 & (1-\Gamma_a)/2 & -(1+\Gamma_a)/2 \\ (1+\Gamma_a) & (1+\Gamma_a) & -\Gamma_a & 0 \\ (1-\Gamma_a) & -(1-\Gamma_a) & 0 & \Gamma_a \end{bmatrix}\begin{bmatrix} V_{p1}^+ \\ V_{p3}^+ \\ V_{ea}^+ \\ V_{oa}^+ \end{bmatrix} \quad (6)$$

(a)

(b)

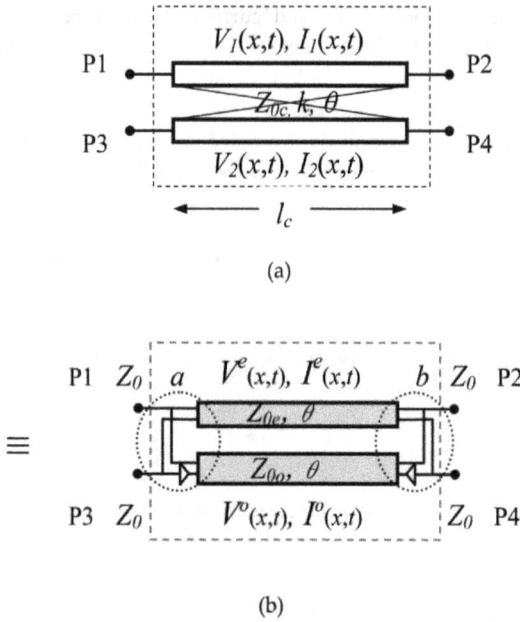

Figure 2. An equivalent circuit model of a basic coupled line.

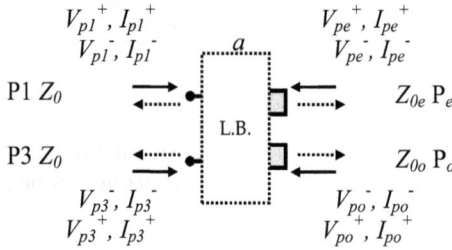

Figure 3. Four ports boundary at left side discontinuity of Figure 2 (b)

$$
\begin{bmatrix} I_{p1}^- \\ I_{p3}^- \\ I_{ea}^- \\ I_{oa}^- \end{bmatrix} = \begin{bmatrix} 0 & -\Gamma_a & -(1+\Gamma_a)/2 & -(1-\Gamma_a)/2 \\ -\Gamma_a & 0 & -(1+\Gamma_a)/2 & (1-\Gamma_a)/2 \\ (1-\Gamma_a) & (1-\Gamma_a) & -\Gamma_a & 0 \\ (1+\Gamma_a) & -(1+\Gamma_a) & 0 & \Gamma_a \end{bmatrix} \begin{bmatrix} I_{p1}^+ \\ I_{p3}^+ \\ I_{ea}^+ \\ I_{oa}^+ \end{bmatrix} \tag{7}
$$

where V_{pi}^+, V_{pi}^-, I_{pi}^+, I_{pi}^- ($i=1, 3, e, o$) denote the incident and reflected voltages, and currents from each line at the left side discontinuity, and the reflection coefficient is

$$\Gamma_a = (Z_{0e} - Z_{0c}) / (Z_{0e} + Z_{0c}) = -(Z_{0o} - Z_{0c}) / (Z_{0o} + Z_{0c}) \qquad (8)$$

Similarly, be done to the right side one. Further, the boundary treatments for each port of the coupled lines with mismatched load are carried out as follows:

$$V_{pk}{}^+ = \Gamma_k V_{pk}{}^+ + (1 + \Gamma_k) V_{pk}{}^- \qquad (9)$$

$$I_{pk}{}^+ = -\Gamma_k I_{pk}{}^+ - (1 - \Gamma_k) I_{pk}{}^- \qquad (10)$$

where $\Gamma_k = (Z_{pk}-Z_{0c})/(Z_{pk}+Z_{0c})$, (k=1 to 4). These boundary treatments need to be performed at each time step.

Here, the initial conditions of each transmission line are given as follows: $V^e(x,t=0)=0$, $I^e(x,t=0)=0$ and $V^o(x,t=0)=0$, $I^o(x,t=0)=0$. Then, for Port1 excitation, the boundary conditions at each port of the circuit component are also given as follows: $V_{port1}(t)= e(t)Z_0/(Z_s+Z_0)$, $I_{port1}(t)=V_{port1}(t)/Z_0$, $V_{port2}(t)=0$, $I_{port2}(t)=0$, $V_{port3}(t)=0$, $I_{port3}(t)=0$, $V_{port4}(t)=0$ and $I_{port4}(t)=0$. The obtained voltage and current solutions of the ring resonator and traveling-wave loop directional filters are demonstrated dynamically.

As a final processing, the line voltage and current solutions of the coupled transmission lines are numerically computed by the following equations.

$$V_1(x,t) = \left(V^e(x,t) + V^o(x,t)\right)/2 \qquad (11)$$

$$I_1(x,t) = \left(I^e(x,t) + I^o(x,t)\right)/2 \qquad (12)$$

$$V_2(x,t) = \left(V^e(x,t) - V^o(x,t)\right)/2 \qquad (13)$$

$$I_2(x,t) = \left(I^e(x,t) - I^o(x,t)\right)/2 \qquad (14)$$

where $V^e(x,t)$, $V^o(x,t)$ and $I^e(x,t)$, $I^o(x,t)$ are the even- and odd-mode line voltages, and currents, respectively.

Algorithm

Preprocessing: Decision of line parameters

Step 1.   Set initial values
             V(x, t=0), I(x, t=0): 0<x<l
Step 2.   t=t +Δt if t > t_n then exit
Step 3.   Calculate update equations
             V(x,t), I(x,t) : 0<x<l
Step 4.   Boundary treatments
Step 5.   Go to Step2

Postprocessing: Visualization of results

The solutions of the basic coupled lines as shown in Figure 2 are computed by the numerical even and odd modes analysis. The voltage and current solutions of the equivalent even- and odd-mode impedance lines of uniform parallel coupled lines are shown in Figure 4 (a)(b)(e)(f). Consequently, by synthesizing these solutions, namely using equations (11)-(14), the solutions for the drive line and sense line are obtained as shown in Figure 4 (c)(d)(g)(h). Subsequently, Figure 5 and Figure 6 show the input and output voltage waveforms extracted from the voltage solutions of the basic coupled lines with and without matched loads, respectively, from 0 to 800 time steps. The used design parameters as follows: Z_0=50Ω, Z_{0c}=50Ω, k=0.707, Z_{0e}=120Ω, Z_{0o}=20.8Ω, l_c=40mm, Δx=1mm, Δt=3.33ps, ε_r=1, for e_i=$\sin(2\pi f_c t)$: f_c =1.875GHz).

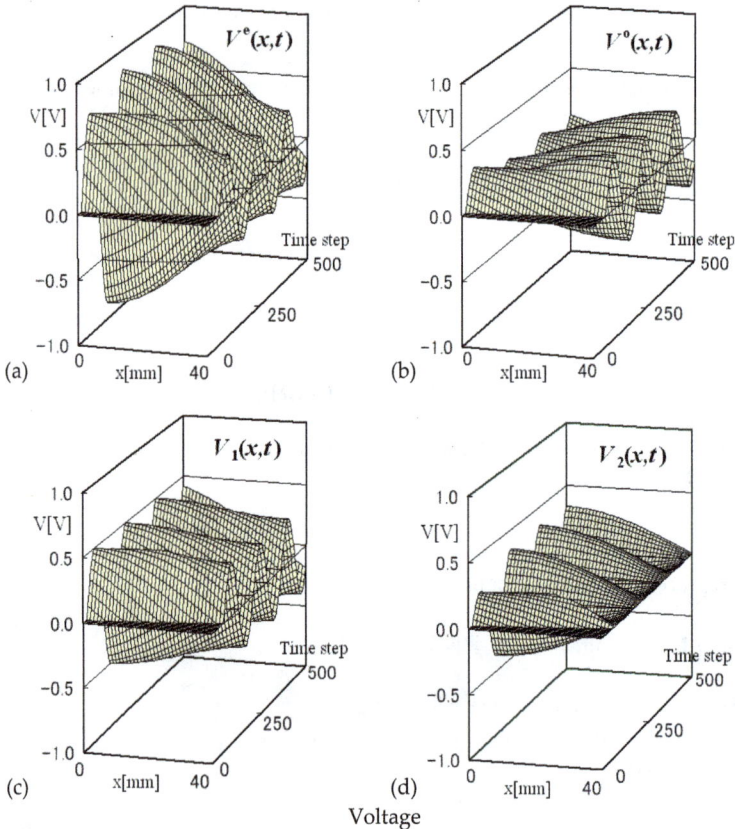

Voltage

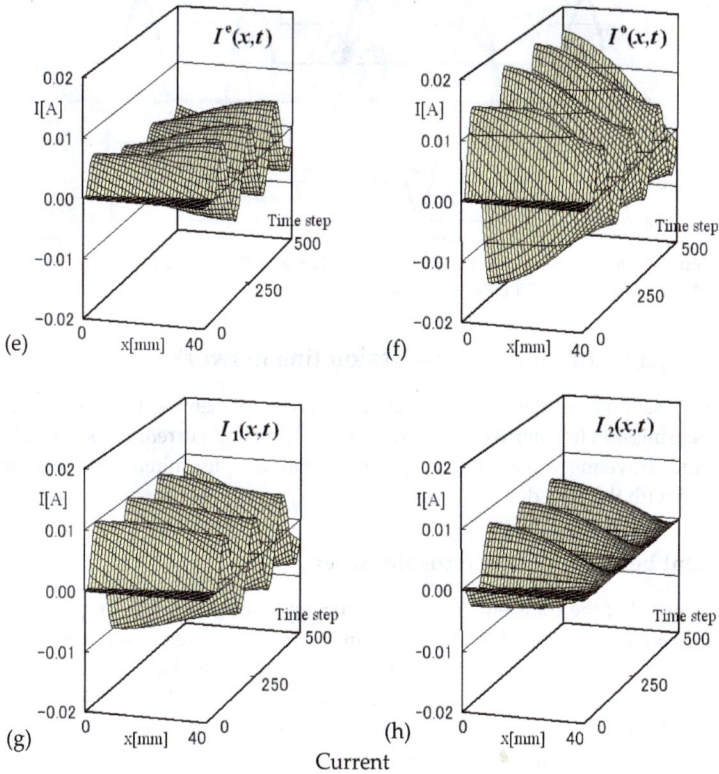

Figure 4. Transient solutions of the even-mode and odd-mode equivalent lines and the solution of the basic coupled line synthesized with their solutions: Voltage (upper) and Current (lower).

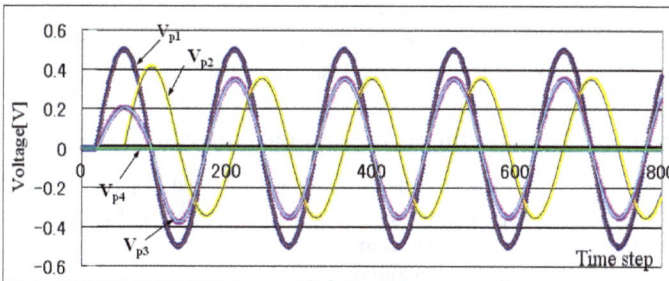

Figure 5. Input and output responses of the basic coupled lines with matched loads: MCD (dotted line) and SNAP (solid line). ($Z_0=50\Omega$, $Z_{0c}=50\Omega$, $k=0.707$, $l_c=40$mm).

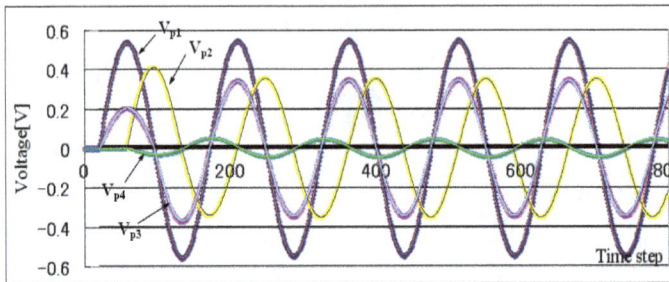

Figure 6. Input and output responses of the basic coupled lines with mismatched loads: MCD (dotted line) and SNAP (solid line). (Z_0=50Ω, Z_{0c}=60Ω, k=0.707, l_c=40mm).

3. Wave propagation in the transmission line networks

In this section, several simulation results are shown for representing the wave propagation along the transmission line network. The transient voltage and current responses of the ring resonator and traveling-wave multi-loop directional coupler filters are analyzed and demonstrated with the time domain analysis.

3.1. Transient behavior of ring resonator filter

First, it is shown that the transient voltage and current responses of the ring resonator filter in Figure 1 (a) are demonstrated in the time domain. The designed parameters for this band stop filter (BSF) were used as follows: Z_0=50Ω, Z_{01}= Z_{03}=55Ω, Z_{02}=35.4Ω, Z_{0e}=130Ω, Z_{0o}=40.9Ω, l=40mm, Δx=1mm, Δt=3.33ps, ε_r=1, for a sinusoidal wave (e_i=$sin(2\pi f_c t)$: f_c center frequency 1.875GHz) excitation. The variations of the instantaneous voltage and current distributions of the transmission line in the transient region are represented by the dynamic expression as shown in Figure 7. Also, Figure 8 shows the voltage and current distributions at steady state after 5000 time steps, to represent the standing wave at the center frequency. This figure represents the difference of the resonance phenomena in loop line between BSF and all-pass ring resonator. Figure 9 shows the input and output responses, in which the transmission zero and full-pass into port 2, respectively, are observed. In the case of the BSF, from the observed responses, it considerably takes much more CPU time up to steady state region. The designed parameters for the case of all-pass were used as follows: Z_0=50Ω, Z_{01}=Z_{03}=50Ω, Z_{02}=50Ω, Z_{0e}=120.7Ω, Z_{0o}=20.8Ω, l=40mm, Δx=1mm, Δt=3.33ps, ε_r=1.

3.2. Time responses of traveling-wave multi-loop directional filters

Next, we show the time responses for the traveling-wave double-loop 3dB directional coupler as shown in Figure 1 (b). Figure 10 shows the variation of instantaneous voltage and current distributions on the directional coupler in the steady state after 1500 time steps. The isolation at port 4 and the equal power division into ports 2 and 3 are observed for the 3-dB coupler at the center frequency. Note that the envelope denotes the standing waves on the transmission lines, in which they appear at only the coupled line area. In Figure 11, the input and output voltage waveforms at each port are represented for 2000 time steps. The CPU time took in a few seconds. The following designed parameters were used: $Z_0=50\Omega$, $Z_{011}=Z_{031}=Z_{012}=Z_{032}=50\Omega$, $Z_{0e1}=Z_{0e2}=Z_{0e3}=120.7\Omega$, $Z_{0o1}=Z_{0o2}=Z_{0o3}=20.8\Omega$, $l=40$mm, $\Delta x=1$mm, $\Delta t=3.33$ps, $\varepsilon_r=1$ at $f_c=1.875$GHz.

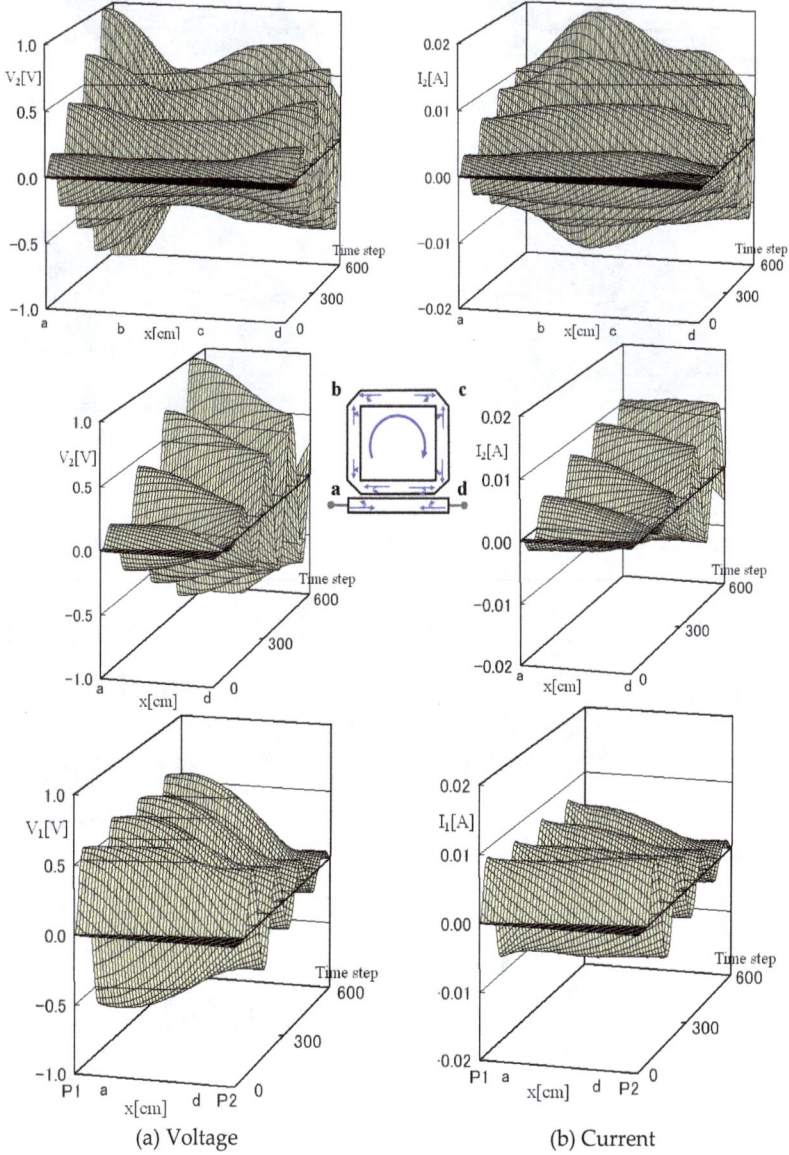

(a) Voltage   (b) Current

Figure 7. Transient behaviors of voltage and current on the ring resonator BSF from 0 to 600 time steps.

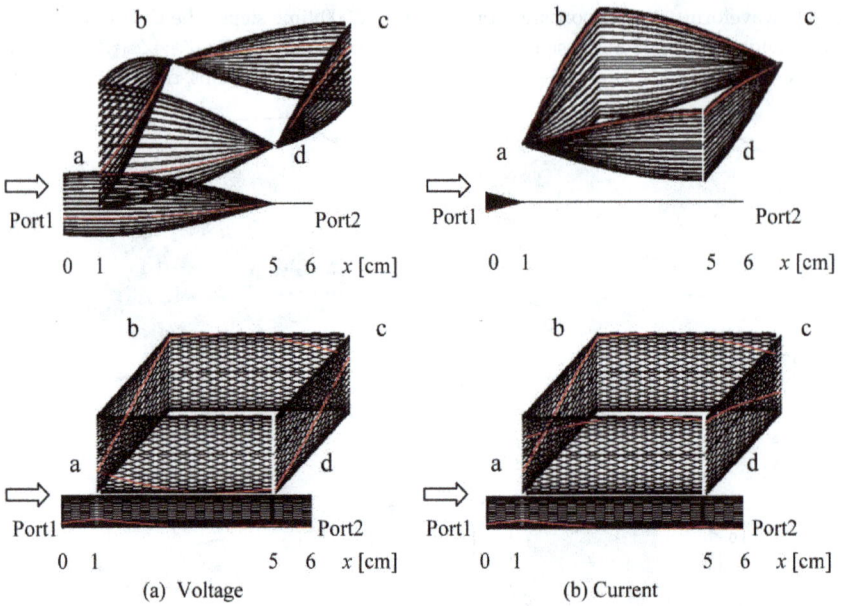

(a) Voltage   (b) Current

Figure 8. Variation of instantaneous voltage and current distributions on the ring resonator BSF (upper) and all-pass (lower) at steady state after 5000 time steps. Red lines denote a snapshot of the plot.

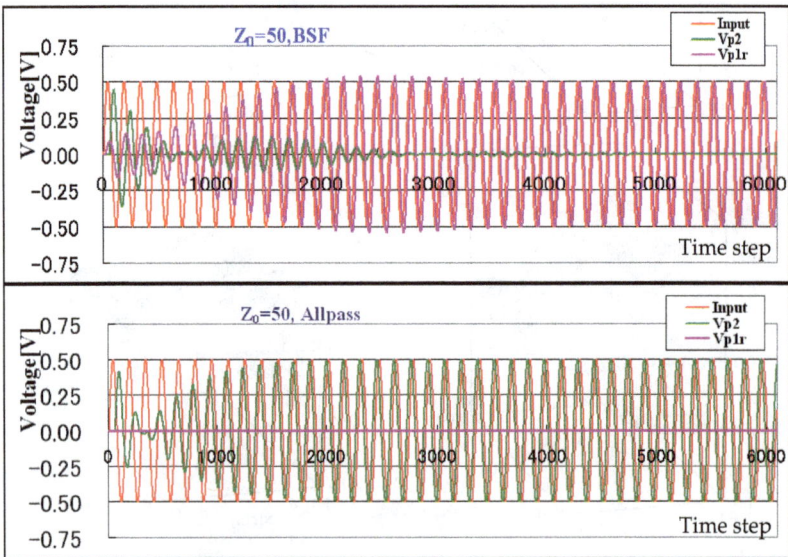

Figure 9. Input and output voltage waveforms of the ring resonator BSF (upper) and all-pass (lower) from 0 to 6000 time steps.

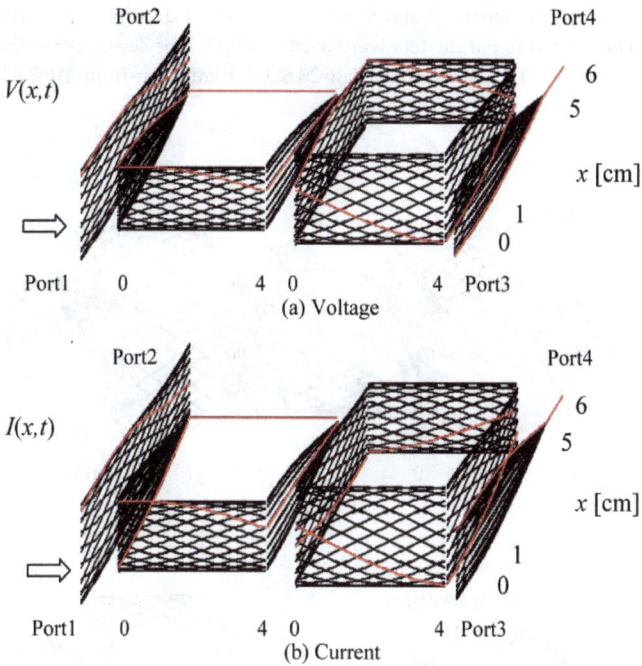

Figure 10. Variation of the instantaneous voltage and current distributions on the double-loop 3-dB directional coupler after 1500 time steps. Red lines denote a snapshot of the plot.

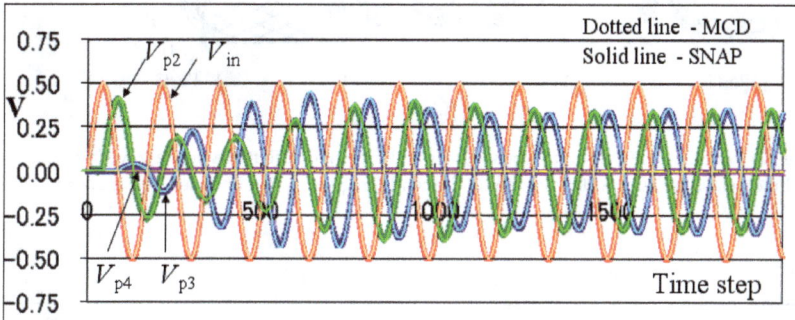

Figure 11. Input and output voltage waveforms of the double-loop directional coupler for sinusoidal wave excitation at the center frequency.

Figure 12 shows the time responses for the traveling-wave triple-loop directional filter. The variation of instantaneous voltage and current distributions on the directional filter in the steady state are illustrated after 3000 time steps. The isolation at ports 4 and 2, and full power into port 3 is observed for the directional filter at the center frequency f_c. The input and output voltage waveforms at each port for 4000 time steps are represented in Figure 13.

Additionally, at dual frequencies f_a and f_b, near equal power division into ports 2 and 3 are represented. The following parameters were used: $Z_0=50\Omega$, $Z_{011}=Z_{031}=Z_{012}=Z_{032}=Z_{013}=Z_{033}=50\Omega$, $Z_{0e1}=Z_{0e2}=Z_{0e3}=Z_{0e4}=120.7\Omega$, $Z_{0o1}=Z_{0o2}=Z_{0o3}=Z_{0o4}=20.8\Omega$, $l=40$mm, $\Delta x=1$mm, $\Delta t=3.33$ps, $\varepsilon_r=1$ at f_c $=1.875$GHz.

(a) Voltage

(b) Current

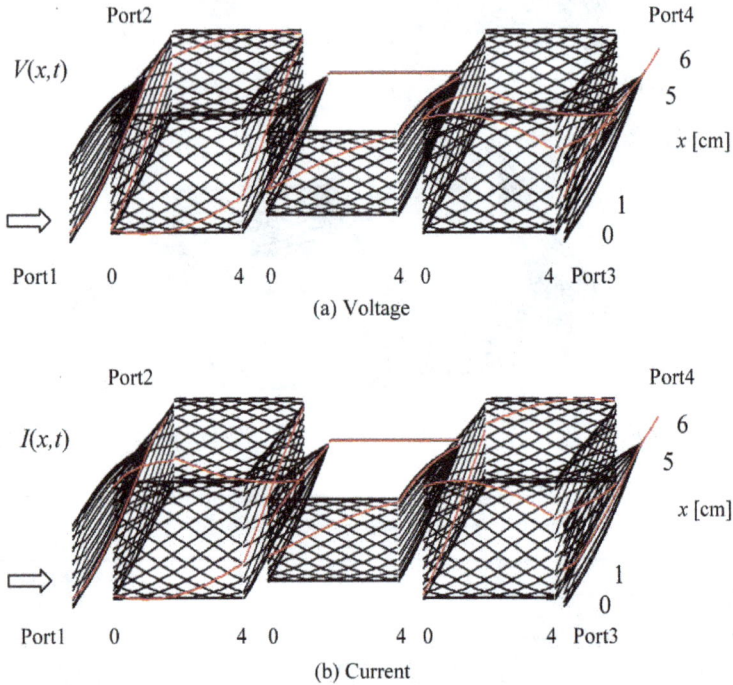

Figure 12. Variation of the instantaneous voltage and current distributions on the traveling-wave triple-loop directional filter after 3000 time steps at the center frequency. Red lines denote a snapshot of the plot.

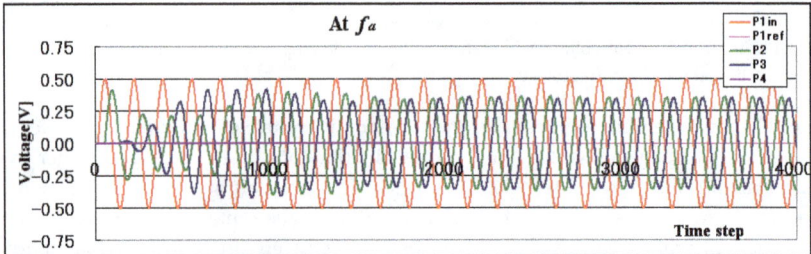

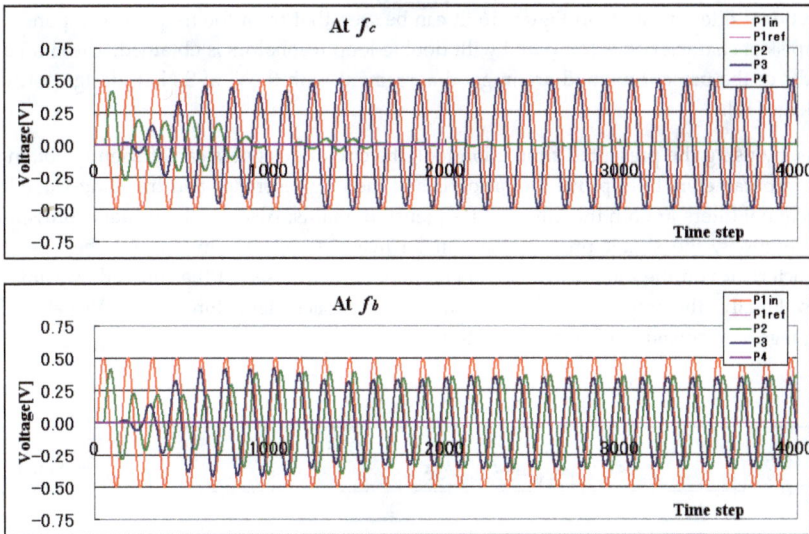

Figure 13. Input and output voltage waveforms of the traveling-wave triple-loop directional filter for f_a (=1.72GHz), f_c(=1.875GHz) and f_b(=2.03GHz) frequencies. (See Figure 16 (b))

4. Frequency responses

As an application of this method, it is shown that frequency responses are obtained from the transient responses by the fast Fourier transform technique. Using the input and output responses extracted from the transient pulse solutions, we obtain each voltage $V_{pn}(t)$ and current $I_{pn}(t)$ as the response quantities at port n. By using the FFT technique, S parameters for port 1 input of the coupler are given by

$$S_{n1} = FFT[V_{pn}(t)]/FFT[V_{in}(t)] \qquad (15)$$

where $V_{in}(t)=(V_{p1}(t)+I_{p1}(t)Z_0)/2$ is the incident voltage at port 1. Then, the return loss S_{11} and insertion loss S_{21} are obtained. This simulation was carried out with a Gaussian pulse $e_i(t)$ =$exp(-((t-t_0)/\tau)^2)$ excitation, where t_0 is the initial delay time and τ is the pulse width parameter.

Figure 14 (a) illustrates the voltage pulse responses at each port for one loop resonator filter. Figure 14 (b) shows the return loss and insertion loss of the frequency characteristics, which has the characteristic of very sharp notch filter at the center frequency. The same line parameters in the previous section were used.

Figure 15 (a) illustrates the voltage pulse responses at each port for the traveling-wave double-loop 3-dB directional coupler. And, Figure 15 (b) shows the return loss and insertion loss of the frequency characteristics, which has the characteristic of band pass filter with 3dB coupling at the center frequency. Similarly, the result of the traveling-wave triple-loop directional filter is shown in Figure 16. It can be seen that from the frequency response, the sharp skirt characteristics compared with double-loop resonators is obtained. The simulated results of the present method are in good agreement with those of S-NAP design software package [10].

Thus, by using the presented simulation tool, the circuit designers can efficiently obtain the design parameters to improve the properties of the ring resonator and traveling-wave loop directional filters in both the time and frequency domains. Also, by the visual expression of the solutions, the signal propagation can be make it easy to understand the operation characteristics on the microwave circuit components composed of the coupled transmission lines. Finally, the implemented program is of compact algorithm by the Visual BASIC language, and considerably saves CPU time.

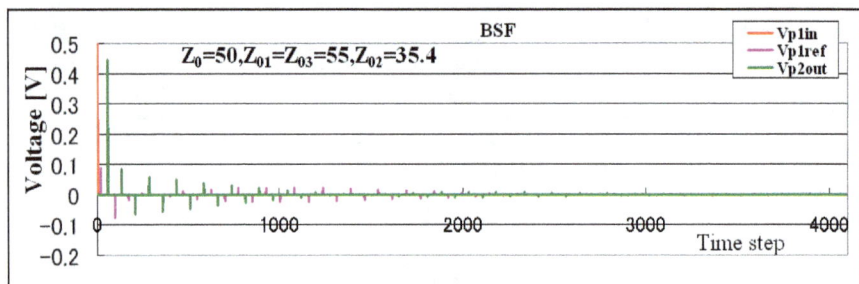

(a)

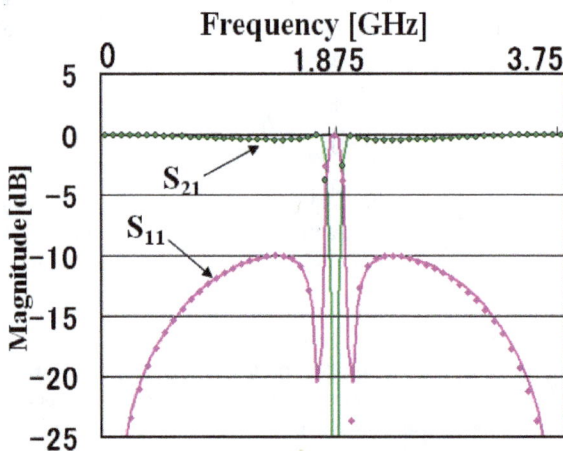

(b)

Figure 14. (a) Input and output voltage responses of the ring resonator BSF from 0 to 4000 time steps for a Gaussian excitation. (b) S-parameters by present method (dotted lines), and SNAP simulator (solid lines).

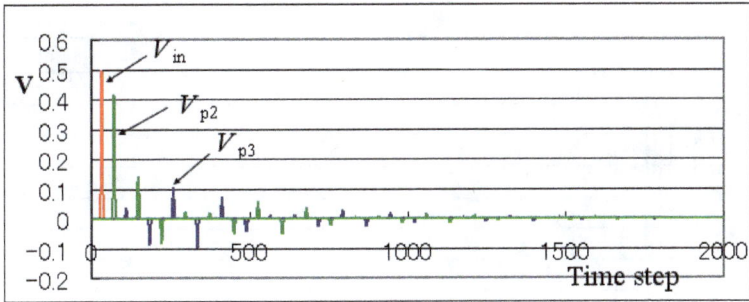

(a)

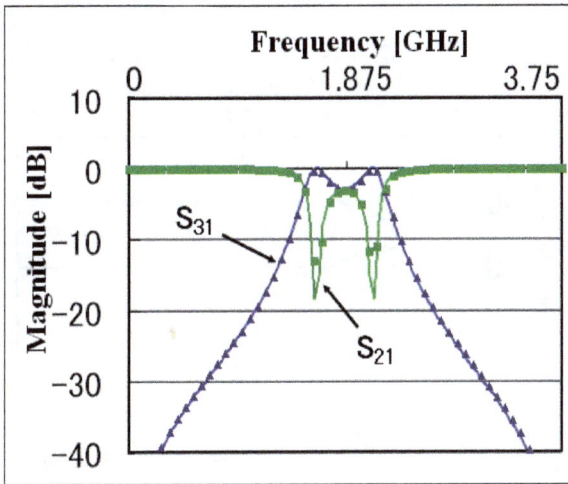

(b)

Figure 15. (a) Input and output responses of the traveling-wave double-loop directional coupler for a Gaussian pulse excitation. (b) S-parameters of this coupler: presented method (dotted lines), and SNAP simulator (solid lines).

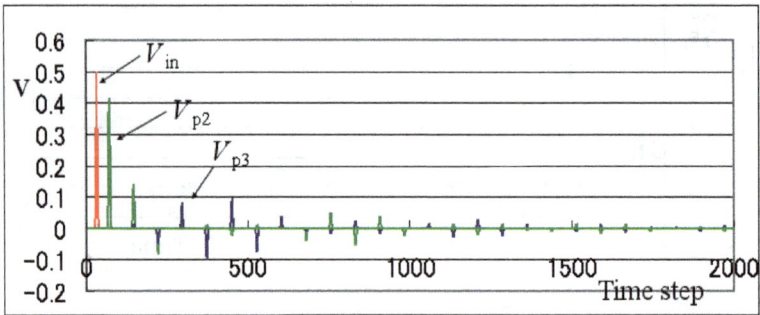

(a)

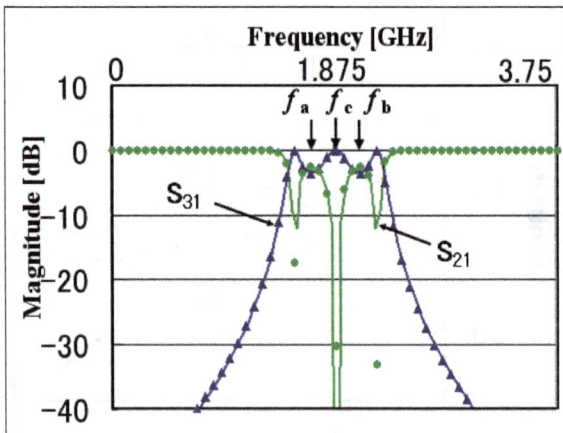

(b)

Figure 16. (a) Input and output responses of the traveling-wave triple-loop directional filter for a Gaussian pulse excitation. (b) S-parameters of this filter: presented method (dotted lines), and SNAP simulator (solid lines).

5. Conclusion

Visualizing the signal propagation in transmission line circuits is very important for understanding the operation mechanism of the circuit systems. It has been shown that the presented time domain simulation method can analyze and demonstrate the transient behaviors of the voltage and current waves of the ring resonator and traveling-wave loop directional filters by dynamic expression. Moreover, we have shown the frequency responses of the multi-loop coupled line filters by using the FFT technique. As a brief test tool, this method introduced the systematic mixed even and odd modes model for the part of parallel coupled lines is useful to confirm the operation characteristics of the circuit components consisted of loop resonator with coupled lines. This run can process on a small size PC system sufficiently.

Author details

Kazuhito Murakami
Computing Center, Kinki University, Japan

6. References

[1] T. Itoh, et al, Numerical techniques for microwave and millimeter-wave passive structures, JOHN WILEY & SONS, 1989.

[2] R. F. Harrington, Field computation by moment method, IEEE Press, 1993.

[3] K. S. Yee, Numerical solutions of initial boundary value problems involving Maxwell's equations in isotropic media, IEEE Trans. Antennas & Propag., vol.AP-14, no.3, May 1966, pp.302-307.

[4] A. Taflove and S. C. Hagness, Computational electrodynamics the Finite-Difference Time-Domain method, ARTECH HOUSE, MA, 2000.

[5] G. Matthaei, L. Young and E.M.T. Jones, Directional, channel-separation filters and traveling-wave ring-resonators, in Microwave filters, impedance-matching networks, and coupling structures, Chapter 14, pp.843-887, ARTECH HOUSE, MA, 1980.

[6] F. S. Coale, A traveling-wave directional filter, IRE Trans. on MTT, vol. 4, no.10, Oct. 1956, pp.256-260.

[7] M. K. M. Sallich, G. Prigent, O. Pigaglio, and R. Crampagane, Quarter-wavelength side-coupled ring resonator for bandpass filters, IEEE Trans. Microwave Theory Tech., vol. MTT-56, no. 1, Jan. 2008, pp.156-162.

[8] K. Murakami, and J. Ishii, Transient analysis of multi-section transmission line using modified central difference method (In Japanese), Trans. IEICE Vol.J78-A, May 1995, pp. 602-609.

[9]   K. Murakami, and S. Kitazawa, Transient responses of voltage and current on ring resonator and traveling-wave loop directional filters, in Proc. of APMC2010, pp. 1185-1188, Dec. 2010, Yokohama, Japan; 2010.

[10] S-NAP Design user's manual, MEL Inc., 1998.

Chapter 12

Ray Launching Modeling in Curved Tunnels with Rectangular or Non Rectangular Section

Émilie Masson, Pierre Combeau, Yann Cocheril,
Lilian Aveneau, Marion Berbineau and Rodolphe Vauzelle

Additional information is available at the end of the chapter

1. Introduction

Since several years, digital wireless transmissions are experiencing a significant increase with the development of digital TV, satellite communications, mobile phones, internet, wireless local area networks and automation in various domains. The transportation field as a whole is a major actor of these changes. Technologies and systems for wireless communications are increasingly used in the field of guided transport (underground, conventional trains, high speed trains, tramways, *etc.*) to ensure communications between trains and between a train and the infrastructure. These systems answer the key operational needs for safety and comfort, such as control and command of the trains, traffic management, maintenance, security and information for passengers and crew members. We can distinguish two main families of systems: low-data rate reliable transmissions, for traffic control and command, and robust high data rate transmissions, for video surveillance, remote diagnostic or embedded multimedia applications. Consequently, the radio coverage prediction of these wireless systems is mandatory to optimize the deployment of the radio access points in order to ensure the robustness and reliability of radio links and to minimize the antennas positioning phase duration. Indeed, minimum field strengths are generally required to ensure Key Performance Indicators (KPI) related to the requirements of dependability and Quality of Service (QoS).

For underground applications, these systems operate in complex environments, such as tunnels, where the usual laws for predicting the propagation in free space are no longer valid. Like the telecommunication operators, railway manufacturers or guided transport operators should invest in planning tools to deploy their radio communication systems. However, to our knowledge, no general model for predicting free propagation in tunnel exists, as easy to use as well known existing models in the world of mobile phones, such as statistical models derived from Okumura-Hata, Cost231, *etc.* The tunnel case is usually solved with heavy measurement campaigns that are time consuming and costly. Thus, the development of specific and efficient propagation models for tunnels is very relevant for railway industry to allow both fast responses to tenders, and quick deployments of their wireless communication systems in these specific environments. The tunnels can be rectangular, circular or arch-shaped. In addition, they may be straight or curved, with only one or two tracks.

Several methods to model radio wave propagation in tunnels have been published in the literature and will be presented in this chapter with their advantages and drawbacks. Among them, only few works are dedicated to non rectangular cross section and curved tunnels. Hence, we focus on a new method recently developed. The structure of the chapter is as follows. Section 2 presents the context of the works and why deployments of wireless telecommunication systems are needed for transport applications. Existing techniques to model radio wave propagation in tunnel are presented in section 3 with their respective advantages and drawbacks. The fourth and fifth sections are respectively devoted to the design and the evaluation of a propagation prediction model for curved tunnel with a rectangular or a circular cross section. Finally, section 6 concludes and presents some perspectives to these works.

2. Wireless telecommunication systems needs for transport applications

The growing of wireless networks including cellular (GSM, GPRS, EDGE, UMTS, *etc.*) saw the arrival on the market of a number of software tools dedicated to radio systems planning, for industrial and telecom operators. These tools allow the identification of sites to set up base stations from the definition of the characteristics of areas to cover, possibly by imposing the location of some sites (favorite sites). Most of these tools are very heavy to handle, dedicated to radio engineers, optimized and enriched by in-house phone operators to take into account the different propagation environments encountered.

Like the general public telecommunications world, the widespread use of wireless communication systems in the field of transport requires the use of software tools for planning and optimization for the efficient deployment of these systems. Some existing tools in the world of telecommunications are specifically adapted, for example for the deployment of GSM-R infrastructures in various European countries. But to our knowledge, no module specifically dedicated to the metro tunnels is currently sold.

Thus, in most cases, the industrials perform measurement campaigns that are time consuming and costly. These experimentations require free access to the operational sites, which is a difficult task to achieve:

- In the case of a line under construction, site access depends on the progress of the construction that may experiences significant delays. Indeed, the development of a wireless communication system is the final step of the overall transportation system;

- In the case of a line already in operation, access to the site has to be done outside the hours of operation of the metro, which leads to very short ranges that occur most often during night.

To limit the measurement campaigns, it is necessary that the railway industrials possess planning tools to determine the field strength regardless of the environment encountered: consideration of tunnels with different cross and longitudinal sections and building materials for example. Propagation models must meet the best compromise between accuracy and computation time and have to be simple to use.

For outdoor areas, predictive models of the radio coverage are the well-known statistical models, such as free space attenuation, Okumura-Hata model or 2-rays model. These models work well because the deployment is in Line-Of-Sight (LOS) and distances between transmitter and receiver are weak. The measurement campaigns are easy to implement and generally there are few constraints for the positioning of access points that are most often located on existing poles (signaling, lighting, *etc.*).

In tunnel areas, the prediction of radio coverage is currently based on intensive use of feedback on measurements and implementation of engineering rules.

Configurations of complex tunnels, such as curved tunnels of circular cross section are more frequently encountered in the modern metro lines and it becomes necessary to refine the prediction process of the radio coverage for deployment based on more sophisticated tools, such as the method presented in this chapter.

3. Existing techniques to model radio wave propagation in tunnel

Methods based on simulation and measurement of radio wave propagation in tunnels were presented in the literature. Analyses based on measurements at 900 MHz and 1800 MHz represent the first approach (Hwang et al., 1998), (Zhang & Hwang, 1998a), (Zhang et. al, 1998), (Zhang & Hong, 2004), in order to characterize propagation in wide rectangular tunnels. In (Zhang & Hwang, 1998b), (Lienard & Degauque, 1998), statistical characteristics of propagation channel in tunnels are performed from measurement results. However, analyses performed from the measurements presented in these papers are specific to a given shape of tunnel. Radio wave propagation modeling in generic tunnels is thus a major research field.

Some authors have proposed methods based on the modal theory to provide all the modes propagating in tunnel. Approximate and exact solutions have been determined for straight rectangular tunnels and straight circular tunnels respectively (Laakman & Steier, 1976), (Emslie et al., 1975), (Mahmoud, 1974), (Dudley & Mahmoud, 2006), (Dudley et al., 2007). The modal theory considers tunnels as oversized waveguides. It provides good results but it is limited to canonical geometries, which are not the main cases according to the generally encountered tunnels.

Several papers present results based on an exact resolution of Maxwell's equations using numerical techniques such as integral methods or the resolution of the Vector Parabolic Equation (Chang et al., 2009), (Reutskiy, 2008), (Popov, 2000), (Bernardi et al., 2009). One more time, these kinds of techniques are limited, mostly due to the computational complexity. Indeed, these methods are based on a volume discretization of the propagation environment with a scale that should be smaller than the wavelength. Consequently, the complexity is prohibitive for operational environments.

Finally, frequency asymptotic techniques based on the ray concept, being able to handle complex tunnel geometries in a reasonable computation time, seem to be a good solution. A first simple approach based on a 2-rays model was proposed in (Zhang, 2003), (Ahmed et al., 2008). This method suffers from lack of accuracy due to approximations on the number of rays considered in the model. In (Mahmoud, 1974), a model based on a Ray Tracing, based on image method, takes into account multiple reflections, but only for the case of straight rectangular tunnels. In (Mariage et al., 1994), the authors use this method and adds diffraction phenomenon in order to analyze the coupling between indoor and outdoor. In (Agunaou et al., 1998), works are purchased by considering changes of tunnel sections, but still only in straight rectangular tunnels. However, tunnels in real environments, such as metro ones, can have non rectangular cross section. Furthermore, they can be curved. Ray Tracing is no longer valid with curved surfaces since the source image is no more unique. Consequently, only a few studies deal with the case of non-rectangular cross sections and curved tunnels.

The first intuitive approach for curved surfaces consists of a tessellation of the curved geometry into multiple planar facets, as proposed in (Chen & Jeng, 1996), (Torres et al., 1999), (Baranowski et al., 1998), (Masson et al., 2010). Unfortunately, the surface curvature is not taken into account in this type of techniques. Furthermore, one of the major drawbacks of this approach is the impossibility to identify an optimal number of facets for a given tunnel cross section and a given frequency (Masson et al., 2010). In (Wang & Yang, 2006), a ray-tube tracing method is used to simulate wave propagating in curved road tunnels. An analytical representation of curved surfaces is proposed. Comparisons with measurement results are performed in straight arch-shaped and curved tunnels. The paper lacks of information. Consequently, we focused on the works presented in (Didascalou et al., 2000), (Didascalou et al., 2001). A Ray Launching combined with a ray-density normalization is presented. The surface curvature is taken into account. Comparisons with measurements are performed respectively in a scaled tunnel (with 20 cm reduced diameter dimension, and then higher frequency at 120 GHz for compensating the former), and in curved subway tunnels at 900 MHz and 1800 MHz. The method provides good results but exhibits two main defaults: the geometric shapes of the tunnels are not all flexible, and the computation time is high due to the large number of rays launched at transmission. To overcome these drawbacks, we developed a method to model the electromagnetic propagation in tunnels with curved geometry, either for the cross section or the main direction. The main advantage of this new method is its best compromise between accuracy and computation time for applications at 5.8 GHz. It is presented in the next section.

4. Design of a propagation prediction model for non rectangular cross section and curved tunnels

The aim of the developed method is to model radio wave propagation in non rectangular cross section and curved tunnels. The problem to be solved consists in taking into account the presence of curved surfaces which induces divergence of rays by reflection on surfaces. An adaptation of the Ray Launching technique and a correction on paths trajectories have to be performed.

The method consists of three main steps. First step consists of an adaptation of a classical Ray Launching technique for the consideration of curved surfaces. Developments on emission, reception and intersection between rays and curved surfaces are performed. Second step is to correct the received paths trajectories after the Ray Launching step. It consists of a minimization of the paths length, a choice of optimized paths and an adapted algorithm for the Identification of Multiple Rays (IMR). Last step consists in taking into account the divergence of rays for the Electric Field calculation in the presence of curved surfaces.

We consider four kinds of tunnel geometry encountered in operational cases, illustrated in Figure 1:

- the straight rectangular tunnel;
- the straight circular tunnel;
- the curved rectangular tunnel;
- the curved circular tunnel.

For tunnels with circular cross section, a floor and/or a roof can be added in the model of the environment to take into account arch-shaped sections for example. All these configurations lead us to consider three kinds of elementary geometry: The plane for the straight rectangular tunnel and the additional floor/roof, the cylinder for the straight circular tunnel and the walls of the curved rectangular one, and finally the torus according to the curved circular tunnel. All these components are quadrics, which is important for the simplicity of intersection computation with rays.

4.1. Ray launching for curved surfaces

4.1.1. Emission

Since no *a priori* information on privileged propagation direction is available, we chose to implement a Ray Launching technique based on a uniform distribution of rays radiated from the transmitter. There are different techniques to obtain a uniform distribution.

We consider a stochastic Monte Carlo method (Didascalou et al., 2000) in order not to skew the results. Instead of using random sequences, the method consists in using quasi-random sequences (Morokoff & Caflisch, 1995) to regularly cover all the space. Indeed, quasi-random sequences allow minimizing the discrepancy, which corresponds to a measure of the gap between a reference situation, generally perfect uniformity, and a given configuration.

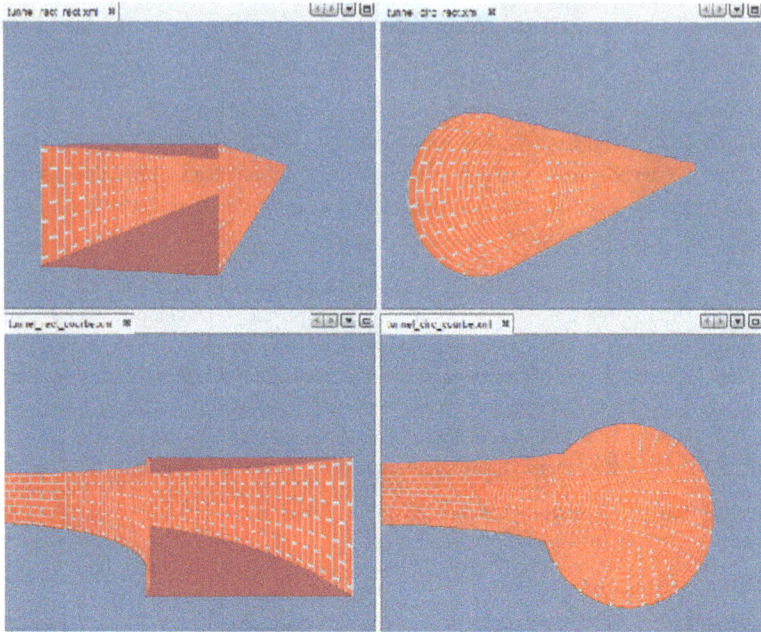

Figure 1. Four kinds of treated tunnels

Figure 2 illustrates the distribution of 512 rays computed from a random sequence compared to ones obtained from two quasi-random sequences, namely Halton (Halton, 1960) and Hammersley (Hammersley, 1964). The Hammersley sequence allows us to obtain the best uniform distribution.

In order to prove it, we present in Figure 3 the histogram of the angles between each ray and its nearest neighbor. So, the most uniform sequence is the one which provides the highest peak for a given angle value in its histogram, it will then minimize the discrepancy. It appears again that the best one is the Hammersley sequence. Therefore, we chose to use it to trace the rays radiated from the transmitter.

Random   Halton   Hammersley

Figure 2. Comparison of the distribution of 512 rays at transmission between the use of a random sequence and 2 quasi-random sequences: Halton and Hammersley

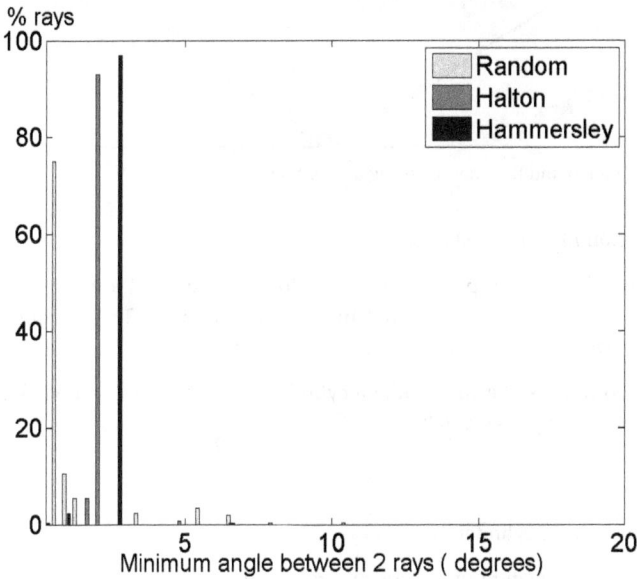

Figure 3. Histogram of minimum angles between two rays launched at transmission

4.1.2. Reception

The Ray Launching technique requires a reception sphere to determine which rays contribute to the received power. Indeed, the probability that a given ray meets a receiver point is always zero. The reception sphere radius, denoted as r_R, depends on the path length r and the transmission angle γ (Seidl & Rappaport, 1994):

$$r_R = \frac{\gamma r}{\sqrt{3}} \qquad (1)$$

The main drawback of a reception sphere is that it should receive a collection of rays whereas an only one is predicted by Geometrical Optics. These rays are called multiple rays, as illustrated in Figure 4. If all of them are added together the result will be an overestimation of the received power. Thus, it is necessary to discriminate them. This can be performed with an Identification of Multiple Rays (IMR) algorithm (Iskander & Yun, 2002). Classical criteria include the number of reflections, the path length and the transmission angle. In case of curved surfaces, multiple rays are much more scattered and distant from each other. A specific treatment of the rays and an adapted IMR are so needed and proposed in section 4.2.

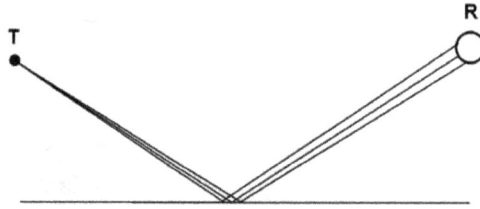

Figure 4. Illustration of multiple rays by using a reception sphere

4.1.3. Intersection ray/curved surface

We have shown that besides planar surfaces, both cylindrical and toroidal surfaces have to be taken into account in the considered tunnel configurations. These geometrical objects belong to the quadrics family and can thus be easily expressed in an analytical form.

In Cartesian coordinates, the equation of a cylinder with the axial direction along the y-axis and centered at the origin is given by:

$$x^2 + z^2 = r_c^2 \tag{2}$$

with r_c the radius of the cylinder.

In a same way, the equation of a torus around the z-axis and centered at the origin is as follows:

$$(x^2 + y^2 + z^2 + R_t^2 - r_t^2)^2 - 4R_t^2(x^2 + y^2) = 0 \tag{3}$$

with R_t and r_t the radiuses of the torus.

The definition of a ray being as follows:

$$x = d_x + t \cdot di_x,$$

$$y = d_y + t \cdot di_y, \tag{4}$$

$$z = d_z + t \cdot di_z$$

with (d_x, d_y, d_z) the origin of the ray and (di_x, di_y, di_z) the direction vector of the ray, the intersection of a ray with a cylinder and a torus leads to the resolution of respectively, a quadratic equation and an equation of degree 4.

4.2. Optimized IMR and final choice of corrected rays

The section 4.1.2 has shown that using a reception sphere leads to take multiple rays into account. The multiple rays are determined by classical IMR algorithms based on similarity criterion according to geometrical characteristics of the rays. Then, the choice of the retained ray is randomly realized among the identified multiple rays, without any physical considerations. So, they lead to a bad approximation of the real ray. To enhance this approach, we propose here to add a correction algorithm of the trajectories of multiple rays that makes them converge to the correct geometrical one. The ambiguous choice of the ray for a correct field calculation is thus avoided.

4.2.1. Correction algorithm of the paths trajectories

The general principle is as follows. Once a ray is launched from the transmitter using the Hammersley quasi-random sequence, its propagation is recursively computed, by calculating its intersection with the curved surfaces, until it reaches the reception sphere. Either it undergoes the fixed maximum number of reflections, or it goes outside the tunnel. So, a contributive ray (a path), *i.e.* one that has reached the reception sphere, presents some geometrical approximations since it does not go through the exact receiver position, as it has been explained in section 4.1.2. Using the Fermat Principle, indicating that the path followed by a wave between two points is always the shortest one, we propose to reduce the geometrical approximation involved in each path: The correction algorithm consists, for a given path, in minimizing its length assuming that it reaches the center of the reception sphere.

While the path length function is not linear, we propose to use the well-known Levenberg-Marquardt algorithm (Marquardt, 1963). Nevertheless, the algorithm is efficient only if the starting point (*i.e.* the initial ray) is near the solution. By using the Ray Launching, it can be assumed that the paths caught by the reception sphere are close to the real paths existing with a Geometrical Optics meaning, *i.e.* the shortest. Thus the received rays represent a good initialization of the algorithm. The principle of the Levenberg-Marquardt algorithm consists in finding the best parameters of a function which minimize the mean square error between the curve to approximate and its estimation.

Applied to propagation in tunnels, the objective becomes a path length minimization. The criterion to minimize is then the total path length J given by:

$$J = \left\| \overrightarrow{EP_1} \right\| + \sum_{k=2}^{N} \left\| \overrightarrow{P_{k-1}P_k} \right\| + \left\| \overrightarrow{P_N R} \right\| \tag{5}$$

With E the transmitter position, R the receiver position and P_k the kth interaction point position of the considered path, as illustrated in Figure 5.

The vectors defined in (5) depend on the coordinates of the interaction points along the path. The iterative algorithm requires the inversion of a Hessian matrix which contains the partial derivatives of the J criterion to minimize with respect to parameters. To keep computation time and numerical errors reasonable, the matrix dimensions and thus the number of parameters have to be minimized. Local parametric coordinates (u, v) from the given curved surface are used instead of global Cartesian coordinates (x, y, z). The parameters vector θ can be written as:

$$\theta = \left[u_1 v_1 \ldots u_N v_N \right] \qquad (6)$$

Where (u_k, v_k) correspond to coordinates of the reflection point P_k.

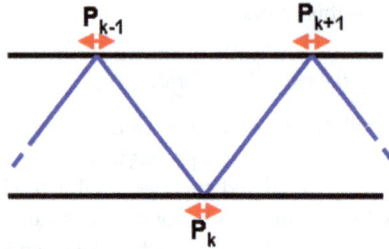

Figure 5. Principle of optimization of paths

4.2.2. Validation of corrected trajectories

Since the Levenberg-Marquardt algorithm is a numerical method, its convergence cannot be guaranteed. So, we have to proceed to a validation test on the corrected paths, which can be incoherent in the case of an algorithm divergence.

This test consists in checking if the Geometrical Optics laws are respected, specifically the Snell-Descartes ones. For each reflection point, we verify if the angle of reflection θ_r equals the angle of incidence θ_i, as illustrated in Figure 6. If not, the path is discarded. For instance an unphysical path $\overrightarrow{d_{r_{LM}}}$ computed by the Levenberg-Marquardt algorithm different from $\overrightarrow{d_{r_{ok}}}$ (by considering $\theta_i = \theta_r$) is shown in figure 6. The path $\overrightarrow{d_{r_{LM}}}$ is thus eliminated.

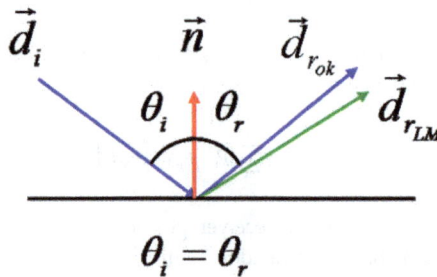

Figure 6. Validation criterion of corrected paths

4.2.3. Final choice of the correct ray

Section 4.1.2 indicates that the Ray Launching algorithm leads to the existence of multiple rays corresponding to the same contribution according to Geometrical Optics. They must be identified and only one ray has to be kept. However, the correction technique previously presented allows obtaining multiple rays very close to the real path, and consequently very close to each other. Nevertheless, due to numerical errors, they cannot be strictly equal to each other. Thus, the choice of the final ray can be done on the base of the reflection points localization: If the reflection points of two candidate paths are at a given maximal inter-distance, the paths are considered to be equal and one of the two is removed, it does not matter which.

4.3. Electric field calculation

In the case of a curved surface, Electric Field can be computed after reflection by classical methods of Geometrical Optics as long as the curvature radiuses of surfaces are large compared to the wavelength (Balanis, 1989).

It can be expressed as follows (Figure 7):

$$\overrightarrow{E^r}(P) = \sqrt{\frac{\rho_1^r \rho_2^r}{(\rho_1^r + r)(\rho_2^r + r)}} e^{-jkr} \overline{\overline{R}} \overrightarrow{E^i}(Q) \tag{7}$$

With ρ_1^r and ρ_2^r the curvature radiuses of the reflected ray, r the distance between the considered point P and the reflection point Q, $k=2\pi/\lambda$ with λ the wavelength, and $\overline{\overline{R}}$ the matrix of dyadic reflection coefficients.

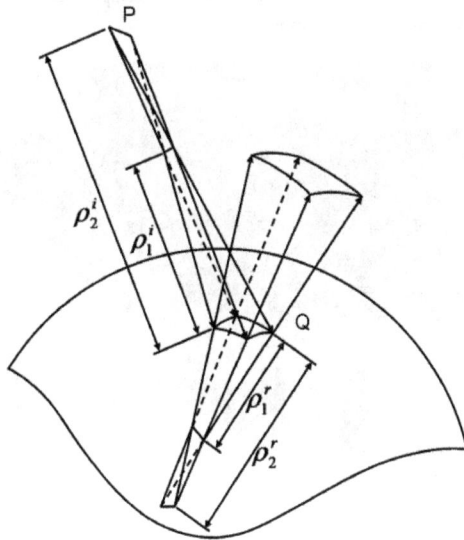

Figure 7. Reflection on a curved surface

Unlike the case of planar surfaces, the curvature radiuses of the reflected ray are different from those of the incident ray. Indeed, the following relation holds (Balanis, 1989):

$$\frac{1}{\rho_{1,2}^{r}} = \frac{1}{2}\left(\frac{1}{\rho_{1}^{i}} + \frac{1}{\rho_{2}^{i}}\right) + \frac{1}{f_{1,2}} \tag{8}$$

With ρ_{1}^{i} and ρ_{2}^{i} the curvature radiuses of the incident ray and $f_{1,2}$ a function depending on ρ_{1}^{i}, ρ_{2}^{i} and the curvature radiuses $R_{1,2}$ of the curved surface (Balanis, 1989).

5. Performance analysis in curved tunnels

This section is dedicated to the performance analysis of the method presented in section 4 in tunnels presenting non rectangular cross section and/or curved longitudinal section. This evaluation is performed from measurement results obtained in real tunnel environments. Performances are evaluated in narrow band by using a comparison of received narrow band powers. First part of the section is dedicated to analyses in a straight arch-shaped tunnel. Second and third parts of the section focus on rectangular and circular curved tunnels.

5.1. Performance analysis in a straight arch-shaped tunnel

5.1.1. Configuration

Measurements were first performed in a straight arch-shaped tunnel, the tunnel of Roux. It consists of a two-way road tunnel located in Ardèche in France, and is illustrated in Figure 8. The straight section has a length of 3325 m. The arch-shaped cross section has a maximal height and width of 5.8 m and 8.3 m respectively.

Figure 8. Illustration of the Tunnel of Roux

The measurements were performed at 5.8 GHz. The transmitter was located at a fixed position at the center of the section, at a height of 4.8 m. It consists of a horn antenna (10.1 dBi gain at 5.8 GHz), vertically polarized. The receiver was moving along the longitudinal axis of the tunnel thanks to a go-kart allowing a very small and regular velocity, compatible with the acquisition rate. It is illustrated in Figure 9. The receiving antenna was vertically polarized and had the same characteristics than the transmitting one. It was located in the middle of the two tracks, 2.4 m from the sidewall, at a height of 4.1 m. Figure 10 illustrates the antennas configuration of measurements.

Figure 9. Go-kart used at reception

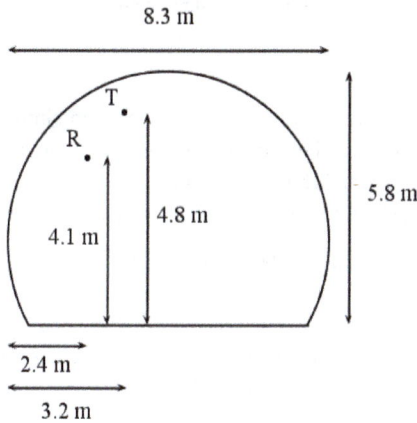

Figure 10. Antennas configuration of measurements in the Tunnel of Roux

5.1.2. Comparison between simulation and measurement loss

Figure 11 presents the comparison between measurements and simulations in terms of normalized received power (P_r) with respect to the transmitted one (P_t), versus the distance between transmitter and receiver along the longitudinal tunnel axis in the Tunnel of Roux.

A quite good concordance between simulations and measurements is highlighted. A detailed analysis is performed in the following.

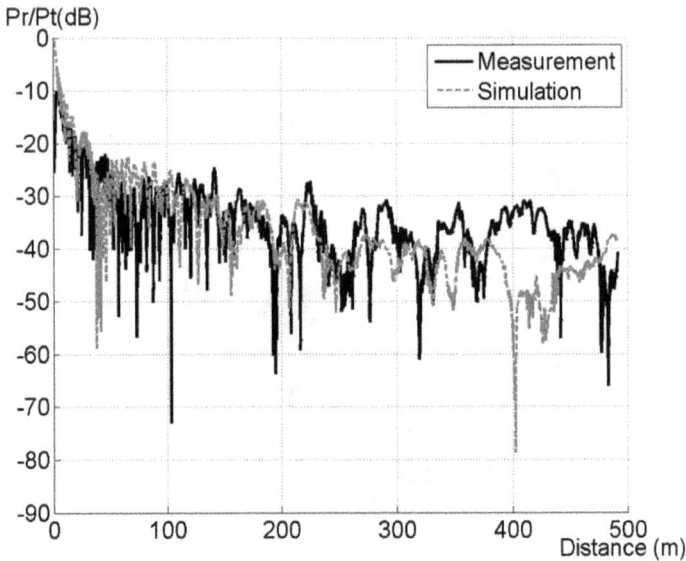

Figure 11. Comparison between measurements and simulations loss in the Tunnel of Roux – 5.8 GHz

5.1.3. Statistical analysis of results

Statistical analysis of slow and fast fading is performed on simulation and measurement results. The procedure to extract these data is as follows. First step consists in smoothing signals by using a running mean. The window's length is 40 λ on the first 50 m, and 100 λ elsewhere, according to the literature (Lienard & Degauque, 1998).

Then, the analysis of slow fading can be performed by the computation of the mean and standard deviation of the error between measurements and simulations. Results are illustrated in Table 1. The slow fading analysis highlights a good agreement between results both for the mean and the standard deviation. Indeed, the values are in the range of 4/6 dB, which generally, from a practical point of view, illustrates good agreement.

	Mean (dB)	Standard deviation (dB)
5.8 GHz	5.35	6.86

Table 1. Mean and standard deviation (in dB) of the error between simulations and measurements in the Tunnel of Roux – 5.8 GHz

Second step consists of an analysis of fast fading. The data are extracted by the smoothing procedure as above. The Cumulative Density Functions (CDF) are computed from measured and simulated data and the Kolmogorov-Smirnov (KS) criterion is applied. Comparisons with theoretical distributions of Rayleigh, Rice, Nakagami and Weibull are performed. Results are given in Table 2. It appears that the Weibull distribution better minimizes the KS criterion, for both measurements and simulations. This study is very useful to reproduce statistically fast fading variations in order to realize studies on system performance.

	Measurements	Simulations
Rayleigh	0.22	0.24
Rice	0.22	0.24
Nakagami	0.05	0.06
Weibull	**0.03**	**0.02**

Table 2. KS criteria of the simulations and measurements compared to theoretical distributions in the Tunnel of Roux – 5.8 GHz

All the analysis carried out according to measurement and simulation results leads us to a quite good matching between measurements and simulations in terms of slow and fast fading. The developed method provides good performances in a straight arch-shaped tunnel.

Furthermore, it has to be noticed that results presented in this part do not depend on the location in the tunnel, while the cross section is similar. This kind of results can be found in literature (Lienard et al., 2007). On the other hand, position of antennas in the cross section can have a big impact on the received power.

5.2. Performance analysis in curved tunnels

5.2.1. Curved rectangular tunnel

5.2.1.1. Configuration

Measurements were performed in a curved rectangular tunnel by an ALSTOM-TIS team. The measurement procedure is as follows. The transmitter is located on a side near the tunnel wall. It is connected to the radio modem delivering a signal at the required frequency. Two receivers, separated by almost 3 m, are placed on the train roof. They are connected to a radio modem placed in the train. Tools developed by ALSTOM-TIS allow us to carry out field measurements and to take into account a simple spatial diversity by keeping the maximum level of the two receivers. Measurements were performed at 5.8 GHz. The measurements configuration is depicted in Figure 12. The curved rectangular tunnel has a curvature radius equal to 299 m, a width of 8 m and a height of 5 m.

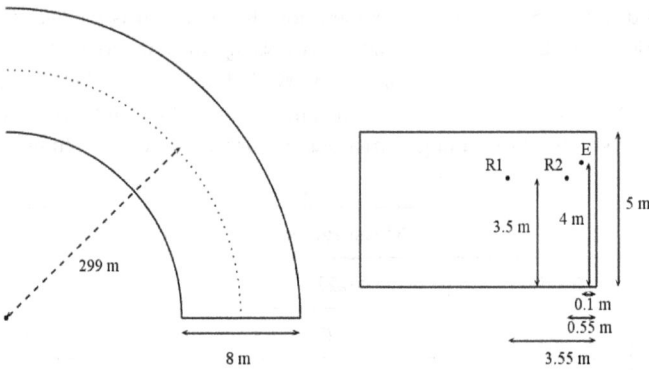

Figure 12. Antennas configuration of measurements in the curved rectangular tunnel

5.2.1.2. Comparison between simulations and measurements

Figure 13 presents the comparison between measurements and simulations in the configuration of Figure 12, in terms of normalized received power (P_r) with respect to the transmitted one (P_t), versus the distance between transmitter and receiver along the longitudinal tunnel axis in the curved rectangular tunnel.

One more time, a quite good concordance between simulations and measurements is highlighted. An analysis is presented below.

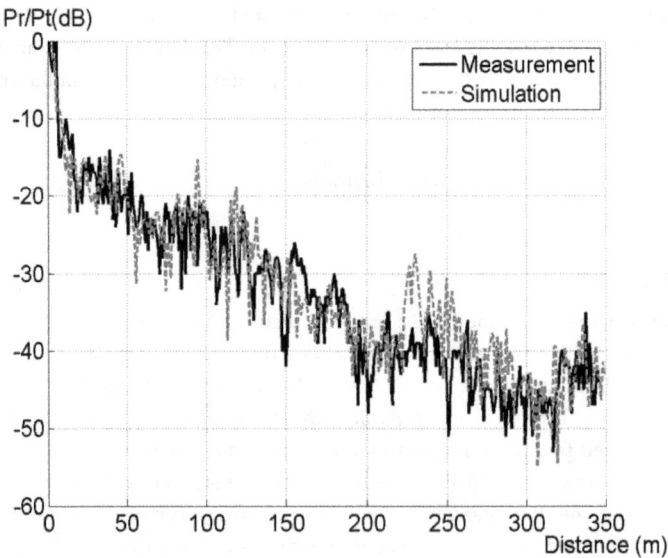

Figure 13. Comparison between measurements and simulations in the curved rectangular tunnel – 5.8 GHz

5.2.1.3. Statistical analysis of results

For these measurements, analysis of fast fading was not possible because of the lake of measured data (spatial sampling greater than $\lambda/2$). We only present a slow fading analysis, as it was performed in the section 5.1.3. Mean and standard deviation values of the error between simulations and measurements are presented in Table 3. It can be observed values around 2 dB which represents a quite good concordance between measurements and simulations in the case of the curved rectangular tunnel.

	Mean (dB)	Standard deviation (dB)
5.8 GHz	2.15	2.55

Table 3. Mean and standard deviation (in dB) of the error between simulations and measurements in the curved rectangular tunnel – 5.8 GHz

5.2.2. Curved circular tunnel

5.2.2.1. Configuration

Measurements were finally performed in a curved circular tunnel. As for the curved rectangular tunnel, the measurements were also realized by an ALSTOM-TIS team. The same measurement procedure was followed. Measurements were performed at 5.8 GHz. The measurements configuration is depicted in Figure 14. The curved circular tunnel has a curvature radius equal to 1000 m and a radius of 2.6 m.

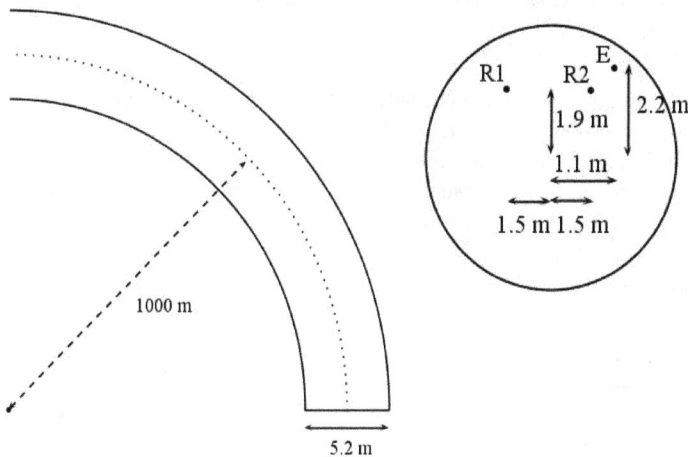

Figure 14. Antennas configuration of measurements in the curved circular tunnel

5.2.2.2. Comparison between simulations and measurements

Figure 15 presents the comparison between measurements and simulations in terms of normalized received power (P_r) with respect to the transmitted one (P_t), versus the distance between transmitter and receiver along the longitudinal tunnel axis in the curved circular tunnel.

It appears a relative good concordance between simulations and measurements. We analyze these results in the following section.

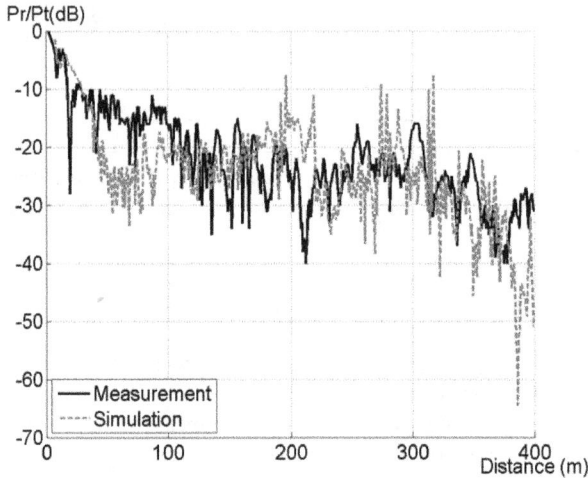

Figure 15. Comparison between measurements and simulations in the curved circular tunnel – 5.8 GHz

5.2.2.3. Statistical analysis of results

As for the curved rectangular tunnel, analysis of fast fading was not performed in this case. Mean and standard deviation values of the error between simulations and measurements are presented in Table 4. It can be observed values around 6 dB which represents a relative good concordance between measurements and simulations in the case of the curved rectangular tunnel. However, some different behaviors can be observed on few areas, such as between 50 and 100 m or 150 and 200 m. These differences are due to problems that can remain in the correction algorithm and lead to a bad estimation of the received power.

	Mean (dB)	Standard deviation (dB)
5.8 GHz	5.35	6.80

Table 4. Mean and standard deviation (in dB) of the error between simulations and measurements in the curved circular tunnel – 5.8 GHz

A global remark can be added for the case of curved tunnels (with rectangular or circular cross section). As for the case of results presented in section 5.1, results are not dependent of the location in the tunnel while the cross section is similar and also while the curvature radius is the same. In this case also, position of antennas in the cross section can have a big impact on the received power.

6. Analysis and discussion

Modeling of radio wave propagation in non rectangular and/or curved tunnels was presented. A first part is dedicated to the presentation of the requirements of wireless telecommunication systems for transport applications. These systems lead to the need for industrial of models to predict radio propagation in particular environments such as tunnels. As presented in details in (Masson, 2010), no such tools are marketed in transport domain and specific tools have to be developed for transport applications in tunnels.

Second part presents the methods to model radio wave propagation in tunnels. Three kinds of models are presented. Methods based on the modal theory provide all the modes propagating in tunnel. These methods provide good results but are limited to canonical geometries, such as rectangular and circular straight tunnels. Some papers are based on the exact resolution of Maxwell's equations. These kinds of techniques are also limited due to the computational complexity. Finally, frequency asymptotic methods based on the ray concept leads to the best compromise between accuracy and computation time. However, tunnels in real environments can have non-rectangular cross section and can also be curved. Only a few studies deal with these cases.

The developed method is presented in the third part where each step of the method is detailed. This one is based on a Ray Launching technique combined with an optimized IMR technique based on a correction of paths trajectories. It consists in a minimization of the total distance of the considered path in order to make it converge to the real one, existing in Geometrical Optics meaning. The technique is based on the well-known Levenberg-Marquardt algorithm. A validation of computed paths after correction is realized by verifying the angles of incidence and reflection. A new IMR algorithm is also developed leading on comparison of reflection points.

Last part is devoted to the evaluation of the method in three different geometries of tunnels: the straight arch-shaped tunnel, the curved rectangular tunnel and the curved circular tunnel. This evaluation is performed from measurements realized in real environments, at 5.8 GHz. For the case of the straight arch-shaped tunnel, a slow fading study was realized and lead to a mean and standard deviation between measurements and simulations of about 5-6 dB, which represents a quite good agreement between simulations and measurements. A fast fading study was also performed. The data were extracted by a smoothing procedure. The CDF were computed from measured and simulated data and the KS criterion was applied. Comparisons with theoretical distributions of Rayleigh, Rice, Nakagami and Weibull were performed. It appeared that the Weibull distribution better minimizes the KS criterion, for both measurements and simulations. This kind of study is very useful to reproduce statistically fast fading variations in order to realize studies on system performance. For the case of curved rectangular and circular tunnels, similar studies on slow fading were realized and lead to a mean and standard deviation between measurements and simulations of about 3-6 dB, which represents a good agreement of results, validating the method.

There exist many perspectives to this work. As it was seen in the previous section, in some areas, bad estimation of the received power is obtained in the case of a curved circular tunnel. These errors can be due to imperfection of correction algorithm that could be improved to obtain reliable results.

Right now, the method is able to treat empty tunnels. The following step will consist in taking into account the presence of trains in the tunnel. The phenomenon of masking train will also be investigated. In this case, diffraction has to be taken into account in the Ray Launching technique. The diffraction phenomenon would also allow taking into account complex geometries such as station, crossing, *etc.*

7. Conclusion

The chapter was dedicated to the radio wave propagation modeling in non rectangular and/or curved tunnels. It was shown that classical methods (modal theory, rigorous methods, asymptotic methods) to model propagation in these kinds of tunnels can not be used. Specific treatments on Ray Launching, based on a correction of paths trajectories, have to be implemented. They were presented, evaluated and discussed.

Author details

Émilie Masson, Yann Cocheril and Marion Berbineau
Univ Lille Nord de France, F-59000, Lille, IFSTTAR, LEOST, France

Pierre Combeau, Lilian Aveneau and Rodolphe Vauzelle
XLIM-SIC laboratory, UMR CNRS 7252, University of Poitiers, France

Acknowledgement

The authors would like to thank the ALSTOM-TIS (Transport Information Solution) who supported this work.

8. References

Agunaou, M.; Belattar, S., Mariage, P. & Degauque, P. Propagation d'ondes électromagnétiques hyperfréquences à l'intérieur d'un métro - Modélisation numérique de l'influence des changements de section. *Recherche Transports Sécurité*, Vol.64, (1998), pp.55-68.

Ahmed, B. T.; Ramon, M. C. & Ariet, L. H. Comments on the Hybrid Model for Propagation Loss Prediction in Tunnels. *IEEE Antennas and Propagation Magazine*, Vol.45, No.5, (2008), pp.156-158.

Baranowski, S.; Bourdier, G. & Degauque, P. Optimisation des règles d'ingénierie radio-modélisation de la propagation d'ondes radioélectriques en tunnels courbes. *Convention* d'études INRETS/USTL 1997/L1, Laboratoire de Radio propagation et Electronique, Villeneuve d'Ascq, 1998.

Bernardi, P., Caratelli, D., Cicchetti, R., Schena, V. & Testa, O. A Numerical Scheme for the Solution of the Vector Parabolic Equation Governing the Radio Wave Propagation in Straight and Curved Rectangular Tunnels. *IEEE Transactions on Antennas and Propagation*, Vol.57, No.10, (2009), pp.3249-3257.

Chang, H. W.; Wu, Y. H., Lu, S. M., Cheng, W. C. & Sheng, M. H. Field Analysis of Dielectric Waveguide Devices based on Coupled Transverse-mode Integral Equation-Numerical Investigation. *Progress In Electromagnetics Research PIER 97*, (2009), pp.159-176.

Chen, S. H. & Jeng, S. K. SBR image approach for radio wave propagation in tunnels with and without traffic. *IEEE Transactions on Antennas and Propagation*, Vol.45, No.3, (1996), pp.570-578.

Didascalou, D.; Schafer, T. M., Weinmann, F. & Wiesbeck, W. Ray-density normalization for ray-optical wave propagation modeling in arbitrarily shaped tunnels. *IEEE Transactions on Antennas and Propagation*, Vol.48, No.9, (2000), pp.1316-1325.

Didascalou, D., Maurer, J. & Wiesbeck, W. Subway tunnel guided electromagnetic wave propagation at mobile communications frequencies. *IEEE Transactions on Antennas and Propagation*, Vol.49, No.11, (2001), pp. 590-1596.

Dudley, D. G. & Mahmoud, S. F. Linear Source in a Circular Tunnel. *IEEE Transactions on Antennas and Propagation*, Vol.54, No.7, (2006), pp.2034-2047.

Dudley, D. G; Lienard, M., Mahmoud, S. F. & Degauque, P. Wireless propagation in tunnels. *IEEE Antennas and Propagation Magazine*, Vol.49, No.2, (2007), pp.11-26.

Emslie, A.; Lagace, R. & Strong, P. Theory of the propagation of UHF radio waves in coal mine tunnels. *IEEE Transactions on Antennas and Propagation*, Vol.23, No.2, (1975), pp. 192-205.

Halton, J. H. On the efficiency of certain quasi-Monte-Carlo sequences of points in evaluating multi-dimensional integrals. *Nummer. Math.*,Vol.2, (1960), pp.84-90.

Hammersley, J. M. Monte Carlo methods, (1964).

Hwang, Y.; Zhang, Y. P. & Kouyoumjian, R. G. Ray-optical prediction of radio-wave propagation characteristics in tunnel environments. 1. Theory. *IEEE Transactions on Antennas and Propagation*, Vol.46, No.9, (1998), pp.1328-1336.

Iskander, M. F. & Yun, Z. Propagation prediction models for wireless communication systems. IEEE *Transaction on Microwave Theory and Techniques*, Vol.50, No.3, (2002), pp.662-673.

Laakmann, K. D. & Steier, W. H. Waveguides: characteristic modes of hollow rectangular dielectric waveguides. *Appl. Opt.*, Vol.15, No.5, (1976), pp.1334-1340.

Lienard, M. & Degauque, P. Propagation in wide tunnels at 2 GHz: a statistical analysis. *IEEE Transactions on Vehicular Technology*, Vol.47, No.4, (1998), pp.1322-1328.

Lienard, M. Nasr A., Garcia Pardo J. M., Degaugue P. Experimental analysis of wave depolarization in arched tunnels. *The 18th Annual IEEE International Symposium on Personal, Indoor and Mobile Radio Communications (PIMRC'07)*, 2007.

Mahmoud S. Characteristics of Electromagnetic Guided Waves for Communication in Coal Mine Tunnels. *IEEE Trans. Commun. Syst.*, Vol.22, No.10, (1974), pp.1547-1554.

Mahmoud, S. F. Characteristics of Electromagnetic Guided Waves for Communication in Coal Mine Tunnels. *IEEE Trans. Commun. Syst.*, Vol.22, No.10, (1974), pp.1547-1554.

Mariage, P.; Lienard, M. & Degauque, P. Theoretical and experimental approach of the propagation of high frequency waves in road tunnels. *IEEE Transactions on Antennas and Propagation*, Vol.42, No.1, (1994), pp.75-81.

Marquardt, D. W. An algorithm for Least-Squares Estimation of Nonlinear Parameters. *J. Soc. Indust. Appl. Math.*, Vol.11, No.2, (1963), pp.431-441.

Masson, E. Etude de la propagation des ondes électromagnétiques dans les tunnels courbes de section non droite pour des applications métro et ferroviaire. PhD thesis of the University of Poitiers, (2010).

Masson, E.; Combeau, P., Cocheril, Y., Berbineau, M., Aveneau, L. & Vauzelle, R. Radio Wave Propagation in Arch-Shaped Tunnels: Measurements and Simulations using Asymptotic Methods. *Comptes Rendus - Physique*, Vol.11, (2010), pp.44-53.

Morokoff, W. J. & Caflisch, R. E. Quasi-monte carlo integration. *Journal of Computational Physics*, Vol.122, No.2, (1995), pp. 218–230.

Popov, A. V. Modeling radio wave propagation in tunnels with a vectorial parabolic equation. *IEEE Transactions on Antennas and Propagation*, Vol.48, No.9, (2000), pp.1403-1412.

Reutskiy, S. The methods of external excitation for analysis of arbitrarily-shaped hollow conducting waveguide. *Progress In Electromagnetics Research PIER 82*, (2008), pp.203-226.

Seidl, S. Y & Rappaport, T. S. Site-specific propagation prediction for wireless in-building personal communication system design. *IEEE Transactions on Vehicular Technology*, Vol.43, No.3, (1994), pp.879-891.

Torres, R. P.; Valle, L., Domingo, M., Loredo, S. & Diez, M. C. CINDOOR: An engineering tool for planning and design of wireless systems in enclosed spaces. *IEEE Antennas and Propagation Magazine*, Vol.41, No.4, (1999).

Wang, T. S. & Yang, C. F. Simulations and Measurements of Wave Propagations in Curved Road Tunnels for Signals From GSM Base Stations. *IEEE Transactions on Antennas and Propagation*, Vol.54, No.9, (2006), pp.2577-2584.

Zhang, Y. P. & Hwang, Y. Theory of the radio-wave propagation in railway tunnels. *IEEE Transactions on Vehicular Technology*, Vol.47, No.3, (1998), pp.1027-1036.

Zhang, Y. P.; Hwang, Y. & Kouyoumjian, R. G. Ray-optical prediction of radio-wave propagation characteristics in tunnel environments. 2. Analysis and measurements. *IEEE Transactions on Antennas and Propagation*, Vol.46, No.9, (1998), pp.1337-1345.

Zhang, Y. P. & Hwang, Y. Characterization of UHF radio propagation channels in tunnel environments for microcellular and personal communications. *IEEE Transactions on Vehicular Technology*, Vol.47, No.1, (1998), pp.283-296.

Zhang, Y. P. Novel model for propagation loss prediction in tunnels. *IEEE Transactions on Vehicular Technology*, Vol.52, No.5, (2003), pp.1308-1314.

Zhang, Y. P. & Hong, H. J. Ray-Optical Modeling of Simulcast Radio Propagation Channels in Tunnels. *IEEE Transactions on Vehicular Technology*, Vol.53, No.6, (2004), pp.1800-1808.

Chapter 13

A Numerical Model Based on Navier-Stokes Equations to Simulate Water Wave Propagation with Wave-Structure Interaction

Paulo Roberto de Freitas Teixeira

Additional information is available at the end of the chapter

1. Introduction

The interaction between waves and structures is a very important subject of the coastal engineering. Many numerical models have been developed over recent decades to analyze these types of cases, which involve phenomena that combine reflection, shoaling, refraction, diffraction, breaking and wave-wave interaction. These non-linear effects provide harmonic generation, including energy transferences with high complexity.

Models based on the Laplace equation assume the potential flow, in which the movement is irrotational and the flow is incompressible. Models based on the boundary element technique [1,2,3] and spectral methods [4,5] are some examples. This theory applies neither to viscous flows nor to situations in which there are flow separations, vortex generations and turbulences.

Other models, called depth-integrated models [6], based on a Boussinesq-type equation for variable depth, consider polynomial approximations for the vertical velocity distribution and vertical integration in the resulting equations at a certain depth. The simplified hypotheses, that include slight non linearity and dispersion, limit the applicability of this type of models to shallow and intermediate waters. Several researches have been developed to extend the applicability of these equations, including high order terms, to deeper water and strong non linearity cases in the last decades. Wave propagation phenomena, such as breaking, bottom friction and run-up, have also been included in these extended Boussinesq equation [7,8,9,10,11,12,13,14]. The accuracy of these type of models has recently been improved by the implementation of the multi-layer concept, in which the water column is divided into layers and a velocity profile is adopted to each one [15,16]. Although the accuracy has been improved significantly, the simplified hypotheses, related to the vertical integration along each layer, limit the use of these models to depths without strong variations.

Many efforts have been carried out to develop non hydrostatic models to prevent the difficulties of the Boussinesq models [17]. The non hydrostatic models capture the free surface movement using a function on the horizontal plane, requiring lower vertical discretization by comparison with those that use classic methods to describe the free surface. In some of these models, the pressure and velocity fields are decomposed by hydrostatic and non-hydrostatic pressure to improve their efficiency.

The numerical solution of the fully Navier-Stokes equations to determine the tridimensional velocity and pressure fields and the free surface position demands high computational cost, due to the large horizontal scale of many coastal engineering problems. However, in cases in which there are flow separation, vortex shedding and turbulence, these models provide more real results. There are several methods to capture the free surface movements, such as the arbitrary lagrangian eulerian (ALE) [22,46], the marked and cell [25], the volume of fluid [26,27] and the level-set methods [28].

This text describes a code (in Fortran 90 language) that integrates the Navier-Stokes equations using a fractional method to simulate 3D incompressible flow problems with free surface, named FLUINCO [29]. The model employs a semi implicit two-step Taylor Galerkin method to discretize the Navier-Stokes equations in time and space; uses the ALE method and a mesh velocity distribution technique to deal with free surface movement.

To show the applicability of the code, two study cases are analyzed: the wave propagation over a submerged horizontal cylinder and submerged trapezoidal breakwaters. In Section 2, the numerical model is described. Section 3 presents study cases, their results and discussion. Finally, Section 4 concludes the analysis.

2. Numerical model

2.1. Governing equations

The algorithm is based on the continuity equation, given by:

$$\frac{\partial \rho}{\partial t} = \frac{1}{c^2}\frac{\partial p}{\partial t} = -\frac{\partial U_i}{\partial x_i} \quad (i=1,2,3), \tag{1}$$

and momentum equations, that are represented by the following equations according to the ALE formulation:

$$\frac{\partial U_i}{\partial t} = -\frac{\partial f_{ij}}{\partial x_j} + \frac{\partial \tau_{ij}}{\partial x_j} - \frac{\partial p}{\partial x_i} + w_j\frac{\partial U_i}{\partial x_j} \quad (i,j=1,2,3), \tag{2}$$

where ρ is the specific mass, p is the pressure, $U_i = \rho v_i$, $f_{ij} = v_j(\rho v_i) = v_j U_i$, v_i are the velocity components, w_i the reference system velocity components and τ_{ij} is the viscous stress tensor $(i,j=1,2,3)$.

2.2. Semi-implicit two-step Taylor-Galerkin method

Basically, the algorithm consists of the following steps [30]:

a. Calculate non-corrected velocity at $\Delta t/2$, obtained by time discretization of Eq. (2), where the pressure term is at t instant, according to Eq. (3).

$$\tilde{u}_i^{n+1/2} = u_i^n - \frac{\Delta t}{2} \left(\frac{\partial f_{ij}^n}{\partial x_j} - \frac{\partial \tau_{ij}^n}{\partial x_j} + \frac{\partial p^n}{\partial x_i} - \rho g_i - w_j^n \frac{\partial u_i^n}{\partial x_i} \right) \quad (i,j = 1,2,3), \tag{3}$$

where g_i are the gravity acceleration components.

b. Update the pressure p at $t+\Delta t$, obtained by time discretization of Eq. (1), given by the Poisson equation:

$$\frac{1}{c^2} \Delta p = -\Delta t \left[\frac{\partial \tilde{u}_i^{n+1/2}}{\partial x_i} - \frac{\Delta t}{4} \frac{\partial}{\partial x_i} \frac{\partial \Delta p}{\partial x_i} \right], \tag{4}$$

where $\Delta p = p^{n+1} - p^n$ and $i = 1,2,3$.

c. Correct the velocity at $t+\Delta t/2$, adding the pressure variation term from t to $t+\Delta t/2$, according to the equation:

$$u_i^{n+1/2} = \tilde{u}_i^{n+1/2} - \frac{\Delta t}{4} \frac{\partial \Delta p}{\partial x_i} \quad (i = 1,2,3), \tag{5}$$

d. Calculate the velocity at $t+\Delta t$ using variables updated in the previous steps as follows:

$$u_i^{n+1} = u_i^n - \Delta t \left(\frac{\partial f_{ij}^{n+1/2}}{\partial x_j} - \frac{\partial \tau_{ij}^{n+1/2}}{\partial x_j} + \frac{\partial p^{n+1/2}}{\partial x_i} - w_j^{n+1/2} \frac{\partial u_i^{n+1/2}}{\partial x_i} - \rho g_i \right) \quad (i,j = 1,2,3), \tag{6}$$

2.3. Space discretization

The classical Galerkin weighted residual method is applied to the space discretization by using a tetrahedron element. In the variables at $t+\Delta t/2$ instant, a constant shape function **PE** is used, and in the variables at t and $t+\Delta t$, a linear shape function **N** is employed. By applying this procedure to Eq. (3), (4), (5) and (6), the following expressions in the matrix form are obtained [29]:

$$\Omega_E^{n+1/2} \tilde{\mathbf{U}}_i^{n+1/2} = \mathbf{C} \, \bar{\mathbf{U}}_i^n - \frac{\Delta t}{2} \left[\mathbf{L}_j \left(\bar{\mathbf{f}}_{ij}^n - \bar{\tau}_{ij}^n \right) + \mathbf{L}_i \bar{\mathbf{P}}^n - \mathbf{T} \, \bar{\mathbf{U}}_i^n - \Omega_E^{n+1/2} \overline{\rho g}_i \right] \tag{7}$$

$$\left(\tilde{\mathbf{M}} + \frac{\Delta t^2}{4} \mathbf{H} \right) \Delta \bar{\mathbf{p}} = \Delta t \left(\mathbf{L}_i^T \tilde{\mathbf{U}}_i^{n+1/2} + \mathbf{f}_a \right) \tag{8}$$

$$\bar{\mathbf{U}}_i^{n+1/2} = \tilde{\bar{\mathbf{U}}}_i^{n+1/2} - \frac{\Delta t}{4\Omega_E} \mathbf{L}_i \Delta \bar{\mathbf{p}} \tag{9}$$

$$\mathbf{M}^{n+1} \bar{\mathbf{U}}_i^{n+1} = \mathbf{M}^n \bar{\mathbf{U}}_i^n + \Delta t \mathbf{L}_j^T \left(\bar{\mathbf{f}}_{ij}^{n+1/2} - \bar{\mathbf{w}}_j^{n+1/2} \bar{\mathbf{U}}_i^{n+1/2} \right)$$

$$-\Delta t \mathbf{Q}_j \bar{\tau}_{ij}^n + \Delta t \mathbf{Q}_i \left(\bar{\mathbf{p}}^n + \Delta \bar{\mathbf{p}}/2 \right) + \Delta t \mathbf{S}_{bi} - \Delta t \mathbf{C}^T \bar{\mathbf{g}}_i \tag{10}$$

where variables with upper bars at n and $n+1$ instants indicate nodal values, while those at $n+1/2$ instant represent constant values in the element. The matrices and vectors from Eq. (7) to (10) are volume and surface integrals that can be seen in detail in [30].

Equation (8) is solved using the conjugated gradient method with diagonal pre-conditioning [31]. In Eq. (10), the consistent mass matrix is substituted by the lumped mass matrix, and then this equation is solved iteratively.

The scheme is conditionally stable and the local stability condition for the element E is given by

$$\Delta t_E \leq \beta h_E / |\mathbf{u}| \tag{11}$$

where h_E is the characteristic element size, β is the safety factor and $\mathbf{u}$ is the fluid velocity.

2.4. Mesh movement

The free surface is the interface between two fluids, water and air, where atmospheric pressure is considered constant (generally the reference value is null). In this interface, the kinematic free surface boundary condition (KFSBC) is imposed. By using the ALE formulation, it is expressed as:

$$\frac{\partial \eta}{\partial t} + {}^{(s)}v_i \frac{\partial \eta}{\partial x_i} = {}^{(s)}v_3 \tag{12}$$

where η is the free surface elevation, ${}^{(s)}v_3$ is the vertical fluid velocity component and ${}^{(s)}v_i$ ($i=1,2$) are the horizontal fluid velocity components in the free surface. The eulerian formulation is used in the x_1 and x_2 directions (horizontal plane) while the ALE formulation is employed in the x_3 or vertical direction.

The time discretization of KFSBC is carried out in the same way as the one for the momentum equations as presented before. After applying expansion in Taylor series, the expressions for η at $n+1/2$ (first step) and $n+1$ (second step) instants are obtained:

$$\eta^{n+1/2} = \eta^n + \frac{\Delta t}{2} \left({}^{(s)}v_3 - {}^{(s)}v_1 \frac{\partial \eta}{\partial x_1} - {}^{(s)}v_2 \frac{\partial \eta}{\partial x_2} \right)^n \tag{13}$$

$$\eta^{n+1} = \eta^n + \Delta t \left({}^{(s)}v_3 - {}^{(s)}v_1 \frac{\partial \eta}{\partial x_1} - {}^{(s)}v_2 \frac{\partial \eta}{\partial x_2} \right)^{n+1/2} \tag{14}$$

Linear triangular elements coincident with the face of the tetrahedral elements on the free surface are used to the space discretization by applying the Galerkin method.

The mesh velocity vertical component w_3 is computed to diminish element distortions, keeping prescribed velocities on moving (free surface) and stationary (bottom) boundary surfaces. The mesh movement algorithm adopted in this paper uses a smoothing procedure for the velocities based on these boundary surfaces. The updating of the mesh velocity at point i of the finite element domain is based on the mesh velocity of the points j that belong to the boundary surfaces, and is expressed in the following way [32]:

$$w_3^i = \frac{\sum\limits_{j=1}^{ns} a_{ij} w_3^j}{\sum\limits_{j=1}^{ns} a_{ij}} \tag{15}$$

where ns is the total number of points belonging to the boundary surfaces and a_{ij} are the influence coefficients between the point i inside the domain and the point j on the boundary surface given by the following expression:

$$a_{ij} = \frac{1}{d_{ij}^4} \tag{16}$$

with d_{ij} being the distance between points i and j. In other words, a_{ij} represents the weight that every point j on the boundary surface has on the value of the mesh velocity at points i inside the domain. When d_{ij} is low, a_{ij} has a high value, favouring the influence of points i, located closer to the boundary surface containing point j.

The free surface elevation, the mesh velocity and the vertical coordinate are updated according to the following steps:

1. Calculate $\eta^{n+1/2}$ and $\tilde{U}_i^{n+1/2}$, Eq. (13) and Eq. (3), respectively.
2. Calculate Δp, Eq. (4).
3. Calculate $U_i^{n+1/2}$, Eq. (3).
4. Calculate U_i^{n+1}, Eq. (6).
5. Calculate η^{n+1}, Eq. (14).
6. Update the mesh velocity w_3 and the vertical coordinate x_3:
    *   Calculate the mesh velocity in the free surface at $t + \Delta t$: ${}^{(S)}w_3^{n+1} = \left(\eta^{n+1} - \eta^n \right) / \Delta t$.
    *   Calculate the mesh velocity in the interior of the domain at $n+1$ e $n+1/2$ by using Eq.

    (13) and $w_3^{n+1/2} = \dfrac{(w_3^{n+1} + w_3^n)}{2}$, respectively.

- Update the vertical coordinates in the interior of the domain:
  $$x_3^{n+1/2} = x_3^n + w_3^n \frac{\Delta t}{2}, \; x_3^{n+1} = x_3^n + w_3^{n+1/2} \Delta t.$$

2.5. Wave generation and radiation conditions

The wave generation is considered imposing the free surface elevation and the fluid velocity components to each time step directly, considering the linear wave equations [33].

The Flather's radiation condition [34] is used to deal with open boundaries. In this method, the Sommerfeld condition to free surface elevation is combined with one-dimensional version of the continuity equation. Then, the normal velocity of the boundary can be expressed by:

$$u = \eta \sqrt{\frac{g}{h}} \, , \tag{17}$$

where g is the gravity acceleration and h is the depth.

3. Study cases

3.1. Submerged cylinder

The interaction among regular waves and submerged circular cylinders, with their axes parallel to the crests of the incident waves, has been studied analytically, experimentally and numerically by many authors. The presence of an obstacle near the free surface may cause reflected and modified transmitted waves. These phenomena depend on the characteristics of the incident wave, the obstacle geometry and the depth. Many studies of this interaction are available to provide a good example to validate numerical codes.

The first study was developed by [35] and, after that, by [36]. Considering a linear behavior, these authors showed that (a) the cylinder does not reflect any energy, regardless of its ray, depth or wave frequency; (b) the transmitted waves are out of phase, but their amplitudes are the same. Chaplin [37] studied the nonlinear forces and characteristics of the reflected and transmitted waves experimentally. He showed that the reflection is negligible up to the third order. This author and Schonberg and Chaplin [38] presented many experimental and numerical studies concerning the nonlinear interaction among waves and submerged cylinders. A detailed review of analyses for this case can be found in Paixão Conde et al. [39].

This case considers a 5.2 m long and 0.425 m deep channel with a submerged cylinder of r = 0.025m positioned 1.60 m from the wave generator (Figure 1). The cylinder center is 0.075 m (3r) from the free surface. The frequency wave is f = 1.4Hz; its amplitude is a = 0.0119 m and its wavelength is L=0.796 m, characterizing a deep water case.

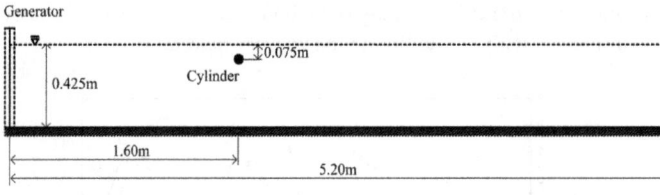

Figure 1. Geometry of the horizontal cylinder case.

Table 1 shows the period, the frequency and the wavelength for the fundamental frequency and its 2nd, 3rd and 4th harmonics, according to the linear wave theory.

	Fundamental	2nd harmonic	3rd harmonic	4th harmonic
T (s)	0.7143	0.3571	0.2381	0.1786
f(Hz)	1.4	2.8	4.2	5.6
L(m)	0.796	0.199	0.0885	0.0498

Table 1. Period, frequency and wavelength of the fundamental frequency and its 2nd, 3rd and 4th harmonics in the horizontal cylinder case.

The mesh, with 173900 nodes and 515623 elements, has one layer of elements in the transversal direction. The average element size on the cylinder boundary is 0.0015 m (105 divisions in the circumference). The element size diminishes from the ends to the region near the cylinder and from the bottom to the free surface. The element sizes on the end where the wave generator is located and on the opposite end are 0.015 m (53 points per fundamental wavelength) and 0.02 m (40 points per fundamental wavelength), respectively. On the bottom, 0.0015m is also used.

The initial conditions are: null velocity components in all domain and hydrostatic pressure (null on the free surface). The wave is generated by imposing the surface elevation and the velocity components. The non-slip condition is imposed to the bottom and to the cylinder wall. The time step is 0.0002s, which satisfies the Courant condition.

Figure 2 shows the free surface elevation obtained by the code and experimental tests, where xc is the horizontal coordinate of the cylinder center. In general, there is agreement between numerical and experimental results [39]. We can notice the free surface disturbance downstream the cylinder. When $(x\text{-}xc)/L$ is above 1.7, the numerical results are smoother than the experimental ones, showing the necessity of a refinement in this region.

Figure 3 shows a comparison among numerical and experimental results in terms of free surface elevation on four gauges located at $(x\text{-}xc)/L$ equal to -0.503, 0.0692, 0.509 and 1.264 (there is only a numerical result on the first gauge). We can observe the similarity among numerical and experimental results.

Figure 4 shows the streamlines and the velocity modulus distribution at the same instant used in Figure 2. Recirculation and separation cannot be observed at downstream. Due to the oscillatory flow behavior, there is no time for recirculation productions. We can notice the flow acceleration near the cylinder due to the boundary layer effect. The viscous effects have only local influence, without modifying the velocity field far from it.

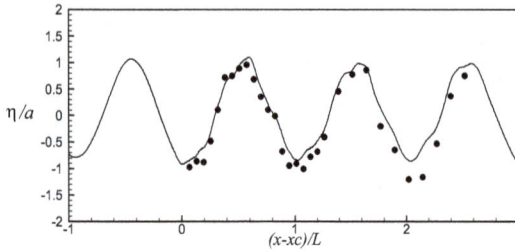

Figure 2. Free surface elevation in the submerged cylinder case (Numerical —; Experimental ■).

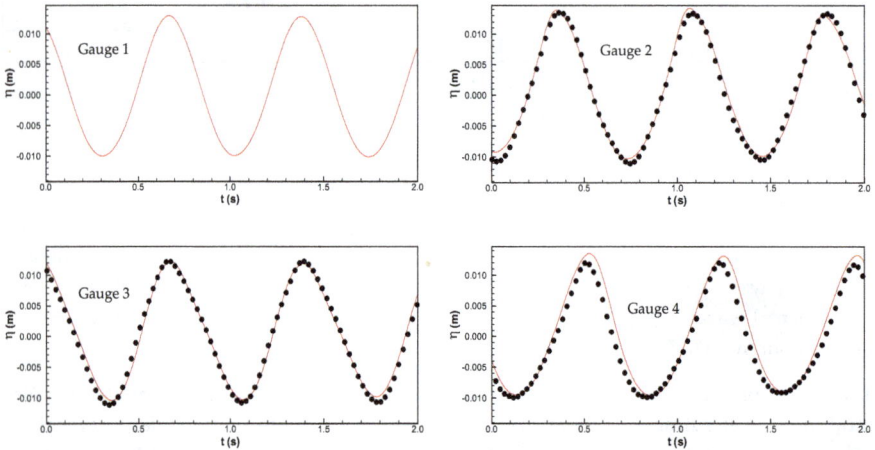

Figure 3. Free surface elevations on the gauges located at (x-xc)/L equal to -0.503, 0.0692, 0.509 and 1.264 in the horizontal cylinder case (Numerical —; Experimental ●).

In Figures 5 and 6, velocity component profiles, u and v, on the same gauge positions are presented. These profiles were constructed at the same instant as that used in Figure 2. According to the linear theory, the maximum value for both horizontal and vertical components is equal to 0.105 m/s. For horizontal components, these values occur on the crest and the trough, while for vertical ones, these values occur on upward and downward zero-crossings. When one component is the maximum, another is null, because the phase difference is 90 degrees.

Gauge 1 ((x-xc)/L = -0.503) is located upstream, near the wave crest; no significant disturbance in u and v profiles is observed. The horizontal velocity component is positive and its maximum value is similar to the theoretical value in the crest. The wave trough passes by gauge 2; the vertical velocity component presents low values and the horizontal velocity component has negative values, reaching the maximum absolute value close to the theoretical ones (0.105 m/s). Gauge 3 is located near the first crest upstream, resulting in high horizontal component values. Finally, gauge 4 is on a region between the trough and upward zero-crossing. Both component profiles are negative and the vertical component magnitude shows how close the gauge is to upward zero-crossing.

The non-slip boundary condition on the bottom does not change the general behavior of the wave propagation, because this case is considered a deep water one.

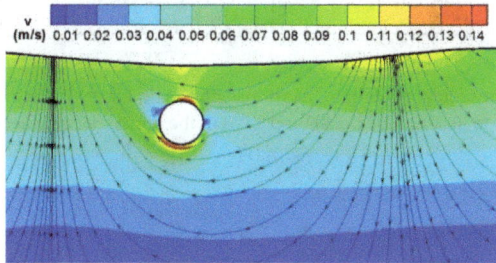

Figure 4. Streamlines and velocity modulus at the instant in which the free-surface elevation was captured (Figure 2) in the horizontal cylinder case.

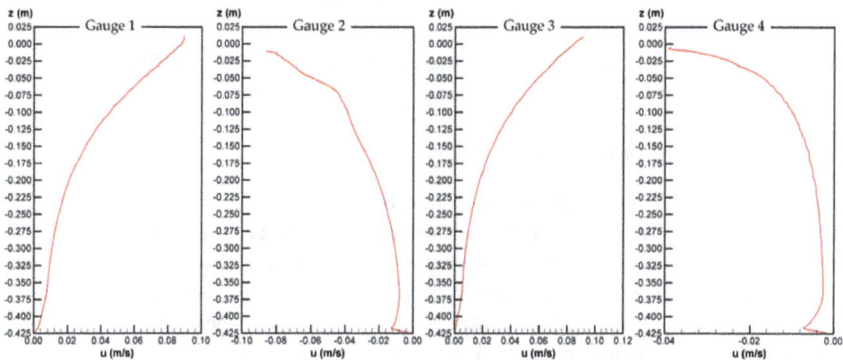

Figure 5. Horizontal velocity components at the same instant used in Figure 2 along the depth on gauges located at $(x-xc)/L$ equal to -0.503, 0.0692, 0.509 and 1.264 in the horizontal cylinder case.

Figure 7 shows the frequency spectra for these four gauges distributed along the channel. In all cases, the energy is concentrated on the fundamental frequency and its harmonic waves. On gauge 1, the fundamental frequency presents most energy and the second harmonic shows a little value. On gauges 2 and 4, located upstream, significant energy up to the third harmonic appears, similar to the experimental results.

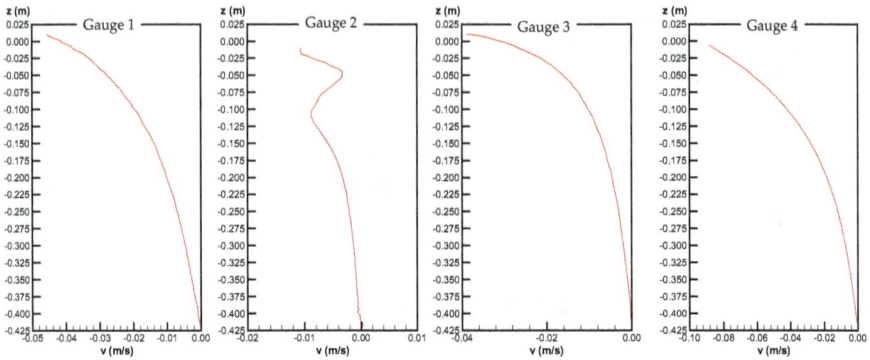

Figure 6. Vertical velocity components at the same instant used in Figure 2 along the depth on gauges located at $(x-xc)/L$ equal to -0.503, 0.0692, 0.509 and 1.264 in the horizontal cylinder case.

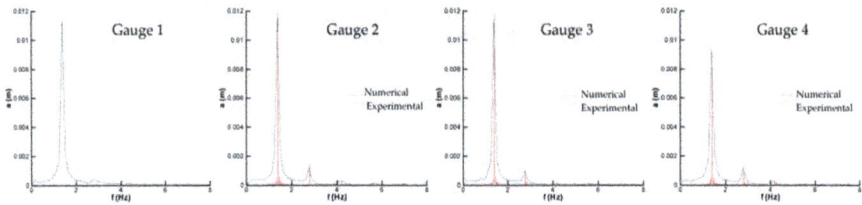

Figure 7. Frequency spectra on gauges located at $(x-xc)/L$ equal to -0.503, 0.0692, 0.509 e 1.264 in the horizontal cylinder case.

3.2. Submerged trapezoidal breakwaters

Whatever the numerical model characteristics, the simulation of wave propagation over submerged breakwaters are important tests to validate wave propagation models. In these cases, the harmonic generation [40,41] and the vortex formation, depending on the geometry [42], also occur. When waves propagate in deep waters over a submerged obstacle, part of the wave energy is transferred from the primary wave component to their harmonics, contributing to increase non linearity. Harmonic generation phenomena that occur when waves propagate over obstacles, such as natural reefs, were studied theoretically [6], experimentally [43,44,45] and numerically [46,23,17,47,48,45,49,50]. In some situations, the correct simulation of the flow can only be figured out considering the viscosity effects [51]. Huang and Dong [42] studied the interaction between solitary waves and rectangular submerged breakwaters using a model based on 2D Navier-Stokes equations and concluded that the flow around the breakwater is laminar, without turbulence. The experimental studies carried out by Ting and Kim [51] and Zhuang and Lee [52] show that velocity fluctuations do not exist around the breakwater.

Two different configurations of the trapezoidal breakwaters, with different level of non-linearity, are used to test the behaviour of the numerical model. In the first case, the downstream and upstream slopes are 1:20 and 1:10, respectively [44]. In the second one, both slopes are 1:2 [45], where the non-linear effects are more significant.

3.2.1. Breakwater with slopes 1:20 and 1:10

Figure 8 shows the channel and the submerged breakwater geometries, and the position of the gauges. The channel is 23m in length, 0.4m and 0.1m are the maximum and the minimum depths, respectively. In the channel entrance, a monochromatic wave is generated with a period of 2.02s and an amplitude of 0.01m.

Table 2 presents some parameters for this case study, in which H is the wave length, h is the depth, k=2π/L is the wave number and Ur = gHT²/h² is the Ursell number, where T is the wave period. H/h, even on the platform, has small values in comparison with breaking limit of approximately 0.8 [33]. The case involves intermediate water for the channel (0.314 < kh < 3.142) and shallow water for the platform (kh < 0.314). Ursell numbers show that the non-linear effects on the platform are more intensive.

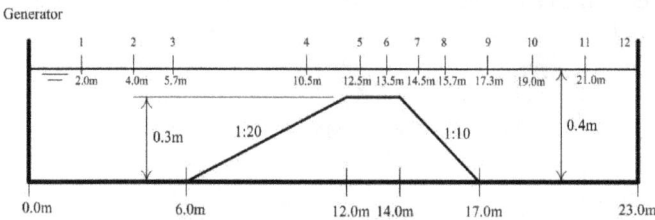

Figure 8. Channel geometry for the 1:20 and 1:10 breakwater

	H/h	kh	Ur
Channel (h = 0.4m)	0.050	0.674	5.0
Platform (h = 0.1m)	0.259	0.318	103.6

Table 2. Wave parameters for the 1:20 and 1:10 breakwater

Table 3 presents periods, frequencies and wavelengths concerning the fundamental frequency and the harmonic components that occur along the wave propagation. The wavelength was estimated according to the dispersion equation of the linear theory. These values are references to determine discretizations in time and space to be used in the modeling.

	Fundamental	2nd harmonic	3rd harmonic	4th harmonic
Period (s)	2.02	1.01	0.67	0.50
Frequency (Hz)	0.50	1.00	1.50	2.00
Wavelength (m)	3.73	1.46	0.70	0.39

Table 3. Period, frequency and wavelength concerning the fundamental frequency, and 2nd, 3rd and 4th harmonics for the 1:20 and 1:10 breakwater

FLUINCO used a mesh with 88700 elements and 37296 nodes. Twenty layers of elements were used in vertical direction, where small elements are located near the bottom and the free surface. Along the channel, the element sizes vary from $\Delta x = 0.08$m in the boundary to $\Delta x = 0.025$m around the platform. In the transversal direction, only one layer of elements is used, because the behavior of the flow is bi-dimensional. In the entrance of the domain, the wave generation condition is imposed while at the end the radiation condition is imposed. The velocity components are null on the bottom and the KFSBC is imposed in the free surface. The velocity component perpendicular to the surface is null for lateral walls (symmetry condition). As an initial condition, the velocity field is null and the pressure one is hydrostatic. The time step is 0.003s, a fact that satisfies the Courant stability condition.

Figure 9 shows the free surface elevations in gauge 3, located downstream the breakwater (x=5.7m); in gauge 6, on the platform (x=13.5m); in gauge 8, in the middle of the upstream slope (x=15.7m); and in gauge 11, on the upstream and far from the breakwater (x=23m). Results obtained by numerical model are compared with the experimental ones presented by Dingemans [44].

In general, there is good agreement between numerical results and experimental ones in gauges 3 and 6. In gauge 6, FLUINCO presents slightly smooth surface deformation. In gauges 8 and 11, corresponding to downstream, the nonlinear effects are more significant. The deformations in gauge 8 are well represented by FLUINCO; although the results get closer to the experimental ones in some regions, there are difficulties in representing the deformations related to higher harmonics, possibly due to the lack of an appropriate discretization to capture the nonlinear phenomena.

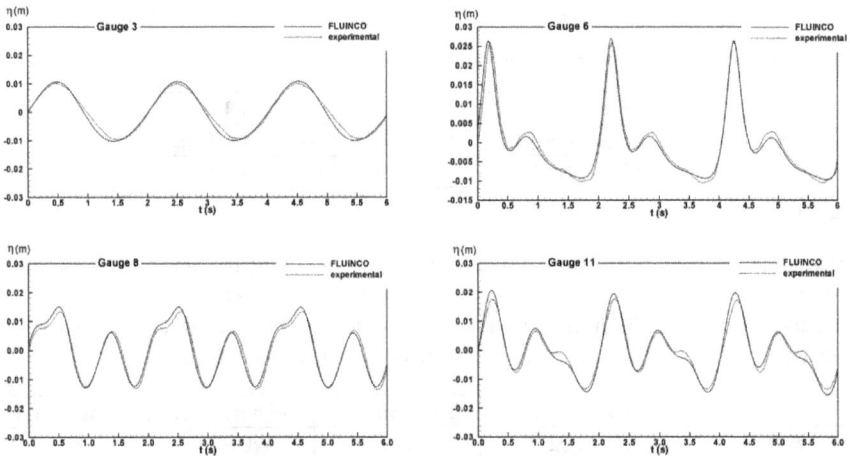

Figure 9. Free surface elevation of the 1:20 and 1:10 breakwater

Figure 10 shows the frequency spectra obtained by the model in the gauges and a comparison with the experimental results. The differences found in the free surface elevation are confirmed in Figure 10, which shows differences in the intensity of harmonic components, mainly in gauges located at the end of the channel. The numerical model adequately simulate the position of the peaks of the fundamental frequency and the harmonic components throughout the domain. However, there are some differences in the amplitude of these peaks, especially in gauges 8 and 11.

Figure 11 presents the streamlines around the upstream slope of the breakwater in eleven instants completing one wave period obtained by FLUINCO. We can observe that the flow separation and the vortex do not exist at all instants, due to the mild inclination of the upstream slope.

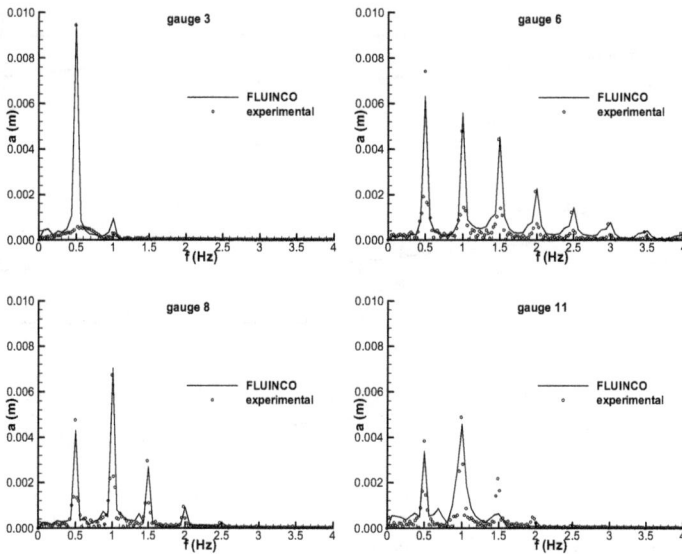

Figure 10. Numerical and experimental frequency spectrum in the gauges of the breakwater 1:20 and 1:10

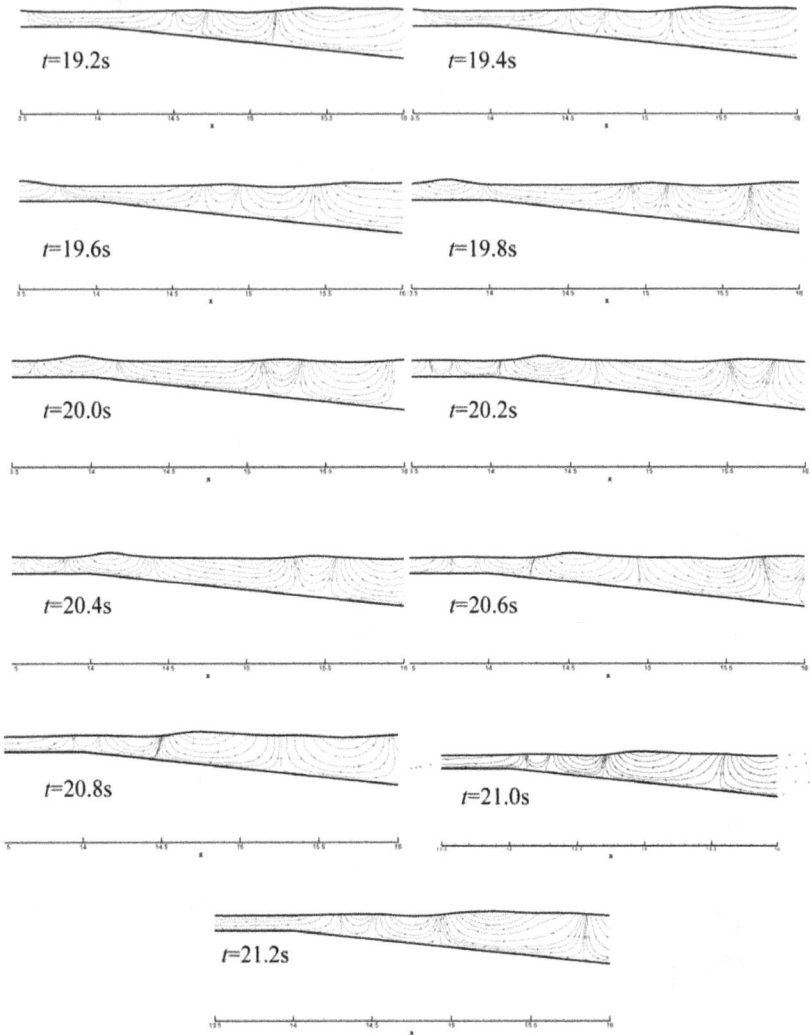

Figure 11. Streamlines of the 1:20 and 1:10 breakwater

3.2.2. Breakwater with slopes 1:2

In this case, the length of the channel is 35m and the maximum and the minimum depths are 0.5m and 0.15m, respectively (See Figure 13). In the entrance of the channel, a monochromatic wave is generated with a period of 2.68s, related to a wavelength of 5.66m in the channel, and an amplitude of 0.025m. This problem is case 6 studied by Ohyama et al.

[45] who analyzed six different types of waves experimentally. Table 4 shows some parameters that characterize the problem, calculated according to the linear theory. The Ursell number on the platform is 210, indicating the strong non-linearity in this region. Parameter H/h shows that breaking does not even occur on the platform.

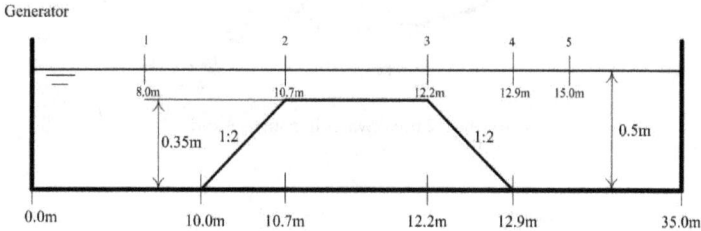

Figure 12. Geometry of the channel for the 1:2 breakwater

	H/h	kh	Ur
Channel (h = 0.5m)	0.100	0.555	14.1
Platform (h = 0.15m)	0.355	0.294	210.0

Table 4. Wave parameters for the 1:2 breakwater

Table 5 shows periods, frequencies and wavelengths concerning the fundamental frequency and the harmonic components that occur along the wave propagation.

A mesh with 120200 elements and 50526 nodes was used for FLUINCO in this simulation. The element sizes along the channel vary between dx=0.08m at the ends and dx=0.01m on the platform. The boundary and the initial conditions are similar to the ones in the previous case, and 0.002s was the time step.

	Fundamental	2nd harmonic	3rd harmonic	4th harmonic
Period (s)	2.68	1.34	0.89	0.67
Frequency (Hz)	0.373	0.746	1.124	1.493
Wavelength (m)	5.66	2.42	1.22	0.70

Table 5. Period, frequency and wavelength related to the fundamental frequency, and 2nd, 3rd, and 4th harmonics for the 1:2 breakwaters

Figure 13 shows the free surface elevations in gauges 3 and 5 (gauge positions are indicated in Figure 5). Numerical results are compared with the experimental ones presented by Ohyama et al. [45]. The FLUINCO model represents the surface deformation recorded in gauge 3 well. The deformations of gauge 5 indicate that the nonlinearity increases. In this case, FLUINCO captures the variation of the surface elevation more accurately.

Figure 14 shows frequency spectra obtained in gauges 3 and 5. The fundamental and the harmonic waves are well represented by the models, but their amplitudes differ. The FLUINCO results are closer for the two gauges.

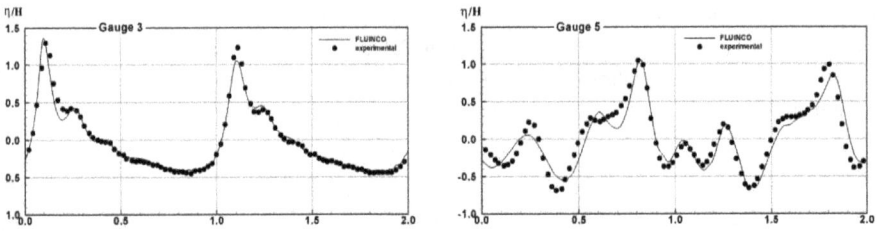

Figure 13. Free surface elevation for the 1:2 breakwater in gauges 3 and 5.

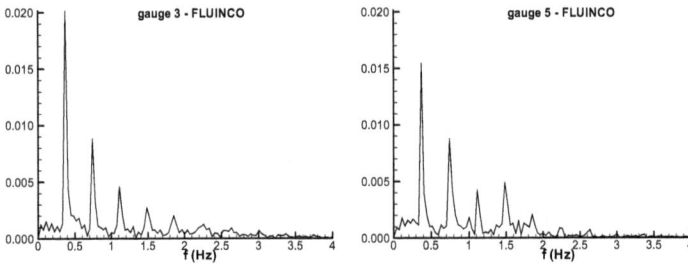

Figure 14. 1:2 Breakwater case. Frequency spectra in gauges 3 and 5.

Streamlines during one wave period obtained by FLUINCO are presented in Figure 15. Unlike the previous case, a vortex, located between the upstream slope and the bottom, occurred during part of the wave period.

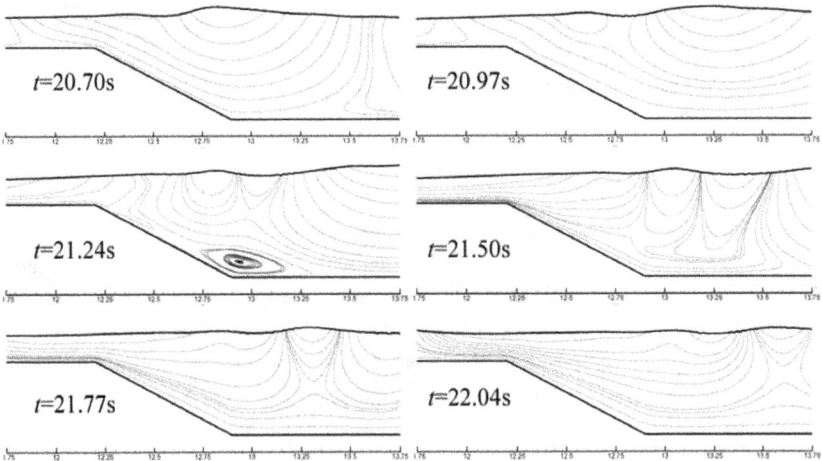

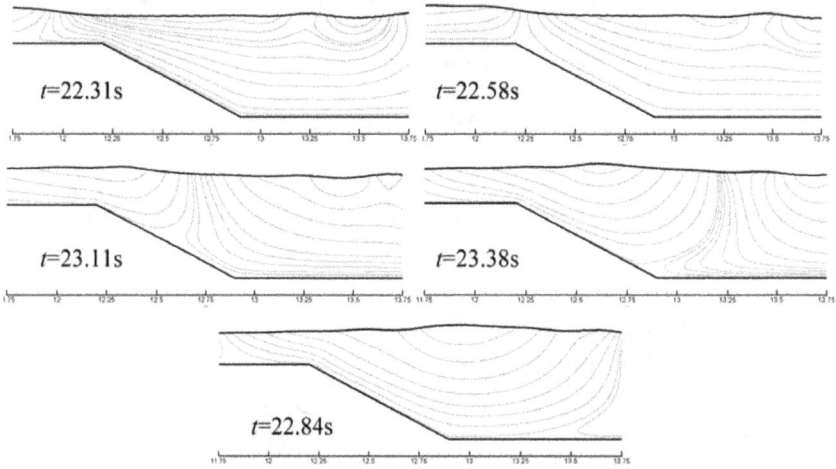

Figure 15. Streamlines of the 1:2 breakwater

4. Conclusions

In this text, we showed a model, named FLUINCO, capable of simulating flows on free surface. It is based on the semi-implicit two-step Taylor- Galerkin method to integrate Navier-Stokes equations in time and space. An ALE formulation is employed to describe the free surface movement. The methodology was validated in two study cases: the wave propagation over a submerged horizontal cylinder and submerged trapezoidal breakwaters. Both study cases showed the application of a Navier-Stokes based code, which considers accurately vertical flow effects, in the wave-submerged structure interaction problems.

In the case of the submerged horizontal cylinder, the free surface elevations and the velocity profiles obtained by the model were similar to experimental ones [39]. The numerical results presented a slight free surface deformation downstream, possibly because of the lack of refinement that caused numerical diffusion. In this case, the viscous effects influenced the flow behavior locally whereas the viscosity was not important far from the cylinder. The non-slip condition on the bottom did not modify the wave propagation significantly because it is a deep water case.

In the case of trapezoidal breakwater, two analyses were carried out for different upstream and downstream slopes. The first analysis deals with upstream and downstream slopes of 1:20 and 1:10, respectively. The results obtained by the model were compared with Dingemans' experimental data [44]. A comparison of the surface elevations and the energy spectrum for some gauges along the channel showed that the model provided good results. Although the FLUINCO results have been somewhat smoothed, they were closer to the experimental ones, including the ones in the gauges placed downstream, where nonlinear effects are more significant. Streamlines over a wave period showed that there was no flow separation in this case.

The second analysis consists of two 1:2 breakwater slopes; it showed a strong influence of nonlinear effects on the results of the surface elevation and the energy spectrum. The numerical results were compared with experimental ones presented by Ohyama et al [45]. The vertical velocity field obtained by FLUINCO showed that a vortex of non-turbulent origin was formed in the flow. The model obtained results closer to the experimental ones, including the ones downstream of the breakwater, where the nonlinearity effects are more significant. Both breakwater analyses showed that FLUINCO captures the nonlinear effects of the flow accurately, due to the fact that this model considers the influence of the vertical circulation in the flow.

Author details

Paulo Roberto de Freitas Teixeira
Engineering School, Federal University of Rio Grande - FURG, Rio Grande, Brazil

Acknowledgement

The author acknowledges Conceição Juana Fortes, Eric Didier and José Manuel Paixão Conde and the partnership between FURG and Laboratório Nacional de Engenharia Civil (LNEC). The author also thanks the support of Conselho Nacional de Desenvolvimento Científico e Tecnológico (CNPq - project 303308/2009-5).

5. References

[1] Longuet-Higgins M S and Cokelet E D (1976) The deformation of steep surface waves on water. I. A numerical method of computation, Proc. R. Soc. London A 95. 1-26.

[2] Isaacson M (1982) Non-linear wave effects on fixed and floating bodies. J. Fluid. Mech., j. 120: 267-281.

[3] Grilli ST, Guyenne P, Dias F (2001) A fully non-linear model for three-dimensional overturning waves over an arbitrary bottom. Int. J. Numer. Meth. Fluids. j. 35 (7): 829-867.

[4] Dommermuth D G, Yue D K P (1987) A high-order spectral method for the study of nonlinear gravity waves. J. Fluid Mech. j. 184: 267-288.

[5] Bateman W J D, Swan C, Taylor P H (2001) On the efficient numerical simulation of directionally spread surface water waves. J. Comput. Phys. j. 174 (1): 277-305.

[6] Peregrine DH (1967) Long waves on a beach. J. Fluid Mech. j. 27: 815-882.

[7] Madsen P A, Murray R, Sorensen O R (1991) A new form of Boussinesq equations with improved linear dispersion characteristics. Coast Engrg. j. 15: 371-388.

[8] Madsen P A, Bingham H B, Liu P L-F (2002) A new Boussinesq method for fully nonlinear waves from shallow to deep water. J. Fluid Mech. j. 462: 1-30.

[9] Nwogu O (1993) An alternative form of Boussinesq equations for nearshore wave propagation. J. Waterway, Port, Coastal Engineering. j. 119: 618-638.

[10] Wei G, Kirby J T, Grilli S T, Subramanya R (1995) A fully nonlinear Boussinesq model for surface waves. Part 1. Highly nonlinear unsteady Wave. J. Fluid Mech. j. 294: 71-92.

[11] Gobbi M F and Kirby J T (1999) Wave evolution over submerged sills: Tests of a high-order Boussinesq model. Coastal Engineering. j. 37: 57-96.

[12] Schäffer H A and Madsen P A (1995) Further enhencements of Boussinesq-type equations. Coastal Engineering. j. 26: 1-14.

[13] Agnon Y, Madsen P A, Schaffer H (1999) A new approach to high order Boussinesq models. Journal of Fluid Mechanics. j. 399: 319-333.

[14] Kennedy A B, Kirby J T, Chen Q, Dalrymple R A (2001) Boussinesq-type equations with improved nonlinear behaviour. Wave Motion. j. 33: 225-243.

[15] Lynett P and Liu PL-F (2004) A two-layer approach to water wave modelling. Proceedings of the Royal Society of London A. j. 460: 2637– 2669.

[16] Hsiao S C, Lynett P, Hwung H H, Liu P L-F (2005) Numerical simulations of nonlinear short waves using a multilayer model. ASCE. J. Eng. Mech. j. 131: 231-243.

[17] Casulli V and Stelling G S (1998) Numerical simulation of 3D quasi-hydrostatic, free-surface flows. J. Hydr. Eng. ASCE. j. 124: 678-686.

[18] Stansby P K and Zhou J G (1998) Shallow-water flow solver with non-hydrostatic pressure: 2D vertical plane problems. Int. J. Numer. Meth. Fluids. j. 28: 514-563.

[19] Zijlema M and Stelling G S (2008) Efficient computation of surf zone waves using the nonlinear shallow water equations with non-hydrostatic pressure. Coastal Eng. j. 55: 780-790.

[20] Mahadevan A, Oliger J, Street R (1996) A non-hydrostatic mesoscale ocean model. Part 1: Well-posedness and scaling. J. Phys. Oceanogr. j. 26: 1868-1880.

[21] Marshall J, Adcroft A, Hill C, Perelman L, Heisey C (1997) A finite-volume, incompressible Navier-Stokes model for studies of the ocean on parallel computers. J. Geophys. Res. j. 102: 5753-5766.

[22] Hodges B R and Street R L (1999) On simulation of turbulent nonlinear free-surface flows. J. Comput. Phys. j. 151: 425-457.

[23] Lin P and Li C W (2002) A σ-coordinate three-dimensional numerical model for surface wave propagation. Int. J. Numer. Meth. Fluids. j. 38: 1045-1068.

[24] Zhou J G, Stansby P K (1999) An arbitrary Lagrangian-Eulerian σ (ALES) model with non-hydrostatic pressure for shallow water flows. Comput. Meth. Appl. Mech. Eng. j. 178: 199-214.

[25] Harlow F H and Welch J E (1965) Numerical calculation of time-dependent viscous incompressible flow. Phys. Fluids. j. 8: 2182-2189.

[26] Lin P and Liu P F L (1998) A numerical study of breaking waves in the surf zone. J. Fluid Mech. j. 359: 239-264.

[27] Hieu P D, Katsutohi T, Ca V T (2004) Numerical simulation of breaking waves using a two-phase flow model. Appl. Math. Model. j. 28: 983-1005.

[28] Iafrati A, Di Mascio A, Campana E F (2001) A level set technique applied to unsteady free surface flows. Int. J. Numer. Meth. Fluids. j. 35: 281-297.

[29] Teixeira P R F (2011) Simulação numérica da interação de escoamentos tridimensionais de fluidos compressíveis e incompressíveis e estruturas deformáveis usando o método de elementos finitos. Doctoral thesis, PPGEC-UFRGS, Porto Alegre, RS (in Portuguese).

[30] Teixeira P R F and Awruch A M (2000) Numerical simulation of three dimensional incompressible flows using the finite element method. 8th Brazilian Congress of Thermal Engineering and Sciences – ENCIT, Porto Alegre.

[31] Argyris J, St. Doltsinis J, Wuestenberg H, Pimenta P M (1985) Finite element solution of viscous flow problems. In: Finite Elements in Fluids.Wiley, New York. Vol. 6: 89-114.

[32] Teixeira P R F. and Awruch A M (2005) Numerical simulation of fluid-structure interaction using the finite element method. Computers & Fluids. j. 34: 249-273.

[33] Dean R G and Dalrymple R A (1994) Water Wave Mechanics for Engineers and Scientists. In: Advanced Series on Ocean Engineering 2, World Scientific Publishing Co. Ltd. 470 p.

[34] Flather R A (1976) A tidal model of the northwest European continental shelf. Mem. Soc. R. Liege, Ser. j. 10 (6): 141-164.

[35] Dean W R (1948) On the Reflection of Surface Waves by a Submerged Circular Cylinder. Proc. Camb. Phil. Soc. j. 44: 483-491.

[36] Ursell F (1950) Surface Waves on Deep Water in presence of a Submerged Circular Cylinder, Proc. Camb. Phil. Soc. j. 46: 141-158.

[37] Chaplin J R (2001) Nonlinear Wave Interactions with a Submerged Horizontal Cylinder. Proc. 11th Int. Offshore and Polar Eng. Conf., Stavanger. j. 3: 272-279.

[38] Schonberg T and Chaplin J R (2003) Computation of Nonlinear Wave Reflections and Transmissions from Submerged Horizontal Cylinder, Int. J. Offshore and Polar Eng. j. 13: 29-37.

[39] Paixão Conde J M, Didier E, Lopes M F P, Gato L M C (2007) Nonlinear wave diffraction by a submerged horizontal circular cylinder. Proc. 17th Int. Offshore and Polar Eng. Conf. (ISOPE 2007), Lisbon.

[40] Johnson J W, Fuchs R A, Morison J R (1951) The damping action of submerged breakwaters. Trans. Am. Geophys. Union. j. 32: 704-718.

[41] Jolas P (1960) Passage de la houle sur un seuil. Houille Blanche. j. 2: 148-152.

[42] Huang C J, Dong C M (1999) Wave deformation and vortex generation in water waves propagating over a submerged dike. Coastal Engrg. j. 37: 123-148.

[43] Beji S, Battjes J A (1993) Experimental investigations of wave propagation over a bar. Coastal Engineering. j. 19: 151-162.

[44] Dingemans M W (1994) Comparison of computations with Boussinesq-like models and laboratory measurements. Report H-1684.12 Delft Hydraulics. 32 p.

[45] Ohyama T, Kioka W, Tada A (1995) Applicability of numerical models to nonlinear dispersive waves. Coastal Engrg. j. 24: 297-313.

[46] Zhou J G, Stansby P K (1999) An arbitrary Lagrangian-Eulerian σ (ALES) model with non-hydrostatic pressure for shallow water flows. Comput. Meth. Appl. Mech. Eng. j. 178: 199-214.

[47] Yuan H L, Wu C H (2004) A two-dimensional vertical non-hydrostatic σ model with an implicit method for free-surface flows. Int. J. Numer. Meth. Fluids. j. 44: 811-835.

[48] Stelling G S, Zijlema M (2005) Further experience in computing non-hydrostatic free-surface flow involving with water waves. Int. J. Numer. Meth. Fluids. j. 48: 169-197.

[49] Beji S and Battjes J A (1994) Numerical simulation of nonlinear wave propagation over a bar. Coast. Eng. j. 23: 1-16.

[50] Shen Y M, Ng C O, Zheng Y H (2004) Simulation of wave propagation over a submerged bar using the VOF method with a two-equation k-epsilon turbulence modeling. Ocean Eng. j. 31: 87-95.

[51] Ting F C K, Kim Y.K (1994) Vortex generation in water-waves propagating over a submerged obstacle. Coast. Eng. j. 24: 23-49.

[52] Zhuang F, Lee J J (1996) A viscous rotational model for wave overtopping over marine structure. Proc. 25th Int. Conf. Coastal Eng. ASCE. 2178-2191.

Chapter 14

Wavelet Based Simulation of Elastic Wave Propagation

Hassan Yousefi and Asadollah Noorzad

Additional information is available at the end of the chapter

1. Introduction

Multiresolution-based studying has rapidly been developed in many branches of science and engineering; this approach allows one to investigate a problem in different resolutions, simultaneously. Some of such problems are: signal & image processing; computer aided geometric design; diverse areas of applied mathematical modeling; and numerical analysis.

One of the multiresolution-based schemes reinforced with mathematical background is the wavelet theory. Development of this theory is simultaneously done by scientists, mathematicians and engineers [1]. Wavelets can detect different local features of data; the properties that locally separated in different resolutions. Wavelets can efficiently distinguish overall smooth variation of a solution from locally high transient ones separated in different resolutions. This multiresolution feature has been interested many researchers, especially ones in the numerical simulation of PDEs [1]. Wavelet based methods are efficient in problems containing very fine and sharp transitions in limited zones of a computation domain having an overall smooth structure. In brief, the most performance of such multiresolution-based methods is obtained in systems containg several length scales.

Regarding wavelet-based simulation of PDEs, two different general approaches have been developed; they are: 1) projection methods, 2) non-projection ones.

In the projection schemes, in general, the wavelet functions are used as solution basis functions. There, all of the computations are performed in the wavelet spaces; the results are finally re-projected to the physical space [2, 3]. In non-projection schemes, the wavelets are only used as a tool to detect high-gradient zones; once these regions are captured, the other common resolving schemes (e.g., finite difference or finite volume method) are employed to simulate considered problems. In this approach, all computations are completely done in the physical domain, and thereby the corresponding algorithms are straightforward and conceptually simple [4]. There are some other schemes that incorporate these two general approaches. They use wavelets as basis functions in a wavelet-based adapted grid points, e.g. [5].

The advantages of the wavelet-based projection methods are:

1. Wavelets provide an optimal basis set; it can be improved in a systematic way. To improve an approximation, wavelet functions can locally be added; such improvements do not lead to numerical instability [3].
2. Most of the kernels (operators) have sparse representation in the wavelet spaces and therefore speed of solutions is high. The band width of the sparse operators can also be reduced by considering a pre-defined accuracy. This leads to inherent adaptation which no longer needs to grid adaptation [6-8].The matrix coefficients can easily be computed considering wavelet spaces relationship [6-8].
3. The coupling of different resolution levels is easy [3]. The coupling coefficients can easily be evaluated considering multiresolution feature of the wavelet spaces [6-8].
4. Different resolutions can be used in different zones of the computation domain.
5. The numerical effort has a linear relationship with system size [3]. In the wavelet system, the fast algorithms were developed [9]. Another considerable property of the wavelet transform is its number of effective coefficients: it is much smaller than data size, itself (in spit of the Fourier transform). These two features leads to fast and accurate resolving algorithms.

The wavelet-based projection methods, however, have two major drawbacks: 1) projection of non-linear operators; 2) imposing both boundary conditions and corresponding geometries [4, 8].

The most common wavelet based projection methods are: the telescopic representation of operators in the wavelet spaces [6-8], wavelet-Galerkin [2, 3,10-19], wavelet-Taylor Galerkin [20-22], and collocation methods [5,23-26] (in this approach, the wavelet-based grid adaptation scheme is incorporated with the wavelet-based collocation scheme). Some efforts have been done to impose properly boundary conditions in these methods. Some of which are: 1) wavelets on an interval [11, 27], 2) fictitious boundary conditions [12, 13, 28, 29], 3) reducing edge effects by proper extrapolation of data at the edges [14], 4) incorporation of boundary conditions with the capacitance matrix method [2, 15].

Regarding non-projection approaches, the common method is to study a problem in accordance with the solution variation; i.e., using different accuracy in different computational domains. In this method more grid points are concentrated around high-gradient zones to detect high variations, the adaptive simulation. In this case, only the important physics of a problem are precisely studied, a cost-effective modeling. Once the grid is adapted, the solution is obtained by some other common schemes, (e.g., the finite difference [4, 30- 38], or finite volume [39-43] method) in the physical space. The wavelet coefficients of considerable values concentrate in the vicinity of high-gradient zones. The coefficients have a one to one correspondence with their spatial grid points, and thereby, by considering points of considerable coefficient values, the grid can be adapted. For this purpose, the points of small enough coefficient values are omitted from the computing grid. In these grid-based adaptive schemes, the degrees of freedom are considered as point values in the physical space; this feature leads to a straightforward and easy method. In some cases the two approaches, projection and non-projection ones, are incorporated; e.g., adaptive collocation methods [5, 23-26], and adaptive Galerkin ones [28, 29].

There is also some other approaches using wavelets only to detect local feature locations, without grid adaptation. In one approach, spurious oscillation locations are captured by the wavelets; thereafter, the oscillations are locally filtered out by a post-processing step [44-45]. The filtering can be done by the conjugate filtering method only in the detected points. In the other approach, to control spurious oscillations the spectral viscosity is locally added in high-gradient/dicontinuous regions; such zones are detcted by the wavelets. This approach is suitable for simulation of hyperbolic systems containing discontinuous solutions. There, artificial diffusion is locally added only in high-frequency components [46]. In these two approaches, the wavelet transforms are used as a tool to detect highly non-uniform localized spatial behaviors and corresponding zones.

The two aforementioned general wavelet based outlooks, projection and non-projection ones, have successfully been implemented for simulation of stress wave propagation problems. The wavelet-based projection methods were successfully used for simulation of wave propagation problems in infinite and semi-infinite medias [12, 47-52]. Another important usage is wave propagation in structural engineering elements; e.g. wave propagation in the nano-composites [53]. The non-projection methods were also employed for wave-propagation problems, one can refer to [35-38].

In brief, it should be mentioned that other powerful and common methods exist for simulation of wave propagation problems for engineering problems; some of which are: the finite difference and finite element schemes, e.g. [54, 55]. These methods are precisely studied and relevant numerical strength and drawbacks are investigated. Regarding these schemes, some of important numerical features are: 1) source of numerical errors: truncation and roundoff errors, [56]; 2) effect of grid/element irregularities on truncation error and corresponding dissipation and dispersion phenomena [57]; 3) internal reflections from grids/element faces [58-65]; 4) the inherent dissipation property [66, 67]. These features lead in general to numerical (artificial) dissipation and dispersion phenomena. In general to control these two numerical drawbacks in wave propagation problems, it is desirable to refine spatio-temporal discritizations [68]. Considering the spatial domain, this can effectively be done by the wavelet theory. In the wavelet-based projection methods, the inherent adaptation is used, while in the non-projection ones, the multiresolution-based grid adaptation is utilized.

This chapter is organized as follows. In section 2, the wavelet-based projction method will be survived. This section includes: 1) a very brief explanation of main concept of multiresolution analysis; 2) in brief review of wavelet-based projection method for solution of PDEs and computation of the spatial derivatives; 3) the issues related to a 2D wave propagation example. In section 3, the wavelet-based non-projection ones will be presented.

It includes: 1) wavelet-based grid adaptation scheme with interpolating wavelets; 2) solution algorithm; 3) smoothing splines; 4) an example: wave propagation in a two layered media. This chapter ends with a brief conclusion about the presented wavelet-based approaches.

2. Wavelet based projection method in wave propagation problem

In the wavelet based projection methods, the wavelets are used as basis functions in numerical simulation of wave equations. This section has following sub-sections: multiresolution analysis and wavelets; representation of operators in the wavelet spaces; the semi group time integration methods; a SH wave propagation problem.

2.1. Multiresolution analysis and wavelet basis

In this subsection, wavelet-based multiresolution analysis and wavelet construction methods will be survived.

2.1.1. Multiresolution analysis

A function or a signal, in general, can be viewed as a set of a smooth background with low frequency component (approximation one) and local fluctuations (local details) of variant high frequency terms. The word "multiresolution" refers to the simultaneous presence of different resolutions in data. In the multiresolution analysis (MRA), the space of functions that belong to square integrable space, $L^2(\mathbb{R})$, are decomposed as a sequence of detail subspaces, denoted by $\{w_k\}$, and an approximation subspace, indicated with v_j. The approximation of $f(t)$ at resolution level j, $f(t)$, is in v_j and the details $d_k(t)$ are in w_k (detail sub-spaces of level k). The corresponding scale of resolution level j is usually chosen to be of order 2^{-j} [69, 70]. In orthogonal wavelet systems, the multiresolution analysis of $L^2(\mathbb{R})$ is nested sequences of the subspaces $\{v_j\}$ such that:

i. $\ldots \subset v_{-1} \subset v_0 \subset v_1 \ldots \subset L^2(\mathbb{R})$

ii. $v_{-\infty} = \{0\}$, $v_{+\infty} = L^2$

iii. $f(t) \in v_j \Leftrightarrow f(2t) \in v_{j+1}$

iv. $f(t) \in v_0 \Rightarrow f(t-k) \in v_0$

v.   Exists a function $\phi(t)$, called the scaling function such that set $\{\phi(t-k)\}_{k \in Z}$ is a basis of
     v_0.

The sub-space v_j denotes the space spanned by family $\{\phi_{j,k}(t)\}$, i.e., $v_j = \overline{span\{\phi_{j,k}(t)\}}_k$ where

$\phi_{j,k}(t) = 2^{j/2} \phi(2^j t - k)$. The $\phi_{j,k}(t)$ is a scaled and shifted version of the $\phi(t)$; thereby the function $\phi(t)$ is known as the father wavelet. The scale functions ($\phi_{j,k}(t)$) are localized in both spatial (or time) and frequency (scale) spaces. The function $\phi(t)$ is usually designed so that: $\int \phi(t)dt = 1$ & $\int |\phi(t)|^2 dt = 1$. The second equation implies that the scaling function ($\phi(t)$) has unit energy and therefore by multiplying it with data, the energy of signals do not alter. The dilated and shifted version of the scale function, $\phi_{j,k}(t)$ is usually normalized with the coefficient $2^{j/2}$ to preserve the energy conservation concept; namely, $\int |\phi_{j,k}(t)|^2 dt = 1$. Since $v_j \subset v_{j+1}$, there exist a detail space w_j that are complementary of v_j in v_{j+1}, i.e., $v_j \oplus w_j = v_{j+1}$. The subspace w_j itself is spanned by a dilated and shifted wavelet function family, i.e. $\{\psi_{j,k}(t)\}$, where $\psi_{j,k}(t) = 2^{j/2}\psi(2^j t - k)$; the function $\psi(t)$ is usually referred as the mother wavelet. The wavelet function, $\psi_{j,k}(t)$ is localized both in time (or space) and frequency (scale); it oscillates in such a way that its average to be zero, i.e.: $\int \psi_{j,k}(t)dt = 0$ .This is because the wavelet function measures local fluctuations; the variations which are assumed to have zero medium. Similar to scaling function and for the same reason, energy of the wavelet functions are unit, i.e., $\int |\psi_{j,k}(t)|^2 dt = 1$. The approximate and detail subspaces satisfy orthogonally conditions as follows: $v_j \perp w_j$ & ($w_j \perp w_{j'}$ for $j \neq j'$). These relations lead to: $\langle \psi_{j,k}, \phi_{j,l} \rangle = 0$ & $\langle \psi_{j,k}, \psi_{j',l} \rangle = \delta_{kl}\delta_{jj'}$ & $\langle \phi_{j,k}, \phi_{j,l} \rangle = \delta_{kl}$ where $< f(t).g(t) > = \int f^*(t)g(t)dt$ (the inner product). Due to the fact that $v_0 \subset v_1$ and $w_0 \subset v_1$, then any function in v_0 or w_0 can be expanded in terms of the basis function of v_1, i.e.: $\phi(t) = \sqrt{2}\sum_k h_k\phi(2t - k)$ & $\psi(t) = \sqrt{2}\sum_k h'_k\phi(2t - k)$. These important equations are known as: dilation equations, refinement equations or two-scale relationships [69-72]. The h_k and h'_k are called filter coefficients, and can be obtained, in general, by the following relationships: $h_k = < \phi_{1,k}, \phi >$ and $h'_k = < \phi_{1,k}, \psi >$. The orthogonality condition of $\phi(x - k)$ and $\psi(x - k)$ leads to relationship: $h'_k = (-1)^k h_{N-1-k}$ where N is length of the scaling coefficient filters, $\{h_k\}, k = 1,...,N$.

As mentioned before, the multiresolution decomposition of $L^2(\mathbb{R})$ leads to a set of subspaces with different resolution levels; i.e., $L^2 = v_j \oplus w_j \oplus w_{j+1} \oplus ...$. In this regard, by using one decomposition level, a function $f(x) \in v_{J_{max}}$ (a space with sampling step $1/2^{J_{max}}$) can be expanded as: $f(x) = \left(\sum_{l=-\infty}^{+\infty} c(J_{max} - 1, k)\phi_{J_{max}-1,,l}(x) \right) + \left(\sum_{n=-\infty}^{\infty} d(J_{max} - 1, n)\psi_{J_{max}-1,,n}(x) \right)$

By following the step by step decomposition of approximation space, if the coarsest resolution level is $J_{min} \leq J_{max-1}$, then the function $f(x) \in v_{J_{max}}$ can be represented as:

$$f(x) = \left(\sum_{l=-\infty}^{+\infty} c(J_{min}, k)\phi_{J_{min},l}(x) \right) + \left(\sum_{j=J_{min}}^{J_{max}-1} \sum_{n=-\infty}^{\infty} d(j,n)\psi_{j,n}(x) \right)$$

This equation shows that the function $f(x)$ is converted into an overall smooth approximation (the first parenthesis), and a series of local fluctuating (high-frequency details) of different resolutions (the second parenthesis). In the above equation $c(j,k)$ and $d(j,k)$ are called the scaling (approximation) and wavelet (detail) coefficients, respectively. These transform coefficients are usually stored in an array as follows: $\left\{ \{d(J_{max}-1,n)\}, \{d(J_{max}-2,n)\}, \ldots, \{d(J_{min},n)\}, \{c(J_{min},n)\} \right\}$; this storing style is commonly referred as the *standard form*.

In the orthogonal wavelet systems, the coefficients $c(j,k)$ and $d(j,k)$ can be determined by: $c(j,k) = < f(x), \phi_{j,k}(x) >$ & $d(j,k) = < f(x), \psi_{j,k}(x) >$. Fast algorithms were developed to these coefficient evaluations and relevant inverse transform [9, 69].

In the following, the multiresolution-based decomposition procedure is qualitatively investigated by an example. Figure 1, illustrates the horizontal acceleration recorded at the El-Centro substation (f_1) and corresponding wavelet-based decompositions. There, the symbol a_0 refers to the approximation space ($a_0 = \sum_{l=-\infty}^{+\infty} c(0,k)\phi_{0,l}(x)$) and d_0 to d_9 denote the detail spaces ($d_j = \sum_{n=-\infty}^{+\infty} d(j,n)\psi_{j,n}(x)$); where, the finest and coarsest resolution levels are $J_{max} = 10$ and $J_{min} = 0$, respectively. The superposition of all projected data, f_2 (i.e., $f_2 = a_0 + \sum_{j=0}^{9} d_j$) and the difference $f_2 - f_1$ are presented as well. It is clear that a_0 approximates the overall smooth behavior; the projections d_0-d_9 include local fluctuations in different resolutions. There, the frequency content of d_j is in accordance with the resolution level j. The wavelet used for the decompositions is the Daubechies wavelet of order 12 (will be explained subsequently).

2.1.2. Derivation of filter coefficients

Considering the above mentioned necessary properties of scaling function and other possible assumptions for scaling/wavelet functions, the filter coefficients can be evaluated.

In orthogonal systems, necessary conditions for the scaling functions are [71]:

1. Normalization condition: $\int \phi(t)dt = 1$ which leads to $\sum_{k=1}^{N} h_k = 1$; N denotes filter length.

2. Orthogonality condition: $\int \phi(t).\phi(t+l)dt = \delta_{0,l}$ or equivalently $\sum_{k=1}^{N} h_k h_{k+2l} = \delta_{0,l}$; the parameter $\delta_{0,l}$ denotes *the Kronecker delta*. For filter set $\{h_k\} : k = 1,2,\ldots,N$ where N is an even number, this condition provides $N/2$ independent conditions.

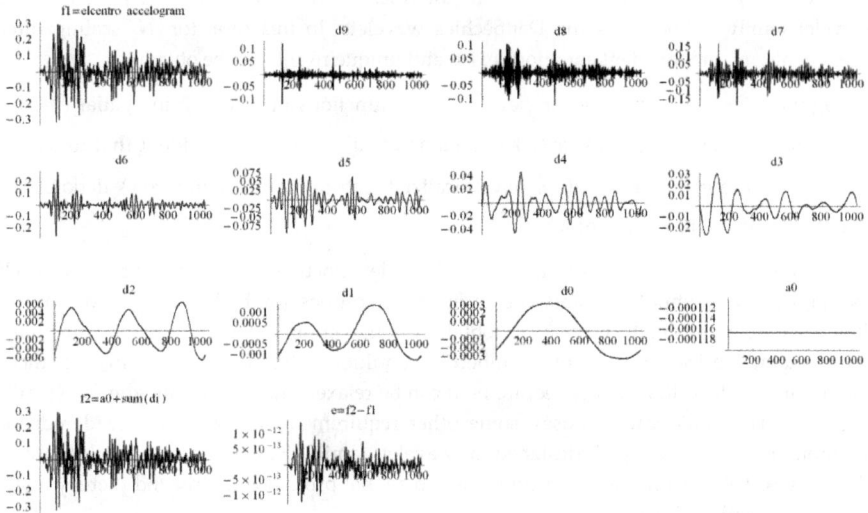

Figure 1. The El Centro acceleration and corresponding multiresolution representation.

Essential conditions 1 & 2 provide $N/2+1$ independent equations. Other requirements can be assumed to obtain the remaining equations.

One choice is the necessity that the set function $\{\varphi(x-k)\}$ can exactly reconstruct polynomials of order upto but not greater than p [71, 72]. The polynomial can be represented as: $f(x) = \alpha_0 + \alpha_1 x + \ldots + \alpha_{p-1}x^{p-1}$; on the other hands: $f(x) = \sum\limits_{k=-\infty}^{\infty} c_k\varphi(x-k)$.

By taking the inner product of the wavelet function $(\psi(x))$ with the above equation, other conditions can be obtained; since: $< f(x),\psi(x) >= \sum\limits_{k=-\infty}^{\infty} c_k < \varphi(x-k),\psi(x) > \equiv 0$, or:

$$< f(x),\psi(x) >= \alpha_0\int\psi(x)dx + \alpha_1\int x\psi(x)dx + \ldots + \alpha_{p-1}\int x^{p-1}.\psi(x)dx = 0.$$

As α_i coefficients are arbitrary, then it is necessary that each of the above integration to be equal to zero: $\int x^l.\psi(x)dx = 0$, $l = 0,1,\ldots,p-1$; these equations lead to p equations where $p-1$ of them are independent [69, 71, 72]. These equations mean that the first p moments of the wavelet function must be equal to zero; this condition in the frequency domain leads to relationship $\left[d^l\hat{\psi}(w)/dw^l\right]_{w=0} = 0$, $l = 0,1,\ldots,p-1$. It can be shown that these conditions

lead to conditions: $\sum\limits_{k}^{N}(-1)^k h_k k^l = 0$, $l = 0,1,\ldots,p-1$.

In case that $p = N/2$, where N is even (for filter coefficients of length N) the resulted wavelet family is known as the Daubechies wavelets. In this case, for N scaling filter coefficients, N independent equations exist, and unique results can be obtained.

In Figure 2 the Daubechies scaling and wavelet functions of order 12 in spatial ($\varphi(x)$ & $\psi(x)$) and frequency ($|\hat{\varphi}(w)|$ & $|\hat{\psi}(w)|$) domains are illustrated. It is evident that functions $\varphi(x)$ & $\psi(x)$, and $|\hat{\varphi}(w)|$ & $|\hat{\psi}(w)|$ have localized feature. In this figure $\varphi'(x)$ denotes the first derivative of the scaling function.

Other choices can be considered for scaling/wavelet functions construction; some of such assumptions are: imposing vanishing moment conditions for both scaling and wavelet functions (e.g., Coiflet wavelets), obtaining maximum smoothness of functions, interpolating restriction, and/or symmetric condition [69]. To fulfill some of these requirements, the orthogonality requirement can be relaxed and the bi-orthogonal system is used [69]. For numerical purposes, some other requirements can also be considered; for example Dahlke et al. [16] designed a wavelet family which is orthogonal to their derivatives. This feature leads to a completely diagonal projection matrix and thereby a fast solution algorithm [14].

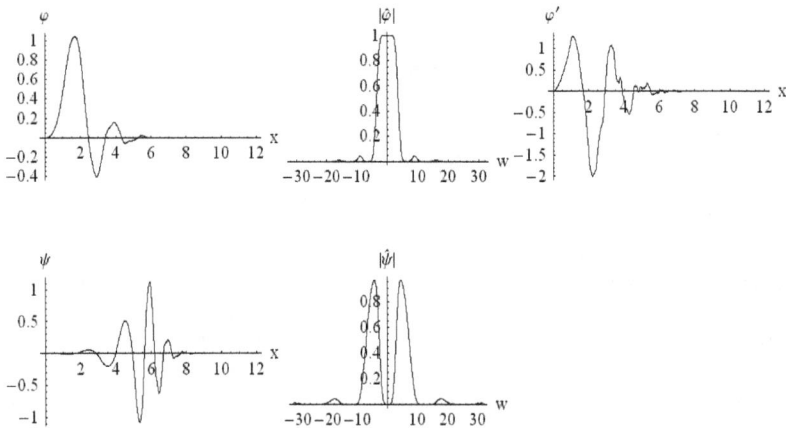

Figure 2. The Daubechies scaling and wavelet functions of order 12 in spatial and frequency domains, as well as first derivative of the scaling function.

2.2. Expressing operators in wavelet spaces

In this subsection, multiresolution analysis of operators will be presented [6-8].

Assume T denotes an operator of the following form: $T : L^2(\mathbb{R}) \rightarrow L^2(\mathbb{R})$. The aim is to represent the operator T in the wavelet spaces; this can be done by projection the operator in the wavelet spaces.

The projection of the operator in the approximation space of resolution level j (v_j) can be represented as: $P_j : L^2(\mathbb{R}) \rightarrow v_j$ where, $(P_j f)(x) = \sum \langle f, \phi_{j,k} \rangle \phi_{j,k}(x)$. In the same way, the projection of the operator in the detail subspace w_j, of resolution j, is: $Q_j : L^2(\mathbb{R}) \rightarrow w_j$; $Q_j = P_{j+1} - P_j$ where, $(Q_j f)(x) = \sum \langle f, \psi_{j,k} \rangle \psi_{j,k}(x)$. The Q_j definition is directly resulted from the multiresolution property, i.e., $v_{j-1} \subset v_j$ and $v_j = v_{j-1} \oplus w_{j-1}$.

For representing the operator T in the multiresolution form, firstly, a signal $x \in v_{J_{max}}$ is considered, where J_{max} denotes the finest resolution level, where $dx = 1/2^{J_{max}}$. The data x can then be projected into the scaling (approximation) and detail spaces of resolution $j = J_{max} - 1$ by one step wavelet transform, i.e.: $x = P_j(x) + Q_j(x)$.

Considering a linear operator (function) T and multiresolution feature, the function $T(x)$ can be presented as follows: $T(x) = T_{J_{max}} = T(P_j + Q_j) = T(P_j) + T(Q_j) = TP_j + TQ_j$.

However $T(P_j)$ & $T(Q_j)$ are no longer orthogonal to each other; so each of them can be re-projected to v_j and w_j as follows: $T(P_j) = P_j TP_j + Q_j TP_j$ & $T(Q_j) = P_j TQ_j + Q_j TQ_j$.

By substituting these relationships in the equation $T(x) = T_{J_{max}}$, we have:

$$T(x) = (Q_j TQ_j + Q_j TP_j + P_j TQ_j) + P_j TP_j$$

Each term of the above equation belongs to either v_j or w_j as follows:

$$A_j = Q_j fQ_j \in w_j; \quad B_j = Q_j fP_j \in w_j; \quad \Gamma_j = P_j fQ_j \in v_j; \quad T_j = P_j fP_j \in v_j$$

In the above equations, B_j and Γ_j represent interrelationship effects of subspaces v_j and w_j. Using these symbols, the operator T can be rewritten as: $T_{J_{max}} = T_{j+1} = (A_j + B_j + \Gamma_j) + T_j$ By continuously repeating the above mentioned procedure for operators T_j, finally, the $T_{J_{max}}$ can be expressed in the multiresolution representation as follows:

$$T_{J_{max}} = \sum_{i=J_{min}}^{J_{max}} (A_i + B_i + \Gamma_i) + T_{J_{min}}$$

where J_{min} denotes the coarsest resolution level (i.e., $dx = 1/2^{J_{min}}$). This representation is the telescopic form of the operator T.

The schematic shape of the operator T in telescopic (multiresolution) form is presented in Figure 3; this form of representation is known as the Non-Standard form (NS form). In this figure, it is assumed that: $J_{\min} = 1$, $J_{\max} = 4$. There, the coefficients d^i and s^i are the scale and detail coefficients, respectively; these coefficients are obtained from the common discrete wavelet transform of data x. The $\hat{d}^i$ & $\hat{s}^i$ are the NS form of the wavelet coefficients, and should be converted to standard form by a proper algorithm (will be discussed).

The projection of the operator T in the wavelet space results to set $\left\{A_j, B_j, \Gamma_j\right\}_{j \in Z}$, where j denotes resolution levels. This form is called NS from, since both of the scale (s^j) and detail (d^j) coefficients are simultaneously appeared in the formulation, see Figure 3.

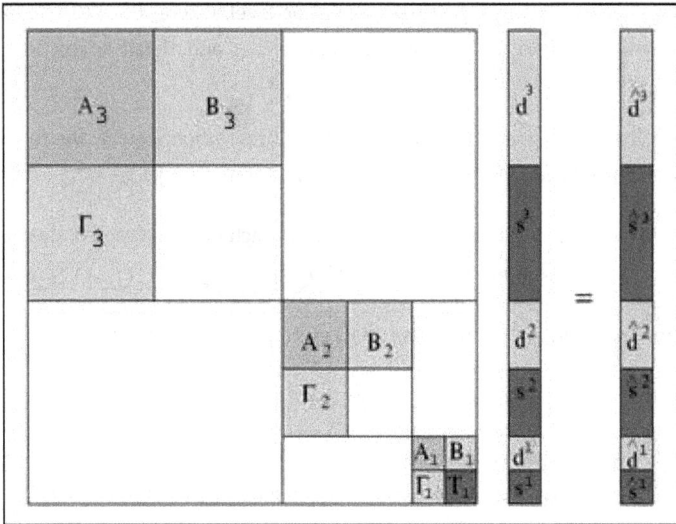

Figure 3. Schematic shape of a NS form of the operator T.

The matrix elements of projected operators A_j, B_j, Γ_j, and T_j are α^j, β^j, γ^j and s^j, respectively; for the derivative operator of order n, d^n / dx^n, the element definitions are:

$$\alpha_{il}^j = \left(2^j\right)^n \int_{-\infty}^{\infty} \psi(2^j x - i)\psi^{(n)}(2^j x - l)\left(2^j\right)dx = \left(2^j\right)^n \alpha_{i-l}$$

$$\beta_{il}^j = \left(2^j\right)^n \int_{-\infty}^{\infty} \psi(2^j x - i)\phi^{(n)}(2^j x - l)\left(2^j\right)dx = \left(2^j\right)^n \beta_{i-l}$$

$$\gamma_{il}^j = \left(2^j\right)^n \int_{-\infty}^{\infty} \phi(2^j x - i)\psi^{(n)}(2^j x - l)\left(2^j\right)dx = \left(2^j\right)^n \gamma_{i-l}$$

$$s_{il}^j = \left(2^j\right)^n \int_{-\infty}^{\infty} \phi(2^j x - i)\phi^{(n)}(2^j x - l)\left(2^j\right)dx = \left(2^j\right)^n s_{i-l}$$

Where:

$$\beta_l = \int_{-\infty}^{+\infty} \psi(x-l)\frac{d^n}{dx^n}\phi(x)dx \quad \alpha_l = \int_{-\infty}^{+\infty} \psi(x-l)\frac{d^n}{dx^n}\psi(x)dx,$$

$$\gamma_l = \int_{-\infty}^{+\infty} \phi(x-l)\frac{d^n}{dx^n}\psi(x)dx, \quad s_l = \int_{-\infty}^{+\infty} \phi(x-l)\frac{d^n}{dx^n}\phi(x)dx$$

The coefficients α_{il}^j, β_{il}^j, γ_{il}^j and s_{il}^j are not independent; the coefficients α_{il}^j, β_{il}^j, γ_{il}^j can be expressed in terms of s_{il}^j. This is because there is the two-scale relationship between the wavelet (detail) and scale functions; for more details see [6, 7]. This fact, leads to a simple and fast algorithm for calculation of A_j, B_j, Γ_j elements. The NS form of the operator d/dx obtained by the Daubechies wavelet of order 12 ($Db12$) is presented in Figure 4; there $J_{min} = 7$ & $J_{max} = 10$. It is clear that the projected operator is banded in the wavelet space.

To convert $\left\{\hat{d}_j, \hat{s}_j\right\}_{J_{min} \le j \le J_{max}-1}$ to the standard form $\left\{\{d_j\}_{J_{min} \le j \le J_{max}-1}, s_{J_{min}}\right\}$, the vector $\hat{s}_j$ is expanded for $J_{min} \le j \le J_{max}-1$ by the following algorithm [73]:

1. Set $\bar{d}_{J_{max}-1} = \bar{s}_{J_{max}-1} = 0$ (the initialization step),

2. For $j = J_{max}-1, J_{max}-2, \ldots, J_{min}$

   (2.1.) If $j \ne J_{max}-1$ then evaluate $\bar{d}_{j-1}$ & $\bar{s}_{j-1}$ from equation $\bar{s}_j + \hat{s}_j = \bar{d}_{j-1} + \bar{s}_{j-1}$, where $\bar{d}_{j-1} = Q_{j-1}(\bar{s}_j + \hat{s}_j)$ and $\bar{s}_{j-1} = P_{j-1}(\bar{s}_j + \hat{s}_j)$.

   (2.2.) evaluate $d_{j-1} = \bar{d}_{j-1} + \hat{d}_{j-1}$,

3. At level $j = J_{min}$, we have $s_{J_{min}} = \bar{s}_{J_{min}} + \hat{s}_{J_{min}}$.

The aforementioned telescopic representation is for 1D data. For higher dimensions, the extension is straightforward: the method can independently be implemented for each dimension.

2.3. The semi-group time discretization schemes

The scheme used here for temporal integration is the semi-group methods [74, 75]. These schemes have a considerable stability property: corresponding explicit methods have a stability region similar to typical implicit ones.

The semi-group time integration scheme can be used for solving nonlinear equations of form: $u_{,t} = Lu + N f(u)$ in $\Omega \in \mathbb{R}^d$

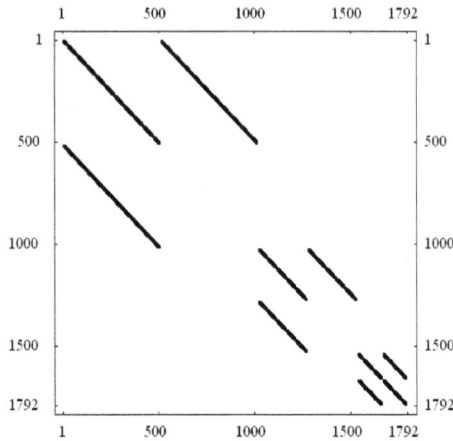

Figure 4. The NS form of operator d/dx obtained by $Db12$; it is assumed: $J_{min} = 7$ & $J_{max} = 10$.

where: L and N represent the linear and non-linear terms, respectively; $u = u(x,t)$; $x \in \mathbb{R}^d$, $d = 1,2,3$; $t \in [0,T]$. The initial condition is: $u(x,0) = u_0(x)$ in Ω, and the linear boundary condition is: $Bu(x,0) = 0$ on $\partial\Omega \in \mathbb{R}^{d-1}$, $t \in [0,T]$.

Regarding the standard semi-group method, the solution of the above mentioned equations is a non-linear integral equation of the form: $u(x,t) = e^{t.L}.u(x,0) + \int_0^t e^{(t-\tau)L} N(u(x,\tau))d\tau$.

For numerical simulations, the $u(x,t)$ should be discretized in time; the discretized value at time $t_n = t_0 + n\Delta t$ (Δt is the time step) will be denoted by $u_n \equiv u(x,t_n)$. In the same way the discrete form of $N(u(x,t))$ at $t = t_n$ is $N_n \equiv N(u(x,t_n))$.

If the linear operator is a constant, i.e., $L = q$, the discretized form of the above equation is [74]: $u_{n+1} = e^{q.l.\Delta t} u_{n+1-l} + \Delta t(\gamma.N_{n+1} + \sum_{m=0}^{M-1} \beta_m.N_{n-m})$, where $M+1$ is the number of time levels considered in the discretization and $l \leq M$; the coefficients γ and β_m are functions of $q.\Delta t$. It is clear that the explicit solution is obtained when $\gamma = 0$; for other choices the scheme is implicit. For case $l = 1$ & $\gamma = 0$ (the explicit method) the coefficients γ and β_m are presented in Table (1). In this table, for linear operator L the coefficient Q_k is [74]:

$$Q_k = Q_k(L.\Delta t); \quad Q_j(L.\Delta t) = \frac{e^{L.\Delta t} - E_j(L.\Delta t)}{(L.\Delta t)^j}; \quad E_j(L.\Delta t) = \sum_{k=0}^{j-1} \frac{(L.\Delta t)^k}{k!}; \quad j = 0,1,...$$

For $j = 0,1,2$ the above mentioned relations yield: $Q_0(L\Delta t) = e^{L\Delta t}$; $Q_1(L\Delta t) = (e^{L\Delta t} - I)(L\Delta t)^{-1}$; $Q_2(L\Delta t) = (e^{L\Delta t} - I - L\Delta t)(L\Delta t)^{-2}$; where I is the identity matrix.

M	β_0	β_1	β_2	order
1	Q_1	0	0	1
2	$Q_1 + Q_2$	$-Q_2$	0	2
3	$Q_1 + 3Q_2/2 + Q_3$	$-2(Q_2 + Q_3)$	$Q_2/2 + Q_3$	3

Table 1. Coefficient values for case $\gamma = 0 \,\&\, l = 1$ (the explicit scheme), where $Q_k = Q_k\left(q\Delta t\right)$.

2.4. Simulation of 2D SH propagating fronts

The governing equation of the SH scalar wave (anti-plane shear wave) is:

$$\rho\frac{\partial^2 u_y}{\partial t^2} = \frac{\partial}{\partial x}\left(\mu\frac{\partial u_y}{\partial x}\right) + \frac{\partial}{\partial z}\left(\mu\frac{\partial u_y}{\partial z}\right) + f_y$$

Where $u_y = u_y(x,z)$ is the out-of plane displacement; μ and ρ are shear modules and density, respectively. By defining a linear operator L_y, the above mentioned equation can

be rewritten as: $\dfrac{\partial^2 u_y}{\partial t^2} = L_y.u_y + \dfrac{f_y}{\rho}$ where $L_y = \dfrac{1}{\rho}\dfrac{\partial}{\partial x}\left(\mu\dfrac{\partial}{\partial x}\right) + \dfrac{1}{\rho}\dfrac{\partial}{\partial z}\left(\mu\dfrac{\partial}{\partial z}\right)$.

For using the semi-group temporal integration scheme, a new variable $v_y = \partial u_y/\partial t$ is introduced and consequently the above equation will be represented as a system of vectors:

$$\frac{\partial u_y}{\partial t} = v_y \quad \& \quad \frac{\partial v_y}{\partial t} = L_y.u_y + \frac{f_y}{\rho}$$

This system can be rewritten in vector notation as follows [48]:

$$U = LU + F \text{ where } U = \begin{pmatrix} u_y \\ v_y \end{pmatrix}; \quad L = \begin{pmatrix} 0 & I \\ L_y & 0 \end{pmatrix}; \quad F = \frac{1}{\rho}\begin{pmatrix} 0 \\ f_y \end{pmatrix}$$

The simplest explicit semi-group time integration scheme is obtained for case $\gamma = 0 \,\&\, M = 1$; in this case the discretized form of the wave equation is: $U_{n+1} = e^{\Delta t L}U_n + \Delta t.\beta_0.F_n$.

For utilizing the semi-group method the non-linear term $e^{\Delta t L}$ is approximated by corresponding Taylor expansion [48]: $e^{\Delta t L} = I + \Delta t.L + \dfrac{\Delta t^2}{2!}L^2 + \dfrac{\Delta t^3}{3!}L^3 + \dfrac{\Delta t^4}{4!}L^4 + \dots$.

The coefficient β_0 can be evaluated as: $\beta_0 = Q_1(L\Delta t) = (e^{L\Delta t} - I)(L\Delta t)^{-1}$. Similarly, the β_0 can be approximated by its Taylor expansion, i.e.: $\beta_0 = I + \dfrac{\Delta t}{2}.L + \dfrac{\Delta t^2}{6}L^2 + \dfrac{\Delta t^3}{24}L^3 + \dots$.

2.4.1. *The absorbing boundary conditions: infinite boundaries*

The absorbing boundaries are usually used for presenting infinite boundaries. The defect of numerical simulations is occurrence of artificial boundaries which reflect incoming energies to the computation domain. In this study, the absorbing boundary introduced in [76] is used to simulate infinite boundaries, where the absorbing boundary condition is considered explicitly. Therefore, the wave equation is modified by a damping term $Q(x,z).\ddot{u}_y(x,z,t)$ where, $Q(x,z)$ is an attenuation factor. This factor is zero in computation domain and increases gradually approaching to the artificial boundaries. Consequently, the waves incoming towards these boundaries are gradually diminished. In general, no absorbing boundary can dissipate all incoming energies, i.e. some small reflections will always remain.

The above mentioned modification, performed for SH wave equation is as follows:

$$\frac{\partial^2 u_y}{\partial t^2} + Q\frac{\partial u_y}{\partial t} = \frac{1}{\rho}\left(\frac{\partial}{\partial x}\left(\mu\frac{\partial u_y}{\partial x}\right) + \frac{\partial}{\partial z}\left(\mu\frac{\partial u_y}{\partial z}\right)\right) + \frac{f_y}{\rho}$$

And the modified vector form of the equation is: $\mathbf{U} = \begin{pmatrix} u_y \\ v_y \end{pmatrix}$; $\mathbf{L} = \begin{pmatrix} 0 & I \\ L_y & -Q \end{pmatrix}$; $\mathbf{F} = \frac{1}{\rho}\begin{pmatrix} 0 \\ f_y \end{pmatrix}$.

2.4.2. *Free boundaries*

There are different approaches for imposing the free boundary conditions in finite-difference methods; some of which are: 1) using equivalent surface forces (explicit implementation) [48]. In this method the equivalent forces will be up-dated in each time step; 2) employing artificial grid points by extending the computing domain (a common method); 3) considering nearly zero properties for continuum domain in simulation of the free ones [77]; in this case the boundary is replaced with an internal one. In the following examples (done by the wavelet based projection method) the third approach will be used. The first method are mostly be used for simple geometries.

2.4.3. *Example*

In the following, a scalar elastic wave propagation problem will be considered. The results confirm the stability and robustness of the wavelet-based simulations.

Example: Here scattering of plane SH waves due to a circular tunnel in an infinite media will be presented. The absorbing boundary is used for simulation of infinite domain; the considered function of $Q(x,z)$ is: $Q(X,Z) = a_x.\left(e^{b_x.X^2} + e^{b_x.(X-n_x)^2}\right) + a_z.\left(e^{b_z.Z^2} + e^{b_z.(Z-n_z)^2}\right)$

where: $X = x/dx$; $Z = z/dz$; $dx = dz = 1/128$; $a_x = a_z = 10000$; $b_x = b_z = -0.02$; $n_x = n_z = 128$; $x \in [0,1]$; $z \in [0,1]$. In simulations it assumed: $J_{max} = 7$ (the finest resolution level);

$J_{min} = 4$. The Daubechies wavelet of order 12 is considered in calculations. The assumed mechanical properties are: $\mu = 1.8 \times 10^4 kPa$ & $\rho = 2$ ton / m^3.

The plane wave condition is simulated by an initial imposed out-of plane harmonic deformation where corresponding wave number is $k = 64 / 12$. For time integration, the simplest form of the semi-group temporal integration method is used. The snapshots of results (displacement $u_y(x,z,t)$) at different time steps are presented in Figure 5; there the light gray circle represents the tunnel. The displacement $u_y(x,z,t = 0.0048)$ is illustrated in Figure 6; the total CPU computation time is 569 sec. for two different uniform grids, this problem is re-simulated by the finite difference method (with accuracy of order 2 in the spatial domain); the grid sizes are 143×143 and 200×200 uniform points. Temporal integrations are done by the 4th Runge-Kutta method. Corresponding displacements at $t = 0.0048$ are illustrated in Figure 7. Considering Figures 6 & 7, it is clear that the dispersion phenomenon occurs in the common finite difference scheme. There, in each illustration, total CPU computational time presented in the below of each figure.

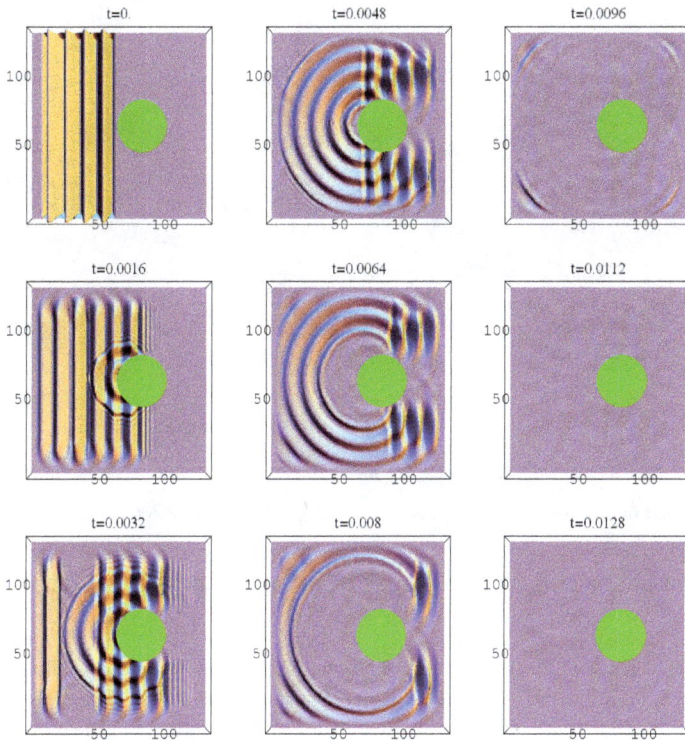

Figure 5. The snapshots of displacement ($u_y(x,z,t)$) at different times.

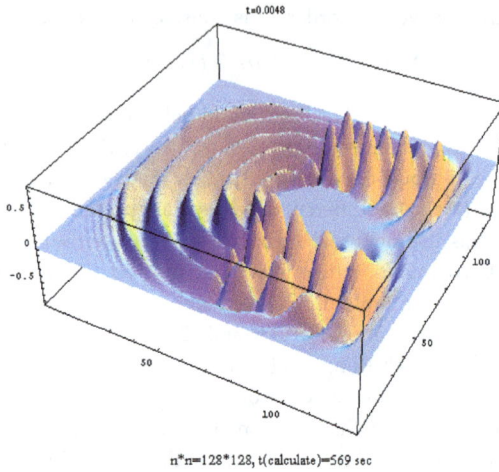

Figure 6. The displacement $u_y(x,z,t)$ at time $t = 0.0048$, obtained by the wavelet-based method.

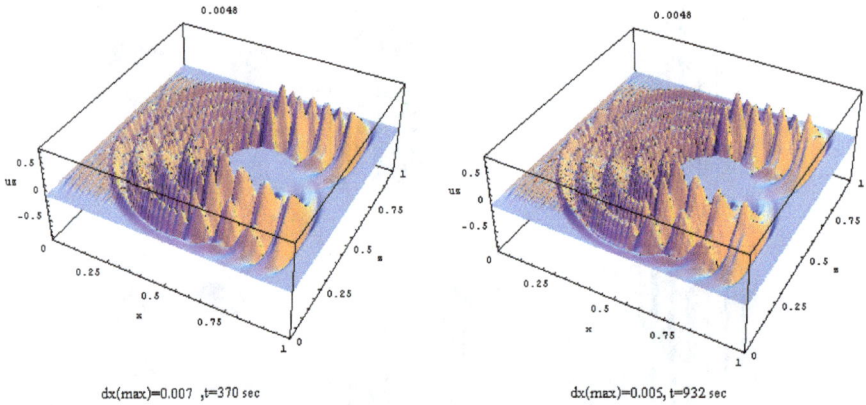

Figure 7. The displacement $u_y(x,z,t)$ at $t = 0.0048$, obtained by the finite difference method; the right and the left figures correspond to grids of size 143×143 and 200×200, respectively.

3. Wavelet based simulation of second order hyperbolic systems (wave equations)

In this section, wavelet-based grid adaptation method is survived for modeling the second order hyperbolic problems (wave equations). The strategy used here is to remove spurious oscillations directly from adapted grids by a post-processing method. The employed stable smoothing method is the cubic smoothing spline, a kind of the Tikhonov regularization method. This section is devoted to the following subsections: interpolating wavelets and corresponding grid adaptation; relevant algorithm for adaptive simulation of wave equations; smoothing splines definition; a 2D P-SV wave propagation example.

3.1. Interpolating wavelets and grid adaptation

In multiresolution analysis, each wavelet coefficient (detail or scale) is uniquely linked to a particular point of underlying grid. This distinctive property is incorporated with compression power of the wavelets and therefore a uniform grid can be adapted by grid reduction technique.

In this method a simple criteria is applied in 1D grid, based on the magnitude of corresponding wavelet coefficients. The existing odd grid points at level j should be removed if corresponding detail coefficients are smaller than predefined threshold (ε); wavelet coefficients and grid points have one-to-one correspondence [69].

In this work, Dubuc-Deslauriers (D-D) interpolating wavelet [69] is used to grid adaptation. The D-D wavelet of order $2M-1$ (with support $Supp(\phi) = [-2M+1, 2M-1]$), is obtained by auto-correlations of Daubechies scaling function of order M (with M vanishing moments).

The D-D scaling function satisfies the interpolating property and has a compact support [69]. In the case of the D-D wavelets, the grid points correspond to the approximation and detail spaces at resolution j are denoted by V_j and W_j, respectively. These sets are locations of the wavelet transform coefficients: the $c(j,k)$ and $d(j,k)$ locations belong to V_j and W_j, respectively. These locations are:

$$V_j = \{x_{j,k} \in [0,1] : x_{j,k} = k / 2^j\}; \quad j \in \mathbb{Z}, \quad k \in \{0,1,\cdots,2^j\}$$

$$W_j = \{x_{j+1,2k+1} \in (0,1) : x_{j+1,2k+1} = (2k+1) / 2^{j+1}\}; \quad j \in \mathbb{Z}, \quad k \in \{0,1,\cdots,2^j - 1\}$$

Regarding interpolation property of D-D scaling functions, the approximation coefficients ($c(J_{min},k)$) at points $x_{J_{min},k} \in V_{J_{min}}$ are equal to sampled values of a considered function $f(x)$ at these points, i.e., $c(J_{min},k) = f(x_{J_{min},k})$. The detail coefficients measured at points $x_{j+1,2k+1} \in W_j$ (of resolution j) is the difference of the function at points $x_{j+1,2k+1}$ (i.e., $f(x_{j+1,2k+1})$) and corresponding predicted values (the estimated ones in the approximation space). The predicted values are those obtained from the approximation space of resolution j (the corresponding points belong to V_j); the estimated values are denoted by $Pf_{j+1}(x_{j+1,2k+1})$. In the D-D wavelets, a simple and physical concept exist for such estimation; the estimation at $x_{j+1,2k+1}$ is attained by the local Lagrange interpolation by the known surrounding grid points $\{x_{j+1,2k} = x_{j,k}\} \in V_j$ (namely, the even-numbered grid points in V_{j+1}). For the D-D wavelet of order $2M-1$, $2M$ most neighbor points, including in V_j, are selected in the vicinity of $x_{j+1,2k+1}$ for interpolation; for points far enough from boundary points, the selected points are: $\{x_{j+1,2k-2n}\}$ $n \in \{-M+1, -M+2, \cdots, M\}$. Using such set, the estimation at the point $x_{j+1,2k+1}$ is denoted by $Pf_{j+1}(x_{j+1,2n+1})$, and the detail coefficients are: $d_{j,n} = f(x_{j+1,2n+1}) - Pf_{j+1}(x_{j+1,2n+1})$.

The above mentioned 1D reduction technique can easily be extended to 2D grid points [30, 34]. The boundary wavelets, introduced by Dohono [78], are also used around edges of finite grid points.

3.2. Wavelet-based adaptive-grid method for solving PDEs

At the time step ($t = t_n$), if the solution of PDE is $f(x,t)$, then the procedure for wavelet-based adaptive solution is:

1. Determining the grids, adapted by adaptive wavelet transform, using $f(x,t_{n-1})$ (step $n-1$). The values of points without $f(x,t_{n-1})$, are obtained by locally interpolation (for example, by the cubic spline method);
2. Computing the spatial derivatives in the adapted grid using local Lagrange interpolation scheme, improved by anti-symmetric end padding method [36]. In this regard, extra non-physical fluctuations, deduced by one sided derivatives, are reduced. Here, five points are locally chosen to calculate derivatives and therefore a high-order numerical scheme is achieved [4, 33];
3. Discretizing PDEs in spatial domain first, and then solving semi-discrete systems. The standard time-stepping methods such as Runge-Kutta schemes can be used to solve ODEs at the time $t=t_n$;
4. Denoising the spurious oscillations directly performed in non-uniform grid by smoothing splines (the post processing stage);
5. Repeating the steps from the beginning.

For 1D data of length n, smoothing spline of degree $2m-1$, needs $m^2.n$ operations [79], and a wavelet transform (employing pyramidal algorithm) uses n operations. Therefore both procedures are fast and effective. However for cost effective simulation, the grid is adapted after several time steps (e.g. 10-20 steps) based on the velocity of moving fronts. In this case, the moving fronts can be properly captured by adding some extra points to the fronts of adapted grid at each resolution level (e.g., 1 or 2 points to each end at each level).

3.3. Smoothing splines

The noisy data are recommended not to be fitted exactly, causing significant distortion particularly in the estimation of derivatives. The smoothing fit is used to remove noisy components in a signal; therefore, interpolation constraint is relaxed. The discrete values of n observations $y_j = y(x_j)$ where $j = 1,2,...,n$ and $x_1 < x_2 < ... < x_n$ are assumed in order to determine a function $f(x)$, that $y_j = f(x_j) + \varepsilon_j$. ε_j are random, uncorrelated errors with zero mean and variance σ_j^2. Here, $f(x)$ is the smoothest possible function in fitting the observations to a specific tolerance. It is well known that the solution to this problem is minimizer, $f(x)$, of the functional:

$$\sum_{j=1}^{n} W_j \left| y_j - f(x_j) \right|^2 + \frac{(1-p)}{p} \int \left| (d^m f(x) / dx^m) \right|^2 dx, \quad 0 \le p \le 1$$

where, $\lambda = (1-p)/p$ ($0 \leq \lambda \leq +\infty$) is a Lagrangian parameter, n is the number of observations, W_j is weight factor at point x_j and m is the derivative order.

It can be shown that spline of degree $k = 2m-1$, having $2m-2$ continuous derivatives, is an optimal solution; where, $n \geq 2m$. In this chapter the cubic smoothing spline is chosen to have a minimum curvature property; hence, $m = 2$ ($2m-1 = 3$) and $f \in C^2[x_1, x_n]$ [80-82].

According to this formula, the natural cubic spline interpolation is obtained by $p = 1$ and the least-squares straight line fit by $p = 0$. In $p < 1$ the interpolating property is vanished while the smoothing property is increased. In the above functional, the errors are measured by summation and the roughness by integral. Therefore, the smoothness and accuracy are obtained simultaneously. In the mentioned equations, the trade-off between smoothness and goodness of fit to the data is controlled by smoothing parameter.

The p should be selected properly, otherwise it leads to over smoothed or under smoothed results. The former are seen in the scheme presented in Reinsch [80], according to Hutchinson-Hoog, [79] and the latter in the scheme offered in Craven-Wahba [83], according to Lee [84].

The smoothness and accuracy in fitting should be incorporated in such a way that the proper adapted grid and accurate solution are obtained simultaneously in adaptive simulations. Hence trial-and-error method is effective in finding appropriate range of p. This study shows that in $\{W_j\} = 1$, the approximated proper values of p are 0.75- 0.95. The lower values of p are applicable for non-uniformly weighed data, i.e. $W_j \geq 1$. The values of $\{W_j\}$ and p can be constant or variable in $\{(x_i, y_i)\}$ sequence [85]. Here, the constant weights and smoothing parameter are studied. The $\{W_j\}$ is assumed as 1 in all considered cases.

Smoothing spline, being less sensitive to noise in the data, has optimal properties for estimating the function and derivatives. The error bounds in estimating the function, belonging to Sobolev space, and its derivatives are presented by Ragozin [86]. He showed that the estimation of function and its corresponding derivatives are converged as the interpolating properties and the sampled points are increased [86].

The smoothing splines work satisfactory for irregular data; this is because the method is a kind of Tikhnov regularization scheme [82, 87, 88].

3.4. Numerical example

The following example is to study the effectiveness of the proposed method concerning some phenomena in elastodynamic problems. Regarding using multiresolution-based adaptive algorithm, the simulation of wave-fields can properly be performed in the media especially one has localized sharp transition of physical properties. The example of such media is solid-solid configurations. In fact, to be analyzed by traditional uniform grid-based methods, these media show major challenges. The main assumptions in the presented example are: 1- applying D-D interpolating wavelet of order 3; 2- decomposing the grid (sampled at $1/2^8$ spatial step in the finest resolution) in three levels; 3- repeating re-adaptation and smoothing processes every ten time steps.

Example: In this example, the wave-fields are presented in inclined two-layered media with sharp transition of physical properties in solid-solid configuration. The numerical methods which do not increase the number of grid points around the interface, have difficulties with the problems of layered media. In such problems, the speeds of elastic waves are largely different. The incident waves, either P or S, can be reflected and refracted from interface in the form of P and S waves.

Schematic shape of considered computational domain is illustrated in Figure 8. It is assumed that the top layer is a soft one, while the other one is a stiff layer. It is considered that at point S, the top layer is subjected to an initial imposed deformation $u_x(x,z,t=0)$ which is: $u_x(x,z,t=0) = \exp(-500((x-0.35)^2 + (z-0.25)^2))$. In the numerical simulation it is assumed that: $p = 0.85$ and $\varepsilon = 10^{-5}$. As mentioned before, the absorbing boundary condition is considered explicitly for simulation of infinite boundaries. This modification, performed for P-SV wave equations, is:

$$\left((\lambda+2\mu)u_{x,xx} + \mu u_{x,zz}\right) + \left((\lambda+\mu)u_{z,xz}\right) = \rho(\ddot{u}_x + Q(x,z).\dot{u}_x)$$
$$\left((\lambda+2\mu)u_{z,zz} + \mu u_{z,xx}\right) + \left((\lambda+\mu)u_{x,xz}\right) = \rho(\ddot{u}_z + Q(x,z).\dot{u}_z)$$

In the above equation, it is assumed that: $Q = a_x(e^{b_x.x^2} + e^{b_x.(1-x)^2}) + a_z(e^{b_z.(1-z)^2})$, where $a_x = a_z = 30$, and $b_x = -110, b_z = -70$. The free boundary is imposed by equivalent force in the free surface boundary [48].

The snapshots of solutions u_x and u_z and corresponding adapted grids are shown in Figures 9-11, respectively. In each figure, the illustrations (a) to (d) correspond to times 0.298, 0.502, 0.658, and 0.886 sec, respectively. It is obvious that, the points are properly adapted and most of the energy is confined in the top layer, the soft one.

4. Conclusion

Multiresolution based adaptive schemes have successfully been used for simulation of the elastic wave propagation problems. Two general approaches are survived: projection and non-projection ones. In the first case the solution grid in not adapted, while in the second one it is done. the results confirm that the projection method is more stable than the common finite differnce schemes; since in the common methods spurious oscillations develop in numerical solutions. In the wavelet-based grid adaptation method, it is shown that grid points concentrate properly in both high-gradient and transition zones. There, for remedy non-physical oscillations the smoothing splines (a regularization method) are used.

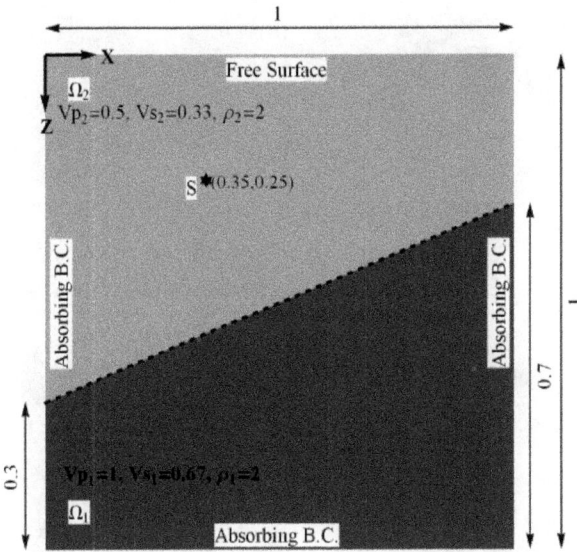

Figure 8. Schematic shape of a inclined two-layered media, solid-solid configuration. The soft layer is above a stiff layer.

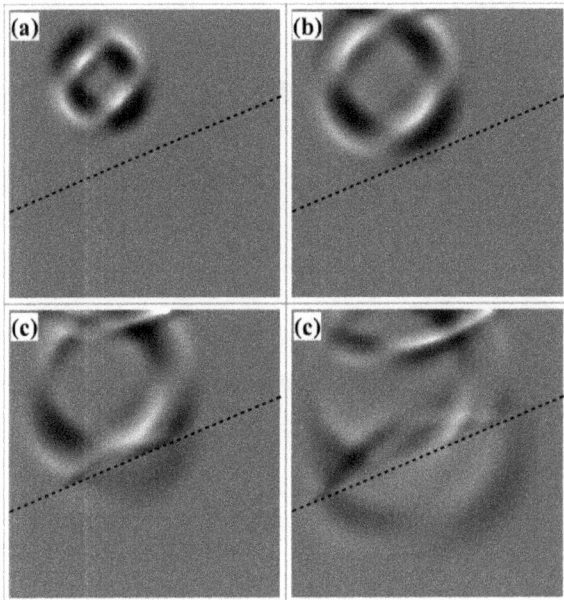

Figure 9. Snapshots of solution u_x at times: a) 0.298, b) 0.502, c) 0.658, d) 0.886.

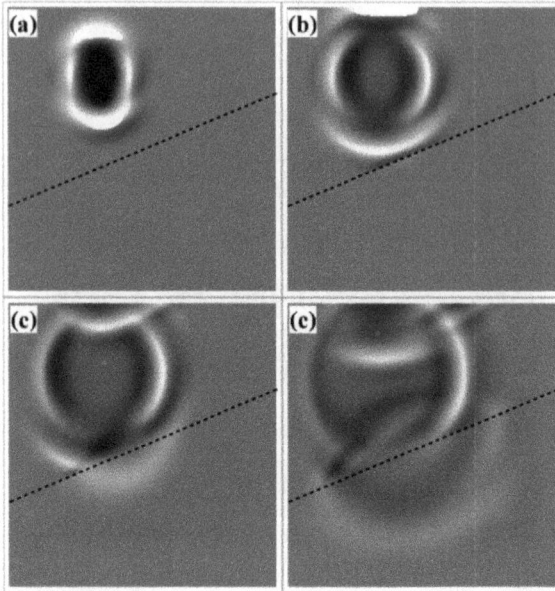

Figure 10. Snapshots of solution u_z at times: a) 0.298, b) 0.502, c) 0.658, d) 0.886.

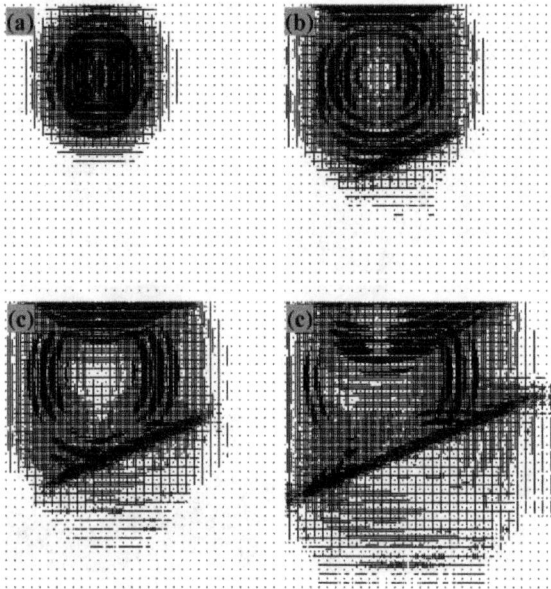

Figure 11. Adapted grid points at times: a) 0.298, b) 0.502, c) 0.658, d) 0.886.

Author details

Hassan Yousefi and Asadollah Noorzad
School of Civil Engineering, University of Tehran, Tehran, Iran

5. References

[1] Sweldens W, Schröder P. Building your own wavelets at home, in: Wavelets in Computer Graphics. ACM SIGGRAPH Course Notes, ACM, 1996; 15-87.

[2] Amaratunga K, Williams JR, Qian S, Weiss J. Wavelet-Galerkin Solutions for One Dimensional Partial Differential Equations. IESL Technical Report No. 92-05, Intelligent Engineering Systems Laboratory, MIT 1992.

[3] Goedecker S, Wavelets and Their Application for the Solution of Partial Differential Equations in Physics. Max-Planck Institute for Solid State Research, Stuttgart, Germany; 2009, goedeck@prr.mpi-stuttgart.mpg.de.

[4] Jameson LM, Miyama T. Wavelet Analysis and Ocean Modeling: a Dynamically Adaptive Numerical Method "WOFD-AHO". Monthly Weather Review 2000; 128(5) 1536-1548.

[5] Cai W, Wang J. Adaptive Multiresolution Collocation Methods for Initial Boundary Value Problems of Nonlinear PDEs. SIAM Journal on Numerical Analysis 1996; 33(3) 937-970.

[6] Beylkin G, Coifman R, Rokhin V. Fast Wavelet Transforms and Numerical Algorithms I. Communications on Pure and Applied Mathematics 1991; 64 141-184.

[7] Beylkin G. On the Representation of Operators in Bases of Compactly Supported Wavelets. SIAM Journal on Numerical Analysis 1992; 6(6) 1716-1740.

[8] Beylkin G, Keiser JM. An Adaptive Pseudo-Wavelet Approach for Solving Nonlinear Partial Differential Equations. In: Dahmen W, Kurdila A, Oswald P (eds.), Multiscale Wavelet Methods for Partial Differential Equations (Wavelet Analysis and Its Applications, V. 6). San Diego: Academic Press; 1997. p137-197.

[9] Mallat SG, A Theory for Multiresolution Signal Decomposition: The Wavelet Representation. IEEE Transactions on Pattern Analysis and Machine Intelligence 1989; 2(7) 647-693.

[10] Pan GW. Wavelets in Electromagnetics and Device Modeling. New Jersey: John Wiley & Sons; 2003.

[11] Xu JC, Shann WC. Wavelet-Galerkin Methods for Two-point Boundary Value Problems. Numerische Mathematik 1994; 37 2703-2716.

[12] Dianfeng LU, Ohyoshi T, Zhu L. Treatment of Boundary Condition in the Application of Wavelet–Galerkin Method to a SH Wave Problem. 1996, Akita Univ. (Japan).

[13] Mishra V, Sabina, Wavelet Galerkin Solutions of Ordinary Differential Equations. International Journal of Mathematical Analysis 2011; 5(9) 407–424.

[14] Amaratunga K, Williams JR. Wavelet-Galerkin Solution of Boundary Value Problems. Archives of Computational Methods in Engineering 1997; 4(3) 243-285.

[15] Williams JR, Amaratunga K, Wavelet Based Green's Function Approach to 2D PDEs. Engineering Computations 1993; 10 349-367.

[16] Dahlke S, Weinreich I. Wavelet-Galerkin Methods: an Adapted Biorthogonal Wavelet Basis. Applied and Computational Harmonic Analysis 1994; 1(3) 237-267.

[17] Williams JR, Amaratunga K. A Multiscale Wavelet Solver with O(n) Complexity. Journal of computational physics 1995;122 30-38.

[18] Qian S, Weiss J. Wavelets and the Numerical Solution of Boundary Value Problems. Applied Mathematics Letter 1993; 6(1) 47-52.

[19] Holmstrom M, Walden J. Adaptive Wavelet Methods for Hyperbolic PDEs. Journal of Scientific Computing 1998; 13(1) 19-49.

[20] Mehra M, Kumar BVR. Time-Accurate Solution of Advection-Diffusion Problems by Wavelet-Taylor-Galerkin Method. Communications in Numerical Methods in Engineering 2005; 21 313-326

[21] Kumar BVR, Mehra M, Wavelet-Taylor Galerkin Method for the Burgers Equation. BIT Numerical Mathematics 2005; 45 543-560.

[22] Mehra M, Kumar BVR. Fast Wavelet Taylor Galerkin Method for Linear and Non-linear Wave Problems. Applied Mathematics and Computation 2007; 189 1292-1299.

[23] Vasilyev OV, Kevlahan NK R. An Adaptive Multilevel Wavelet Collocation Method for Elliptic Problems. Journal of Computational Physics 2005; 206(2) 412-431.

[24] Alam JM, Kevlahan NKR, Vasilyev OV. Simultaneous Space-Time Adaptive Wavelet Solution of Nonlinear Parabolic Differential Equations. Journal of Computational Physics 2006; 214(2) 829-857.

[25] Bertoluzza S, Castro L. Adaptive Wavelet Collocation for Elasticity: First Results. Technical Report 1276, Pub. IAN-CNR de Pavia. 2002.

[26] Griebel, M, Koster, F. Adaptive Wavelet Solvers for the Unsteady Incompressible Navier-Stokes Equations. In: Malek J, Nečas J, Rokyta M (eds.) Advances in Mathematical Fluid Mechanics. Berlin: Springer; 2000. p67-118.

[27] Latto A, Resnikoff HL, Tenenbaum E. The evaluation of connection coefficients of compactly supported wavelets. In: Proceedings of the French-USA Workshop on Wavelets and Turbulence, 1991, Princeton, New York: Springer-Verlag; 1992.

[28] Jang GW, Kim JE, Kim YY. Multiscale Galerkin Method Using Interpolation Wavelets for Two-Dimensional Elliptic Problems in General Domains. International Journal for Numerical Methods in Engineering 2004; 59 225-253.

[29] Kim JE, Jang GW, Kim YY. Adaptive Multiscale Wavelet-Galerkin Analysis for Plane Elasticity Problems and Its Applications to Multiscale Topology Design Optimization. International Journal of Solids and Structures 2003; 40 6473-6496.

[30] Santos JC, Cruz P, Alves MA, Oliveira PJ, Magalhães FD, Mendes A. Adaptive Multiresolution Approach for Two-Dimensional PDEs. Computer Methods in Applied Mechanics and Engineering 2004; 193(3) 405-425.

[31] Cruz P, Mendes A, Magalhães FD. Wavelet-Based Adaptive Grid Method for the Resolution of Nonlinear PDEs. AIChE Journal 2002; 48(4) 774-785.

[32] Cruz P, Mendes A, Magalhães FD. Using Wavelets for Solving PDEs: an Adaptive Collocation Method. Chemical Engineering Science 2001; 56(10) 3305-3309.

[33] Jameson LM. A Wavelet-Optimized, Very High Order Adaptive Grid and Order Numerical Method. SIAM Journal on Scientific Computing 1998; 19(6) 1980-2013.

[34] Holmstrom M. Solving Hyperbolic PDEs Using Interpolating Wavelets. SIAM Journal on Scientific Computing 1999; 21(2) 405-420.

[35] Operto S, Virieux J, Hustedt B, Malfanti F. Adaptive Wavelet-Based Finite-Difference Modeling of SH-Wave Propagation. Geophysical Journal International 2002; 148 476-498.

[36] Yousefi H, Noorzad A, Farjoodi J. Simulating 2D Waves Propagation in Elastic Solid Media Using Wavelet Based Adaptive Method. Journal of Scientific Computing 2010; 42 404-425.

[37] Yousefi H, Noorzad A, Farjoodi J, Vahidi M. Multiresolution-Based Adaptive Simulation of Wave Equation. Applied Mathematics & Information Sciences 2012; 6(1S) 47-58.

[38] Pei ZhL, Fu LY, Yu GX, Zhang LX. A Wavelet-Optimized Adaptive Grid Method for Finite Difference Simulation of Wave Propagation. Bulletin of the Seismological Society of America 2009; 99(1) 302-313.

[39] Alves MA, Cruz P, Mendes A, Magalhães FD, Pinho FT, Oliveira PJ. Adaptive Multiresolution Approach for Solution of Hyperbolic PDEs. Computer Methods in Applied Mechanics and Engineering 2002; 191 3909 - 3928.

[40] Cruz P, Alves MA, Mendes A, Magalhães FD, Mendes A. Solution of Hyperbolic PDEs Using a Stable Adaptive Multiresolution Method. Chemical Engineering Science 2003; 58 1777-1792.

[41] Harten A. Adaptive Multiresolution Schemes for Shock Computations. Journal of Computational Physics 1994; 115(2) 319-338.

[42] Cohen A, Kaber SM, Muller S, Postel M. Fully Adaptive Multiresolution Finite Volume Schemes for Conservation Laws. Mathematics of Computation 2003; 72 183-225.

[43] Muller S, Stiriba Y. Fully Adaptive Multiscale Schemes for Conservation Laws Employing Locally Varying Time Stepping. Journal of Scientific Computing 2007; 30(3) 493-531.

[44] Wei GW, Gu Y. Conjugate Filter Approach For Solving Burgers' Equation. Journal of Computational and Applied Mathematics 2002; 149(2) 439–456.

[45] Gu Y, Wei GW. Conjugate Filter Approach for Shock Capturing. Communications in Numerical Methods in Engineering 2003; 19(2) 99–110.

[46] Diez DC, Gunzburger M, Kunoth A. An Adaptive Wavelet Viscosity Method for Hyperbolic Conservation Laws. Numerical Methods for Partial Differential Equations 2008; 24(6) 1388–1404.

[47] Hong TK, Kennett BLN. A Wavelet-Based Method for Simulation of Two-Dimensional Elastic Wave Propagation. Geophysical Journal International 2002; 150 610–638.

[48] Hong TK, Kennett BLN. On a Wavelet-Based Method for the Numerical Simulation of Wave Propagation. Journal of Computational Physics 2002; 183 577–622.

[49] Hong TK, Kennett BLN. Scattering Attenuation of 2D Elastic Waves: Theory and Numerical Modeling Using a Wavelet-Based Method. Bulletin of the Seismological Society of America 2003; 93(2) 922–938.

[50] Hong TK, Kennett BLN. Modelling of Seismic Waves in Heterogeneous Media Using a Wavelet-Based Method: Application to Fault and Subduction Zones. Geophysical Journal International 2003; 154 483–498.

[51] Hong TK, Kennett BLN. Scattering of Elastic Waves in Media with a Random Distribution of Fluid-Filled Cavities: Theory and Numerical Modeling. Geophysical Journal International 2004; 159 961–977.

[52] Wu Y, McMechan GA. Wave Extrapolation in the Spatial Wavelet Domain with Application to Poststack Reverse-Time Migration. Geophysics 1998; 63(2) 589-600

[53] Gopalakrishnan S, Mitra M. Wavelet Methods for Dynamical Problems with Application to Metallic, Composite, and Nano Composite Structures. New York: CRC Press; 2010.

[54] Moczo P, Kristek J, Halada L. The Finite-Difference Method for Seismologists, an Introduction. Comenius University Bratislava; 2004.

[55] Moczo P, Kristeka J, Galisb M, Pazaka P, Balazovjech M, The Finite-Difference and Finite-Element Modeling of Seismic Wave Propagation and Earthquake Motion. Acta Physica Slovaca 2007; 57(2) 177 – 406.

[56] Prentice JSC. Truncation and Roundoff Errors in Three-Point Approximations of First and Second Derivatives. Applied Mathematics and Computation 2011; 217 4576–4581.

[57] Yamaleev NK. Minimization of the Truncation Error by Grid Adaptation. Journal of Computational Physics 2001; 170 459–497.

[58] Vichnevetsky R. Propagation through Numerical Mesh Refinement for Hyperbolic Equations. Mathematics and Computers in Simulation 1981; 23 344–353.

[59] Vichnevetsky R. Propagation and Spurious Reflection in Finite Element Approximations of Hyperbolic Equations. Computers & Mathematics with Applications 1985; 11(7-8) 733–746.

[60] Vichnevetsky R. Wave Propagation Analysis of Difference Schemes for Hyperbolic Equations: A Review. International Journal for Numerical Methods in Fluids 1987a; 7 409–452.

[61] Vichnevetsky R. Wave Propagation and Reflection in Irregular Grids for Hyperbolic Equations. Applied Numerical Mathematics 1987b; 3 133–166.

[62] Grotjahn R, Obrien JJ. Some Inaccuracies in Finite Differencing Hyperbolic Equations. Monthly Weather Review 1976; 104(2) 180–194.

[63] Trefethen LN. Group Velocity of Finite Difference Schemes. SIAM Review 1982; 23 113–136.

[64] Bazant ZP. Spurious Reflection of Elastic Waves in Nonuniform Finite Element Grids. Computer Methods in Applied Mechanics and Engineering 1978; 16 91-100.

[65] Bazant ZP, Celep Z. Spurious Reflection of Elastic Waves in Nonuniform Meshes of Constant and Linear Strain Finite Elements. Computers & Structures 1982; 15(4) 451–459.

[66] Gottlieb D, Hesthaven JS. Spectral Methods for Hyperbolic Problems. Journal of Computational and Applied Mathematics 2001, 128(1-2) 83–131.

[67] Engquist B, Kreiss HO. Difference and Finite Element Methods for Hyperbolic Differential Equations. Computer Methods in Applied Mechanics and Engineering 1979; 17-18 581–596.

[68] Chin RCY. Dispersion and Gibbs Phenomenon Associated with Difference Approximations to Initial Boundary-Value Problems for Hyperbolic Equations. Journal of Computational Physics 1975; 18 233–247.

[69] Mallet S. A Wavelet Tour of Signal Processing. San Diego: Academic Press; 1998.

[70] Jawerth B, Sweldens W. An Overview of Wavelet Based Multiresolution Analysis. SIAM Review 1994; 36(3) 377-412.

[71] Williams JR, Amaratunga K. Introduction to Wavelets in Engineering. IESL Technical Report No. 92-07, Intelligent Engineering Systems Lab oratory, MIT, 1992.

[72] Daubechies I. Orthonormal Bases of Compactly Supported Wavelets. Communications on Pure and Applied Mathematics 1988; 41 909-996.

[73] Gines D, Beylkin G, Dunn J. LU Factorization of Non-Standard Forms and Direct Multiresolution Solvers. Applied and Computational Harmonic Analysis 1998; 5(2) 156-201.

[74] Beylkin G, Keiser JM, Vozovoi L. A New Class of Time Discretization Schemes for the Solution of Nonlinear PDEs. Journal of Computational Physics 1998;147 362-387.

[75] Beylkin G, Keiser JM. On the Adaptive Numerical Solution of Nonlinear Partial Differential Equations in Wavelet Bases. Journal of Computational Physics 1997; 132 233-259.

[76] Sochacki J, Kubichek R, George J, Fletcher WR, Smithson S. Absorbing Boundary Conditions and Surface Waves. Geophysics 1987; 52(1) 60-71.

[77] Boore DM. Finite Difference Methods for Seismic Wave Propagation in Heterogeneous Materials. In: Bolt BA (ed.) Methods in Computational Physics, volume 11, Seismology: SurfaceWaves and Earth Oscillations. New York: Academic Press; 1972. p1-37.

[78] Donoho DL. Interpolating Wavelet Transforms. Technical Report 408. Department of Statistics, Stanford University. 1992

[79] Hutchinson MF, de Hoog FR. Smoothing Noisy Data with Spline Functions. Numerische Mathematik 1985;47(1) 99–106.

[80] Reinsch CH. Smoothing by Spline Functions. Numerische Mathematik 1967;10 177–183.

[81] Reinsch CH. Smoothing by Spline Functions. II. Numerische Mathematik 1971;16 451–454.

[82] Unser M. Splines: A Perfect Fit for Signal/Image Processing. IEEE Signal Proc Mag 1999;16(6) 22–38.

[83] Craven P, Wahba G. Smoothing Noisy Data with Spline Functions: Estimating the Correct Degree of Smoothing by the Method of Generalized Cross Validation. Numerische Mathematik 1979;31 377–403.

[84] Lee TCM. Smoothing Parameter Selection for Smoothing Splines: A Simulation Study. Computational Statistics & Data Analysis 2003;42(1-2) 139-148

[85] Lee TCM. Improved Smoothing Spline Regression by Combining Estimates of Different Smoothness. Statistics & Probability Letters 2004; 67(2) 133–140.

[86] Ragozin DL. Error Bounds for Derivative Estimates Based on Spline Smoothing of Exact or Noisy Data. Journal of Approximation Theory 1983;37 335–355.

[87] Petrov YP, Sizikov VS. Well-Posed, Ill-Posed, and Intermediate Problems with Applications. VSP; 2005.

[88] Hansen PC. Rank-Deficient and Discrete Ill-Posed Problems, Philadelphia: SIAM; 1998.

Chapter 15

Wave Iterative Method for Electromagnetic Simulation

Somsak Akatimagool and Saran Choocadee

Additional information is available at the end of the chapter

1. Introduction

Presently, the microwave circuit design is the most critical and necessary for modern communication systems. The problems with the design of circuits often include a lack of equipment and tools used to design and build them. Numerical methods related to electromagnetic waves have caused a revolution in microwave engineering; techniques such as FDTD (Finite Differential Time Domain) [1], TLM (Transmission Line Matrix) [2], and the Moment method [3] have been developed. Each method has respective disadvantages and limitations of usage foe both research and education. To perform the electromagnetic wave analysis of waveguide structures, several classic approaches can be found in the literature. A practical example of computer aid design is that the inductive iris in a rectangular waveguide has been analyzed with high accuracy through approximate modeling utilizing the full wave Mode Matching Method (MMM) [4] and Transmission Line Matrix (TLM) [5]. Therefore, the development and optimization of numerical methods are important for an efficient electromagnetic simulation tool [6-7].

In this chapter, we study and introduce an efficient electromagnetic simulation tool for analysis of inductive and capacitive obstacles and rectangular window in waveguides. The Wave Iterative Method (WIM) based on iterative procedure and wave propagation theory is proposed. This method has been combined with mode matching technique to characterize the obstacle in a rectangular waveguide. Also, the compact CAD tool and presenting the electromagnetic field distributions included in this topic.

2. Wave propagation and waveguide

Several years ago, the characteristics of waves in source-free, homogeneous regions of space have been discussed. Accordingly, we present this section by considering the reflection, transmission and incident experienced by electromagnetic waves on stratified dielectric surface.

There are many situations where a wave passes through several layers of different materials. One example is the passage of electromagnetic through a free space with transmission line mode or TEM. There are several ways use to analyze the correct results, but a very common method is an impedance transformation that integrates reflections into a single parameter to generate the numerical results.

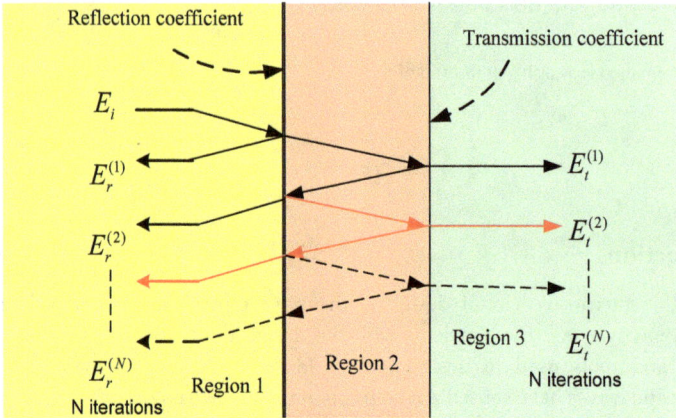

Figure 1. (a) The wave propagation in stratified interface

Fig.1 shows a stratified dielectric region. When a wave is normally incident from the region 1 and transmitted into region 2 and region 3 respectively, an infinite number of transmitted, reflected waves in the same wavelength are produced at the boundary interface. At a time, the all incident waves (E_i), reflected waves (E_r) and transmitted waves (E_t) can be considered to be a single of each wave. We can develop the reflection coefficient expression by following the progression of the incident waves and its reflections. Similarly, the transmission coefficient expression can be presented by following the progression of the incident wave and its transmissions. Finally, we can use the relationship of expressions to calculate the properties of boundary interface.

Waveguides are used to transport electromagnetic energy along a fixed path that carries non-TEM modes, often called waveguide modes. Most important of waveguide properties is that they can support an infinite number mode of filed generated by diffraction form each interface within waveguide [8]. Waveguides are almost operated so that only a single propagation mode is present because the presence of more than one propagating mode causes dispersion. Typically, they must be operated over smaller bandwidths and smaller losses than transmission line, which makes them attractive for many applications. One of several examples is to place obstacle element in waveguides to make devices such as lumped element, filters, couplers or antenna. Fig.2(a) shows two thin metal fins placed on the top and bottom walls of a waveguide, called a capacitive window. Considering in the gap of obstacle, when the electric filed within the window increases, the energy stored between two conducting surfaces will increase. We observed that the equivalent circuit of obstacle is a shunt capacitor.

Similarly, Fig.2(b) shows an inductive window which the most magnetic fields can flow through in the obstacle width. The equivalent circuit of obstacle is a shunt inductor.

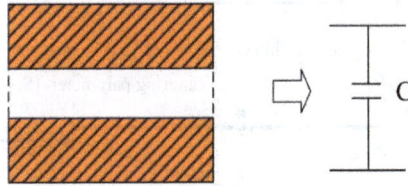

(a) capacitive window

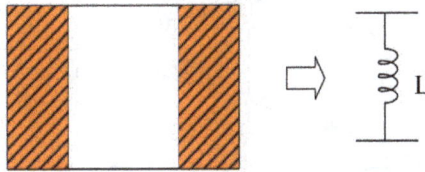

(b) inductive window

Figure 2. (a) The capacitive and inductive windows

3. Wave iterative method (WIM)

An analysis of the electromagnetic wave properties within a waveguide consists of TE and TM field components. Most wave analyses are calculated in the spectral domain based on the series integration equation to present the electromagnetic field and to analyze the two ports network parameters such that the result of the frequency response reflects the characteristic of various planar circuits. In this section, we will present the cooperation of waves between the real domain and the spectrum domain.

The calculating concept for the electromagnetic wave propagation in a waveguide is based on the Wave Iterative Method (WIM) [9-11]. The operating process, as shown in Fig.3, present the amplitude and direction of the incident, reflected, and transmitted waves what propagate in the waveguide obstacle. On the obstacle, the waves are calculated in the real domain (pixels) and the waves in the free space are calculated in the spectrum domain (modes). To alternate between both domains, we use the Fast Fourier Transform (FFT) to reduce the computation time and show directly the electromagnetic field in the real domain.

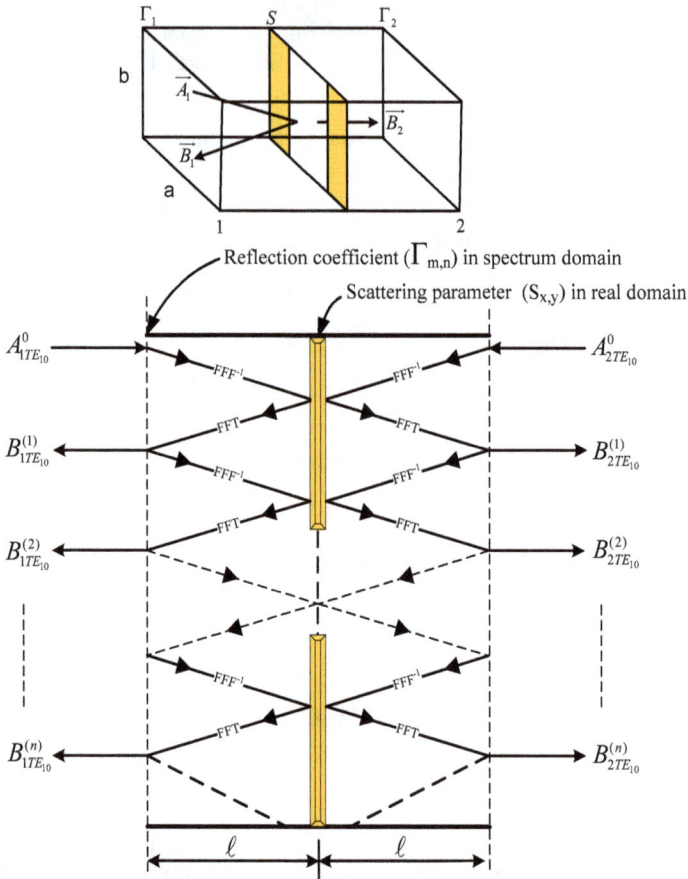

Figure 3. (a) The wave propagation in the waveguide

In Fig.3, the first step, the TE10 field which is the dominant mode, propagate into the obstacle as iris circuit, and then the higher-mode waves caused by the scatter on the conductor surface will be incident and will reflect within the waveguide. An infinite numbers of reflected and transmitted waves are produced at the obstacle interface. Finally the calculation of waves will use the principle of mode matching at the input and the output ports of the waveguide.

The initial value of dominant mode, $TE_{1,0}$ is

$$A_{x,y}^{(0)} = -\sqrt{\frac{2}{ab}} \sin(\frac{\pi x}{a}), \, ,$$

(1)

where a and b are the dimension of the waveguide. The reflected waves $\left(B_{x,y}\right)$ in n^{th} iteration on the obstacle are

$$B^{(n)}_{x,y} = S_{x,y} A^{(n-1)}_{x,y}. \tag{2}$$

The scattering parameter $\left(S_{x,y}\right)$ of waves in the real domain for a two-port network is defined as

$$\left[S_{x,y}\right] = \begin{bmatrix} -S_{C(x,y)} & S_{D(x,y)} \\ S_{D(x,y)} & -S_{C(x,y)} \end{bmatrix}, \tag{3}$$

where S_C is equal to 1 in the conductor area and S_D is equal to 1 for the free space. To transform the wave from the real domain to the spectrum domain, we use the Fast Fourier Transform (FFT) of the TE/TM modes as

$$B^{(n)}_{m,n} = Modal\_FFT\left(B^{(n)}_{x,y}\right). \tag{4}$$

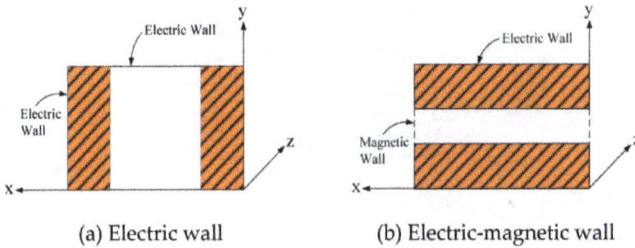

(a) Electric wall   (b) Electric-magnetic wall

Figure 4. The electric and magnetic walls around the waveguide

The basic functions of TE field components in the electric wall [12], as shown in Fig.4(a) for an inductive obstacle, are

$$\vec{E}^{\alpha}_{x(x,y)} = K^{\alpha}_{x} \cos\left(\frac{m\pi x}{a}\right) \sin\left(\frac{n\pi y}{b}\right), \tag{5}$$

$$\vec{E}^{\alpha}_{y(x,y)} = -K^{\alpha}_{y} \sin\left(\frac{m\pi x}{a}\right) \cos\left(\frac{n\pi y}{b}\right), \tag{6}$$

where α refers to the TE mode and K is the constant value with respect to the x and y directions. The equations are defined as

$$K_x^{TE} = \frac{n\sqrt{\frac{2\tau}{ab}}}{b\sqrt{\frac{m^2}{a^2} + \frac{n^2}{b^2}}}, \quad K_x^{TM} = \frac{m\sqrt{\frac{2\tau}{ab}}}{a\sqrt{\frac{m^2}{a^2} + \frac{n^2}{b^2}}}$$

$$K_y^{TE} = \frac{-m\sqrt{\frac{2\tau}{ab}}}{a\sqrt{\frac{m^2}{a^2} + \frac{n^2}{b^2}}}, \quad K_y^{TM} = \frac{n\sqrt{\frac{2\tau}{ab}}}{b\sqrt{\frac{m^2}{a^2} + \frac{n^2}{b^2}}}$$

(7)

where $\tau = 1$ if $m, n = 0$ and $\tau = 2$ if $m, n \neq 0$.

The basic functions of TM field components in the electric-magnetic wall [12], as shown in Fig.4(b) for a capacitive obstacle, are

$$\vec{E}_{x(x,y)}^{\alpha} = -K_x^{\alpha} \sin\left(\frac{m\pi x}{a}\right)\sin\left(\frac{n\pi y}{b}\right),$$

(8)

$$\vec{E}_{y(x,y)}^{\alpha} = K_y^{\alpha} \cos\left(\frac{m\pi x}{a}\right)\cos\left(\frac{n\pi y}{b}\right),$$

(9)

where α refers to the TM mode and K is the constant value respect to the x and y directions. The equations are defined as

$$K_x^{TE} = \frac{n\sqrt{\frac{2\tau}{ab}}}{b\sqrt{\frac{m^2}{a^2} + \frac{n^2}{b^2}}}, \quad K_x^{TM} = \frac{m\sqrt{\frac{2\tau}{ab}}}{a\sqrt{\frac{m^2}{a^2} + \frac{n^2}{b^2}}}$$

$$K_y^{TE} = \frac{m\sqrt{\frac{2\tau}{ab}}}{a\sqrt{\frac{m^2}{a^2} + \frac{n^2}{b^2}}}, \quad K_y^{TM} = \frac{-n\sqrt{\frac{2\tau}{ab}}}{b\sqrt{\frac{m^2}{a^2} + \frac{n^2}{b^2}}}$$

(10)

where $\tau = 1$ if $m, n = 0$ and $\tau = 2$ if $m, n \neq 0$.

At the input port of the waveguide, the incident waves $\left(A_{x,y}\right)$ of higher-order modes will feedback into the obstacle of the waveguide. The equation for incident waves $\left(A_{m,n}\right)$ is

$$A_{m,n}^{(n)} = \Gamma_{m,n} B_{m,n}^{(n)} + A_{1,0}^{TE/TM},$$

(11)

where $A_{1,0}$ is an initial exciton source. The reflection coefficient $\left(\Gamma_{m,n}\right)$ of waves in the spectrum domain at the input and the output ports of the waveguide can be written as

$$\Gamma_{m,n} = \Gamma_{m,n}^{TE/TM} = \frac{1 - Z_0 Y_{m,n}^{TE/TM}}{1 + Z_0 Y_{m,n}^{TE/TM}},$$

(12)

where Z_0 is the intrinsic impedance of the dominant mode and $Y_{m,n}^{TE/TM}$ is the TE/TM modes admittance with the orders of m and n which can be expressed as

$$Y_{m,n}^{TE} = \frac{\left(\dfrac{m\pi}{a}\right)^2 + \left(\dfrac{n\pi}{b}\right)^2 - \omega^2\mu\varepsilon}{j\omega\mu}, \tag{13}$$

$$Y_{m,n}^{TM} = \frac{j\omega\varepsilon}{\left(\dfrac{m\pi}{a}\right)^2 + \left(\dfrac{n\pi}{b}\right)^2 - \omega^2\mu\varepsilon}. \tag{14}$$

Therefore, considering the waves on the obstacle, the waves $\left(A_{m,n}\right)$ will be transformed by using the Inverse Fast Fourier Transform (IFFT) to analyze the basic functions of TE/TM field components by using Eq.(5)-(8), to come back to the real domain as

$$A_{x,y}^{(n)} = Pixel\_IFFT\left(B_{m,n}^{(n)}\right). \tag{15}$$

The implementation of the Wave Iterative method consists of a recurrence relationship of wave between the propagation in both sides of the waveguide and the propagation on the obstacle. From Eq.(2) and Eq.(11), the total waves at n^{th} iteration are

$$A_{m,n}^{(n)} = \Gamma_{m,n}\hat{S}A_{m,n}^{(n-1)} + A_{1,0}^{TE/TM}, \tag{16}$$

where $\hat{S}$ is the spectrum operator of scattering coefficient.

Considering the wave propagation in the rectangular waveguide with zero thickness obstacles, the electromagnetic equivalent circuit of the obstacle section is presented to identify the impedance (Z) element and the input and the output sections of the waveguide are presented by dominant mode admittance, as shown in Fig.5.

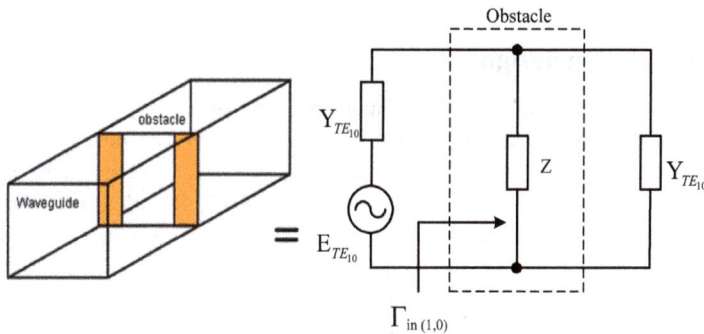

Figure 5. (a) The electromagnetic equivalent circuit of an obstacle

At the convergence condition, the reflected wave on the obstacle at n^{th} iteration, is tend toward zero, the total waves will be the steady state field. So, we can obtain the input reflection coefficient of obstacle circuit of dominant mode, $\Gamma_{in(1,0)}$ as,

$$\Gamma_{in(1,0)} = \frac{B_{1,0}^{TE/TM}}{A_{1,0}^{TE/TM}} = \frac{\sum_{n=1}^{N} B(n)_{1,0}^{TE/TM}}{A_{1,0}^{TE/TM}}. \tag{17}$$

The input impedance, $Z_{in(1,0)}$ of zero thickness obstacles in the rectangular waveguide can be written as

$$Z_{in(1,0)} = Z_0 \left(\frac{\Gamma_{in(1,0)} - 1}{2\Gamma_{in(1,0)}} \right). \tag{18}$$

Then, the input reactance will be a positive imaginary value for an inductive element and a negative imaginary value for a capacitive element. If the input impedance is equal to zero or infinity, the obtained element will be a LC series or parallel resonant circuit respectively. The net 2.5D electric field distribution on the obstacle by summing the amplitude of the incident and the reflected waves of N iterations can be expressed as

$$E_{x,y} = \sqrt{Z_0} \sum_{n=1}^{N} \left(A_{x,y}^{(n)} + B_{x,y}^{(n)} \right). \tag{19}$$

Similarly, the net 2.5D magnetic fields can be expressed by subtracting the amplitude of the incident and the reflected waves of N iterations series as can be expressed as

$$H_{x,y} = \frac{1}{\sqrt{Z_0}} \sum_{n=1}^{N} \left(A_{x,y}^{(n)} - B_{x,y}^{(n)} \right). \tag{20}$$

4. WCD simulation design

In this topic, we present a developed simulation program that conducts a numerical analysis by using the Wave Iterative Method (WIM), shown in the Fig.6. This developed simulation program, called WCD (Waveguide Circuit Design), consists of a main menu, parameter setup, and a design and display window, as shown in Fig.7. The WCD is constructed by using the GUI (Graphic User Interface) of MATLAB®. The user can setup the initial values that are used for calculating of the two ports network, calculate the obstacle characteristics and select the display windows of the simulated results. With the simulation program, it is possible to analyze and design the waveguide iris, waveguide filter and also visualize the 2.5D electromagnetic field distribution.

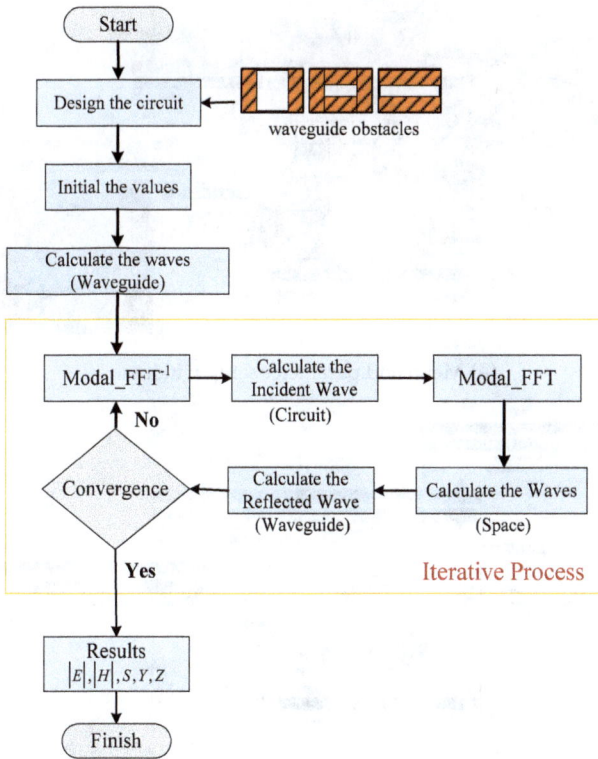

Figure 6. Flowchart of WCD simulated process

The simulated process of the WCD, as shown in Fig.6, is following:

1. Start the simulation program form the main menu which consists of an inductive, a capacitive and a window obstacle bottom, as shown in Fig.7 (a).
2. Design the waveguide obstacle structure what the user can determinate the desired obstacle dimensions, as shown in Fig.7(a).
3. Set the initial values; such as the size, length, frequency operating, dielectric constant, etc., as shown in Fig.7(a).
4. Calculate the incident wave of the dominant mode.
5. Calculate the reflected and the transmitted waves on the obstacle surface in the real domain and in the free space in the spectrum domain. In this process, an infinite number of the transmitted and the reflected waves are produced on the obstacle interface. However the reasonable numbers of iteration are determined, the process will result in termination.
6. Present the simulated results; such as the electric and the magnetic field distribution, the impedance, the scattering and the admittance parameters, as shown in Fig.7(b).

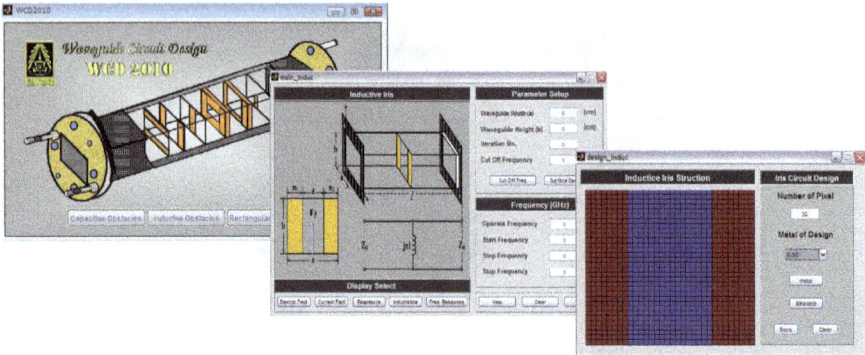

(a) Menu and parameter setup windows

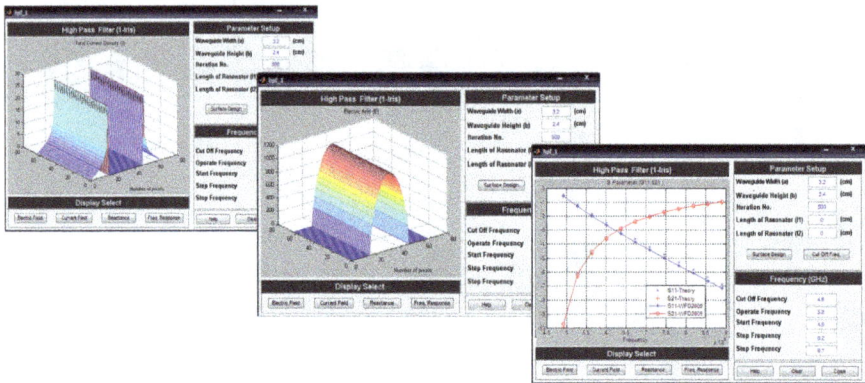

(b) Display windows

Figure 7. Waveguide Circuit Design (WCD) Tool

5. Simulation results

In this chapter, we present the analysis of the inductive, the capacitive obstacles and the rectangular window in the waveguide by using the developed WCD simulation and also introduce the electromagnetic distribution in the waveguide. The simulated WCD results will be compared with the CST simulation.

5.1. Inductive obstacle analysis

The vertical obstacle section transforms the inductive equivalent circuit, shown in Fig.8. The dimensions of rectangular waveguide consists of a width (a) equal to 6.4 cm., a height (b) equal to 3.2 cm. and the usable obstacle width (d) equal to 3.2 cm. and 4 cm. respectively. The cutoff frequency of the waveguide is 3.24 GHz.

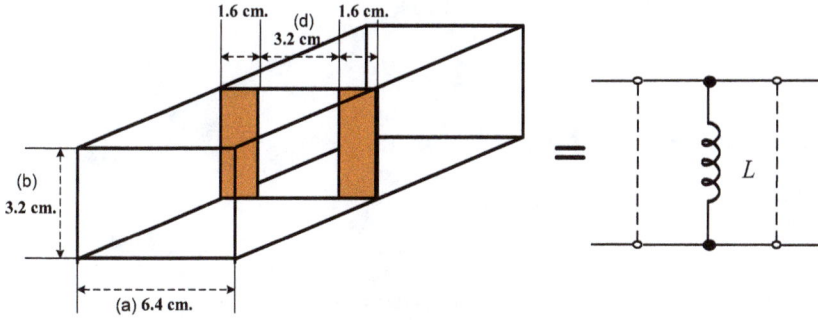

Figure 8. (a) Inductive obstacle structure

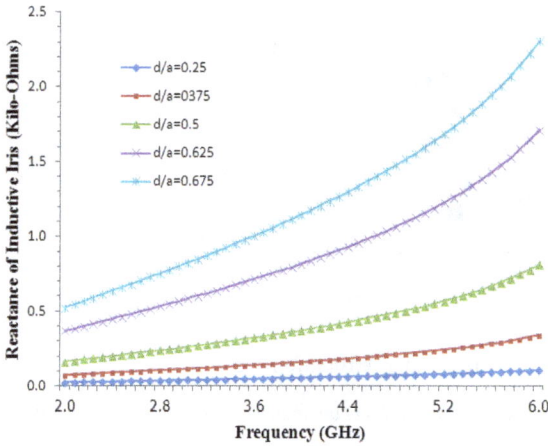

Figure 9. The inductance of inductive obstacle

Fig.9 presents the input reactance of the inductive obstacle in frequency range from 2-6 GHz following the variation the ratio of obstacle and the waveguide width. We observed that the obstacle width increased, the reactance of the inductive obstacle will increase that is consistent with the properties of the inductor.

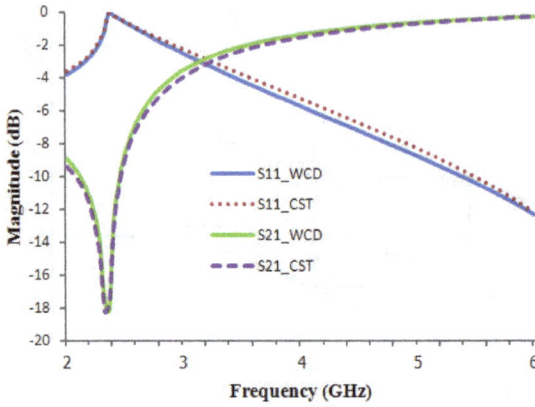

(a) Obstacle structure ratio of d/a = 0.5

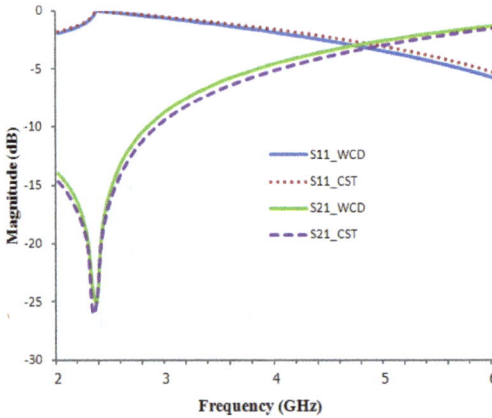

(b) Obstacle structure ratio of d/a = 0.375

Figure 10. Comparison of dB(S11), dB(S21) of the inductive obstacle between the WCD and the CST simulation

Fig.10(a) presents the comparison of dB(S11) and dB(S21) of the inductive obstacle structure between the WCD and the CST simulation In the case of a obstacle width equal to 3.2 cm., we can obtain the -3 dB cutoff frequency at 3.15 GHz. and Fig.10(b) presents the comparison of dB(S11) and dB(S21) of the inductive obstacle structure between the WCD and the CST simulation, at a obstacle width equal to 4.0 cm. We can obtain the -3 dB cutoff frequency at 4.78 GHz. This comparison revealed good agreement. We observed that the shunt inductive circuit is presented as a high pass filter.

(a) Electric field

(b) Magnetic field

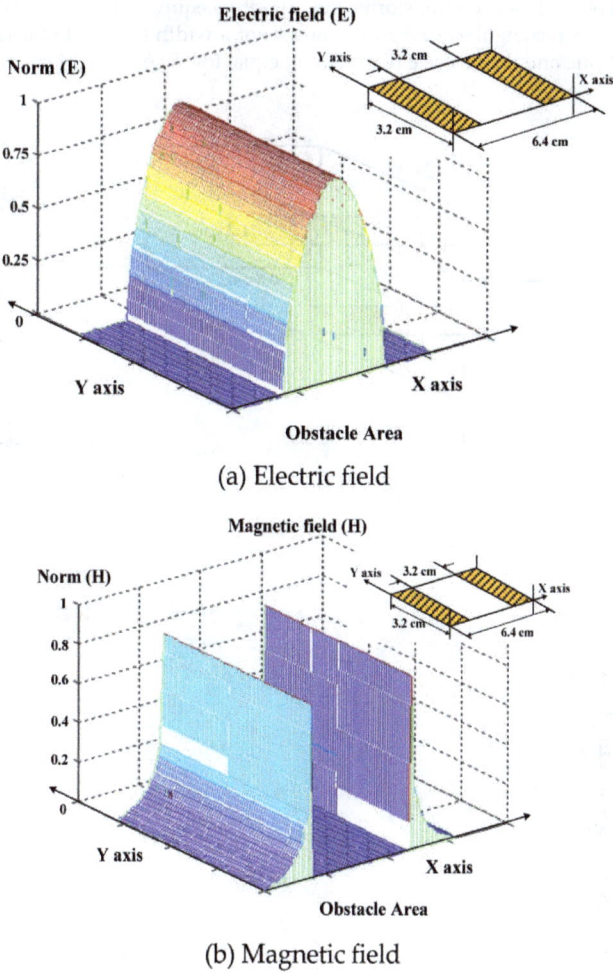

Figure 11. The 2.5D electromagnetic field illustration

Fig.11 shows the 2.5D normalized electromagnetic field distributions on the inductive obstacle that are agreements with the electromagnetic field theory. The electric field density of TE10 fundamental mode, as shown in Fig.11(a) is maximum at the center obstacle and is minimum at the discontinuity of boundary obstacle. We observed that the electric field on the conducting surface is null. In Fig.11 (b), the amplitude of magnetic field on discontinuity of boundary obstacle is peak, and decreases until to zero following the distance on the both sides of the conductor.

5.2. Capacitive obstacle analysis

The horizontal obstacle section transforms the capacitive equivalent circuit, shown in Fig.12. The dimensions of rectangular waveguide consists of a width (a) equal to 6.4 cm., a height (b) equal to 3.2 cm., and the obstacle of iris circuit equal to 0.8 cm. and 0.4 cm. respectively.

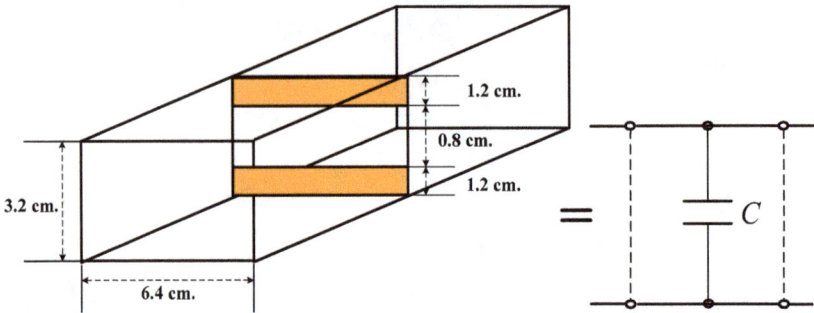

Figure 12. (a) Capacitive obstacle structure

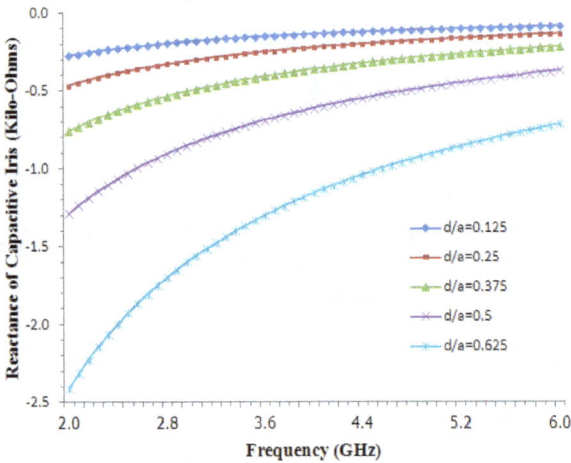

Figure 13. The capacitance of capacitive obstacle

Fig.13 presents the input reactance of the capacitive obstacle in frequency range from 2-6 GHz following the variation the ratio of obstacle and waveguide width. We observed that the obstacle width increased, the reactance of the capacitive obstacle will increase that is consistent with the properties of the capacitor.

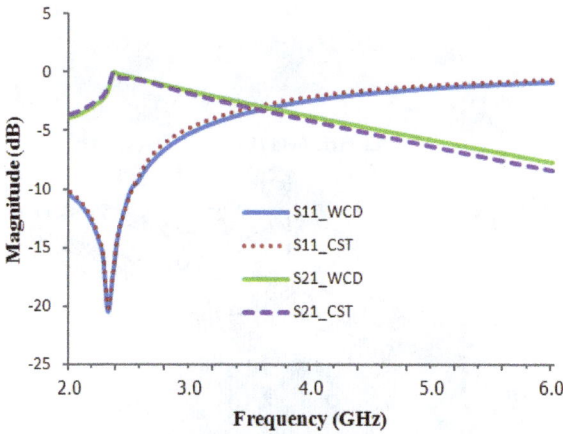

(a) Obstacle structure ratio of $d/a = 0.25$

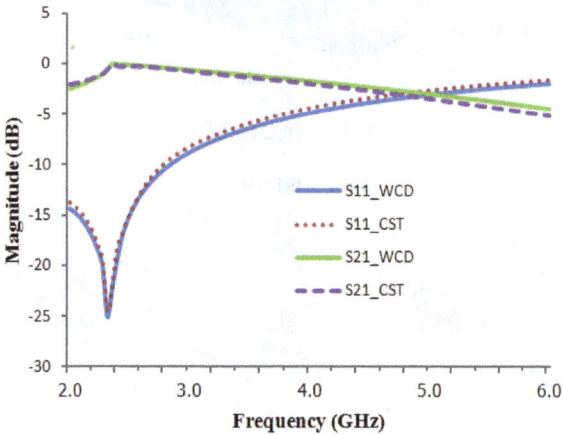

(b) Obstacle structure ratio of $d/a = 0.125$

Figure 14. Comparison of dB(S11), dB(S21) of capacitive obstacle between the WCD and the CST simulation

Fig.14(a) presents the comparison of dB(S11) and dB(S21) of the capacitive obstacle structure between the WCD and the CST simulation. In the case of a width of obstacle structure is equal to 0.8 cm., we can obtain the -3 dB cutoff frequency at 3.43 GHz. Fig.14(b) presents the comparison of dB(S11) and dB(S21) of capacitive obstacle structure between The WCD and the CST simulation, of the obstacle structure width equal to 0.4 cm. We can obtain the -3 dB cutoff frequency at 4.82 GHz. This comparison revealed good agreement. We observed that the shunt capacitive circuit is presented as a low pass filter.

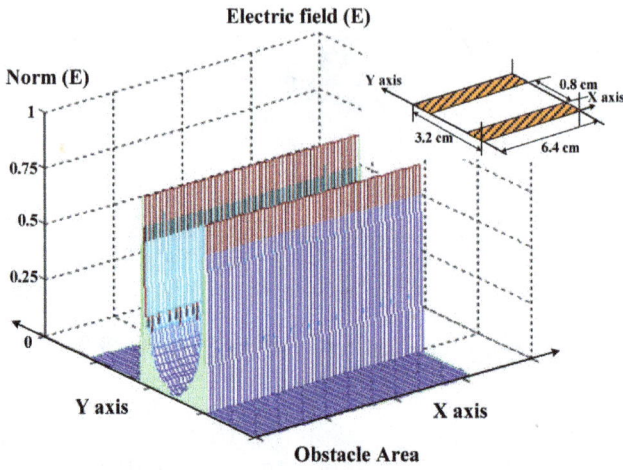

(a) Electric field

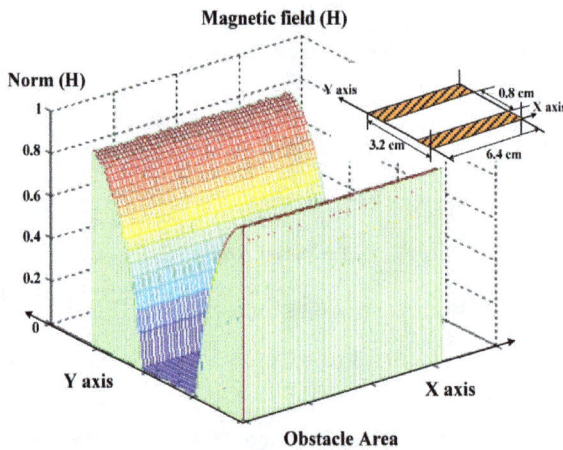

(b) Magnetic field

Figure 15. (a) The 2.5D electromagnetic field illustration

Fig.15 shows the 2.5D normalized electromagnetic field distributions on the capacitive obstacle that are agreements with the electromagnetic field theory. The electric field density of TE_{01} fundamental mode, as shown in Fig.15(a), is minimum at the center obstacle and is maximum at the discontinuity of boundary obstacle, and we observed that the electric field on the conducting surface is null. In Fig.15(b), the amplitude of magnetic field on the both sides of the conducting surface in y direction is maximal and decreases until to zero at the obstacle internal edges.

5.3. Rectangular window analysis

Finally, we study the rectangular window with a width of window equal to 0.9 cm. and a height equal to 0.675 cm., then we compare the analyzed results between the WCD and the CST simulation. The rectangular window section in the waveguide transforms the LC shunt parallel resonant circuit, as shown in Fig.16. We use the dimensions of rectangular waveguide consisting of a width (a) equal to 2.4 cm., a height (b) equal to 1.8 cm., the rectangular window dimensions are determine as in Fig.16. The cutoff frequency of the waveguide is 6.25 GHz. After simulating by the WCD tool, the input reactance of the rectangular window can be obtained at a resonant frequency of 15.5 GHz. An example of the reactance variation on the WCD display window is shown in Fig.17.

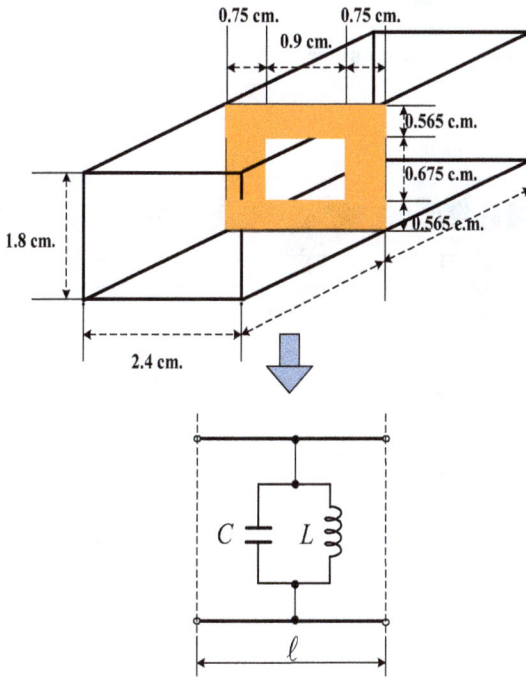

Figure 16. (a) The rectangular window structure

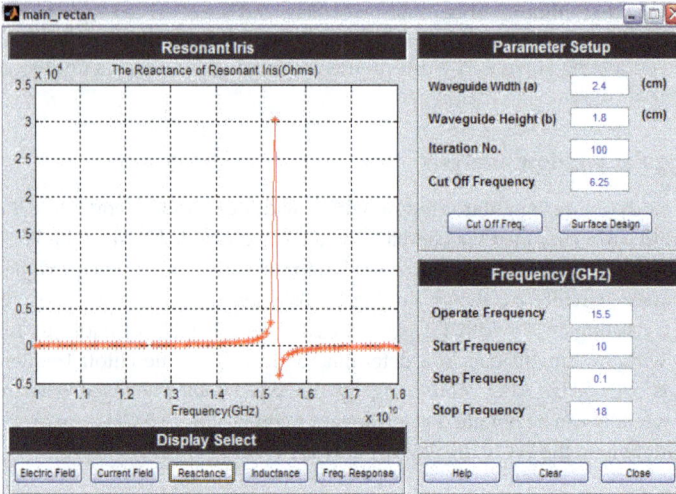

Figure 17. The reactance of rectangular window

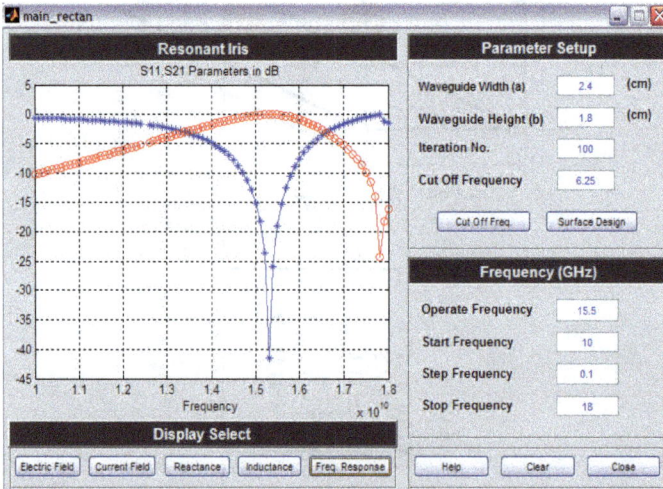

Figure 18. The dB(S11) and dB(S21) parameters of rectangular window

Fig.18 shows the dB(S11) and dB(S21) parameters of the operating frequency of 10-18 GHz, the rectangular windows is presented as a band pass filter, the -3 dB lower and upper cutoff frequency is at 15.25 GHz and at 16.12 GHz respectively. We observed that the center frequency is equal to 11 GHz and the bandwidth is equal to 2.98 GHz.

Fig.19 presents the comparison of dB(S11) and dB(S21) of the rectangular obstacle structure between the both simulations, we can obtain the -3 dB lower cutoff frequency at 13.25 GHz. and -3 dB upper cutoff frequency at 16.35 GHz. This comparison revealed good agreement.

We observed that the equivalent circuit at the resonant frequency is a shunt LC circuit.

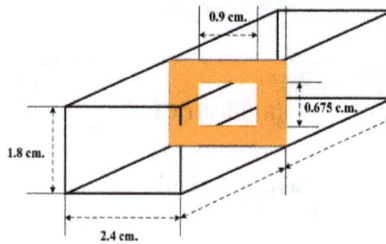

(a) Rectangular obstacle structure

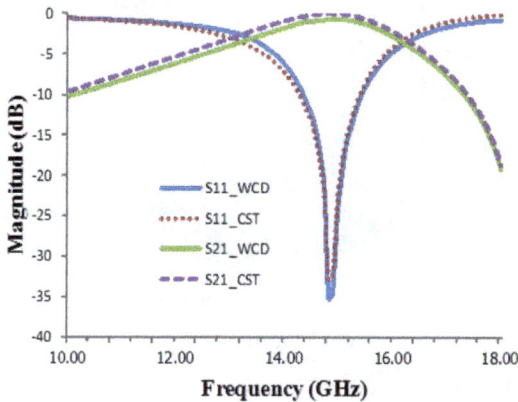

(b) Comparison of dB(S11), dB(S21) of rectangular obstacle

Figure 19. (a) The rectangular obstacle structure and the comparison of dB(S11), dB(S21) of rectangular obstacle between the WCD and CST simulation

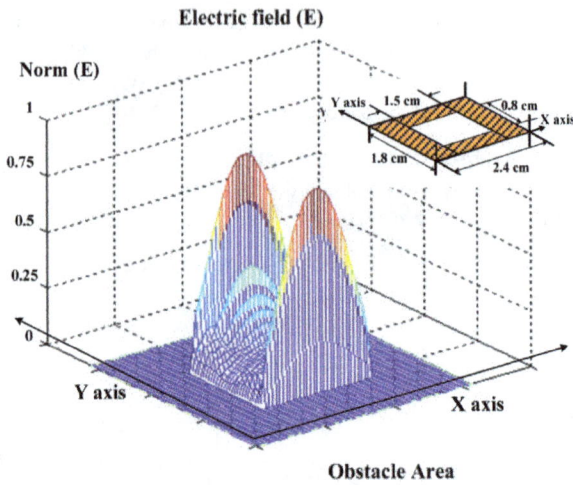

(a) Electric field

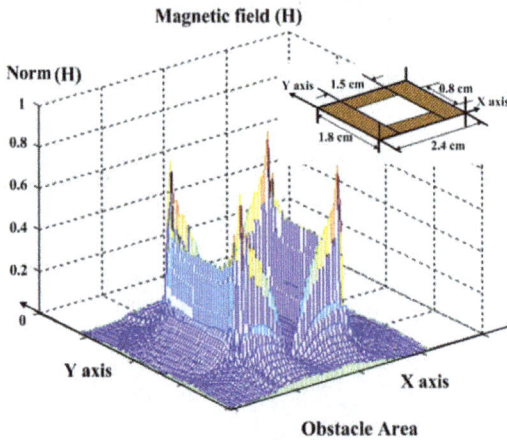

(b) Magnetic field

Figure 20. (a) The 2.5D electromagnetic field illustration

Fig.20 shows the 2.5D electric and current distributions of TE01 fundamental mode on the rectangular window which are agreements with electromagnetic field theory. In Fig.20 (a), the normalized electric field peak is at the obstacle edge in x direction and minimum values are at the rectangular center window. In Fig.20(b), most current density distributions are presented at around conducting edge. The field density distribution can be used to lean qualitatively about the obstacle's operating behavior.

For efficiency of calculation, the computation time of the WCD simulation depends on the EM algorithm associating a pixel of obstacle structure to a class in an iterative process. In this simulation, the calculation at the 10 frequencies steps with 32*64 of cell numbers and 1 mm. of cell size will be equal to approximately 45 seconds of computing time, which is less than 2 times of the CST simulation (run on Pentium IV Processor, 1 GB of RAM at 3.0 GHz of speed).

6. Conclusion

The capability of developed WCD simulation based on the Wave Iterative Method (WIM) has been presented to the complex description of the 2.5D electromagnetic field and the characteristic analysis of the obstacle structures. The WCD simulation is properly conceived to apply for the inductive and the capacitive element in the rectangular waveguide without resorting to extensive computation times; it appears as an efficient alternative to the CST simulation. In the case of wave propagation, the WCD simulation can use shown the electromagnetic field distributions on the conducting obstacle. In conclusion, the WCD simulation can be used in the waveguide lumped circuit analysis and has proven to be with high accuracy and efficacy a useful tool for microwave circuit design and an educational aid.

Author details

Somsak Akatimagool and Saran Choocadee
Department of Teacher Training in Electrical Engineering,
King Mongkut's University of Technology North Bangkok, Thailand

Acknowledgement

The authors would like to thank the Faculty of Technical Education, Rajamangala University of Technology Thanyaburi, Thailand, for providing the CST Microwave Studio® software for data analysis.

7. References

[1] T. Hiraoka, C. P. Chen, T. Anada, J. P. Hsu, Z. W. Ma, and C. C. Christopoulos, "Electric Field Distributions in Microwave Plana Circuits by Small Coaxial Probe and Comparison with FDTD Method," in Microwave Conference, European, pp.269-272, 2005.

[2] J. W. Bandler, A. S. Mohamed, and M. H. Bakr, "TLM-based Modeling and Design Exploiting Space Mapping," Microwave Theory and Techniques, IEEE Transactions on, vol.53, pp. 2801-2811, 2005.

[3] S. Seunghyun, K. Hyeong-Seok, J. Hyun-Kyo, and M. Un-Chul, "Frequency Domain Analysis of Microstrip Filters and Antennae using an Adaptive Frequency Sampling Moment Method," Magnetics, IEEE Transactions on, vol.42, pp.607-610, 2006.

[4] V. K. Chaudhary, P. Verma, and U. Balaji, "Field Theory Based CAD of Inductive Iris Waveguide Filter," in Asia-Pacific Microwave Conference, APMC 2001, vol.1, pp.318-321, 2001.

[5] A. Sallier, J. Bornemann, and W. J. R. Hoefer, "Field-Based Waveguide Filter Synthesis in the Time Domain," AEU - International Journal of Electronics and Communications, vol.57, pp.119-127, 2003.

[6] I. C. Hunter, L. Billonet, B. Jarry, P. Guillon, "Microwave Filters-Applications and Technology", Microwave Theory and Techniques, IEEE Transactions on, vol.50, pp. 794 – 805, 2002.

[7] Y. P. Zhang, H. Guorui, Z. Wenmei and J. H. Sheng, "Electromagnetic Mode Theory of Periodically-Loaded Oversized Imperfect Waveguide and its Application to the Propagation of Radio Waves in Long Wall Coal Mining Face Tunnels", Antennas and Propagation, IEEE Transactions on, vol.58, no.5, pp.1816-1822, 2010.

[8] N.Marcuvitz, "Waveguide Handbook", Short Run Press Ltd., London, UK, 1986.

[9] S. Akatimagool, D. Bajon, and H. Baudrand, "Analysis of Multi-layer Integrated Inductors with Wave Concept Iterative Procedure (WCIP)," in Microwave Symposium Digest, 2001 IEEE MTT-S International, vol.3, pp.1941-1944, 2001.

[10] S. Khamkleang and S. Akatimagool, "Microwave Planar Circuit Design Tool in the teaching of Microwave Engineering," in Electrical Engineering/Electronics, Computer, Telecommunications and Information Technology, ECTI-CON 2009, 6th International Conference on, pp.830-833, 2009.

[11] S. Choocadee and S. Akatimagool, "Development of Efficiency EM Simulation Tool for Capacitive and Inductive Obstacle Analysis," in Electrical Engineering/Electronics Computer Telecommunications and Information Technology (ECTI-CON), International Conference on, pp.1154-1158, 2010.

[12] H.Baudrand,"Introduction of Microwave Circuit ", Toulouse, France, 1993.

Permissions

The contributors of this book come from diverse backgrounds, making this book a truly international effort. This book will bring forth new frontiers with its revolutionizing research information and detailed analysis of the nascent developments around the world.

We would like to thank Yi Zheng, for lending his expertise to make the book truly unique. He has played a crucial role in the development of this book. Without his invaluable contribution this book wouldn't have been possible. He has made vital efforts to compile up to date information on the varied aspects of this subject to make this book a valuable addition to the collection of many professionals and students.

This book was conceptualized with the vision of imparting up-to-date information and advanced data in this field. To ensure the same, a matchless editorial board was set up. Every individual on the board went through rigorous rounds of assessment to prove their worth. After which they invested a large part of their time researching and compiling the most relevant data for our readers. Conferences and sessions were held from time to time between the editorial board and the contributing authors to present the data in the most comprehensible form. The editorial team has worked tirelessly to provide valuable and valid information to help people across the globe.

Every chapter published in this book has been scrutinized by our experts. Their significance has been extensively debated. The topics covered herein carry significant findings which will fuel the growth of the discipline. They may even be implemented as practical applications or may be referred to as a beginning point for another development. Chapters in this book were first published by InTech; hereby published with permission under the Creative Commons Attribution License or equivalent.

The editorial board has been involved in producing this book since its inception. They have spent rigorous hours researching and exploring the diverse topics which have resulted in the successful publishing of this book. They have passed on their knowledge of decades through this book. To expedite this challenging task, the publisher supported the team at every step. A small team of assistant editors was also appointed to further simplify the editing procedure and attain best results for the readers.

Our editorial team has been hand-picked from every corner of the world. Their multi-ethnicity adds dynamic inputs to the discussions which result in innovative outcomes. These outcomes are then further discussed with the researchers and contributors who give their valuable feedback and opinion regarding the same. The feedback is then collaborated with the researches and they are edited in a comprehensive manner to aid the understanding of the subject.

Apart from the editorial board, the designing team has also invested a significant amount of their time in understanding the subject and creating the most relevant covers. They scrutinized every image to scout for the most suitable representation of the subject and create an appropriate cover for the book.

The publishing team has been involved in this book since its early stages. They were actively engaged in every process, be it collecting the data, connecting with the contributors or procuring relevant information. The team has been an ardent support to the editorial, designing and production team. Their endless efforts to recruit the best for this project, has resulted in the accomplishment of this book. They are a veteran in the field of academics and their pool of knowledge is as vast as their experience in printing. Their expertise and guidance has proved useful at every step. Their uncompromising quality standards have made this book an exceptional effort. Their encouragement from time to time has been an inspiration for everyone.

The publisher and the editorial board hope that this book will prove to be a valuable piece of knowledge for researchers, students, practitioners and scholars across the globe.

List of Contributors

Alexey Androsov, Sven Harig, Annika Fuchs, Antonia Immerz, Natalja Rakowsky, Wolfgang Hiller and Sergey Danilov
Alfred Wegener Institute for Polar and Marine Research, POB 120161, 27515 Bremerhaven, Germany

Yi Zheng and Aiping Yao
Department of Electrical and Computer Engineering, St. Cloud State University, St. Cloud, Minnesota, USA

Xin Chen, Haoming Lin, Yuanyuan Shen, Ying Zhu, Minhua Lu, Tianfu Wang and Siping Chen
Department of Biomedical Engineering, School of Medicine, ShenZhen Univeristy, ShenZhen, China
National-Regional Key Technology Engineering Laboratory for Medical Ultrasound, ShenZhen, China

Mohamad Abed A. LRahman Arnaout
Toulouse University, Plasma and Energy Conversion Laboratory, Toulouse, France

Alexey Pavelyev, Alexander Pavelyev, Stanislav Matyugov and Oleg Yakovlev
FIRE RAS, Fryazino, Russia

Yuei-An Liou
CSRSR, NCU, Taiwan

Kefei Zhang
RMIT University School of Mathemathical & Geospatial Sciences, Melbourne, Australia

Jens Wickert
GeoForschungsZentrum Potsdam (GFZ-Potsdam), Telegrafenberg, Potsdam, Germany

Mir Ghoraishi
Tokyo Institute of Technology, Japan
University of Surrey, United Kingdom

Jun-ichi Takada
Tokyo Institute of Technology, Japan

Tetsuro Imai
NTT DOCOMO Inc., Kanagawa, Japan

Hitendra K. Malik
Department of Physics, Indian Institute of Technology Delhi, New Delhi, India

Michal Čada
Dalhousie University, Halifax, Canada

Montasir Qasymeh
Abu Dhabi University, Abu Dhabi, United Arab Emirates

Jaromír Pištora
VŠB - Technical University of Ostrava, Ostrava, Czech Republic

Z. Menachem and S. Tapuchi
Department of Electrical Engineering, SCE-Shamoon College of Engineering, Israel

Jorge Avella Castiblanco, Divitha Seetharamdoo and Marion Berbineau
French institute of science and technology for transport, France

Michel Ney and François Gallée
Institute Telecom Bretagne, France

Shahrooz Asadi
University of Ottawa, Canada

Kazuhito Murakami
Computing Center, Kinki University, Japan

Émilie Masson, Yann Cocheril and Marion Berbineau
Univ Lille Nord de France, F-59000, Lille, IFSTTAR, LEOST, France

Pierre Combeau, Lilian Aveneau and Rodolphe Vauzelle
XLIM-SIC laboratory, UMR CNRS 7252, University of Poitiers, France

Paulo Roberto de Freitas Teixeira
Engineering School, Federal University of Rio Grande - FURG, Rio Grande, Brazil

Hassan Yousefi and Asadollah Noorzad
School of Civil Engineering, University of Tehran, Tehran, Iran

Somsak Akatimagool and Saran Choocadee
Department of Teacher Training in Electrical Engineering, King Mongkut's University of Technology North Bangkok, Thailand